Essential Physics
Volume II
2013

Andrew Duffy

Table of Contents for *Essential Physics, Volume 2*

Accompanying web site: <u>http://physics.bu.edu/~duffy/EssentialPhysics/</u>

The web site accompanying the book has several features, including:

- Simulations and animations to accompany the material in the printed text. Some things, such as the connection between uniform circular motion and simple harmonic motion, are made significantly clearer when the entire motion is shown as opposed to showing still frames from the motion as can only be done in a book.

- Additional examples. Example problems are very important for students, but they also take up a significant fraction of the pages in a textbook. To keep the number of pages down in this book, some examples will appear only on the web site.

- Answers to most of the odd-numbered problems for each chapter.

- Additional topics. The book itself will include the core topics. Additional sections will be included on the web site, an example being a chapter on Alternating Current (AC) Circuits.

This is a close-up photograph, taken by Ansel Adams, of the electric transmission lines at the Hoover Dam (originally called the Boulder Dam). Electricity is transmitted from the dam's hydroelectric power plant to locations in Nevada, Arizona, and California. Sights like this one, which are common now because of how heavily we rely on electricity in our daily lives, were non-existent just 120 years ago. Photo credit: public-domain image by Ansel Adams.

Chapter 16 – Electric Charge and Electric Field

CHAPTER CONTENTS

16-1 **Electric Charge**
16-2 **Charging an Object**
16-3 **Coulomb's Law**
16-4 **Applying the Principle of Superposition**
16-5 **The Electric Field**
16-6 **Electric Field: Special Cases**
16-7 **Electric Field Near Conductors**

With this chapter, we begin a new sequence of topics, falling under the broad heading of Electricity and Magnetism. In this chapter, our focus will be on electric charge. Electric charge sounds quite different from what we have done previously, but there are many parallels between this new material and gravity, which we investigated in Chapter 8. We will make as many connections as we can between the behavior of objects with charge and the behavior of objects with mass, so we can learn by analogy.

The interactions between charged objects play a major role in our daily lives. To begin with, all the atoms that make us up contain protons and electrons, which are charged particles. Forces that we know of as normal forces arise from interacting charges. The reason toaster elements and the coils in hair dryers heat up is because of the charged particles (electrons, generally) that pass through them. Anything electronic relies on flowing charges for its operation. You can probably think of many other situations in which charge is important. In this chapter, we will start by considering the very basics of how charges interact, to lay the groundwork for understanding the wide variety of applications of charge.

16-1 Electric Charge

In previous chapters, we have been concerned with several properties of objects, such as mass, momentum, energy, and angular momentum. These last three properties are associated with the object's motion, but the first property, mass, we often view as being an inherent property of the object itself. We often think of charge in a similar way, particularly the charge of an electron. Larger objects, such as ourselves, generally acquire a charge when they either lose electrons or acquire some extra electrons (we do this when we scuff our feet across a carpet, for instance). Table 16.1 shows the masses and charges of three basic constituents of atoms.

Particle	Mass (kg)	Charge
Electron	9.11×10^{-31} kg	$-e = -1.602 \times 10^{-19}$ C
Proton	1.672×10^{-27} kg	$+e = +1.602 \times 10^{-19}$ C
Neutron	1.674×10^{-27} kg	0

Table 16.1: The masses and charges of the electron, proton, and neutron, the basic building blocks of the atom.

EXPLORATION 16.1 – Experimenting with charge
One way to charge an object is to rub it with a cloth made from a different material. In this Exploration, we will investigate what happens when we do this with various combinations of materials. Such investigations go back as far as the ancient Greeks.

Step 1 – *For this experiment, we need a piece of silk, two glass rods, and one piece of string. Suspend one of the glass rods from a string tied around the middle of the rod so that the rod is balanced. Rub one end of the rod with the silk. Rub one end of a second glass rod with the silk, and then bring the rubbed end close to, but not touching, the rubbed end of the rod that is suspended from the string. What do you observe?*
What you should observe in this case is that the end of the suspended rod moves away from the other rod – the suspended rod is repelled by the second rod. By Newton's third law, we know that the rods must be repelling one another with equal-and-opposite forces.

Step 2 – *For this experiment, we need a piece of fur, two rubber rods, and one piece of string. Suspend one of the rubber rods from a string tied around the middle of the rod so that the rod is balanced. Rub one end of the rod with the fur. Rub one end of a second rubber rod with the fur, and then bring the rubbed end close to, but not touching, the rubbed end of the rod that is suspended from the string. What do you observe?*
Again, what you should observe in this case is that the end of the suspended rod moves away from the other rod – the suspended rod is repelled by the second rod. By Newton's third law, we know that the rods must be repelling one another with equal-and-opposite forces.

Step 3 – *Now, bring the rubbed end of a glass rod (rubbed with silk) close to, but not touching, the rubbed end of the rubber rod (rubbed with fur) that is suspended from the string. What do you observe?*
In this situation, what you should observe is that the rods attract one another.

Step 4 – *Repeat the experiments with a number of other types of rod material rubbed with different materials. What do you observe?*
In general, you should observe that all rubbed rods tend to act either like a glass rod rubbed with silk, or like a rubber rod rubbed with fur.

Step 5 – *Can you explain these observations using a simple model involving charge? If so, describe the features of the model.*

The model we use to explain the observations with the rods is to first say that rubbing one material with a different material generally causes a transfer of charge from one material to the other. All glass rods rubbed with silk, for instance, should acquire charge of the same sign. To account for the observation that identical charged rods repel one another, our model states that like charges repel. We also build into the model that unlike charges attract, explaining why a glass rod rubbed with silk will attract a rubber rod rubbed with fur – a rubber rod rubbed with fur must acquire charge of the opposite sign to a glass rod rubbed with silk. Our model also accounts for two types of charge, which we call positive and negative.

Key ideas about interacting rubbed rods: The observations we make with the charged rods enable us to construct a basic model of charge. In this model, there are two types of charge, positive and negative. However, both types of charge can be obtained from the transfer of electrons, which have a negative charge. Rubbing a glass rod with silk generally transfers electrons from the glass to the silk, leaving the glass with a positive charge. Rubbing a rubber rod with fur generally transfers electrons from the fur to the rubber, giving the rubber a negative charge.
Related End of Chapter Exercise: 40.

Acquiring Charge

Everyday objects contain large numbers of electrons (negative charges) and protons (positive charges). In many instances the number of electrons is the same as the number of protons, so the object has no net charge. It is quite easy to give an object a net charge, however.

As we have learned, one way to charge an object is to rub it with a different material. For instance, rubbing a glass rod with silk transfers electrons from the glass to the silk, leaving the glass rod with a positive charge and giving the silk a negative charge. How effective this process is, and which material ends up with the negative charge, depends on where the two materials fit in the triboelectric series, shown in Table 16.2. "Tribos" is a Greek word meaning "rubbing", so triboelectricity is all about giving objects net electric charges by rubbing. Many centuries ago, the Greeks themselves did experiments with charge, rubbing amber with wool. It is no coincidence that the Greek word for amber is "electron."

Rubbing promotes charge transfer, but all that is necessary is to bring the two materials into contact, causing chemical bonds (which involve electrons) to form between them. Upon separation the atoms in one material tend to keep some of the electrons while atoms in the other material tend to give them up. In general, the farther apart the materials in the triboelectric series, the more charge is transferred, with the material farther down the list acquiring electrons and ending up with a negative charge.

MOST POSITIVE
Leather
Rabbit's fur
Glass
Nylon
Wool
Silk
Paper
Cotton
NEUTRAL
Amber
Polystyrene
Rubber balloon
Hard rubber
Saran wrap
Polyethylene
Vinyl (PVC)
MOST NEGATIVE

Table 16.2: The triboelectric series. When one material is brought into contact with another and then separated, some electrons can be transferred from one to the other. The material further down the list generally becomes negative.

Essential Question 16.1: 1 coulomb (1 C) represents a large amount of charge. If –1.0 C worth of electrons is transferred from a glass rod to a piece of silk, how many electrons are involved? By how much does the mass of each object change? *(The answer is at the top of the next page.)*

Answer to Essential Question 16.1: Dividing the total charge by the charge on one electron gives the number of electrons involved:

$$n = \frac{q}{-e} = \frac{-1.0 \text{ C}}{-1.602 \times 10^{-19} \text{ C/electron}} = 6.2 \times 10^{18} \text{ electrons}.$$

Because electrons are transferred from the glass rod to the silk, the mass of the glass rod decreases while the mass of the silk increases. Multiplying the number of electrons transferred by the mass of each electron (see Table 16.1) gives the total mass involved, which is very small:

$$m = (6.242 \times 10^{18} \text{ electrons}) \times (9.11 \times 10^{-31} \text{ kg/electron}) = 5.7 \times 10^{-12} \text{ kg}.$$

16-2 Extending our Model of Charge

Key facts about charge: keep the following in mind when dealing with electric charge.

- The symbol for charge is q or Q. The MKS unit for electric charge is the coulomb (C), although we will also use units of e, the magnitude of the charge on the electron.
- Unlike mass, which is always positive, charges can be either positive (+) or negative (−).
- Like charges (both + or both −) repel one another; unlike charges (a + and a −) attract.
- Charge is quantized – it can only be particular values. Charging an object generally involves a transfer of electrons, so the charge on an object is an integer multiple of e.
- Charge is conserved. This is another of the fundamental conservation laws of physics, that the net charge of a closed system remains constant. See Example 16.2 for an application.

Conductors and Insulators

When an electric appliance, such as a television or refrigerator, is on, electric charges flow through the wires connecting the appliance to a wall socket, and through the wires within the appliance itself. Even so, it is generally safe to touch the cable connecting the appliance to the wall socket as long as the metal wires inside that cable are completely covered by rubber. This exploits the different material properties of metal and rubber, specifically the differences in their conductivity. Metals (which we classify as conductors) generally have conductivities that are orders of magnitude larger than the conductivities of materials like rubber and plastic – those materials we call insulators. The major difference between these two classes of materials is that, in an insulator, each electron is closely tied to its molecule, while some fraction of the electrons in a conductor (these are known as the conduction electrons) are free to move around.

Charge is Quantized

When something is quantized it can not take on just any value – only particular values are possible. An example is money, which is quantized in units of pennies (in the USA and Canada, at least). It is possible to have $1.27, the equivalent of 127 pennies, but it is not possible to have 2/7 of a dollar. For something to be quantized does not necessarily mean that its allowed values are integer multiples of its smallest unit, but that is how things work with money and charge.

For now, we can say that the smallest unit of charge is $e = 1.602 \times 10^{-19}$ C, the magnitude of the charge on the electron and the proton. Expressing charge quantization as an equation:

$$q = ne, \qquad \text{(Eq. 16.1: \textbf{Charge is quantized, and comes in integer multiples of } } e)$$

where n is any positive or negative integer.

Charging by Induction

An uncharged conducting object like a metal sphere can be charged by rubbing it with a charged rod, acquiring charge of the same sign as that of the rod. However, it can also be charged without touching it with a charged rod, in the process known as charging by induction:

1. ***Bring a charged insulating rod close to the conductor, without allowing the rod to touch the conductor.*** Bringing the charged rod close causes conduction electrons in the conductor to move in response to the presence of the charge. The electrons move toward the rod, if the rod is positive, or (as shown in Figure 16.1), they move away from the rod if the rod is negative. The conductor is now polarized, but it still has no net charge.

2. ***Ground the conductor.*** A ground is a large object, like the Earth, that can accept or give up electrons without being affected. We can ground the conductor by connecting a wire from the conductor to a metal pipe. This allows electrons to be transferred from ground to the conductor, if the rod is positive, or from the conductor to ground if the rod is negative. The conductor now has an excess charge with a sign opposite to that of the charge on the rod.

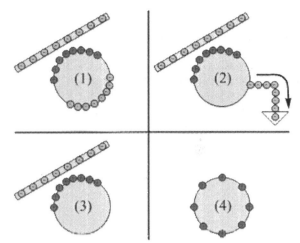

3. ***Remove the ground connection.*** This strands the transferred charge.

4. ***Remove the charged rod.*** The charge on the conductor redistributes itself, but the conductor keeps the net charge it has by the end of step 2.

Figure 16.1: The process of charging by induction.

In the charging by induction process, the conducting object ends up with a net charge of the opposite sign as the charge on the rod. Note that steps 1 and 2 in the process can be reversed – the charge is only transferred after both these steps have been done. However, steps 3 and 4 must be done in the order above. Removing the charged rod before removing the ground connection allows charge transferred between the conductor and ground to return to where it started. The presence of the rod keeps the net charge on the conductor until the ground connection is removed.

EXAMPLE 16.2 – Two spheres touch

Two identical conducting spheres sit on separate insulating stands. Sphere A has a net positive charge of $+8Q$, while sphere B has a net negative charge of $-2Q$. The spheres are touched together briefly and then separated again. How much charge is on each sphere now?

SOLUTION

Each sphere has a charge of $+3Q$. The net charge in the system is $+6Q$, so, if each sphere ends up with half of this, we satisfy the law of Conservation of Charge and also get the symmetry we expect because the spheres are identical. Even though sphere B has a net charge of $-2Q$, it transfers $-5Q$ to sphere A. The extra $-3Q$ is made up of some of sphere B's conduction electrons.

Related End-of-Chapter Exercises for this section: 2 and 3.

Essential Question 16.2: In Example 16.2, the spheres are conducting. Would we get the same result if the spheres were both made of rubber, an insulating material? Explain.

Answer to Essential Question 16.2: No. Charge does not flow on an insulator, so touching charged rubber spheres together could transfer a small amount of charge between the spheres, but most of the excess charge would stay where it was. This is the big difference between conductors and insulators – charge flows easily in a conductor but does not flow through an insulator.

16-3 Coulomb's Law

To qualitatively measure the charge on an object, we can use an electroscope (see Figure 16.2). When the electroscope is charged (such as by rubbing it with a charged rod), the charge distributes itself over the entire electroscope because the electroscope is made from conducting material. Like charges repel, so the arm of the electroscope swings out, as in the electroscope on the right. The larger the charge, the more the arm swings out. To get a more quantitative measure of charge than we can get with an electroscope, we use Coulomb's law.

Figure 16.2: An uncharged electroscope, on the left, and a charged electroscope, on the right. Photo courtesy A. Duffy.

Point charges are charged objects that are so small that all the charge is effectively at one point. The force between point charges of charge q and Q, separated by a distance r, is given by:

$$\vec{F}_E = \frac{kqQ}{r^2}\hat{r} \qquad \text{(Eq. 16.2: \textbf{Coulomb's Law for the force between point charges})}$$

where $k = 8.99\times10^9 \text{ N m}^2/\text{C}^2$ is a constant. The unit vector $\hat{r}$ tells us the force is directed along the line joining the charges. If the charges have opposite signs, qQ is negative and the force is attractive, directed toward the object applying the force. If the charges have the same sign, qQ is positive and the force is repulsive, directed away from the object applying the force.

Comparing Coulomb's law to Newton's law of universal gravitation, which gives the force between two objects with mass, we see that they have the same form:

$$\vec{F}_G = -\frac{GmM}{r^2}\hat{r} \qquad \text{(Equation 8.1: \textbf{Newton's Law of Universal Gravitation})}$$

where $G = 6.67\times10^{-11}\text{N m}^2/\text{kg}^2$ is the universal gravitational constant. For Coulomb's law, k takes the place of G, and the charges q and Q take the place of the masses, m and M. Note that objects with mass always attract one another, while charged objects can attract or repel.

EXAMPLE 16.3 – Inside the hydrogen atom
A hydrogen atom contains an electron and a proton. In the Bohr model of the atom, the electron follows a circular orbit around the proton. Determine the ratio of the magnitudes of the electrostatic force to the gravitational force between the proton and electron.

SOLUTION
Note that we don't need to know the radius, because the radius cancels out in the ratio.

$$\frac{F_E}{F_G} = \frac{k|qQ|/r^2}{GmM/r^2} = \frac{k|qQ|}{GmM} = \frac{\left(9.0\times10^9 \text{ N m}^2/\text{C}^2\right)\times\left(1.6\times10^{-19} \text{ C}\right)\times\left(1.6\times10^{-19} \text{ C}\right)}{\left(6.67\times10^{-11} \text{ N m}^2/\text{kg}^2\right)\times\left(9.11\times10^{-31} \text{ kg}\right)\times\left(1.67\times10^{-27} \text{ kg}\right)} = 2.3\times10^{39}$$

The gravitational interaction is negligible, being 39 orders of magnitude smaller!

EXPLORATION 16.3 – The principle of superposition

Three balls, with charges of $+q$, $-2q$, and $-3q$, are equally spaced along a line. The spacing between the balls is r. We can arrange the balls in three different ways, as shown in Figure 16.3. In each case, the balls are in an isolated region of space very far from anything else.

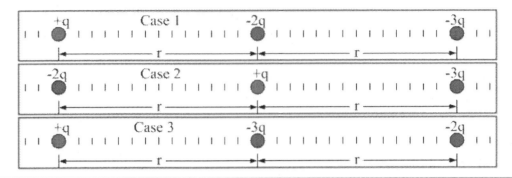

Figure 16.3: Three different arrangements of three balls of charge $+q$, $-2q$, and $-3q$ placed on a line with a distance r between neighboring balls. Each ball experiences two electrostatic forces, one from each of the other balls. We can neglect any other interactions.

Step 1 – *Consider Case 1. Is the electrostatic force that the ball of charge $+q$ exerts on the ball of charge $-3q$ affected by the fact that the ball of charge $-2q$ lies between the other two balls?*
No, we can apply the principle of superposition – the force from the interaction between any pair of charged objects is unaffected by the presence of any other charged object in the vicinity.

Step 2 – *In which case does the ball of charge $-2q$ experience the largest-magnitude net force? Argue qualitatively.* Let's attach arrows to the ball of charge $-2q$, as in Figure 16.4, to represent the two forces it experiences in each case. The length of each arrow is proportional to the force.

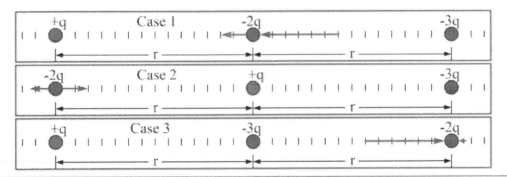

Figure 16.4: Attaching force vectors to the ball of charge $-2q$, which is attracted to the $+q$ ball and repelled by the $-3q$ ball. The length of each vector is drawn in units of kq^2/r^2.

The net force is largest in case 1. In cases 2 and 3, the forces partly cancel, while only in case 1 are the directions of the two forces acting on the ball of charge $-2q$ the same, and the net force case 1 is clearly larger than it is in the other two cases.

Key ideas about adding electrostatic forces: The net force acting on an object can be found using the principle of superposition, remembering that each individual force is unaffected by the presence of other forces. **Related End of Chapter Exercises: 28, 29.**

Essential Question 16.3: An object with a charge of $+5Q$ is placed a distance r away from an object with a charge of $+2Q$. Which object exerts a larger electrostatic force on the other?

Answer to Essential Question 16.3: Newton's third law tells us that the electrostatic forces the two objects exert on one another are equal in magnitude (and opposite in direction). This follows from Coulomb's law, because whether we look at the force exerted by the first object or the second object the factors going into the equation are the same in both cases.

16-4 Applying the Principle of Superposition

EXPLORATION 16.4 – Three objects in a line

Let's return again to the situation of three different arrangements of three balls that we looked at in Exploration 16.3. The balls, with charges of $+q$, $-2q$, and $-3q$, are equally spaced along a line. The spacing between the balls is r. In each case, the balls are in an isolated region of space very far from anything else.

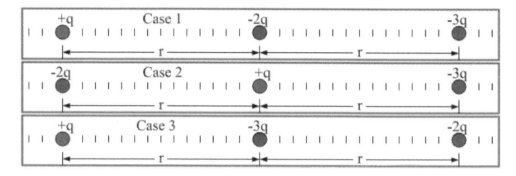

Figure 16.5: Three different arrangements of three balls of charge $+q$, $-2q$, and $-3q$ placed on a line with a distance r between neighboring balls.

Step 1 – *Calculate the force experienced by the ball of charge $-2q$ in each case.*

To do this, we will make extensive use of Coulomb's law. Let's define right to be the positive direction, and use the notation $\vec{F}_{21}$ for the force that the ball of charge $-2q$ experiences from the ball of charge $+q$. The $+$ and $-$ signs in the equation come from the direction of the force, not the signs on the charges. In each case:

$$\vec{F}_{2,net} = \vec{F}_{21} + \vec{F}_{23}$$

Case 1: $\vec{F}_{2,net} = \vec{F}_{21} + \vec{F}_{23} = -\dfrac{kq(2q)}{r^2} - \dfrac{k(2q)(3q)}{r^2} = -\dfrac{2kq^2}{r^2} - \dfrac{6kq^2}{r^2} = -\dfrac{8kq^2}{r^2}$

Case 2: $\vec{F}_{2,net} = \vec{F}_{21} + \vec{F}_{23} = +\dfrac{kq(2q)}{r^2} - \dfrac{k(2q)(3q)}{(2r)^2} = +\dfrac{2kq^2}{r^2} - \dfrac{3kq^2}{2r^2} = +\dfrac{kq^2}{2r^2}$

Case 3: $\vec{F}_{2,net} = \vec{F}_{21} + \vec{F}_{23} = -\dfrac{kq(2q)}{(2r)^2} + \dfrac{k(2q)(3q)}{r^2} = -\dfrac{kq^2}{2r^2} + \dfrac{6kq^2}{r^2} = +\dfrac{11kq^2}{2r^2}$

The ball of charge $-2q$ does experience the largest-magnitude net force in case 1.

Key ideas about adding electrostatic forces: Again, we see that the net force acting on an object can be found using the principle of superposition, remembering that each individual force is unaffected by the presence of other forces. In addition, $+$ and $-$ signs should be based on the direction of the force, rather than the signs of the charges.
Related End of Chapter Exercises: 30, 31.

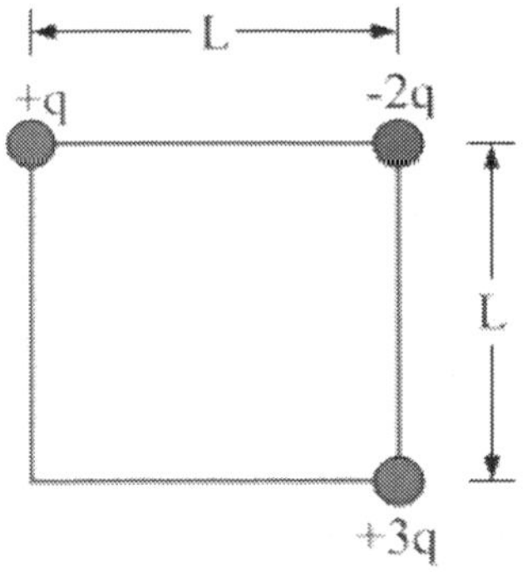

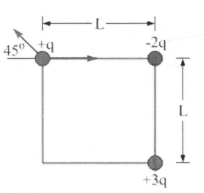

Figure 16.6: Three charged balls placed at the corners of a square.

EXAMPLE 16.4 – A two-dimensional situation

Compare this example to Example 8.1. Three balls, with charges of $+q$, $-2q$, and $+3q$, are placed at the corners of a square measuring L on each side, as shown in Figure 16.6. Assume this set of three balls is not interacting with anything else in the universe, and assume that gravitational interactions are negligible. What is the magnitude and direction of the net electrostatic force on the ball of charge $+q$?

SOLUTION

Let's attach force vectors (see Figure 16.7) to the ball of charge $+q$, which is attracted to the $-2q$ ball and repelled by the $+3q$ ball. The length of each vector is proportional to the magnitude of the force it represents.

We can find the two individual forces acting on the ball of charge $+q$ using Coulomb's law. Let's define $+x$ to the right and $+y$ up.

From the ball of charge $-2q$: $\vec{F}_{21} = \dfrac{kq(2q)}{L^2}$ to the right.

From the ball of charge $+3q$: $\vec{F}_{31} = \dfrac{kq(3q)}{L^2 + L^2}$ at 45° above the $-x$-axis.

Finding the net force is a vector addition problem.

Figure 16.7: Attaching force vectors to the ball of charge $+q$.

In the x-direction, we get:

$$\vec{F}_{1x} = \vec{F}_{21x} + \vec{F}_{31x} = +\frac{2kq^2}{L^2} - \frac{3kq^2}{2L^2}\cos 45° = \left(2 - \frac{3}{2\sqrt{2}}\right)\frac{kq^2}{L^2}.$$

Note that the signs on each term come not from the signs on the charges, but from comparing the direction of the forces to the directions we chose to be positive above.

In the y-direction, we get: $\vec{F}_{1y} = \vec{F}_{21y} + \vec{F}_{31y} = 0 + \dfrac{3kq^2}{2L^2}\sin 45° = \left(+\dfrac{3}{2\sqrt{2}}\right)\dfrac{kq^2}{L^2}.$

The Pythagorean theorem gives the magnitude of the net force on the ball of charge $+q$:

$$F_1 = \sqrt{F_{1x}^2 + F_{1y}^2} = \sqrt{\left(4 - \frac{6}{\sqrt{2}} + \frac{9}{8} + \frac{9}{8}\right)\frac{kq^2}{L^2}} = 1.42\frac{kq^2}{L^2}.$$

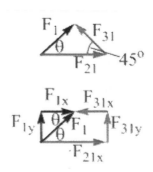

The angle is given by: $\tan\theta = \dfrac{F_{1y}}{F_{1x}} = \dfrac{\dfrac{3}{2\sqrt{2}}}{\dfrac{4\sqrt{2}-3}{2\sqrt{2}}} = \dfrac{3}{4\sqrt{2}-3}.$

So, the angle is 48.5° above the $+x$-axis.

Related End-of-Chapter Exercises: 4, 5, 16, 17, 41, 49, 50, 53.

Figure 16.8: The triangle representing the vector addition problem above.

Essential Question 16.4: In Exploration 16.4, on the previous page, which ball experiences the largest-magnitude net force in (i) Case 1, (ii) Case 2, and (iii) Case 3?

Answer to Essential Question 16.4: The object experiencing the largest-magnitude net force is the ball of charge −2*q* in case 1, and the ball of charge −3*q* in cases 2 and 3.

16-5 The Electric Field

There are important parallels between the electric field $\vec{E}$ and the gravitational field $\vec{g}$, so many that you may find it helpful to review Section 8-3.

The electric field, $\vec{E}$, at a particular point can be defined in terms of the electric force, $\vec{F}_E$, that an object of charge q would experience if it were placed at that point:

$$\vec{E} = \frac{\vec{F}_E}{q}, \quad \text{or} \quad \vec{F}_E = q\vec{E}. \quad \text{(Eq. 16.3: \textbf{Connecting electric field and electric force})}$$

The units for electric field are N/C.

A special case is the electric field from a point charge with a charge Q:

$$\vec{E} = \frac{kQ}{r^2}\hat{r}, \quad \text{(Equation 16.4: \textbf{Electric field from a point charge})}$$

where r is the distance from the charge to the point in space where we are finding the field. The magnitude of the field is kQ/r^2. The electric field points away from a positive charge and toward a negative charge.

One way to think about an electric field is the following: it is a measure of how a charged object, or a set of charged objects, influences the space around it.

Visualizing the electric field

It can be useful to draw a picture that represents the electric field near an object, or a set of objects, so we can see at a glance what the field in the region is like. In general there are two ways to do this, by using either field lines or field vectors. The field-line representation is shown in Figure 16.9. Figure 16.9(a) represents a uniform electric field directed down, while Figure 16.9 (b) represents the electric field near a negative point charge. Figure 16.9(c) shows the field from an electric dipole, which consists of two objects with equal-and-opposite charge.

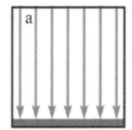
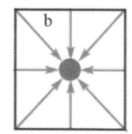
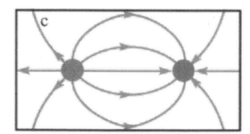

Figure 16.9: Field-line diagrams for various situations. Diagram *a* represents a uniform electric field directed down, while diagram b represents the electric field near a negative point charge. Diagram c shows the electric field near an electric dipole, which is a pair of charges of equal magnitude and opposite sign separated by some distance.

Question: How is the direction of the field at a particular point shown on a field-line diagram? What indicates the relative strength of the field at a point on the field-line diagram?

Answer: As with gravitational field lines, each field line has a direction marked on it with an arrow that shows the direction of the electric field at all points along the field line. The relative strength of the field is indicated by the density of the field lines (that is, by how close the lines are). The more lines there are in a given area, the larger the field.

A second method of representing a field is to use field vectors. A field vector diagram reinforces the idea that every point in space has an electric field associated with it, because a grid made up of equally spaced dots is superimposed on the picture and a vector is attached to each of these grid points. All the vectors are the same length. The situations represented by the field-line patterns in Figure 16.9 are now re-drawn in Figure 16.10 using the field-vector representation.

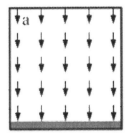 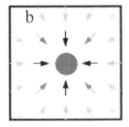 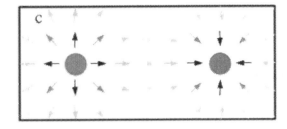

Figure 16.10: Field-vector diagrams for various situations. In figure *a*, the field is uniform and directed down. Figure *b* represents the non-uniform but symmetric field found near a negative point charge, while figure *c* shows field vectors for an electric dipole.

Question: How is the direction of the field at a particular point shown on a field-vector diagram? What indicates the relative strength of the field at a point on the field-vector diagram?

Answer: The direction of the field at a particular point is represented by the direction of the field vector at that point (or the ones near it if the point does not correspond exactly to the location of a field vector). The relative strength of the field is indicated by the darkness of the arrow. The larger the magnitude of the field the darker the arrow.

Related End of Chapter Exercises: 25, 26, 58.

We often use a test charge to sample the electric field near a charge or a set of charges. A test charge is a positive charge of such a small magnitude that it has negligible impact on the field it is sampling. Based on the relationship $\vec{F}_E = q\vec{E}$, the force on a positive test charge is in the direction of the electric field at the point where the test charge is, and the size of the force is proportional to the magnitude of the field at that point.

Essential Question 16.5: Figure 16.11 shows the force a positive test charge feels when it is placed at the location shown, near two charged balls. The ball on the left has a positive charge +Q, while the ball on the right has an unknown sign and magnitude. Based on the force experienced by the test charge, what is the sign of the charge on the ball on the right? How does the magnitude of that charge compare to Q?

Figure 16.11: The arrow shows the force experienced by a test charge when it is placed at the position shown near two charged balls.

Answer to Essential Question 16.5: The $+Q$ charge exerts a force on the test charge that points directly away from the $+Q$ charge. The unknown charge shifts the direction of the force on the test charge towards the unknown charge, so the unknown charge must be negative. If the unknown charge was $-Q$, by symmetry the test charge would experience a net force directed right, with no vertical component. Decreasing the magnitude of the unknown charge gives the net force an upward component; so the magnitude of the unknown charge is less than Q.

16-6 Electric Field: Special Cases

Let's consider two important special cases. The first involves a charged object in a uniform electric field, while the second involves the electric field from point charges.

EXPLORATION 16.6A – Motion of a charge in a uniform electric field

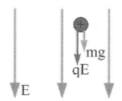

Step 1 – Sketch the free-body diagram for a proton placed in a uniform electric field of magnitude E = 250 N/C that is directed straight down. Then apply Newton's Second Law to find the acceleration of the proton. How important is gravity in this situation? The free-body diagram in Figure 16.12 shows the two forces applied to the proton, the downward force of gravity and the downward force applied by the electric field.

Figure 16.12: Free-body diagram for a proton in a uniform electric field directed down.

Taking down to be positive, applying $\Sigma \vec{F} = m\vec{a}$, gives: $+qE + mg = m\vec{a}$.

The values of the mass and charge of the proton are given in Table 16.1. Solving for the acceleration gives:

$$\vec{a} = +\frac{qE}{m} + g = +\frac{\left(1.60\times10^{-19}\ \text{C}\right)\left(250\ \text{N/C}\right)}{\left(1.67\times10^{-27}\ \text{kg}\right)} + 9.8\ \text{m/s}^2 = +2.40\times10^{10}\ \text{m/s}^2.$$

Note that gravity is negligible in this case, because g is orders of magnitude less than qE/m.

Step 2 – How far has the proton traveled 25 μs after being released from rest in this field? The electric field is uniform, so the acceleration is constant. We can apply the constant acceleration equations from one-dimensional motion (Chapter 2) or projectile motion (Chapter 4).

From Eq. 2.11: $\vec{x} = \vec{x}_i + \vec{v}_i t + \frac{1}{2}\vec{a}t^2 = 0 + 0 + \frac{1}{2}\left(2.395\times10^{10}\ \text{m/s}^2\right)\left(25\times10^{-6}\ \text{s}\right)^2 = 0.75\ \text{m}.$

Key idea: The acceleration of a charged particle in a uniform electric field is constant, so the constant-acceleration equations from Chapters 2 and 4 apply. The scale of the accelerations and times are different but the physics is the same. **Related End-of-Chapter Exercises: 27 and 28.**

EXAMPLE 16.6A – Where is the electric field equal to zero?
Two point charges, with charges of $q_1 = +2Q$ and $q_2 = -5Q$, are separated by a distance of 3.0 m, as shown in Figure 16.13. Determine all locations on the line passing through the two charges where their individual electric fields combine to give a net electric field of zero.

(a) For the net field to be zero, what condition(s) must the individual fields satisfy?

Figure 16.13: Two point charges separated by 3.0 m.

(b) Using qualitative arguments, can the electric field be zero at:
    I.   a point on the line, to the left of the $+2Q$ charge?
    II.   a point on the line, in between the charges?
    III. a point on the line, to the right of the $-5Q$ charge?
(c) Calculate the locations of all points near the charges where the net electric field is zero.

SOLUTION
    (a) The two individual fields add as vectors, so they must cancel one another for the net field to be zero. The two fields must be of equal magnitude and point in opposite directions.

    (b) Figure 16.14 shows three representative points on the line through the charges. In region I, to the left of the $+2Q$ charge, the two fields are in opposite directions. There is only one point at which the fields cancel in that region because close to the $+2Q$ charge the field from that charge dominates, and far from the charges the field from the $-5Q$ charge dominates because that charge is larger. At some point in between, the fields exactly balance. In region II, between the charges, there is no such point because both fields are directed right, and can not cancel.

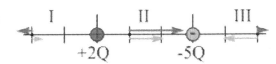

Figure 16.14: Choosing a point in each region allows us to do a qualitative analysis to determine where the net field is zero. At each point we draw two field vectors, one from each charge.

    We also see from Figure 16.14 that at a point to the right of the $-5Q$ charge the two fields point in opposite directions. However, because such points are closer to the $-5Q$ charge (the larger-magnitude charge), the field from the $-5Q$ charge is always larger in magnitude than the field from the $+2Q$ charge. Thus, there are no points in region III where the net field is zero.

    (c) Thus, we can conclude that there is only one point at which the net electric field is zero. Let's say this point is a distance x to the left of the $+2Q$ charge. Equating the magnitude of the field from one charge at that point to the magnitude of the field from the second charge gives:

$$\frac{k|q_1|}{x^2} = \frac{k|q_2|}{(x+3.0 \text{ m})^2} .$$

Canceling factors of k leads to: $\dfrac{2Q}{x^2} = \dfrac{5Q}{(x+3.0 \text{ m})^2} .$

Canceling factors of Q and re-arranging gives: $\dfrac{(x+3.0 \text{ m})^2}{x^2} = \dfrac{5}{2} .$

Taking the square root of both sides gives: $\dfrac{x+3.0 \text{ m}}{x} = \pm\sqrt{2.5} .$

Using the $+$ sign gives, when we solve for x: $\quad x = \dfrac{3.0 \text{ m}}{\sqrt{2.5}-1} = 5.16 \text{ m} .$

    Thus, the electric field is zero at the point 5.16 m to the left of the $+2Q$ charge.

Related End-of-Chapter Exercises: 57, 59, and 60.

Essential Question 16.6: The method in the previous example gives two solutions. What is the second solution in this case and what is its physical meaning?

Answer to Essential Question 16.6: We can find the other solution by using the – sign. This

gives: $x' = \dfrac{3.0 \text{ m}}{-\sqrt{2.5}-1} = -1.16 \text{ m}$. This is a subtle point, but because x was defined as the distance

of the point to the left of the origin, the negative answer gives us a point 1.16 m to the right of the origin, between the point charges. This is actually the other point on the line joining the charges where the fields from the two charges have the same magnitude. However, between the charges those two fields have the same direction (both pointing right), so they add rather than canceling.

16-7 Electric Field Near Conductors
At equilibrium, the conduction electrons in a conductor move about randomly, somewhat like atoms of ideal gas, but there is no net flow of charge in any direction. If there is a change in the external electric field the conductor is exposed to, however, the conduction electrons respond by redistributing themselves, very quickly coming to a new equilibrium distribution. At equilibrium a number of conditions apply:

1. There is no electric field inside the solid part of the conductor.
2. The electric field at the surface of the conductor is perpendicular to the surface.
3. If the conductor is charged, excess charge lies only at the surface of the conductor.
4. Charge density is highest, and electric field is strongest, on pointy parts of a conductor.

Let's investigate each of these conditions in more detail.

At equilibrium, $E = 0$ within solid parts of a conductor.
If electric field penetrates into a conductor, conduction electrons immediately respond to the field. Because $\vec{F} = q\vec{E}$, and electrons are negative, electrons feel a force that is opposite to the field. As shown in Figure 16.15, there is a net movement of electrons to the region where the field enters the conductor. The field lines end at the electrons at the surface, so $E = 0$ within the conductor. This redistribution of electrons leaves positive charge at the other side of the conductor, so field lines start up again there and go away from the conductor.

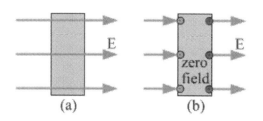
(a)   (b)

Figure 16.15: Conduction electrons in a conductor quickly redistribute themselves until the field is zero inside the conductor.

At equilibrium, electric field lines are perpendicular to the surface of a conductor.
If the electric field lines end at the surface of a conductor but are not perpendicular to the surface, as in Figure 16.16(a), the charges at the surface feel a force from the field. As Figure 16.16(b) shows, the component of the force parallel to the surface ($F_\parallel$) causes the charges to flow along the surface, carrying the field lines with them. The charges are in equilibrium when the electric field lines are perpendicular to the surface, as in Figure 16.16(c).

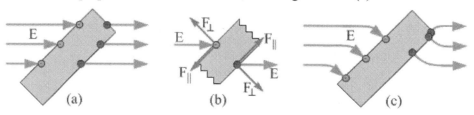
(a)   (b)   (c)

Figure 16.16: If electric field lines are not perpendicular to the surface of a conductor, the charges at the surface redistribute themselves until the field lines are perpendicular.

Electrons at the surface still feel a force component perpendicular to the surface that is trying to remove the electrons from the conductor. In most cases this will not happen because the conductor is surrounded by insulating material (such as air), but if the field is strong enough electrons will jump off the surface. When this happens there is a spark from the conductor.

At equilibrium, any excess charge lies only at the surface of a conductor.

This statement is a consequence of the fact that at equilibrium $E = 0$ within the conductor. If there was excess charge in the bulk of the conductor field lines would either start there, if it was positive, or end there, if it was negative. This non-zero field inside the conductor would cause the charge to move to the surface to bring the field to zero within the conductor.

Charge tends to accumulate on pointy parts of a conductor.

Figure 16.17 shows three different situations. In Figure 16.17(a) a metal sphere has a net positive charge. At equilibrium the excess charge is distributed uniformly over the surface of the sphere. Moving any of the charges around results in forces that act on these charges, driving them back to the equilibrium distribution. In contrast, Figure 16.17(b) shows excess charge (negative in this case, but our analysis is equally valid for positive charge) distributed evenly along a conducting rod. The charge at the center experiences no net force from the other charges, but the other charges experience net forces that push them toward the ends of the rod: each charge on the right experiences a net force pushing it further right, while each charge on the left experiences a net force pushing it further left. The equilibrium situation for the rod is more like that shown in Figure 16.17(c), where there is a much larger charge density at the ends than in the middle.

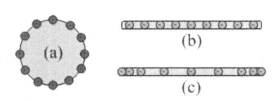

Figure 16.17: In (a), excess positive charge is uniformly distributed over the surface of a metal sphere. The charge is at equilibrium because there are no forces acting on the charges to move them around the sphere. In (b), however, uniformly distributing charge along the length of a conducting rod results in net forces on the charges that shift them toward the ends of the rod, as in (c).

This helps us to understand how a lightning rod works. Lightning occurs when charge builds up, increasing the local electric field to a large enough value that charge can travel between a cloud and, say, your house. Without a lightning rod this can take a long time, requiring a lot of charge, so that when the discharge finally happens it can involve a great deal of energy and cause significant damage. With a sharply-pointed lightning rod, attached to ground, on your house, however, charge and field builds up quickly at the tip of the lightning rod. This causes a slow and steady drain of charge from the cloud to the rod and then the ground, much safer than one sudden large discharge. The lightning rod was invented by Benjamin Franklin.

Related End-of-Chapter Exercises: 62 – 64.

Essential Question 16.7: A point charge with a charge of $+Q$ is placed at the center of a hollow thick-walled metal sphere. The sphere itself has no net charge on it. Which of the three pictures in Figure 16.18 correctly shows the equilibrium charge distribution on the metal sphere?

Figure 16.18: Three possible equilibrium situations for when a charge of $+Q$ is placed at the center of a hollow thick-walled metal sphere that has no net charge. In (a) the sphere has a total charge $+Q$ on its outer surface and $-Q$ on its inner surface; in (b) the sphere has a charge $-Q$ on both its inner and outer surfaces, and in (c) the sphere has a charge of $-Q$ on its inner surface.

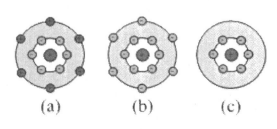

Answer to Essential Question 16.7: Figure 16.18(a) shows the correct result. Enough conduction electrons in the sphere are attracted to the inner surface that the field lines from the center $+Q$ charge do not penetrate into the sphere. This takes $-Q$ worth of charge. That leaves a net positive charge $+Q$ on the outer surface, as shown. In addition, (a) is the only picture that shows no net charge on the sphere itself.

Chapter Summary

Essential Idea: Electric Charge.
In many situations the interaction between objects that have a net electric charge, or the behavior of a charged object in a uniform electric field, is analogous to that of objects with mass that interact with each other via the force of gravity, or that are in a uniform gravitational field.

Key Facts about Charge
- The symbol for charge is q or Q. The MKS unit for electric charge is the Coulomb (C), although we will also use units of e, the magnitude of the charge on the electron.
- Unlike mass, which is always positive, charges can be either positive (+) or negative (−).
- Like charges (i.e, both + or both −) repel one another; unlike charges (a + and a −) attract.
- Charge is quantized – it can only be particular values. Since charging an object generally involves a transfer of electrons, the charge on an object is an integer multiple of e.
- Charge is conserved. This is another of the fundamental conservation laws of physics, that the net charge of a closed system remains constant.
- Most materials fall into one of two categories. Charge can flow easily through **conductors**, while it does not readily flow through **insulators**. Metals are good conductors, while plastic and rubber are examples of good insulators.
- When an insulating rod of one material is rubbed by another material, electrons can be transferred from one to the other. Which material ends up with a negative charge, and how much charge is transferred, depends on where the two materials lie in the **triboelectric series** (see Table 16.1).

Coulomb's Law for the Force Between Charged Objects
The force that an object with a charge q exerts on a second object with a charge Q that is a distance r away is given by Coulomb's Law:

$$\vec{F}_E = \frac{kqQ}{r^2} \hat{r} \qquad \text{(Equation 16.2)}$$

where $k = 8.99 \times 10^9 \ \text{N m}^2 / \text{C}^2$ is a constant. The unit vector $\hat{r}$ indicates that the force is directed along the line joining the two charged objects. If the charges are of opposite signs then the product qQ is negative and the force is attractive, directed back toward the object applying the force. If the charges have the same sign then qQ is positive and the force is repulsive, directed away from the object applying the force.

The principle of superposition applies – the net force on a charged object is the vector sum of the individual forces acting on that object.

Electric Field

The electric field, $\vec{E}$, at a particular point can be defined in terms of the electric force, $\vec{F}_E$, that an object of charge q would experience if it were placed at that point:

$$\vec{E} = \frac{\vec{F}_E}{q}, \quad \text{or} \quad \vec{F}_E = q\vec{E} \qquad \text{(Eq. 16.3: \textbf{Connecting electric field and electric force})}$$

The units for electric field are N/C. An object with a positive charge experiences a force in the direction of the field, while an object with a negative charge experiences a force opposite in direction to the electric field.

Visualizing the Electric Field

We can visualize the electric field by drawing field lines or field vectors. Field lines are continuous lines that start on positive charges, or come from infinity, and that end on negative charges, or go off to infinity. The direction of the field at a point is tangent to the field line passing through that point. The magnitude of the field increases as the density of the field lines increases. In a field vector pattern equal-length arrows are drawn at equally spaced points. The direction of an arrow shows the direction of the field at that point, and the darkness of an arrow indicates the strength of the field at that point.

A Charged Object in a Uniform Electric Field

A uniform field has the same magnitude and direction at all points. A charged object in a uniform electric field experiences a constant acceleration that is given by $\vec{a} = \vec{F}/m = q\vec{E}/m$.

Because the acceleration is constant we can apply the constant-acceleration equations from Chapter 2 (for one-dimensional motion) or from Chapter 4 (for projectile motion in two dimensions) to determine equations of motion for the object.

Electric Field from a Point Charge

Electric fields are set up by charges. One special case is the electric field from a point charge, which is a charged object so small that it can be considered to be a point. If the point charge has a charge Q the magnitude of the electric field a distance r away from it is given by:

$$E = \frac{kQ}{r^2}. \qquad \text{(Equation 16.4: \textbf{Electric field from a point charge})}$$

Electric field is a vector. The electric field points away from a positive charge and toward a negative charge. The net electric field from multiple point charges can be found by superposition, adding the fields from each charge as vectors.

Electric Field near a Conductor, at Electrostatic Equilibrium

When a conducting object has a net charge, and/or when the conducting object is placed in a region in which there is an electric field, a number of conditions apply when electrostatic equilibrium is reached. Electrostatic equilibrium is defined as there being no net flow of charge within or on the conducting object.

1. There is no electric field inside the solid part of the conductor.
2. The electric field at the surface of the conductor is perpendicular to the surface.
3. If the conductor is charged, excess charge lies only at the surface of the conductor.
4. Charge density is highest, and electric field is strongest, on pointy parts of a conductor.

End-of-Chapter Exercises

Exercises 1 – 14 are mainly conceptual questions designed to see if you have understood the main concepts of the chapter. Treat all charged balls as point charges.

1. While you are solving a physics problem, you calculate that the charge on a particular object has a value of $+2.5 \times 10^{-22}$ C. Can this be correct? Choose the one correct statement about this from the set of three options below. Note that e represents the magnitude of the charge on the electron.
   A – Yes, this answer could be correct.
   B – No, this answer cannot be correct because the charge represents a small fraction of e.
   C – No, this answer cannot be correct. The value has a magnitude larger than e, but it does not represent an integer multiple of e.

2. You have three identical metal spheres that have different initial net charges. Sphere A has a net charge of $+5Q$; sphere B has a net charge of $-3Q$; and sphere C has a net charge of $+6Q$. You first touch sphere B to sphere A, and then separate them; you then touch sphere A to sphere C, and then separate them; and finally you touch sphere C to sphere B, and then separate them. (a) Assuming no charge is transferred to you, what is the total combined charge on the three spheres at the end of the process? (b) What is the charge on each one of the spheres at the end of the process?

3. Consider again the system of three charged metal spheres in Exercise 2. You can set their initial charges to be whatever you wish, but you touch them together as described in the previous problem. (a) Is it possible for each sphere to end up with the same non-zero net charge? If so, give an example. (b) Is it possible for each sphere to end up with a different amount of charge? If so, give an example. (c) Is it possible for the sign of the charge on one sphere to be opposite to the charge on the other two spheres, at the end of the process? If so, give an example.

4. A small charged ball with a charge of $+5Q$ is located at a distance of 2.0 m from a charged ball with a charge of $+Q$. Which ball exerts a larger-magnitude force on the other? Justify your answer.

5. Ball A is charged, and so is ball B. The two balls are separated by a distance of d, and they can be treated as point charges. Which of the following changes, done individually, would cause the force that ball B exerts on ball A to double? If the change does not cause a doubling of the force state explicitly what effect the change has. (a) Double the charge on ball B. (b) Double the charge on ball A. (c) Double the charge on both balls. (d) Decrease the separation between the balls to $d/2$.

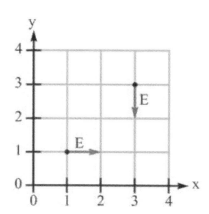

6. The electric field in the region shown in Figure 16.19 is produced by a single point charge, but the location of that point charge is unknown. At the point ($x = 1$, $y = 1$), we know that the electric field is directed to the right. (a) If this is all we knew about the field, what could we say about the location and sign of the point charge? (b) We also know that, at the point ($x = 3$, $y = 3$), the electric field is directed down. With this additional information, what can we say about the location and sign of the point charge?

Figure 16.19: The electric field in this region is produced by a single point charge. The location of the point charge is not shown. For Exercise 6.

7. In a particular uniform electric field, the electric field lines are directed to the right. You draw a diagram to reflect this field, showing a number of equally spaced parallel arrows, separated by 1 cm, that are directed to the right. What would be the spacing between the arrows on your field-line diagram if the field was reduced in magnitude by a factor of 2?

8. You want to sketch a field-line pattern for a situation involving two point charges separated by some distance. One charge has a magnitude of $+3Q$ while the other has a magnitude of $-Q$. (a) If you draw 15 lines emerging from the $+3Q$ charge, how many should you draw ending on the $-Q$ charge? (b) Where do the remaining lines go? (c) At a point quite far from the two charges, the electric field looks like the electric field from a single point charge. What is the charge of this single point charge?

9. As shown in Figure 16.20, a positive test charge experiences a net force directed right when it is placed exactly halfway between a ball of charge $+Q$ and a second ball of unknown charge. (a) What is the direction of the electric field at the point where the test charge is? (b) What, if anything, can you conclude about the sign and/or magnitude of the charge on the second ball?

Figure 16.20: A positive test charge is located halfway between a ball of charge $+Q$ and a second ball of unknown charge. For Exercise 9.

10. As shown in Figure 16.21, a positive test charge experiences a net force directed right when the test charge is placed twice as far from a ball of unknown charge as it is from a ball of charge $+Q$. (a) What is the direction of the electric field at the point where the test charge is? (b) What, if anything, can you conclude about the sign and/or magnitude of the charge on the second ball?

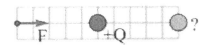

Figure 16.21: A ball of charge $+Q$ is located halfway between a positive test charge and a second ball of unknown charge. For Exercise 10.

11. Figure 16.23 shows three charged balls, which are equally spaced along a line. Each ball has a non-zero charge, but the signs of the charges on balls 1 and 3 are not shown. Ball 2 has a negative charge. The figure also shows the net force acting on each ball, because of its interaction with the other two balls. Assume that the only forces acting are electrostatic forces. (a) Ball 1 experiences no net force. What, if anything, does this tell us about the sign and magnitude of the charge on ball 3? Explain. (b) The net force on ball 2 is directed to the left. What, if anything, does this tell us about the sign and magnitude of the charge on ball 1? Explain. (c) Rank the balls, from largest to smallest, based on the magnitude of the charge on them.

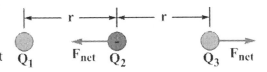

Figure 16.22: Three charged balls are equally spaced along a line. The sign of the charge on ball 2 is negative, but the signs of the charges on the other two balls are not shown. The net force on each ball is also shown – ball 1 experiences no net force due to its electrostatic interaction with the other two balls. For Exercise 11.

12. Three balls, with charges of $+q$, $+2q$, and $+3q$, are arranged so there is one ball at each corner of an equilateral triangle. Rank the balls based on the magnitude of the net electrostatic force they experience, from largest to smallest.

13. Five balls, two of charge $+q$ and three of charge $-2q$, are arranged as shown in Figure 16.23. What is the magnitude and direction of the net electrostatic force on the ball of charge $+q$ that is located at the origin?

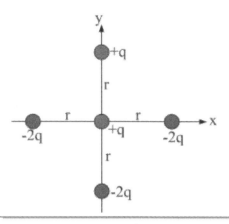

Figure 16.23: An arrangement of five charged balls, for Exercise 13.

14. Three charged balls are placed so that each is at a different corner of a square, as shown in Figure 16.24. Balls 1 and 3 both have positive charges, but the sign of the charge on ball 2 is not shown. The figure also shows the net force acting on each of the balls – the only forces that matter here are those associated with the interactions between the charges. (a) What, if anything, does the direction of the net force acting on ball 2 tell us about the sign of the charge on ball 2? Explain. (b) What, if anything, does the direction of the net force acting on ball 2 tell us about how the magnitude of the charge on ball 1 compares to the magnitude of the charge on ball 3? Explain. (c) What, if anything, does the direction of the net force acting on ball 3 tell us about how the magnitude of the charge on ball 1 compares to the magnitude of the charge on ball 2? Explain. (d) Rank the balls, from largest to smallest, based on the magnitude of the charge on them.

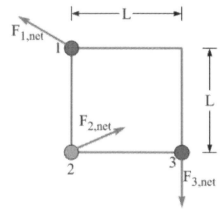

Figure 16.24: An arrangement of three charged balls, for Exercise 14. The arrows represent the net electrostatic force acting on each ball.

Exercises 15 – 20 deal with Coulomb's Law. Treat all charged balls as point charges.

15. Two point charges, one with a charge of $+Q$ and the other with a charge of $+16Q$, are placed on the x-axis. The $+Q$ charge is located at $x = +6a$, and experiences a force of magnitude kQ^2/a^2 because of its electrostatic interaction with the second charge. Where is the second point charge located? State all possible solutions.

16. Two small identical conducting balls have different amounts of charge on them. Initially, when they are separated by 75 cm, one ball exerts an attractive force of 1.50 N on the second ball. The balls are then touched together briefly, and then again separated by 75 cm. Now, both balls have a positive charge, and the force that one ball exerts on the other is a repulsive force of 1.10 N. What was the charge on the two balls originally?

17. Two charged balls are placed on the x-axis, as shown in Figure 16.25. The first ball has a charge $+q$ and is located at the origin, while the second ball has a charge $-4q$ and is located at $x = +4a$. A third ball, with a charge of $+2q$, is then brought in and placed somewhere on the x-axis. Assume that each ball is influenced only by the other two balls, and neglect gravitational interactions. (a) Could the third ball be placed so that all three balls simultaneously experience no net force due to the other two? (b) Could the third ball be placed so that at least one of the three balls experiences no net force due to the other two? Briefly justify your answers.

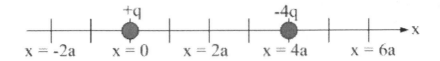

Figure 16.25: Two charged balls on the x-axis. For Exercises 17 – 19.

18. Return to the situation described in Exercise 17, and find all the possible locations where the third ball could be placed so that at least one of the three balls experiences no net force due to the other two.

19. Two charged balls are placed on the x-axis, as shown in Figure 16.25. The first ball has a charge +q and is located at the origin, while the second ball has a charge −4q and is located at x = +4a. Could a third ball, with an appropriate charge, be brought in and placed somewhere on the x-axis so that all three balls simultaneously experience no net force due to the other two? If so, find the charge and location of the third ball.

20. Three balls, each with the same magnitude charge, are arranged so there is one ball at each corner of an equilateral triangle. Each side of the triangle is exactly 1 meter long. (a) If each ball experiences a net force of 8.00 x 10 ^{6} N because of the other two balls, what is the charge of each ball? (b) Must the sign of the charge on each ball be the same, or could the charge on one ball be opposite to that of the charge on the other two balls? Explain.

Exercises 21 – 31 deal with electric field. Treat all charged balls as point charges.

21. An electron with an initial velocity of 1500 m/s directed straight up is in a uniform electric field of 200 N/C that is also directed straight up. (a) The electron is near the surface of the Earth. Is it reasonable to neglect the influence of gravity in this situation? Justify your answer. (b) How long does it take for the electron to come instantaneously to rest? (c) How far does the electron travel in this time?

22. An electron with an initial velocity of 7.5×10^5 m/s directed horizontally is in a uniform electric field of 400 N/C that is directed straight up. The electron starts 2.0 m above a flat floor. (a) How long does it take the electron to reach the floor? (b) How far does the electron travel horizontally in this time? (c) What is the speed of the electron as it runs into the floor?

23. A single point charge is located at an unknown point on the x-axis. There are no other charged objects nearby. You measure the electric field at the origin to be 600 N/C in the positive x-direction, while the electric field on the x-axis at x = +4.0 m is 5400 N/C in the negative x-direction. What is the sign and magnitude of the point charge, and where is it located?

24. The net electric field at the center of a square is directed to the right, as shown in Figure 16.26. This net field is the vector sum of electric fields from four point charges, which are located so that there is one point charge at each corner of the square. The charges have identical magnitudes, but may be positive or negative. Which are positive and which are negative? Is there more than one possible answer?

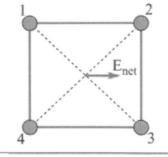

Figure 16.26: Four point charges are placed so that there is one charge at each corner of a square. The charges all have the same magnitude, but they may have different signs. The net electric field at the center of the square, due to these four charges, is directed right. For Exercise 24.

25. A ball of with a charge of $+6q$ is placed on the x-axis at $x = -2a$. There is a second ball of unknown charge at $x = +a$. If the net electric field at the origin due to the two balls has a magnitude of $\dfrac{kq}{a^2}$, what is the charge of the second ball? Find all possible solutions.

26. Repeat the previous problem, but now the net electric field at the origin has a magnitude of $\dfrac{3kq}{a^2}$.

27. A ball with a charge of $-2q$ is placed on the x-axis at $x = -a$. There is a second ball with a charge of $+q$ that is placed on the x-axis at an unknown location. If the net electric field at the origin due to the two balls has a magnitude of $\dfrac{6kq}{a^2}$, what is the location of the second ball? Find all possible solutions.

28. Three balls, with charges of $+4q$, $-2q$, and $-q$, are equally spaced along a line. The spacing between the balls is r. We can arrange the balls in three different ways, as shown in Figure 16.27. In each case the balls are in an isolated region of space very far from anything else. (a) In which case does the ball with the charge of $+4q$ experience a larger-magnitude net force? Give a qualitative argument. (b) Calculate the magnitude and direction of the net force experienced by the $+4q$ charge in each case.

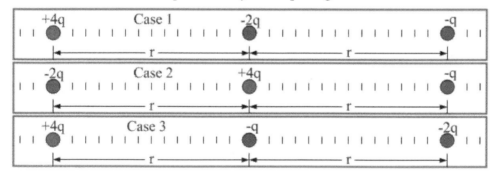

Figure 16.27: Three different arrangements of three balls of charge $+4q$, $-2q$, and $-q$ placed on a line with a distance r between neighboring balls. For Exercises 28 – 31.

29. Return to the situation described in Exercise 28, and shown in Figure 16.27. (a) Duplicate the diagram, and then draw in two force vectors on each ball in each case, to represent the force each ball experiences due to the other two balls. Figure 16.4 shows an example of this process. (b) Rank the three cases, from largest to smallest, based on the magnitude of the net force exerted on the ball in the middle of the set of three balls.

30. Return to the situation described in Exercise 28, and shown in Figure 16.27. Which ball experiences the largest-magnitude net force in (a) Case 1, (b) Case 2, and (c) Case 3? Calculate the magnitude and direction of the force applied to the ball that is experiencing the largest magnitude force in (d) Case 1, (e) Case 2, and (f) Case 3.

31. Consider the situation shown in case 3 in Figure 16.27. Each charged ball experiences a net force because of the other two balls – if you did the previous problem you would have calculated the three different net forces already. (a) If you add these three net forces as vectors, what do you get? Why? (b) Would you get the same result in all similar situations, including cases 1 and 2 in Figure 16.27? Why or why not?

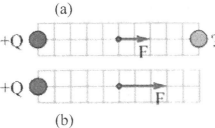

Exercises 32 – 36 deal with test charges. Treat all charged balls as point charges.

32. As shown in Figure 16.28 (a), a positive test charge placed exactly halfway between a ball of charge $+Q$ and a second ball of unknown charge experiences a net force directed right. When the second ball is removed from the situation, as in Figure 16.28 (b), the force experienced by the test charge increases by a factor of 3/2. What is the sign and magnitude of the charge on the second ball in Figure 16.28 (a)?

Figure 16.28: When the second ball shown in (a) is removed, as in (b), the force on the test charge increases by a factor of 3/2. For Exercise 32.

33. Figure 16.29 shows the net force experienced by a positive test charge located at the center of the diagram. The force comes from two nearby charged balls, one with a charge of $+Q$ and one with an unknown charge. (a) What is the sign of the charge on the second ball? (b) Is the magnitude of the charge on the second ball more than, less than, or equal to Q? (c) Find the sign and magnitude of the charge on the second ball.

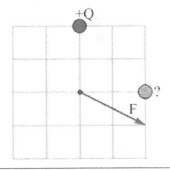

34. Figure 16.30 shows the net force experienced by a positive test charge located at the center of the diagram. The force comes from two nearby charged balls, one with a charge of $+Q$ and one with an unknown charge. (a) What is the sign of the charge on the second ball? (b) Is the magnitude of the charge on the second ball more than, less than, or equal to Q? (c) Find the sign and magnitude of the charge on the second ball.

Figure 16.29: The two charged balls produce a net force directed down and to the right, as shown, on the test charge at the center of the diagram. For Exercise 33.

Figure 16.30: The two charged balls produce a net force at a 45° angle directed down and to the left, as shown, on the test charge at the center of the diagram. For Exercise 34.

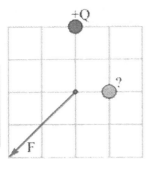

35. Figure 16.31 shows the net force experienced by a positive test charge located at the center of the diagram. The force comes from two nearby charged balls, one with a charge of $+Q$ and one with an unknown charge. What is the sign and magnitude of the charge on the second ball?

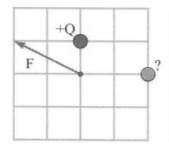

Figure 16.31: The two charged balls produce a net force directed up and to the left, as shown, on the test charge at the center of the diagram. For Exercise 35.

36. Two identical test charges are located at different positions, as shown in Figure 16.32. The test charges experience forces of the same magnitude, and in the directions shown. Could these forces be produced by a single nearby point charge? If so, state where that point charge would be and what you know about it. If not, explain why not.

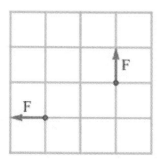

Figure 16.32: Two identical positive test charges experience the forces shown in the diagram. For Exercise 36.

General problems and conceptual questions. Treat all charged balls as point charges.

37. Benjamin Franklin made significant scientific contributions to our understanding of electric charge. Do some research on these contributions, and write two or three paragraphs describing them.

38. The SI unit of charge, the coulomb, is named after Charles-Augustin de Coulomb. Do some research on Coulomb (the scientist) and write a short biographical sketch of him.

39. A photocopier relies on the basic principles of charge. Do some research on how a photocopier works, and write a step-by-step explanation of the photocopying process.

40. At the laundromat, you put your silk pajamas in the dryer with your woolen sweater. When you take them out again, you find they are stuck together, because of static cling. Explain why this happens.

41. What is the speed of an electron in the ground state of a hydrogen atom? See Example 16.3A for relevant data, and use the Bohr model of the hydrogen atom, in which the electron follows a circular orbit around the proton.

42. Two small identical conducting balls have different amounts of charge on them. When they are first separated by 35 cm, one ball exerts a force with a magnitude of 4.50 N on the second ball. The balls are then touched together briefly, and then again separated by 35 cm. Now the force that one ball exerts on the other has a magnitude of 7.50 N. What was the charge on the two balls originally? Is there more than one possible solution?

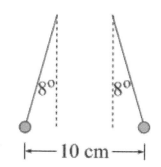

43. Two identical balls each have a charge of -2.5×10^{-6} C. The balls hang from identical strings that are at 8.0° from the vertical because of the repulsive force between the charged balls. The balls are separated by a distance of 10 cm, as shown in Figure 16.33. What is the mass of each ball?

Figure 16.33: Two identical charged balls hang from strings, for Exercise 43.

44. Three balls, with charges of $+q$, $-2q$, and $+3q$, are arranged so there is one ball at each corner of an equilateral triangle. Each side of the triangle is exactly 2 meters long. (a) Find the magnitude of the net electrostatic force acting on the ball of charge $+3q$. (b) What is the magnitude of the electric field at the center of the triangle?

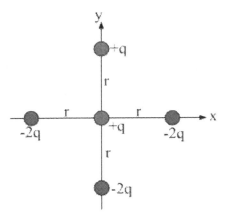

45. (a) Referring to Figure 16.34, which of the two balls of charge $+q$ experiences the largest net electrostatic force? Justify your answer. (b) What is the magnitude of the net electrostatic force experienced by the three different balls of charge $-2q$?

Figure 16.34: An arrangement of five charged balls, for Exercise 45.

46. A ball of charge $+3q$ is placed on the x-axis at $x = -a$. There is a second ball with an unknown charge that is placed on the x-axis at an unknown location. If the electrostatic force the second ball exerts on the first ball has a magnitude of $\dfrac{kq^2}{2a^2}$ and the net electric field at $x = 0$ due to these balls is $\dfrac{69kq}{25a^2}$ in the positive x direction, what is the charge and location of the second ball? Find all possible solutions.

47. Consider the three cases shown in Figure 16.35. Rank these cases, from largest to smallest, based on the (a) magnitude of the electrostatic force experienced by the ball of charge $-q$; (b) magnitude of the electric field at the origin.

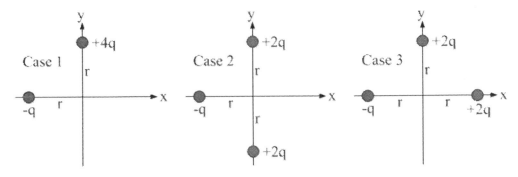

Figure 16.35: Three different configurations of charged objects, for Exercises 47 – 49.

48. Consider the three cases shown in Figure 16.35. Determine the magnitude and direction of the electrostatic force experienced by the ball of charge $-q$ in (a) case 1; (b) case 2; (c) case 3.

49. Consider the three cases shown in Figure 16.35. Determine the magnitude and direction of the electric field at the origin in (a) case1; (b) case 2; (c) case 3.

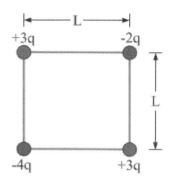

50. Four small charged balls are arranged at the corners of a square that measures L on each side, as shown in Figure 16.36. (a) Which ball experiences the largest-magnitude force due to the other three balls? (b) What is the direction of the net force acting on the ball with the charge of $-4q$? (c) If you doubled the length of each side of the square, so neighboring charges were separated by a distance of $2L$ instead, what would happen to the magnitude of the force experienced by each charge?

51. Four small charged balls are arranged at the corners of a square that measures L on each side, as shown in Figure 16.36. To what value can you adjust the charge on the ball of charge $+3q$ in the lower right corner so that the ball of charge $+3q$ in the upper left corner experiences (a) a net force that points directly toward the ball of charge $-4q$, or (b) a net force that points directly away from the ball of charge $-2q$? (c) Determine the magnitude of the net force (in terms of k, q, and L) acting on the $+3q$ charge in the upper left corner in these two situations.

Figure 16.36: Four charged balls at the corners of a square, for Exercises 50 – 53.

52. Four small charged balls are arranged at the corners of a square that measures L on each side, as shown in Figure 16.36. (a) If you adjust the charge on the ball with the $-4q$ charge at the lower left, could you bring the net force acting on the ball with the $-2q$ charge to zero? (b) If so, calculate the sign and magnitude of the charge on the ball in the lower left corner that would be required. If not, explain why not.

53. Four small charged balls are arranged at the corners of a square that measures L on each side, as shown in Figure 16.36. (a) Calculate the magnitude and direction of the electric field at the center of the square. (b) Could you change the amount of charge on one of the balls to produce a net electric field at the center that is directed horizontally to the left? If so, which ball would you change the charge of and what would you change it to? If not, explain why not.

54. A ball with a weight of 10 N hangs down from a string that will break if its tension is greater than or equal to 25 N. The ball has a charge of $+5.0\times10^{-6}$ C. You want to break the string by introducing a uniform electric field. What is the magnitude and direction of the minimum electric field required to cause the string to break?

55. A small charged ball with a weight of 10 N hangs from a string. When the ball is placed in a uniform electric field of 800 V/m directed left, the string makes a 40° angle with the vertical, as shown in Figure 16.37. What is the sign and magnitude of the charge on the ball?

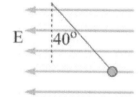

56. Return to Example 16.4, in which we calculated the magnitude and direction of the net force exerted on the $+q$ charge by the other two charges. Now determine the magnitude and direction of the net force exerted on (a) the $-2q$ charge, and (b) the $+3q$ charge.

Figure 16.37: The equilibrium position of a ball in a uniform electric field directed left, for Exercise 55.

57. Return to the situation of Example 16.4. Determine the magnitude and direction of the net electric field at (a) the center of the square, and (b) the unoccupied corner.

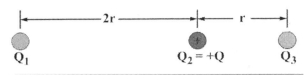

58. Three charged balls are placed in a line, as shown in Figure 16.38. Ball 1, which has an unknown charge and sign, is a distance 2r to the left of charge 2. Ball 2 is positive, with a charge of +Q. Ball 3 has an unknown non-zero charge and sign, and is a distance r to the right of ball 2. Ball 3 feels no net electrostatic force because of the other two balls. (a) Is there enough information given here to find the sign of the charge on ball 1? If so, what is the sign? (b) Can we find the magnitude of the charge on ball 1? If so, what is it? (c) Can we find the sign of the charge on ball 3? If so, what is the sign? (d) Can we find the magnitude of the charge on ball 3? If so, what is it?

Figure 16.38: Three charges in a line. Only the sign and magnitude of charge 2 are known, although we also know that charge 3 is in equilibrium. For Exercise 58.

59. A single point charge is located at an unknown point on the x-axis. There are no other charged objects nearby. You measure the electric field at x = +2.0 m to be 6000 N/C, directed in the +x direction, while the field at x = +5.0 m has a magnitude of 1500 N/C. What is the sign and magnitude of the point charge? State all possible answers.

60. In Example 16.6A we found one point on the line passing through two unequal charges at which the net electric field is zero. Are there any such points that are off this line, a finite distance from the charges? Use one or more diagrams to support your answer.

61. Consider the field-line diagram shown in Figure 16.39. The arrows show field lines emerging from charge 1, on the left, and ending at charge 2, on the right. (a) What is the sign of each of these charges? (b) If the charge on charge 1 has a magnitude of 10 μC, what is the charge on charge 2?

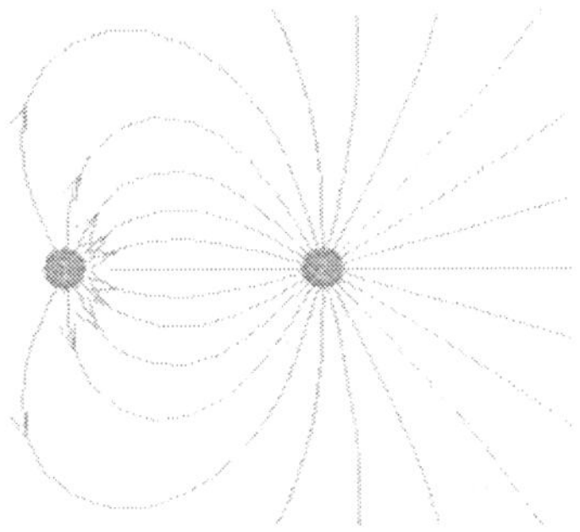

62. One point charge is located at the origin, and has a charge of +5.0 μC. A second point charge is located at x = +2.0 m, and has a charge of –9.0 μC. (a) Analyze the situation qualitatively to determine approximate locations of any points where the net electric field due to these two point charges is zero. (b) Determine the location of all such points.

Figure 16.39: A field-line pattern near two charged objects, for Exercise 61.

63. Repeat Exercise 62, except now the second point charge has a charge of +18.0 μC.

64. A charge of unknown sign and magnitude is located halfway between a small ball with a charge of +Q and a positive test charge. The test charge experiences a net force directed right, as shown in Figure 16.40. (a) What, if anything, can you conclude about the sign and/or magnitude of the unknown charge? (b) If you moved the test charge to the point halfway between the +Q charge and the unknown charge, in which direction would the force be on the test charge? (c) You return the test charge to the position shown in Figure 16.40. You then observe that when you shift the position of the unknown charge a little to the right, the force experienced by the test charge decreases in magnitude a little. What, if anything, can you conclude about the sign and/or magnitude of the unknown charge?

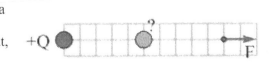

Figure 16.40: A charge of unknown sign and magnitude is located halfway between a small ball of charge +Q and a positive test charge. For Exercise 64.

65. One of the safest places to be in a lightning storm is inside a car, as long as the car is made of metal. Even if lightning strikes the car you should be safe inside. Explain why this is.

66. (a) Sketch a field-line pattern for a situation in which there is a uniform electric field directed straight down. (b) Re-draw the pattern for when a neutral metal sphere is placed into the field. Draw the sphere large enough that it covers a region on the diagram where at least 5 field lines pass through on your original diagram.

67. Figure 16.41 shows possible equilibrium distributions of charge on a hollow, thick-walled metal sphere that has a net charge of −Q. Which is most correct, assuming there are no other charged objects in the vicinity?

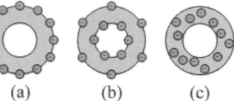

(a)   (b)   (c)

> **Figure 16.41**: Possible equilibrium distributions of charge on a hollow, thick-walled metal sphere that has a net charge of −Q. In (a) the charge is spread uniformly over the outer surface; in (b) half the charge is distributed over the outer surface and half is on the inner surface; and in (c) the charge is randomly distributed throughout the bulk of the sphere. For Exercise 67.

68. Three students are having a conversation. Explain what you think is correct about what they say, and what you think is incorrect. In particular, how would you respond to Brenda's questions in the last statement?

Brenda: So, the question says that we have two objects, one with a +5Q charge and the other with a +Q charge, and it asks us for which one exerts a larger magnitude force on the other. Well, that's the +5Q object, right – it exerts 5 times as much force as the other one.

Paul: Let's think about Coulomb's law – it does say that when you increase one of the charges that the force goes up.

Lauren: Thinking about Coulomb's law makes sense, except that, in Coulomb's law, the force is proportional to the product of the two charges. You can apply it to each charge, and you get the same answer. So, I think the forces are the same.

Paul: That's consistent with Newton's third law, too – the objects have to exert equal and opposite forces on one another. That sounds right.

Brenda: That just doesn't make sense to me – shouldn't the bigger charge exert more force? I kind of got Newton's third law when we were talking about colliding carts a few months ago, but how can it apply for things that don't even touch each other, like these little charges?

This photograph shows a lightning strike in Independence, Missouri. Lightning requires a large build-up of charge in a thundercloud, and an associated large difference in potential between the cloud and the ground. The basic principles of physics that explain the phenomenon of lightning will be the theme of this chapter. Photo credit: Mark Coldren, from publicdomainpictures.net.

Chapter 17 – Electric Potential Energy and Electric Potential

CHAPTER CONTENTS

In this chapter, we will continue to draw on concepts we examined when we learned about gravity. We will then extend our knowledge by defining a new concept, potential, which applies to situations involving charge as well as situations involving gravitational interactions. We will start, however, by looking at the familiar concept of energy. The concept of energy conservation can be applied to charged objects in much the same way that we applied it to understanding carts on inclines and on masses interacting via gravity.

17-1 Electric Potential Energy

Whenever charged objects interact with one another, there is an energy associated with that interaction. In general, we have two special cases to consider. The first is the energy associated with a charged object in a uniform electric field, and the second is the energy associated with the interaction between point charges. These are analogous to the two situations we examined earlier for gravity.

The potential energy associated with the interaction between one object with a charge q and a second object with a charge Q that is a distance r away is given by:

$$U_E = \frac{kqQ}{r}$$ (Eq. 17.1: **Potential energy for the interaction between two charges**)

where $k = 8.99 \times 10^9 \ \text{N m}^2/\text{C}^2$ is a constant. If the charges are of opposite signs, then the potential energy is negative – this indicates an attraction. If the charges have the same sign, the potential energy is positive, indicating a repulsion.

Once again, we can see the parallel with gravity, for which the equivalent expression is:

$$U_G = -\frac{GmM}{r}$$ (Eq. 8.4: **Potential energy for the interaction between two masses**)

where $G = 6.67 \times 10^{-11} \text{N m}^2/\text{kg}^2$ is the universal gravitational constant. For the interaction between charges k takes the place of G, and the values of the charges q and Q take the place of the values of the masses, m and M.

Case 1 – A charged object in a uniform electric field

This situation is directly analogous to the situation of an object with mass in a uniform gravitational field. When we raise a ball of mass m through a height h in a uniform gravitational field g directed down, the change in potential energy is $\Delta U_G = mgh$. Note that we take h to be the component of the displacement parallel to, and opposite in direction to, the field. If the ball experiences a displacement $\Delta \vec{r}$ then an equivalent equation is $\Delta U_G = -m\vec{g} \bullet \Delta \vec{r} = -mg\Delta r \cos\theta$, where θ is the angle between the gravitational field and the displacement.

The equivalent expression for a charge object in a uniform electric field is:

$$\Delta U_E = -q\vec{E} \bullet \Delta \vec{r} = -qE\Delta r \cos\theta, \qquad \text{(Eq. 17.2: \textbf{Change in potential energy in a uniform field})}$$

where θ is the angle between the electric field $\vec{E}$ and the displacement $\Delta \vec{r}$.

We are free to define the zero level of potential energy, but only in a uniform field.

Related End-of-Chapter Exercise: 6.

Case 2 – The electric potential energy of a set of point charges

In this situation, we look at the interaction between pairs of charges. We use equation 17.1 to calculate the energy of each pair of interacting objects, and then simply add up all these numbers because potential energy is a scalar. Note that we are not free to define the zero level. The zero is defined by equation 17.1, in fact, because the potential energy goes to zero as the distance between the charges approaches infinity.

Compare Exploration 17.1 to Exploration 8.4.

EXPLORATION 17.1 – Calculate the total potential energy in a system

Three charged balls, of charge $-q$, $+2q$, and $-3q$, are placed in a line, as shown in Figure 17.1. What is the total electric potential energy of this system?

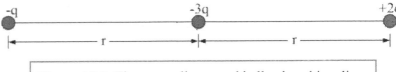

Figure 17.1: Three equally spaced balls placed in a line.

To determine the total potential energy of the system, consider the number of interacting pairs. In this case, there are three ways to pair up the objects, so there are three terms to add together to find the total potential energy. Because energy is a scalar, we do not have to worry about direction. Using a subscript of 1 for the ball of charge $-q$, 2 for the ball of charge $+2q$, and 3 for the ball of charge $-3q$, we get:

$$U_{Total} = U_{13} + U_{23} + U_{12} = \frac{k(-q)(-3q)}{r} + \frac{k(+2q)(-3q)}{r} + \frac{k(-q)(+2q)}{2r} = -\frac{4kq^2}{r}.$$

When a system has a negative total energy (including the total kinetic energy, of which there is none in this situation), that is indicative of a bound system. In general, there is a greater degree of attraction in the system than repulsion.

Key ideas for electric potential energy: Potential energy is a scalar. The total electric potential energy of a system of objects can be found by adding up the energy associated with each interacting pair of objects.   **Related End-of-Chapter Exercises: 4, 42, 46.**

Work – an equivalent approach

Consider again the system shown in Figure 17.1. If the three charged balls start off infinitely far away from their final positions, and infinitely far from one another, how much work do we have to do to assemble the balls into the configuration shown in Figure 17.1? Assume that, aside from their interactions with us, the balls interact only with one another, electrostatically.

Pick one ball to bring into position first. Let's start with the ball with the $-3q$ charge. Because the other charged balls are still infinitely far away, it takes no work to bring the first ball into position. There are no other interactions to worry about.

Now, let's bring the ball with the $-q$ charge into position. The potential energy changes from 0, when the two balls are infinitely far away, to $+3kq^2/r$, when those two balls are in their final positions. This potential energy comes from work we do – we do $+3kq^2/r$ worth of work.

Finally, bring the ball with the $+2q$ charge into position. The potential energy associated with this ball changes from 0, when that ball is at infinity, to $-2kq^2/(2r) + (-6kq^2)/r = -7kq^2/r$, when the ball with the $+2q$ charge is in its final position.

Adding the two individual work values to find the total work to assemble the system gives $+3kq^2/r -7kq^2/r = -4kq^2/r$, the same result we got for the potential energy of the system. The work done in assembling the system is equal to the system's potential energy.

Essential Question 17.1: Return to Exploration 17.1. If we replace the ball of charge $+2q$ by a ball of charge $-2q$, does the potential energy of the system increase, decrease, or stay the same?

Answer to Essential Question 17.1: The potential energy of the system would increase and, in fact, would become $+10kq^2/r$. This is because the balls would all have charge of the same sign, and, overall, the particles of the system would repel one another.

17-2 Example Problems Involving Potential Energy

EXAMPLE 17.2A – Stopping an electron

An electron with a speed of 1.2×10^5 m/s enters a region where there is a uniform electric field with $E = 500$ N/C, as shown in Figure 17.2.

(a) Show that gravitational influences can be neglected in this situation.

(b) If, after traveling for some distance d in the field, the electron comes to rest for an instant, in which direction is the field?

(c) Calculate d.

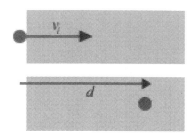

Figure 17.2: The top diagram shows the electron entering the electric field. The bottom diagram shows the point where the electron comes instantaneously to rest.

SOLUTION

(a) Let's assume the electron is at the surface of the Earth, where the acceleration due to gravity is the familiar $g = 9.8$ m/s². What is the acceleration associated with the electric field? The magnitude of the force applied to the electron by the field is $F = qE$. Dividing by the mass gives the magnitude of the acceleration of the electron:

$$a = \frac{F}{m} = \frac{qE}{m} = \frac{\left(1.60 \times 10^{-19} \text{ C}\right)\left(500 \text{ N/C}\right)}{9.11 \times 10^{-31} \text{ kg}} = 8.78 \times 10^{13} \text{ N/kg}.$$

This is 13 orders of magnitude larger than g, which justifies neglecting gravity.

(b) To stop the electron, the force on the electron from the field must be directed opposite to the electron's initial velocity. Because the electron has a negative charge, the field and force are in opposite directions. Thus, the electric field must be in the same direction as the electron's initial velocity.

(c) Let's apply energy conservation here, using the usual energy equation:
$$U_i + K_i + W_{nc} = U_f + K_f.$$

In this case, there are no non-conservative forces acting, and the final state is the point at which the electron comes instantaneously to rest. Thus, the final kinetic energy is zero. In addition, we can define the initial electric potential energy to be zero. This gives $K_i = U_f$.

The final potential energy is given by: $U_f = U_i + \Delta U = 0 + (-(-e)Ed) = +eEd$.

Putting everything together gives: $\frac{1}{2}mv_i^2 = +eEd$.

Solving for the distance gives: $d = \frac{mv_i^2}{2eE} = \frac{\left(9.11 \times 10^{-31} \text{ kg}\right)\left(1.2 \times 10^5 \text{ m/s}\right)^2}{2\left(1.60 \times 10^{-19} \text{ C}\right)\left(500 \text{ N/C}\right)} = 8.2 \times 10^{-5} \text{ m}.$

Compare the following Example to Example 8.4.

EXAMPLE 17.2B – Applying conservation ideas
Two identical balls, of mass 1.0 kg and charge +2.0 μC, are initially separated by 2.0 m in a region of space in which they interact only with one another. When they are released from rest, they accelerate away from one another. When they are 4.0 m apart, how fast are they going?

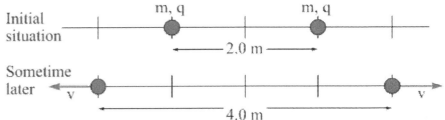

SOLUTION

Figure 17.3 shows the balls at the beginning and when they are separated by 4.0 m. Analyzing forces, we find that the force on each ball decreases as the distance between the balls increases. This makes it difficult to apply a force analysis. Energy conservation is a simpler approach. Our energy equation is:

Figure 17.3: The balls are initially at rest. Repulsion from their like charges causes them to accelerate away from one another.

$$U_i + K_i + W_{nc} = U_f + K_f .$$

In this case, no non-conservative forces act and, in the initial state, the kinetic energy is zero because both objects are at rest. This gives $U_i = U_f + K_f$. The final kinetic energy represents the kinetic energy of the system, the sum of the kinetic energies of the two objects.

Do we need to account for gravity here? In this situation we do not, which we can tell by calculating that the gravitational potential energy is much less than the electric potential energy. The change in potential energy is really what is important, but that is of the same order of magnitude as the potential energy.

$$U_{iG} = -\frac{Gmm}{r} = -\frac{(6.67\times10^{-11} \text{ N m}^2/\text{kg}^2)(1.0 \text{ kg})^2}{r} \qquad U_{iE} = \frac{kqq}{r} = -\frac{(8.99\times10^9 \text{ N m}^2/\text{C}^2)(2.0\times10^{-6} \text{ C})^2}{r}$$

Let's solve this generally, using a mass of m and a final speed of v for each ball.

The energy equation becomes: $\dfrac{kqq}{2.0 \text{ m}} = \dfrac{kqq}{4.0 \text{ m}} + \dfrac{1}{2}mv^2 + \dfrac{1}{2}mv^2 .$

Simplifying: $\dfrac{kq^2}{4.0 \text{ m}} = mv^2 .$

Solving for v gives:

$$v = \sqrt{\frac{kq^2}{m(4.0 \text{ m})}} = \sqrt{\frac{(8.99\times10^9 \text{ N m}^2/\text{C}^2)(2.0\times10^{-6} \text{ C})^2}{(1.0 \text{ kg})(4.0 \text{ m})}} = 9.5\times10^{-2} \text{ m/s.}$$

Related End-of-Chapter Exercises: Problems 1, 2, 3, 32, 33.

Essential Question 17.2: How would we solve Example 17.2B if the balls had different masses? To be specific, let's say the ball on the left has a mass of 1.0 kg while the ball on the right has a mass of 3.0 kg. What are the speeds of the two balls in that case?

Answer to Essential Question 17.2: Because there is no net external force, the system's momentum is conserved. There is no initial momentum. For the net momentum to remain zero, the two momenta must always be equal-and-opposite. Defining right to be positive, and using 1 as a subscript for the ball on the left and 2 for the other ball, momentum conservation gives:

$$0 = -mv_1 + 3mv_2 \text{, which we can simplify to } v_1 = 3v_2.$$

The energy conservation equation is: $+\dfrac{kq^2}{2.0 \text{ m}} = m(3v_2)^2 + 3mv_2^2 = 12mv_2^2$

This gives $v_2 = \sqrt{\dfrac{kq^2}{m(24.0 \text{ m})}}$, and $v_1 = 3v_2 = 3\sqrt{\dfrac{kq^2}{m(24.0 \text{ m})}}$.

Using $m = 1.0$ kg, we get $v_2 = 3.9 \times 10^{-2}$ m/s and $v_1 = 12 \times 10^{-2}$ m/s.

17-3 Electric Potential

Gravitational potential, V_g, is related to gravitational potential energy in the same way that field is related to force, by a factor of the mass. Like field, gravitational potential is a way of measuring how an object with mass, or a set of objects with mass, influences the region around it.

The electric potential, V, at a particular point can be defined in terms of the electric potential energy, U, associated with an object of charge q being placed at that point:

$$V = \frac{U}{q}, \quad \text{or} \quad U = qV. \qquad \text{(Eq. 17.3: } \textbf{Connecting electric potential and potential energy)}$$

The unit for electric potential is the volt (V). 1 V = 1 J/C.

A special case is the electric potential from a point charge with a charge Q:

$$V = \frac{kQ}{r}, \qquad \text{(Equation 17.4: } \textbf{Electric potential from a point charge)}$$

where r is the distance from the charge to the point in space where we are finding the potential. Electric potential is a scalar, so it has a sign but not a direction.

We can visualize potential by drawing **equipotentials**, lines or surfaces that connect points of the same potential. On a 2-D picture like Figure 17.4 equipotentials are lines; in a three-dimensional world we have equipotential surfaces. Around a point charge, for instance, the equipotential surfaces are spheres centered on the charge. As Figure 17.4 suggests, we can define a gravitational potential analogous to electric potential.

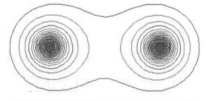

Figure 17.4: This diagram could represent the electric potential near two like charges, or the gravitational potential near two hills.

This is a little abstract, but you have probably seen gravitational equipotential lines, or contour lines, on a topographical map, as in Figure 17.5. Contour lines connect points at the same height – such points have the same gravitational potential. Where the lines are close together the terrain is steep; where the lines are far apart the terrain is flatter. If you want to go for an easy hike you have two choices. You can stick to flat terrain where the contour lines are far apart, or you can walk along a trail along a contour line, where the height is constant.

Figure 17.5: Contour lines on a topographical map, showing the terrain around the summit of Mt. Rainier, in Washington State. Photo credit: NASA/USGS.

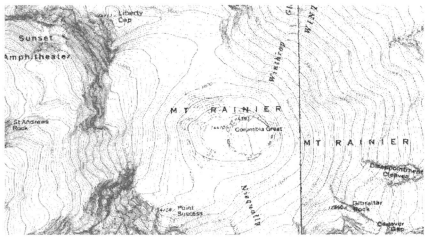

Moving along a contour line requires no work, while moving from one contour line to another does involve work, because the gravitational potential energy changes. Equipotential lines give us the same information. Moving a charged object at constant speed from one place to another along an equipotential requires no work; but moving from one equipotential to another does involve work. Note that equipotential lines are always perpendicular to field lines, and the field points in the direction of decreasing potential.

Potential in a Uniform Field

In a uniform field, we can define the zero for potential to be anywhere we find to be convenient. The value of the potential at a particular point is not what is important. What is of primary importance is how the potential changes from one point to another. In a uniform electric field we define the potential difference (the difference in potential between two points) as:

$$\Delta V = -E\,\Delta r\cos\theta,$$ (Equation 17.5: **Potential difference in a uniform field**)

where θ is the angle between the electric field $\vec{E}$ and the displacement $\Delta\vec{r}$.

Electric potential has units of volts, or J/C. We see that the electric potential difference between points A and B in Figure 17.6 is +20 J/C (starting at A and going to B), and between points C and B (starting at C and going to B) it is also +20 J/C. If we move an object that has a charge of +2 mC from A to B, or from C to B, what is the change in electric potential energy? This equals the work we do to move the object, assuming its kinetic energy remains the same.

$$\Delta U = q\,\Delta V = q\left(V_B - V_A\right) = 2\ \text{mC}\left(+10\ \text{J/C} - (-10\ \text{J/C})\right) = 2\ \text{mC}\times\left(+20\ \text{J/C}\right) = +40\ \text{mJ}.$$

Moving the 2 mC object from C to B also produces a +40 mJ change in potential energy; in moving it from A to C there is no change in potential energy (and no work), and moving it from B to either A or C changes the potential energy by –40 mJ.

If we change objects, using an object with a charge of +5 mC, for instance, the change in potential energy is easy to find. It is simply the new charge multiplied by the change in potential.

Related End-of-Chapter Exercises: Problems 13, 14, 36, 43.

Essential Question 17.3: If we moved an object with a charge of –2 mC from point A to point B in Figure 17.6, is the change in potential energy still +40 mJ? Explain.

                                                    +20 J/C
                                          B
                                                    +10 J/C

                                                    0

                                                    -10 J/C
                                          A   C
                                                    -20 J/C

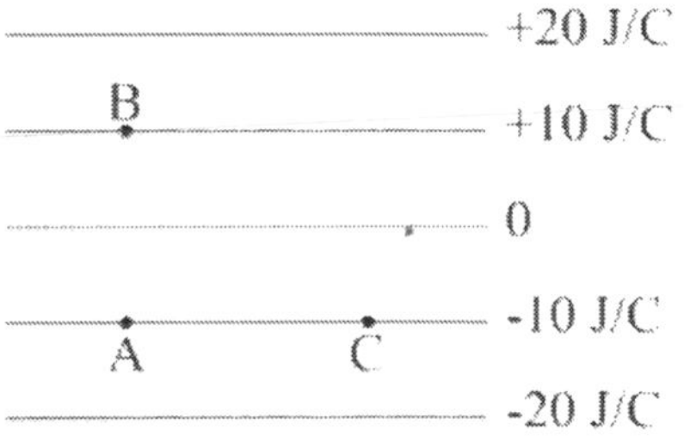

Figure 17.6: Equally spaced equipotentials are a hallmark of a uniform field. Field points in the direction of decreasing potential, so the field is directed down in this case.

Answer to Essential Question 17.3: No. If we re-do the calculation $\Delta U = q\Delta V$ we find that the change in potential energy is –40 mJ. Because potential energy is a scalar, –40 mJ is quite different from +40 mJ. We have to force a positively charged object to move to a region of higher potential (and higher potential energy). A negatively charged object already feels a force toward higher potential, and moving in that direction lowers the potential energy, so we have to pull back on it to cause it to stop at the point of higher potential.

17-4 Electric Potential for a Point Charge

A point charge is generally an object like a small charged ball. Such objects contribute to both the electric field and the electric potential in the space around them. The contribution of an object of charge Q to the electric potential at a point a distance r away is given by:

$$V = \frac{kQ}{r}.$$ (Equation 17.4: **Electric potential from a point charge**)

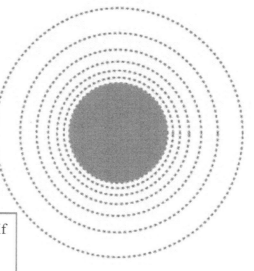

Interestingly, the potential outside a charged sphere that has a spherically symmetric charge distribution is also given by Equation 17.6, where Q is the total charge on the sphere and r is the distance from the center of the sphere to a point outside the sphere. In a two-dimensional diagram of this situation, such as that in Figure 17.7, the equipotential lines are circles centered on the object. Note that the potential difference (which is more important than the value of the potential at a point) between one equipotential line and the next is a constant value, but the lines get further apart as you move away from the object because the field decreases - the farther you are from the object, the smaller its influence.

Figure 17.7: Equipotentials near a charged conducting sphere. If the equipotential at the surface of the sphere connects all the points where the potential is –10 J/C, the equipotentials at gradually increasing distances from the sphere connect all points of potential -9 J/C, -8 J/C, -7 J/C, -6 J/C, -5 J/C, and -4 J/C, respectively. The field points in the direction of decreasing potential, toward the center of the sphere.

EXAMPLE 17.4 – Calculating the electric potential
Two balls are on the *x*-axis, a ball of charge $+q$ at $x = 0$ and a ball of charge $+9q$ at $x = +4a$.
　　(a) Find the electric potential at: (i) the point halfway between them, and (ii) $x = +a$, where the net electric field is zero.
　　(b) Graph the electric potential at points along the axis between $x = -4a$ and $x = +8a$.

SOLUTION
　　As usual, let's begin with a diagram of the situation. This is shown in Figure 17.8.

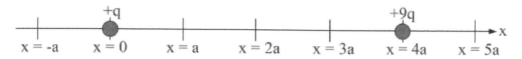

Figure 17.8: The position of the two charged balls on the *x*-axis.

To solve part (a) we use superposition, the idea that the potential at any point is simply the sum of the potential at that point from each of the objects.

(a)(i) The point halfway between the objects is $x = +2a$. Remembering that the r in the equation represents the distance from an object with mass to the point we're considering, which also happens to be $2a$ for both objects in this case, we get:

$$V_{net} = V_{+q} + V_{+9q} = +\frac{kq}{2a} + \frac{k(9q)}{2a} = +\frac{10kq}{2a} = +\frac{5kq}{a}.$$

(a)(ii) The point $x = +a$ is a distance a from the object of charge $+q$ and $3a$ from the object of charge $+9q$. The net potential is again the sum of the two individual contributions:

$$V_{net} = V_{+q} + V_{+9q} = +\frac{kq}{a} + \frac{k(9q)}{3a} = +\frac{4kq}{a}.$$

(b) To plot a graph of the potential as a function of position, we should have a general expression for the net potential. This is:

$$V_{net} = V_{+q} + V_{+9q} = +\frac{kq}{|x|} + \frac{k(9q)}{|4a - x|}.$$

The graph of this function between $x = -4a$ and $x = +8a$ is shown in Figure 17.9.

Related End-of-Chapter Exercises: 10-12, 56, 57.

There is a connection between the electric potential and the electric field that the graph in Figure 17.9 can help us to understand. The electric field is zero at $x = +a$. Does anything special happen at $x = +a$ on the potential graph? It turns out that, at $x = +a$, the graph reaches a local minimum, so the slope of the potential-vs.-position graph is zero at that point.

For $0 < x < +a$, the electric field along the axis points in the positive x direction while the slope of the potential is a negative value. Just the opposite happens in the range $+a < x < +4a$. The electric field is actually the negative of the slope

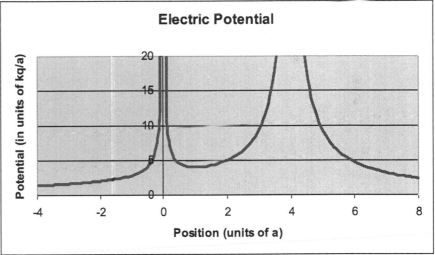

Figure 17.9: Graph of the electric potential along the axis near the balls. Note how the potential approaches positive infinity as you approach either of the balls.

of the potential-vs.-position graph. So, the potential graph is rather powerful, telling us at a glance something about where the electric field is strongest and in what direction it is, while at the same time giving us information about how potential energy would change if we placed another object with charge at a particular location.

Essential Question 17.4: What do we have to account for in determining net electric field that we do not have to account for in determining the net electric potential?

Answer to Essential Question 17.4: A fundamental difference between field and potential is that field is a vector while potential is a scalar. To find the net field we need to account for direction, adding the various fields as vectors. Finding potential simply involves adding numbers.

17-5 Working with Force, Field, Potential Energy, and Potential

Let's consider two important special cases. The first involves a charged object in a uniform electric field, while the second involves the electric field from point charges.

EXAMPLE 17.5A – Charged particles at the corners of a square

Consider the four situations shown in Figure 17.10, involving four charged particles of charge $+q$ or $-q$ placed so that there is one charge at each corner of the square.

(a) In which case does the charge at the top right corner of the square experience the net force of the largest magnitude?
(b) Is the net electric field at the center of the square zero in any of the configurations? If so, which?
(c) Is the net electric potential at the center of the square zero in any of the configurations? If so, which?
(d) Is the electric potential energy of the system equal to zero in any of the configurations? If so, in which configuration(s)?

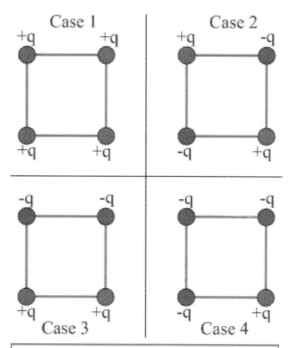

SOLUTION

(a) The charged particle at the top right corner experiences the largest magnitude net force in case 1, where all the forces have a component directed away from the center. In the other cases the particle at the top right experiences a mix of attractive and repulsive forces, which in this case causes some partial cancellation of the forces.

(b) The net electric field is zero at the center of the square in cases 1 and 2. If we pair up the charges that are diagonally across from one another, in cases 1 and 2 the fields within each pair cancel, giving no net field at the center.

Figure 17.10: Four different arrangements of equal-magnitude point charges placed so that there is one charged particle at each corner of a square.

(c) Each positive charge makes a positive contribution to the potential, while each negative charge makes a negative contribution. To produce a zero net potential at the center in this situation we need two positive charges and two negative charges, which is true in case 2 and case 3. The potential at the center of the square is positive in case 1 and negative in case 4.

(d) To find the total potential energy, we can work out the energy associated with each interacting pair of charges (with four charges there are six such interactions) by applying Equation 17.1, and add up these six terms to find the total potential energy. When we do this in case 4, we find that the total potential energy is zero. The total potential energy is non-zero in all other cases. If the square measures $L \times L$, the addition of the six terms gives:

$$U_{net,case4} = \frac{k(-q)(+q)}{L} + \frac{k(-q)(-q)}{L} + \frac{k(+q)(-q)}{L} + \frac{k(-q)(-q)}{L} + \frac{k(-q)(-q)}{\sqrt{2}\,L} + \frac{k(-q)(+q)}{\sqrt{2}\,L} = 0$$

Related End-of-Chapter Exercises for Example 17.5A: 19 - 23.

EXAMPLE 17.5B – Asking questions
Two charged particles, one with a charge of $+q$ and the other with a charge of $-3q$, are placed on the x-axis at $x = 0$ and $x = +4a$, respectively. This is shown in Figure 17.11. Give an example of a question pertaining to this situation that involves:

(a) Force; (b) Electric field; (c) Electric field, with a follow-up question on force;
(d) Electric potential energy; (e) Electric potential; (f) Electric potential, with a follow-up question on potential energy.

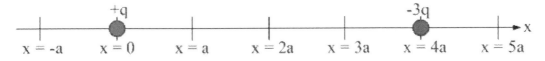

Figure 17.11: Two charged particles are separated by a distance of $4a$ on the x-axis.

SOLUTION
(a) Because force involves an interaction between objects, and we only have two objects in the system, there are a limited number of questions we could ask involving force unless we add another charge to the system. One is: "What is the magnitude and direction of the force exerted on the $+q$ charge by the $-3q$ charge?"

(b) "What is the magnitude and direction of the net electric field at $x = +5a$?" Because the two charged particles create an electric field at all points around them, we can ask an infinite number of questions involving field – we have an infinite number of points to choose from.

(c) "Consider the point $x = +2a$, $y = +2a$. (i) What is the magnitude and direction of the net electric field at that point? (ii) If a third charged particle with a charge of $+2q$ is placed at that point, what is the magnitude and direction of the force it experiences because of the two charges?"

(d) The pattern for (a) – (c) can be repeated for potential energy and potential. "What is the electric potential energy associated with this system of two charges?"

(e) "What is the net electric potential at $x = +5a$ due to the two charges?" As with field, we can ask about the potential at any point.

(f) "Consider the point $x = +2a$, $y = +2a$. (i) What is the net electric potential at that point? (ii) If a third charged particle with a charge of $+2q$ is placed at that point, what is the change in potential energy for the system?"

Related End-of-Chapter Exercises: 16 - 18.

Essential Question 17.5: Return to the situation described in Example 17.5A. In which configurations, could you bring a fifth charged particle from infinitely far away and place it at the center of the square without doing any net work? In which configurations would you do negative net work in bringing a fifth charged particle from infinitely far away to the center of the square?

Answer to Essential Question 17.5: The net work you do to bring a charge Q from infinity to the center is $W_{You} = Q\Delta V = Q(V_f - V_i)$. V_i, the potential at infinity, is zero, so $W_{You} = QV_f$. This is zero if V_f, the potential at the center of the square from the four charges, is zero, which is true in cases 2 and 3. V_f is positive in case 1, so your net work in that case is negative if Q is negative. In case 4 V_f is negative, so your net work is negative if Q is positive. Negative work by you means that the fifth charge experiences a net attraction, and would accelerate unless you held it back.

17-6 Capacitors and Dielectrics

A capacitor is a device for storing charge. One example is a parallel-plate capacitor, consisting of two identical metal plates, each of area A, placed parallel to one another and separated by a distance d. The space between the plates is sometimes filled with insulating material. If it is we say that the capacitor contains a **dielectric**.

One way to charge a capacitor is to connect it to a battery, connecting one wire from the positive terminal of the battery to one plate of the capacitor, and a second wire from the negative terminal of the battery to the other plate. A diagram of this is shown in Figure 17.12, along with a circuit diagram. In the circuit diagram a capacitor is shown as two parallel lines of equal length, looking very much like a parallel-plate capacitor, and a battery is represented by two parallel lines of different length. The longer line on the battery symbol represents the positive terminal.

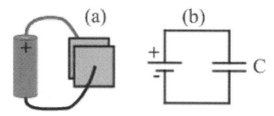

Figure 17.12: (a) A diagram, and (b) a circuit diagram, showing a battery connected to a capacitor (C) by two wires.

It is useful to think of the battery as a charge pump. The battery pumps electrons from the plate attached to the positive terminal, leaving that plate with a charge of $+Q$, to the plate attached to the negative terminal, giving that plate a charge of $-Q$. The capacitor then stores a charge Q. The battery pumps charge until the potential difference, ΔV, across the capacitor equals the potential difference across the battery. Another word for potential difference is voltage - the capacitor is charged when its voltage equals the battery voltage.

The charge Q stored in a capacitor is proportional to the capacitor's potential difference, ΔV:

$$Q = C\,\Delta V .$$ (Eq. 17.6: **Charge on a capacitor is proportional to its voltage**)

where C is the capacitance of the capacitor, a measure of how much charge is stored for a particular voltage. The capacitance depends on the geometry of the capacitor, as well as on what, if anything, is between the plates of the capacitor. For a parallel-plate capacitor made of two plates of equal area A, separated by a distance d, the capacitance is given by:

$$C = \frac{\kappa \varepsilon_0 A}{d} ,$$ (Eq. 17.7: **Capacitance of a parallel-plate capacitor**)

where κ is the dielectric constant of the material between the capacitor plates, and

$$\varepsilon_0 = \frac{1}{4\pi k} = 8.85 \times 10^{-12} \text{ C}^2 / \left(\text{N m}^2\right) \text{ is the permittivity of free space.}$$

The MKS unit for capacitance is the farad (F). 1 F = 1 C/V.

A charged capacitor, with a charge $+Q$ on one plate and $-Q$ on the other, has an electric field directed from the positive plate to the negative plate. The field is uniform if the distance d between the plates is much smaller than the dimensions of the plate itself. In this case the magnitude of the field, E, is connected to ΔV, the potential difference across the capacitor, by:

$$E = \frac{|\Delta V|}{d}.$$ (Eq. 17.8: **Magnitude of the electric field in a parallel-plate capacitor**)

Dielectrics.

A material in the space between capacitor plates is known as a **dielectric**. Usually this is an insulating material, because filling the space with a conductor would discharge the capacitor. It is interesting to review what happens with a conductor, however. As in Figure 17.13, when a conductor is placed in an external electric field the conduction electrons re-distribute themselves, giving an induced electric field E_{ind} inside the conductor equal in magnitude but opposite in direction to the external field. This gives a net field of zero inside the conductor.

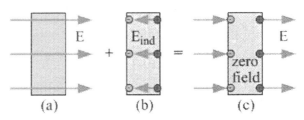

Figure 17.13: Conduction electrons distribute themselves so the field is zero inside a conductor.

The process is similar in an insulator. If the molecules in the insulator are unpolarized, the electric field polarizes them by shifting the electron clouds around the nuclei in the direction opposite to the external field; if the molecules are polarized to begin with, and randomly aligned as in Figure 17.14, the molecules experience a torque from the field that tends to align them with the field. In either case the positive and negative charges cancel one another in the bulk of the insulator, but there is a net positive charge along one face of the insulator and a net negative charge along the other face. This gives rise to an induced electric field E_{ind} inside the conductor that is opposite in direction to the external field. The induced field has a smaller magnitude than the external field, leading to a partial cancellation of fields inside the insulator, as in Figure 17.15.

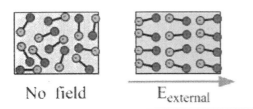

No field $E_{external}$

Figure 17.14: Polar molecules align with the field, when placed in an electric field.

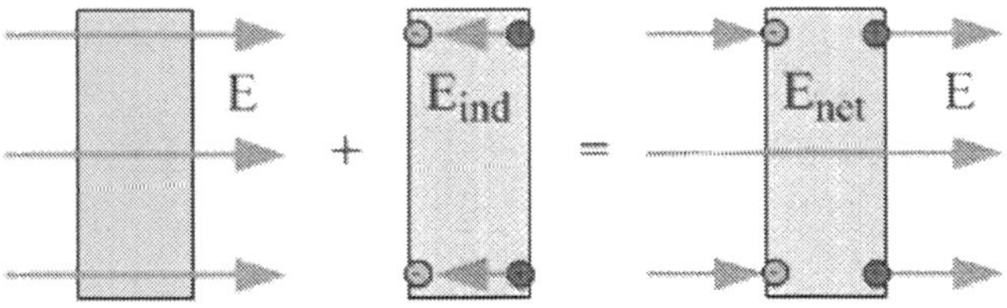

Figure 17.15: The induced field partly cancels the external field in the dielectric.

Related End-of-Chapter Exercises: 31, 61.

The dielectric constant of an insulator is the ratio of the external field to the net field inside the insulator:

$$\kappa = \frac{\vec{E}_0}{\vec{E}_{net}} \geq 1.$$ (Eq. 17.9: **The dielectric constant**)

Selected dielectric constants are shown in Table 17.1.

Material	Dielectric constant
Vacuum	1.00000
Dry air	1.00056
Polystyrene	2.6
Mylar	3.1
Paper	3.6
Water	80

Table 17.1: Dielectric constants of various materials.

Essential Question 17.6: What is the dielectric constant of a conductor?

Answer to Essential Question 17.6: The net field inside a conductor is zero. Using $E_{net} = 0$ in equation 17.9 gives $\kappa = $ infinity. Thus, conductors have infinite dielectric constants.

17-7 Energy in a Capacitor, and Capacitor Examples

When a capacitor stores charge, creating an electric field between the plates, it also stores energy. The energy is stored in the electric field itself. The energy density (the energy per unit volume in the field) is proportional to the square of the magnitude of the field:

$$\text{Energy density} = \frac{1}{2}\kappa\varepsilon_0 E^2.$$ (Eq. 17.10: **Energy per unit volume in an electric field**)

U, the total potential energy stored in the capacitor is thus the energy density multiplied by the volume of the region between the capacitor plates, $A \times d$. The energy can be written as:

$$U = \frac{1}{2}C(\Delta V)^2 = \frac{1}{2}Q\Delta V = \frac{Q^2}{2C}.$$ (Equation 17.11: **Energy stored in a capacitor**)

Let's now explore the two main ways we set up capacitor problems.

EXPLORATION 17.7A – The capacitor stays connected to the battery

A parallel-plate capacitor is connected to a battery that has a voltage of V_0. The capacitor has an initial capacitance of C_0, with air filling the space between the plates. The capacitor stores a charge Q_0, has an electric field of magnitude of E_0, and stores an energy U_0. With the battery connected to the capacitor, the spacing between the plates is doubled, and then a dielectric with a dielectric constant 5 times that of air is inserted, completely filling the space between the plates.

Step 1 – *Do any of the parameters given above stay constant during this process?* Yes, the potential difference is constant. Because the capacitor is connected to the battery at all times, the potential difference across the capacitor matches the battery voltage.

Step 2 – *Complete Table 17.2 to show the values of the various parameters after each step.*

Situation	Potential difference	Capacitance	Charge	Electric field	Energy
Initially	V_0	C_0	Q_0	E_0	U_0
Spacing is doubled					
Dielectric inserted					

Table 17.2: Fill in the table to show what happens to the various parameters in this situation.

The completed table is Table 17.3. The potential difference is constant, so we start with that column. We then use Equation 17.7 to find the capacitance. Doubling the distance between the plates reduces the capacitance by a factor of 2; increasing the dielectric constant by a factor of 5 then increases C by a factor of 5, for an overall increase by a factor of $5/2 = 2.5$.

Knowing ΔV and C, we can use Equation 17.6, $Q = C\Delta V$, to find the charge on the capacitor. Because the potential difference is constant the charge changes in proportion to the capacitance. Similarly, Equation 17.8, $E = |\Delta V|/d$, tells us what happens to the magnitude of the electric field, and the form of equation 17.11 that says the energy is proportional to $C(\Delta V)^2$ tells us what happens to the energy. It is interesting that inserting the dielectric has no effect on the electric field – this is because the battery increases the charge by a factor of 5 to keep the potential difference, and therefore the field, the same when the dielectric is inserted.

Situation	Potential difference	Capacitance	Charge	Electric field	Energy
Initially	V_0	C_0	Q_0	E_0	U_0
Spacing is doubled	V_0	$0.5\,C_0$	$0.5\,Q_0$	$0.5\,E_0$	$0.5\,U_0$
Dielectric inserted	V_0	$2.5\,C_0$	$2.5\,Q_0$	$0.5\,E_0$	$2.5\,U_0$

Table 17.3: Keeping track of the various parameters as changes are made to the capacitor.

Key idea: When a capacitor remains connected to a battery the capacitor voltage is constant – it equals the battery voltage. If changes are made we first determine how the capacitance changes, and then use the various equations to determine what happens to other parameters.

EXPLORATION 17.7B – The battery is disconnected from the capacitor before the changes
Let's return our parallel-plate capacitor to the same initial state as in the previous Exploration. The wires connecting the capacitor to the battery are then removed. After this the spacing between the plates is doubled, and then a dielectric with a dielectric constant 5 times that of air is inserted into the capacitor, completely filling the space between the plates.

Step 1 – What stays constant during this process? The charge is constant. With the capacitor disconnected from the battery, the charge is stranded on the plates.

Step 2 – Complete Table 17.2 to show the values of the various parameters after each step. The completed table is Table 17.4. The charge is constant, so we start with that column. Applying Equation 17.7, we can then determine what happens to the capacitance. Doubling the distance between the plates reduces the capacitance by a factor of 2; increasing the dielectric constant by a factor of 5 then increases C by a factor of 5, for an overall increase by a factor of $5/2 = 2.5$.

Knowing Q and C, we can use Equation 17.6, $Q = C\Delta V$, to find the potential difference across the capacitor. Note that ΔV is inversely proportional to the capacitance. Equation 17.8, $E = |\Delta V|/d$, tells us what happens to the electric field, and the form of equation 17.11 that says the energy is proportional to $Q\,\Delta V$ tells us what happens to the energy. In this case, inserting the dielectric decreases the field, as we would expect from our previous discussion of dielectrics.

Situation	Potential difference	Capacitance	Charge	Electric field	Energy
Initially	V_0	C_0	Q_0	E_0	U_0
Spacing is doubled	$2\,V_0$	$0.5\,C_0$	Q_0	E_0	$2\,U_0$
Dielectric inserted	$0.4\,V_0$	$2.5\,C_0$	Q_0	$0.2\,E_0$	$0.4\,U_0$

Table 17.4: Keeping track of the various parameters as changes are made to the capacitor.

Key idea: When a capacitor is not connected to anything the charge on the capacitor remains constant. If changes are made we first determine how the capacitance changes, and then use the various equations to determine what happens to other parameters.

Related End-of-Chapter Exercises: 25 – 30.

Essential Question 17.7: In Exploration 17.7B, the energy stored by the capacitor doubles when the spacing between the plates doubles. Where does this extra energy come from?

Answer to Essential Question 17.7: The capacitor plates, being oppositely charged, attract one another. Positive work is required to pull the plates farther apart. The energy associated with that process is the extra energy stored by the capacitor.

Chapter Summary

Essential Idea: Electric Potential Energy and Electric Potential.
In this chapter we continued looking at parallels between how charged particles interact and how objects with mass interact. As with gravitational situations, conservation of energy can be applied to many situations involving charged particles. We also went beyond what we had done with gravity, defining electric potential. The analogy still holds – we can define a gravitational potential for objects with mass that is much like electric potential for charged objects.

Electric Potential Energy
The potential energy associated with the interaction between one object with a charge q and a second object with a charge Q that is a distance r away is given by:

$$U_E = \frac{kqQ}{r}$$ (Eq. 17.1: **Potential energy for the interaction between two charges**)

where $k = 8.99 \times 10^9 \text{ N m}^2 / \text{C}^2$ is a constant. If the charges are of opposite signs then the potential energy is negative – this indicates an attraction. If the charges have the same sign the potential energy is positive, indicating a repulsion.

Potential energy is a scalar. The total electric potential energy of a system of charged objects can be found by adding up the energy associated with each interacting pair of objects. For a charged particle in a uniform electric field we use the change in potential energy:

$$\Delta U_E = -q\vec{E} \bullet \Delta \vec{r} = -qE\Delta r \cos\theta ,$$ (Eq. 17.2: **Change in potential energy in a uniform field**)

where θ is the angle between the electric field $\vec{E}$ and the displacement $\Delta \vec{r}$.

In a uniform field, we are free to define the zero level of potential energy. With equation 17.1, however, the potential energy is zero when the charges are separated by an infinite distance.

Electric Potential
Electric potential helps us understand how a charged object, or a set of charged objects, affects the region around it. The electric potential, V, at a particular point can be defined in terms of the electric potential energy, U, associated with an object of charge q placed at that point:

$$V = \frac{U}{q}, \quad \text{or} \quad U = qV .$$ (Eq. 17.3: **Connecting electric potential and potential energy**)

The unit for electric potential is the volt (V). $1V = 1$ J/C.

A special case is the electric potential from a point charge with a charge Q:

$$V = \frac{kQ}{r} ,$$ (Equation 17.4: **Electric potential from a point charge**)

where r is the distance from the charge to the point in space where we are finding the potential. Electric potential is a scalar, so it has a sign but not a direction.

Equipotentials connect points of the same potential. Equipotentials are perpendicular to field lines, because field points in the direction of decreasing potential. In a uniform field, or for a small displacement in a non-uniform field, we can connect the potential difference to the field:

$$\Delta V = -E\,\Delta r\cos\theta\,,\qquad\text{(Equation 17.5: \textbf{Potential difference in a uniform field})}$$

where θ is the angle between the electric field $\vec{E}$ and the displacement $\Delta\vec{r}$.

Working with Force, Field, Potential Energy, and Potential

Keep two points in mind when working with force, field, potential energy, and potential:
- Force and field are vectors, while potential energy and potential are scalars.
- Forces and potential energies arise from interactions between charges, requiring at least two charges. Field and potential exist with a single charged object.

Capacitors and Dielectrics

The charge Q stored in a capacitor is proportional to the capacitor's potential difference, ΔV:

$$Q = C\,\Delta V\,,\qquad\text{(Eq. 17.6: \textbf{Charge on a capacitor is proportional to its voltage})}$$

where C is the capacitance of the capacitor, a measure of how much charge is stored for a particular voltage. The capacitance depends on the geometry of the capacitor, as well as on what, if anything, is between the plates of the capacitor. For a parallel-plate capacitor made of two plates of equal area A, separated by a distance d, the capacitance is given by:

$$C = \frac{\kappa\varepsilon_0 A}{d}\,,\qquad\text{(Eq. 17.7: \textbf{Capacitance of a parallel-plate capacitor})}$$

where κ is the dielectric constant of the material between the capacitor plates, and

$$\varepsilon_0 = \frac{1}{4\pi k} = 8.85\times10^{-12}\ \mathrm{C^2/(N\,m^2)}\ \text{is the \textbf{permittivity of free space.}}$$

The MKS unit for capacitance is the farad (F). 1 F = 1 C/V.

The dielectric constant of an insulator is the ratio of the external field to the net field inside:

$$\kappa = \frac{\vec{E}_0}{\vec{E}_{net}} \geq 1\,.\qquad\text{(Eq. 17.9: \textbf{The dielectric constant})}$$

$$E = \frac{|\Delta V|}{d}\,.\qquad\text{(Eq. 17.8: \textbf{Magnitude of the electric field in a parallel-plate capacitor})}$$

$$U = \frac{1}{2}C(\Delta V)^2 = \frac{1}{2}Q\Delta V = \frac{Q^2}{2C}\,.\qquad\text{(Equation 17.11: \textbf{Energy stored in a capacitor})}$$

The energy is stored in the electric field. The energy density (the energy per unit volume) of an electric field is:

$$\text{Energy density} = \frac{1}{2}\kappa\varepsilon_0 E^2\,.\qquad\text{(Eq. 17.10: \textbf{Energy per unit volume in an electric field})}$$

When a capacitor is connected to a battery the capacitor voltage (i.e., potential difference) is equal to the battery voltage. When a capacitor is charged and then all connections to the capacitor are removed, however, the charge on the capacitor is constant.

End-of-Chapter Exercises

Exercises 1 – 12 are primarily conceptual questions designed to see whether you understand the main concepts of the chapter.

1. (a) If the electric field at a particular point is zero, does that imply that the electric potential is also zero at that point? If so, explain why. If not, give an example involving two or more point charges where the electric field is zero at a point but the electric potential is not. (b) If the electric potential at a particular point is zero, does that imply that the electric field is also zero at that point? If so, explain why. If not, give an example involving two or more point charges where the electric potential is zero at a point but the electric field is non-zero there.

2. For interactions between point charges, the electric potential energy is defined to be zero when the charges are separated by an infinite distance. Is it possible for a collection of two or more point charges to have an electric potential energy of zero when the charges are finite distances from one another? If not, explain why not. If so, give an example in which this happens.

3. An electric dipole consists of two charged particles, one with a charge of $+Q$ and the other with a charge of $-Q$, separated by some distance. In a particular dipole, the positive charge is located on the x-axis at $x = +10$ cm, and the negative charge is located on the x-axis at $x = -10$ cm. Assume non-conservative forces, and gravitational interactions, are negligible for this situation. (a) How much net work do you need to do to bring a third particle, with a charge of $+2Q$, from very far away to the origin if you bring the charge toward the origin along the positive y-axis? Justify your answer. (b) How much net work do you need to do if you bring the third particle to the origin by a more circuitous route, via a path that takes it quite close to the particle of charge $+Q$? Justify your answer.

4. (a) Sketch field lines that represent a uniform field directed down. Add some equipotential lines to your sketch. (b) Make a second diagram that shows a second electric field with twice the magnitude and the same direction as the first field. Add equipotential lines to this diagram so that the potential difference between the lines is the same as that in your first diagram.

5. Figure 17.16 shows electric field lines and equipotentials in three different regions. Comment on each diagram, stating either that it is physically possible or, if not, what is wrong with the situation shown.

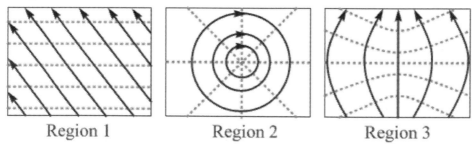

Region 1   Region 2   Region 3

Figure 17.16: State whether the field lines (black arrows) and equipotential lines (dashed lines) are physically correct. For Exercise 5.

6. Two charged balls are placed on the x-axis, as shown in Figure 17.17. The first ball has a charge $+q$ and is located at the origin, while the second ball has a charge $+3q$ and is located at $x = +4a$. A third charged ball is now brought in and placed nearby. Your goal is to determine the sign of the charge on the third ball, and to narrow down its location. (a) The force on the third ball is in the $+x$ direction, with no component in any other direction. What, if anything, does this tell you about the third ball? (b) When the third ball is added to the system, the potential at $x = 2a$ decreases. What, if anything, do you know about the third ball now? (c) When the third ball is added to the system, the force on the ball of charge $+q$ increases, while the force on the ball of charge $+3q$ decreases. Both these forces maintain their original directions. What, if anything, do you know about the third ball now?

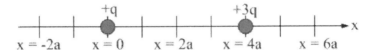

Figure 17.17: Two charged balls on the x-axis. For Exercise 6.

7. Four charged balls, with charges of $+4q$, $-5q$, $-6q$, and $-q$, are placed on either the x or y axes as shown in Figure 17.18. Each charge is the same distance from the origin. You will now remove one of the charged balls from the system. Which charged ball should you remove to cause (a) the largest decrease in the magnitude of the electric field at the origin? (b) the largest increase in the potential at the origin? (c) the largest increase in the potential energy of the system?

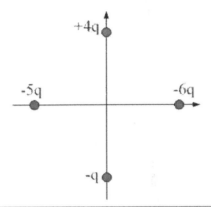

Figure 17.18: Four charged balls are placed on either the x or y axis so that each ball is the same distance from the origin. For Exercises 7 and 8.

8. Return to the system of four charged balls described in Exercise 7 and shown in Figure 17.18. The system is now changed by reversing the sign of each of the charges. Describe what effect, if any, that change has on the answers to Exercise 7.

9. A parallel-plate capacitor is connected to a battery that has a voltage of V_0. The capacitor has an initial capacitance of C_0, with a dielectric of dielectric constant three times larger than that of air filling the space between the plates. The capacitor stores a charge Q_0, has an electric field of magnitude of E_0, and stores an energy U_0. With the battery still connected to the capacitor, the dielectric is removed, leaving air between the plates. The distance between the plates is then decreased by a factor of 2. Make a table to show the potential difference, capacitance, charge, magnitude of the field inside the capacitor, and energy stored by the capacitor, in terms of their initial values, after each change is made.

10. Repeat Exercise 9 but, this time, the wires connecting the capacitor to the battery are removed before the changes are made.

11. Return to the situation described in Exploration 17.7B. When the dielectric is inserted, the energy stored by the capacitor is reduced by a factor of 5. If the dielectric is removed, the energy returns to its original value. (a) Where does the energy go when the dielectric is inserted? (b) Where does the energy come from when the dielectric is removed?

12. A parallel-plate capacitor with air between the plates is connected to a 12-volt battery. (a) What is the potential difference across the capacitor? (b) Without bringing in additional batteries, is it possible to increase the potential difference across the capacitor to 36 V, by making changes such as removing the battery and/or changing the separation between the plates? If so, describe how you could do it. If not, explain why not.

Exercises 13 – 19 deal with electric potential energy.

13. In the Bohr model of the hydrogen atom, the electron in the ground state is a distance of 5.3×10^{-11} m from the proton in the nucleus. Assuming the proton remains at rest, what is the escape speed of the electron? In other words, if the electron was given an initial velocity directed away from the proton, what is the minimum speed it has to have to completely escape from the proton? Assume nothing but the proton influences the electron.

14. Two charged objects are released from rest when they are a particular distance apart. As they accelerate away from one another, their velocities are always equal in magnitude and opposite in direction. Assuming that the only thing acting on one object is the other object, are the following statements true or false? Justify each answer. Statement 1 – The two objects have charges of the same sign. Statement 2 – The two objects have charges of the same magnitude. Statement 3 – The two objects have equal mass.

15. Return to the situation described in Exercise 14. Let's say that the objects are identical small balls with the same mass and charge. If each ball has a mass of 25 grams, a charge of 5.0×10^{-6} C, and is originally separated by 8.0 cm, how fast is each ball moving when the separation between the balls has doubled?

16. Three balls, with charges of $+4q$, $-2q$, and $-q$, are equally spaced along a line. The spacing between the balls is r. We can arrange the balls in three different ways, as shown in Figure 17.19. In each case the balls are in an isolated region of space very far from anything else. (a) Rank the arrangements according to their potential energy, from most positive to most negative. See if you can do this without explicitly calculating the potential energy in each case. (b) Verify your answer to part (a) by calculating the potential energy in each case.

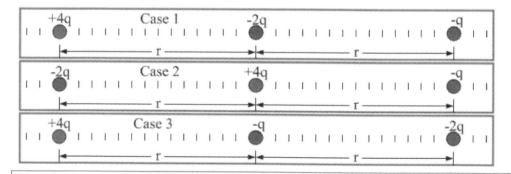

Figure 17.19: Three different arrangements of three balls of charge $+4q$, $-2q$, and $-q$ placed on a line with a distance r between neighboring balls. For Exercises 16 and 17.

17. Consider the three arrangements of charged balls shown in Figure 17.19. When assembling these arrangements, the middle ball is the last one added to the system in each case. How much work do you have to do, against the forces applied by the other two balls, to bring the middle ball from infinity to the point halfway between the other balls in (a) Case 1? (b) Case 2?, (c) Case 3?

18. An electron and a proton are each released from rest in a uniform electric field that has a magnitude of 500 N/C. The energy and speed of each particle is measured after it has moved through a distance of 25 cm. Assume the particles do not influence one another, but are influenced only by the electric field. Without doing any calculations, determine which particle (a) has more kinetic energy, (b) has a higher speed, (c) takes more time to cover 25 cm. Justify your answers. Now calculate the kinetic energy, speed, and elapsed time for (d) the electron, and (e) the proton.

19. In an electron beam in a cathode ray tube television, the electrons are accelerated from rest through a potential difference of 15 kV on their way to the screen. What is the speed of the electrons?

Exercises 20 – 26 deal with electric potential.

20. A ball of with a charge of $+6q$ is placed on the x-axis at $x = -2a$. There is a second ball of unknown charge at $x = +a$. If the net electric potential at the origin due to the two balls is $+\dfrac{2kq}{a}$, what is the charge of the second ball? Find all possible solutions.

21. A single point charge is located at an unknown point on the x-axis. There are no other charged objects nearby. You measure the electric potential at $x = +2.0$ m to be –200 volts, while the potential at $x = +5.0$ m is –400 volts. What is the sign and magnitude of the point charge, and where is it located? State all possible answers.

22. A single point charge is placed on the x-axis at $x = -2.0$ m. If the electric potential at $x = 0$ because of this charge is –500 volts, determine the sign and magnitude of the charge of the point charge.

23. Two charged balls are placed on the x-axis, as shown in Figure 17.20. The first ball has a charge $+q$ and is located at the origin, while the second ball has a charge $-3q$ and is located at $x = +4a$. Your goal is to find all the points on the axis, a finite distance from the charges, where the net electric potential due to these two balls is zero. Start qualitatively. Provide a justification for whether or not there are any such points (a) to the left of the ball of charge $+q$, (b) between the balls, and (c) to the right of the ball of charge $-3q$. (d) Find the locations of all such points.

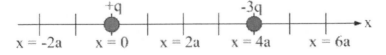

Figure 17.20: Two charged balls on the x-axis. For Exercises 23 and 24.

24. Return to the situation described in Exercise 23 and shown in Figure 17.20. (a) If we reverse the sign of the charge on both the balls, do the answers to Exercise 23 change? If so, describe how. (b) If we reverse the sign of the charge on just one of the balls, do the answers to the previous exercise change? If so, describe how.

25. The potential difference between two points, A and B, in a uniform electric field has a magnitude of 30 volts. Assume gravity is negligible in this situation. When an electron is released from rest at point A, it passes through B a short time later. If the potential at point A is +50 volts, what is the potential at point B?

26. (a) Make a sketch on a piece of paper showing three points, A, B, and C. Point B is located a distance of 20 cm from, and directly above, point A, while point C is located a distance of 20 cm, and horizontally to the right, from point A. The points are in a uniform electric field. All you know about its direction is that it is in the plane of the piece of paper. (b) You measure the electric potential at point A to be +40 volts, while the potential at point B is +50 volts. What, if anything, does this tell you about the magnitude and/or the direction of the electric field? (c) You then measure the electric potential at point C to be +50 volts. What, if anything, can you say about the magnitude and/or the direction of the electric field now? (d) If possible, sketch the field lines and show the +30 V, +40 V, +50 V, and +60 V equipotentials, and find the magnitude of the electric field.

Exercises 27 – 32 involve force, field, potential, and potential energy.

27. Two charged balls are placed on the x-axis, as shown in Figure 17.21. The first ball has a charge $+q$ and is located at the origin, while the second ball has a charge $+3q$ and is located at $x = +4a$. (a) Which of the balls experiences a larger magnitude force because of its interaction with the other ball? Why? (b) Determine the magnitude and direction of the force experienced by the ball of charge $+q$. (c) Calculate the magnitude and direction of the net electric field at $x = +2a$. (d) Determine the electric potential energy of this system. (e) Calculate the electric potential at $x = +2a$, relative to $V = 0$ at infinity.

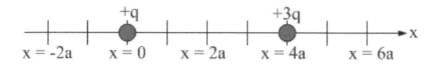

Figure 17.21: Two charged balls on the x-axis. For Exercises 27 and 28.

28. Two charged balls are placed on the x-axis, as shown in Figure 17.21. The first ball has a charge $+q$ and is located at the origin, while the second ball has a charge $+3q$ and is located at $x = +4a$. (a) Determine the location of all points on the x-axis, a finite distance from the balls, where the net electric field is zero. (b) Determine the location of all points on the x-axis, a finite distance from the balls, where the total electric potential is zero, relative to $V = 0$ at infinity.

29. Three charged balls, with charges of $+q$, $-3q$, and $+2q$, are placed at the corners of a square that measures L on each side, as shown in Figure 17.22. (a) What is the magnitude and direction of the force experienced by the ball at the top right corner? (b) What is the magnitude and direction of the net electric field at the lower left corner? (c) What is the electric potential energy of this set of charged objects? (d) What is the electric potential at the lower left corner?

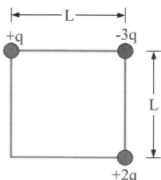

Figure 17.22: Three charged balls, with charges of $+q$, $-3q$, and $+2q$, are placed at the corners of a square that measures L on each side. For Exercise 29.

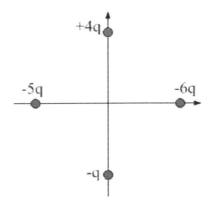

30. Four charged balls, with charges of $+4q$, $-5q$, $-6q$, and $-q$, are placed on either the x or y axes as shown in Figure 17.23. Each ball is a distance d from the origin. (a) Calculate the magnitude and direction of the force exerted on the ball of charge $+4q$ by the other three balls. (b) Calculate the magnitude and direction of the net electric field at the origin. (c) Calculate the electric potential energy of this set of charged balls. (d) Calculate the electric potential at the origin, relative to $V = 0$ at infinity.

Figure 17.23: Four charged balls are placed on either the x or y axis so that each ball is the same distance from the origin. For Exercises 30 and 31.

31. Four charged balls, with charges of $+4q$, $-5q$, $-6q$, and $-q$, are placed on either the x or y axes as shown in Figure 17.23. Each ball is a distance d from the origin. (a) If you remove one ball from the system, which ball requires you to do the largest positive work to remove it? (b) Calculate the value of the work required to remove that ball.

32. A point object with a charge of $+Q$ is placed on the x-axis at $x = -a$. A second point object of unknown charge is placed at an unknown location on either the x-axis or the y-axis. The potential energy associated with the point charges is $U = +2kQ^2/a$; the electric potential at the origin due to the charges is $V = +4kQ/a$; and the net electric field at the origin due to the charges points in the negative x direction. (a) What is the sign of the charge on the second object? Justify your answer. (b) Qualitatively, what, if anything, can you say about the location of the second object? Explain. (c) Determine the sign and magnitude of the charge on the second object, and determine its location. (d) What is the magnitude of the net electric field at the origin due to the two charges?

Exercises 33 – 35 deal with capacitors.

33. A parallel-plate capacitor with air between the plates is connected to a battery that has a voltage of V_0. The capacitor has an initial capacitance of C_0. The capacitor stores a charge Q_0, has an electric field of magnitude of E_0, and stores an energy U_0. The following steps are then carried out. Step 1, the distance between the plates is doubled; step 2, the wires connecting the capacitor to the battery are removed; step 3, a dielectric with a dielectric constant 4 times that of air is inserted, completely filling the space between the plates; step 4, the capacitor is re-connected to the battery. Make a table to show the potential difference, capacitance, charge, magnitude of the field inside the capacitor, and energy stored by the capacitor, in terms of their initial values, after each step.

34. A parallel-plate capacitor with air between the plates is connected to a battery that has a voltage of V_0. The capacitor has an initial capacitance of C_0. The capacitor stores a charge Q_0, has an electric field of magnitude of E_0, and stores an energy U_0. The following steps are then carried out. Step 1, a second capacitor, identical to the first but initially uncharged, is placed so it is touching the first capacitor, effectively doubling the area of the capacitor plates; step 2, the wires connecting the capacitor to the battery are removed; step 3, the distance between the plates is doubled; step 4, the capacitor is re-connected to the battery. Make a table to show the potential difference, capacitance, charge, magnitude of the field inside the capacitor, and energy stored by the capacitor, in terms of their initial values, after each step.

35. The membrane of a living cell can be treated as a parallel-plate capacitor, with a plate separation of 10 nm and a dielectric constant of 5. The area of the plates is approximately 5×10^{-9} m^2. How much energy is stored in the cell membrane? Assume that the potential difference across the membrane has a magnitude of 100 millivolts.

General problems and conceptual questions

36. Two small charged balls are released from rest when they are 6.0 cm apart. One ball has a mass of 50 grams and a charge of 5.0×10^{-5} C, while the second ball has three times the mass and twice the charge as the first ball. Assuming the balls are influenced only by one another, how fast is each ball moving when they are very far apart?

37. Three identical balls, each with a mass of 75 grams and a charge of 8.0×10^{-4} C, are arranged so there is one ball at each corner of an equilateral triangle. Each side of the triangle measures 25 cm. (a) Assuming the balls are influenced only by one another, describe what will happen when the balls are released from rest. (b) When the balls are 50 cm from one another, how fast are they moving? (c) How fast are they moving when they are very far from one another?

38. A charged particle is given an initial speed of 80 cm/s in a uniform electric field. The initial velocity is the same direction as that of the field. Assume gravity can be neglected in this situation. After covering a distance of 30 cm, the particle has a velocity of 40 cm/s, directed in the same direction as its initial velocity. (a) What is the sign of the particle's charge? (b) How much time did the particle take to cover 30 cm? (c) What is the additional distance covered by the particle before it comes instantaneously to rest?

39. The electric potential in a particular region is due solely to a nearby point charge. You find that the potential at a location 50 cm from the charge is 60 volts higher than the potential at a location only 10 cm from the charge. (a) Is this possible? Explain. (b) If it is possible, determine the sign and magnitude of the charge on the point charge.

40. A ball of charge $+2q$ is placed on the x-axis at $x = -2a$. A second ball of charge $-q$ is placed nearby so that the net electric potential at the origin because of the two balls is $-\dfrac{2kq}{a}$. Where is the second ball?

41. Three small balls, with charges of $+q$, $-2q$, and $+3q$, can be placed on the x-axis in three different configurations, as shown in Figure 17.24. In each case one charge is at $x = -a$, one is at $x = +a$, and the third is at $x = +2a$. Rank the configurations based on (a) the magnitude of the force experienced by the ball of charge $-2q$, (b) the net electric field at $x = 0$, (c) the potential energy of the configuration, (d) the total electric potential at $x = 0$.

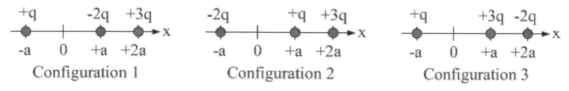

Figure 17.24: Three different configurations of three charged balls, for Exercises 41 and 42.

42. Return to the system described in Exercise 41 and shown in Figure 17.24. Calculate the potential energy in (a) configuration 1, (b) configuration 2, and (c) configuration 3. (d) Is it possible to reverse the locations of two of the charged balls in any configuration without affecting the potential energy of the configuration? Explain.

43. A particle with a mass of 24 grams and a charge of $+3.0 \times 10^{-5}$ C has a speed of 0.75 m/s when it passes through a point at which the potential is +1200 volts. What is the particle's speed when it passes through a second point at which the potential is −1200 volts? Assume that the only force acting on the particle comes from the electric field.

44. The electric potential a distance r from the center of a charged sphere with a spherically symmetric charge distribution is the same as that from a point charge that has the same total charge as the sphere, as long as the point is outside the sphere. Consider a conducting sphere with a net charge of +30 μC and a radius of 4.0 cm. Assume that the system is in electrostatic equilibrium, and that there are no other objects nearby. (a) What is the electric potential at a point 10 cm from the center of the sphere? (b) What is the electric potential at the surface of the sphere? (c) What is the electric field at the center of the sphere? (d) What is the electric potential at the center of the sphere?

45. Consider the three cases shown in Figure 17.25. Rank these cases, from most positive to most negative, based on the (a) electric potential at the origin; (b) electric potential energy of the set of charges.

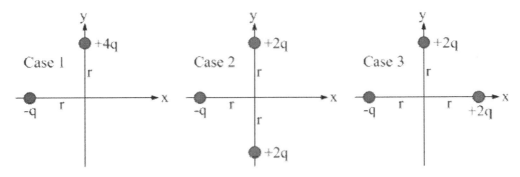

Figure 17.25: Three different configurations of charged objects, for Exercises 45 – 47.

46. Consider the three cases shown in Figure 17.25. Calculate the electric potential energy of the set of charges in (a) case 1; (b) case 2; (c) case 3.

47. Consider the three cases shown in Figure 17.25. Determine the electric potential at the origin in (a) case 1; (b) case 2; (c) case 3.

48. Four small charged balls are arranged at the corners of a square that measures L on each side, as shown in Figure 17.26 (a) Find the electric potential energy associated with this set of balls. (b) What is the electric potential at the center of the square, relative to $V = 0$ at infinity? (c) If you doubled the length of each side of the square, so neighboring charges were separated by a distance of $2L$ instead, what would happen to your answer to part (a)?

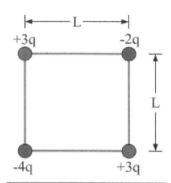

49. Four small charged balls are arranged at the corners of a square that measures L on each side, as shown in Figure 17.26. How much work would you have to do to remove one of the balls with a charge of $+3q$ from the system?

Figure 17.26: Four charged balls at the corners of a square, for Exercises 48 – 50.

50. Four small charged balls are arranged at the corners of a square that measures L on each side, as shown in Figure 17.26. If you could place a fifth ball, having any charge you wish, at the center of the square, could you give the system a total electric potential energy of zero? If not, explain why not. If so, determine the sign and magnitude of the fifth ball.

51. Three charged balls are placed in a line, as shown in Figure 17.27. Ball 1 has an unknown charge and sign, and is a distance $2r$ to the left of ball 2. Ball 2 has a charge of $+Q$. Ball 3 has an unknown non-zero charge and sign, and is a distance r to the right of ball 2. Ball 3 was the last ball brought into the system, and it required zero net work to bring ball 3 from very far away to the location shown. (a) Is there enough information here to find the sign of the charge on ball 1? If so, what is the sign? (b) Can we find the magnitude of the charge on ball 1? If

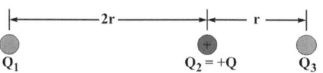

Figure 17.27: Three charges in a line. Only the sign and magnitude of charge 2 are known, although we also know that it took no net work to bring charge 3 into the system from far away. For Exercise 51.

so, what is it? (c) Can we find the sign of the charge on ball 3? If so, what is the sign? (d) Can we find the magnitude of the charge on ball 3? If so, what is it?

52. A single point charge is located at an unknown point on the x-axis. There are no other objects nearby. You measure the electric potential at $x = +2.0$ m to be $+600$ volts, while the potential at $x = +5.0$ m is $+150$ volts. What is the sign and magnitude of the point charge? State all possible answers.

53. A point charge with a charge of $+5.0$ μC is located at the origin. A second point charge is located at $x = +2.0$ m, with a charge of -9.0 μC. (a) Analyze the situation qualitatively to determine approximate locations of any points along the straight line that passes through both charges where the net electric potential due to these two point charges is zero. (b) Determine the location of all such points that are a finite distance from the charges.

54. Repeat Exercise 53, except now the second point charge has a charge of $+18.0$ μC.

55. Return to the situation described in Exercise 53. Are there any points a finite distance from the charges at which the net electric potential due to the two charges is zero, that are off the line passing through the charges? If so, explain how you would locate them.

56. Two point charges are placed at different locations on the x-axis. The graph of the electric potential as a function of position on the axis is shown in Figure 17.28. The position is given in units of a. Note that the electric potential is zero at $x = +a$. (a) Where are the two charges located? (b) If one of the charges has a charge of $+3q$, what is the charge of the other point charge? (c) Is the electric field equal to zero at any point on the x-axis in the region shown on the graph, $-4a \le x \le +4a$? Justify your answer by referring to the graph.

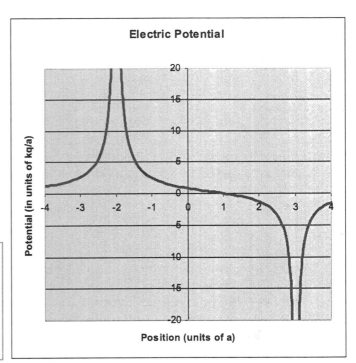

Figure 17.28: The graph of the electric potential along the x-axis in the region $-4a \le x \le +4a$. The potential is due to two point charges that are located on the x-axis. The electric potential is zero at $x = +a$. For Exercise 56.

57. Two point charges are placed at different locations on the x-axis. The graph of the electric potential as a function of position on the axis is shown in Figure 17.29. The position is given in units of a. (a) Where are the two charges located? (b) If one of the charges has a charge of $-2q$, what is the charge of the other point charge? (c) Is the electric field equal to zero at any point on the x-axis in the region shown on the graph, $-4a \le x \le +4a$? Justify your answer by referring to the graph.

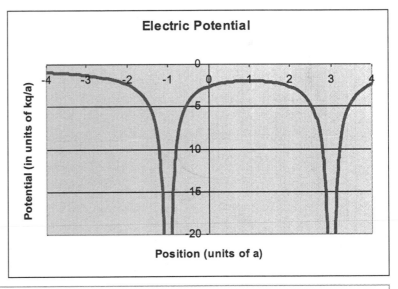

Figure 17.29: The graph of the electric potential along the x-axis in the region $-4a \le x \le +4a$. The potential is due to two point charges that are located on the x-axis. For Exercise 57.

58. A parallel-plate capacitor that has air between the plates is initially charged by being connected to a battery. With the battery still connected, one of the following changes is made. Change 1: the distance between the plates is tripled. Change 2: a dielectric with a dielectric constant 2 times that of air is inserted, completely filling the space between the plates. Change 3: another identical, but initially uncharged, capacitor is placed next to the first so they are touching, effectively doubling the area of each plate. Rank each of these changes, from largest to smallest, based on (a) the final potential difference across the capacitor, (b) the final charge stored on the capacitor, (c) the magnitude of the final electric field with the capacitor, (d) the final energy stored by the capacitor.

59. Repeat Exercise 58, but now the wires connecting the capacitor to the battery are removed before the changes take place.

60. (a) For a parallel-plate capacitor made up of two plates of area A separated by a distance d, what is the volume of the space between the plates? (b) Combine Equation 17.10, for the energy density, with Equations 17.7 and 17.8 to derive one form of Equation 17.11, for the energy stored in a capacitor. (c) Bring in one additional relationship to show how the other forms of Equation 17.11 follow from the expression you obtained in (b).

61. Comment on this part of a conversation between two of your classmates.

Paul: So, let's say we have a charged parallel-plate capacitor that has nothing between the plates. If I then insert a dielectric between the plates, the electric field in the capacitor is reduced, right?

Mary: I don't think so. Sometimes that's true but not always.

The photo shows a part of a circuit on a circuit board, such as that found inside a computer or other electronic device. Among other components, there is a large integrated circuit at the far left, several capacitors, diodes, and three resistors, which we will investigate in some detail in this chapter. Photo credit: Petr Kratochvil, from publicdomainpictures.net.

Chapter 18 – DC (Direct Current) Circuits

CHAPTER CONTENTS

We will now move from the more abstract concepts of field and potential to concrete applications of these ideas, in electric circuits. Humans have learned to exploit electric circuits in a variety of ways to make our lives easier. Electric circuits lie at the heart of flashlights, cell phones, televisions, and iPods, not to mention the central nervous system within each of us. Over the last hundred years, the number and complexity of the circuits in use has increased tremendously. Our goal in this chapter is simply to understand the basic rules on which many of these circuits are based.

18-1 Current, and Batteries

In Chapters 16 and 17, our main focus was on static charge. We turn now to look at situations involving flowing charge. Controlling that flow is the basis of many electric circuits.

> Current is the rate at which charge flows. The symbol we use for current is I:
>
> $$I = \frac{\Delta Q}{\Delta t}$$ (Equation 18.1: **Current, the rate of flow of charge**)
>
> The unit for current is the ampere (A). 1 A = 1 C/s.
>
> The direction of current is the direction positive charges flow, a definition adopted by Benjamin Franklin before it was determined that in most cases the charges that flow in a circuit are electrons (negative charges). However, in a circuit positive charge flowing in one direction is equivalent to an equal amount of negative charge flowing at the same rate in the opposite direction.

In general, circuits in which charge flows in one direction are **direct current (DC)** circuits. In **alternating current (AC)** circuits (such as those in your house) the current direction continually reverses direction. Despite this difference, many of the concepts addressed in this chapter apply to both DC and AC circuits.

Batteries

What causes charge to flow? In general, charge flows from one point to another when there is a potential difference between the points. A battery can create such a potential difference. This potential difference gives rise to an electric field within the wires and the other elements of a circuit. Charged objects, such as electrons, in this field experience a force which can cause them to move. Note that, previously, we discussed how the electric field within a conductor is zero when the conductor is in static equilibrium. In this case, when we have a non-zero electric field within the conductor, we have more of a dynamic equilibrium – a steady flow of charge may be established as the charges respond to the non-zero field.

A battery can be thought of as an electron manufacturing and recycling system. It does not create electrons. Rather, a chemical reaction that liberates electrons takes place at the negative terminal of the battery. If the battery is connected in a circuit the electrons travel through the circuit, giving up energy along the way (such as to a light bulb a toaster element), to the positive terminal of the battery. At the positive terminal a different chemical reaction takes place that recycles the electrons, binding them into waste products. It is also important that charge flows within the battery between the positive and negative terminals – this charge is often positive ions.

Consider a lead-acid car battery as an example. The battery's positive terminal is made from lead dioxide, the negative terminal from lead, and both terminals are immersed in a solution of dilute sulphuric acid. The sulphuric acid contains water molecules as well as ions of H^+ and HSO_4^-. The chemical reaction that takes place at the negative terminal liberates two electrons:

$$Pb + HSO_4^- \Rightarrow PbSO_4 + H^+ + 2e^-.$$

After traveling through the circuit, these electrons are recycled at the positive terminal:

$$PbO_2 + HSO_4^- + 3H^+ + 2e^- \Rightarrow PbSO_4 + 2H_2O.$$

To keep the system going, there must be a net flow of positive ions from the negative terminal to the positive terminal.

As in any manufacturing process, there are raw materials (the electrodes, and the acid solution), there is a product (the electrons), and there are waste products (the $PbSO_4$ and water). A battery runs out when its raw materials are used up, or when enough waste products build up to inhibit the reactions. In a rechargeable battery, the battery is recharged by running the chemical reactions in the opposite direction, re-creating the electrodes and removing waste products.

Fuel cells use a similar process as batteries but, whereas a battery is a closed system in which its raw materials and waste products are sealed in a container, in a fuel cell everything is open so that raw materials can be continually fed into the system and waste products removed. A number of manufacturers are researching ways to run portable electronic devices, such as laptops and cell phones, from fuel cells instead of from batteries. The advantage offered by the fuel cell is that you could run the device for significantly longer than you could run it off a battery. Also, instead of plugging the device into an electrical outlet for a few hours to re-charge it, you could just take a few seconds to top it up with, say, methanol, and the device would be good to go again.

Figure 18.1 has three views of a circuit involving a battery, two wires, and a light bulb. Figure 18.1(a) shows conventional current, where the charges flowing are positive. Figure 18.1(b) shows the actual situation, showing that the charges flowing through the wires and the light-bulb filament are actually electrons, while positive H^+ ions flow within the battery itself. The two situations look different, but the light bulb would be equally bright in either case. Figure 18.1(c) shows the circuit diagram. The current I is in the direction of conventional current.

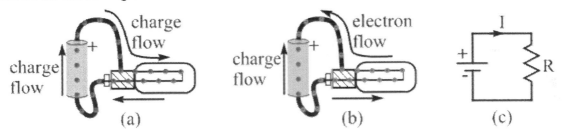

Figure 18.1: Three views of a battery-powered circuit. Figure (a) shows conventional current, in which the charge that flows is always positive. Figure (b) shows the actual situation, in which electrons flow through the circuit and positive ions flow within the battery. Figure (c) shows a circuit diagram for this circuit. R stands for resistor, which we cover in the next section. The arrow under the I shows the direction of the conventional current. The two parallel lines, the shorter one marked with a – and the longer one with a plus, represent the standard symbol for a battery in a circuit diagram. The + is the positive terminal, and the – is the negative terminal.

Every battery has an associated potential difference: for instance, a 9-volt battery provides a potential difference of around 9 volts. This is the potential difference between the battery terminals when there is no current, and is known as the battery emf, ε (emf stands for electromotive force, and you say emf as it is spelled, e-m-f).

Related End-of-Chapter Exercises: 13, 15.

Essential Question 18.1: Which is closer to the truth – a battery is a source of constant potential difference, or a battery is a source of constant current? Explain.

Answer to Essential Question 18.1: The statement that a battery is a source of constant potential difference (until it runs out, at least) is closer to the truth. As we will see in Section 18-2, the current provided by a battery depends on what the battery is connected to.

18-2 Resistance and Ohm's Law

For many objects we find that the current through them is proportional to the potential difference across them. Such objects are said to be ohmic, because they obey Ohm's Law.

The current I through an ohmic device is proportional to the potential difference ΔV across it:

$$\Delta V = IR \qquad \text{(Equation 18.2: Ohm's Law)}$$

R here is known as the electrical resistance. The smaller the resistance of an object the more current flows the object for a given potential difference. An object in a circuit that contributes a relatively large resistance to a circuit is known as a resistor.

Many resistors are simply wires of length L and cross-sectional area A. The electrical resistance of such an object is given by:

$$R = \frac{\rho L}{A}. \qquad \text{(Equation 18.3: Electrical resistance)}$$

The unit of resistance is the ohm (Ω). $1\ \Omega = 1$ V/A.

ρ in Equation 18.3 is the resistivity, a parameter that depends on the material the resistor is made from. Table 18.1 shows some resistivity values for different materials.

Material	Resistivity	Material	Resistivity	Material	Resistivity
Copper	$1.7\times10^{-8}\ \Omega$ m	Nichrome	$1.0\times10^{-6}\ \Omega$ m	Hard rubber	$1\times10^{13}\ \Omega$ m
Tungsten	$5.6\times10^{-8}\ \Omega$ m	Silicon	$640\ \Omega$ m	Teflon	$1.0\times10^{16}\ \Omega$ m

Table 18.1: Resistivity values for various materials.

EXAMPLE 18.2 – Factors contributing to resistance
Three cylindrical resistors are made from the same material and have the same volume. The first resistor has a length L, a cross-sectional area A, and a resistance R. The second has a length $L/2$ and cross-sectional area $2A$. The third has length $2L$ and cross-sectional area $A/2$.
(a) Draw a picture of these three situations.
(b) Rank the three cases in terms of the total resistance.
(c) Determine the resistance, in terms of R, of the second and third resistors.

SOLUTION

(a) The three situations are shown in Figure 18.2.

(b) The ranking by resistance goes as $3 > 1 > 2$. Resistance increases as length increases and cross-sectional area decreases, so case 3, with the longest length and smallest cross-sectional area, has the largest resistance. Case 2 has the smallest resistance because it has the smallest length and the largest area.

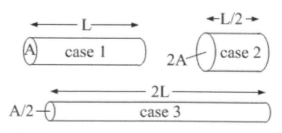

Figure 18.2: The three situations of the resistors described in Example 18.2.

(c) Applying Equation 18.2, we have the resistance in case 1 is $R = \rho L / A$. In case 2, applying Equation 18.2 gives

$$R_2 = \rho(L/2)/2A = \rho L / 4A = R/4.$$

Doing a similar analysis in case 3 gives

$$R_3 = \rho(2L)/(A/2) = 4\rho L / A = 4R.$$

On a circuit diagram, a resistor is usually represented as shown in Figure 18.3(a), although if the resistor is a light bulb the symbol can look like that in Figure 18.3(b), which looks a lot like the actual light bulb.

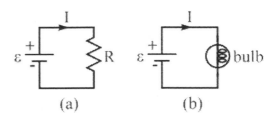

(a)   (b)

Figure 18.3: (a) A circuit diagram showing a battery, at left, connected to a resistor, such as a toaster element, at right. (b) A similar circuit, but this time the resistor is a light bulb.

Temperature dependence of resistance

An object's resistance also generally depends on its temperature. To understand this, consider a simple model of resistance. When there is no potential difference between the ends of a wire the conduction electrons move about at random, much like atoms in an ideal gas. With a potential difference, however, a net drift of electrons toward the higher potential end is superimposed on the random motion, as in Figure 18.4. The drift speed is of the order of millimeters/second, compared to the typical electron speeds of a thousand kilometers per second.

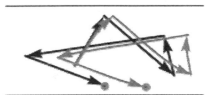

Figure 18.4: With no potential difference a conduction electron follows the random motion illustrated by the darker path. With a potential difference, the electron follows the lighter path, drifting to the right in this case.

A potential difference between the ends of a wire creates an electric field within the wire. Conduction electrons respond to this field, accelerating along the wire. However, as electrons bump into atoms in the wire the collisions transfer energy from the electrons to the atoms. This is where the resistance of the wire comes from – the energy lost by electrons in these collisions.

At higher temperatures atoms vibrate more energetically, making it harder for electrons to move past – thus, resistance generally increases with temperature. In some materials, however, such as semiconductors, increasing the temperature frees more electrons to serve as conduction electrons and is thus associated with a decrease in resistance. We can use a simple model to describe the change in resistivity with temperature:

$$\rho = \rho_i(1 + \alpha \Delta T), \qquad \text{(Equation 18.4: \textbf{Temperature dependence of resistivity})}$$

where α is the temperature coefficient of resistivity, which depends on the material. Equation 18.4 is reminiscent of Equation 13.5, describing an object's change in length when it changes temperature. When a resistor changes temperature both the resistivity and the dimensions change, but the change in size is generally negligible compared to the change in the resistivity.

Related End-of-Chapter Exercises: 16 – 19.

Essential Question 18.2: In what direction is the electric field that gives rise to the drift velocity to the right, shown by the lighter path in Figure 18.4? Which end of the wire is at a higher potential?

Answer to Essential Question 18.2: To cause the electron to drift to the right, as shown, the electric field must be directed to the left. Because the electron has a negative charge, the force it feels is opposite in direction to that of the electric field. Electric field points in the direction of decreasing potential, so the right end of the wire must be at a higher potential.

18-3 Circuit Analogies, and Kirchoff's Rules

Analogies can help us to understand circuits, because an analogous system helps us build a model of the system we are interested in. For instance, there are many parallels between fluid being pumped through a set of pipes and charge flowing around a circuit. There are also useful parallels we can draw between a circuit and a ski hill, in which skiers are taken to the top of the hill by a chair lift and then ski down via various trails. Let's investigate these in turn.

A particular fluid system is shown in Figure 18.5. The fluid is enclosed in a set of pipes, and a water wheel spins in response to the flow. The fluid circulates through the system by means of a pump, which creates a pressure difference (analogous to potential difference in the circuit) between different sections of the system.

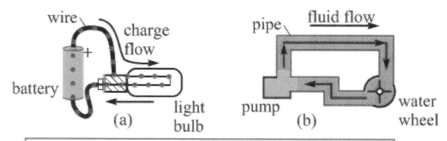

Figure 18.5: Fluid flows through a set of pipes like charges flow through a circuit.

The pump in the fluid system is like the battery in the circuit; the water is like the charge; and the water wheel is like the light bulb in the circuit. The large pipes act like the wires, one pipe carrying water from the pump to the water wheel, and another carrying the water back to the pump, much as charge flows through one wire from the battery to the light bulb, and through a second wire back to the battery. The pressure difference in the fluid system is analogous to the potential difference across the resistor in the circuit.

EXPLORATION 18.3 – Analogies between a circuit and a ski hill

A basic ski hill consists of a chair lift, like that shown in Figure 18.6, that takes skiers up to the top of the hill, and a trail that skiers ski down to the bottom. A short and wide downward slope takes the skiers from the top of the lift to the top of the trail, and another takes the skiers from the bottom of the trail to the bottom of the lift.

Step 1 – *Identify the aspects of the circuit that are analogous to the various aspects of the ski hill. In particular, identify what for the ski hill plays the role of the battery, the flowing charge, and the resistor.* The chair lift is like the battery in the circuit, while the skiers are like the charges. The chair lift raises the gravitational potential energy of the skiers, and the skiers dissipate all that energy as they ski down the trail (which is like the resistor in the circuit) to the bottom. Similarly, the battery raises the electric potential energy of the charges, and that energy is dissipated as the charges flow through the resistor.

Figure 18.6: A chair lift that takes skiers up a hill. Photo credit: Petr Kratochvil, from publicdomainpictures.net.

Step 2 – Does the analogy have limitations? Identify at least one difference between the ski hill and the electric circuit. What happens when the chair lift / battery is turned off? On the ski hill the skiers keep skiing down to the bottom, but in the circuit if a switch is opened the net flow of charge stops. This difference stems from the fact that with the ski hill the potential difference is imposed by something external to the system, the Earth's gravity, while in the circuit the battery provides the potential difference. Another difference is that in the circuit the charges are identical and obey basic laws of physics, while on a ski hill the skiers are not identical, and make choices regarding when to stop for lunch or to enjoy the view, and which route to take down the hill.

Step 3 - Let's use our ski hill analogy to understand two basic rules about circuits, which we will make use of throughout the rest of the chapter. Figure 18.7 shows one ski trail dividing into two, trails A and B. The same thing happens in a circuit, with one path dividing into paths A and B. *What is the relationship between N, the total number of skiers, and N_A and N_B, the number of skiers choosing trails A and B, respectively? What is the analogous relationship between the current I in the top path in the circuit, and the currents I_A and I_B in paths A and B?*

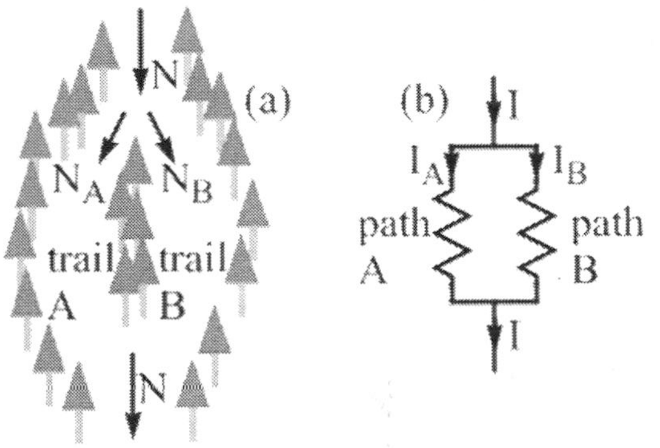

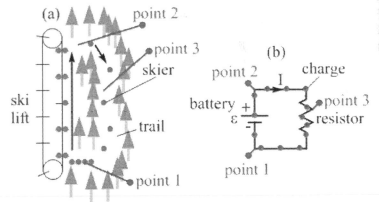

We do not lose or gain skiers, so some skiers choose trail A and the rest choose B, giving $N = N_A + N_B$. The skiers come back together when the trails re-join, and N skiers continue down the trail. Similarly, a certain number of charges flow through the top path in the circuit, and the charges take either path A or path B through the circuit before re-combining. The rate of flow of charge is the current, so we can say that $I = I_A + I_B$. This is the **junction rule**.

Figure 18.7: (a) One trail splits temporarily into two on the ski hill; (b) one path splits into two in the circuit.

Step 4 - *Figure 18.8 shows various points on a ski hill and in a circuit. For a complete loop, say from point 3 back to point 3, if we add up the changes in gravitational potential as we move around the loop, what will we get? If we do the analogous process for the circuit, adding up the electric potential differences as we move around a complete loop, what will we get?*

In both cases we get zero. There is as much up as down on the ski hill, so when we return to the starting point the net change in potential is zero. The same applies to the circuit. This is the **loop rule**.

Figure 18.8: Equivalent points marked on a ski hill and on a circuit.

Key Ideas for Analogies: Analogies, particularly the gravitation-based ski-hill analogy, can give us considerable insight into circuits. In this case we used the analogy to come up with what are known as Kirchoff's Rules. The **Junction Rule** – *the total current entering a junction is equal to the total current leaving a junction*. The **Loop Rule** – *the sum of all the potential differences for a closed loop is zero.* **Related End-of-Chapter Exercises: 1 and 2.**

Essential Question 18.3: In Figure 18.7, let's say the resistance of path A is larger than that of path B. Which current is larger? What is the blanket statement summarizing this idea?

Answer to Essential Question 18.3: If path A has a larger resistance then path B we would expect the current in path A to be smaller than that in path B, much as we might expect more skiers to choose trail B if it is an easier trail than trail A. The blanket statement about this is – **current prefers the path of least resistance**.

18-4 Power, the Cost of Electricity, and AC Circuits

A typical light bulb has two numbers on it. One is the power, in watts, and the other is the voltage, which is typically 120 V in North America, matching something about the voltage you obtain from a typical household electrical outlet. With these numbers you can determine the current through the bulb and the resistance of the bulb. Let's understand how this is done.

A change in electrical potential energy is given by the equation $\Delta U = q \Delta V$. Power is the time rate of change of energy. If we divide electrical potential energy by time we get:

$$P = \frac{\Delta U}{t} = \frac{q \Delta V}{t} = \frac{q}{t} \Delta V = I \Delta V.$$

Using Ohm's Law, $\Delta V = IR$, we can write the power equation in three ways:

$$P = I \Delta V = I^2 R = \frac{(\Delta V)^2}{R}.$$ (Equation 18.5: **Electrical power**)

EXAMPLE 18.4A – Calculating resistance
(a) What is the resistance of a bulb stamped with "100 W, 120 V?"
(b) What is the resistance of a bulb stamped with "40 W, 120 V?"
(c) If you pick up such a light bulb and measure its resistance when the bulb is not lit, the measured value is much less than what you calculate in parts (a) or (b). Why is this?

SOLUTION
(a) We can re-arrange one form of the power equation to solve for resistance of the 100 W bulb:

$$R = \frac{(\Delta V)^2}{P} = \frac{120 \text{ volts} \times 120 \text{ volts}}{100 \text{ W}} = \frac{120 \text{ volts} \times 120 \text{ volts}}{(10 \times 10) \text{ W}} = 12 \times 12 \ \Omega = 144 \Omega.$$

(b) The resistance of the 40 W bulb can be found in a similar way:

$$R = \frac{(\Delta V)^2}{P} = \frac{120 \text{ volts} \times 120 \text{ volts}}{40 \text{ W}} = (120 \times 3.0) \Omega = 360 \Omega.$$

(c) What we calculated above is the resistance of a bulb when the bulb has 120 volts across it. This is when it glows brightly because the filament is a few thousand kelvin. If you measure the resistance when the bulb is off, with the filament at room temperature, the resistance is much less because of the temperature dependence of resistance. As we discussed earlier, resistance generally increases as temperature increases. This is what is going on with the bulbs.

Related End-of-Chapter Exercises: 20, 28.

The Cost of Electricity

On your electric bill you are charged about 20 cents for every kilowatt-hour of electricity you used in a month. What kind of unit is the kilowatt-hour?

$$1 \, kW - hour = 1000 \, W \times 3600 \, s = 3.6 \times 10^6 \, J.$$

The kilowatt-hour is a unit of energy. Note that you are charged only about 20 cents for each 3.6 million joules of energy delivered to your residence. To find the total cost associated with running a particular device, you do the following multiplication:

$$(\text{Number of hours it is on}) \times (\text{power rating in kW}) \times (\text{cost per kW-hr}) \quad \text{(Eq. 18.6: Electricity cost)}$$

The power rating of a device is generally stamped on it. Clock radios and televisions usually show this on the back or the bottom, either on a sticker or printed directly on the device.

EXAMPLE 18.4B – The cost of watching television (the electrical cost, that is)

In a typical American household the television set is on for three hours a day. If the power company charges 20 cents for each kW-hr, what is the daily cost of having the TV on for that length of time if the TV's power rating is 300 W?

SOLUTION

First, let's convert the TV's power rating to kW by dividing by 1000, to get 0.30 kW. Then let's simply apply equation 18.6. The daily cost is:

$$(3 \, hr) \times (0.30 \, kW) \times (20 \, cents/kW\text{-}hr) = 18 \, cents.$$

This is amazingly cheap, particularly compared to what it costs to go to the movies.

Related End-of-Chapter Exercises: 48 – 50.

AC Circuits

In many situations we can treat an electrical outlet, in North America, as acting like a 120-volt battery (in Europe it would be like a 220-volt battery). In reality, however, the voltage signal in North America is a sine wave that oscillates between +170 volts and –170 volts with a frequency of 60 Hz (60 cycles/s), as in Figure 18.9. For a sine wave it turns out that the root mean square value of the voltage is the peak value divided by $\sqrt{2}$. This is where the 120 V comes from, it is the root-mean-square value of the voltage signal.

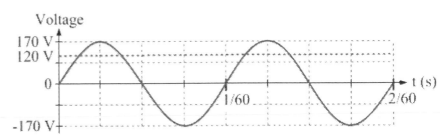

Figure 18.9: The voltage signal from an electrical outlet in North America.

When connected to a wall socket, an incandescent light bulb flickers, but it does so at a rate much faster than is observable to us. The average power dissipated in the light bulb, however, is the root-mean-square voltage multiplied by the root-mean-square current.

Essential Question 18.4: When connected to a wall socket, an incandescent light bulb flickers at a frequency that is not 60 Hz. At what frequency does it flicker? Why?

Answer to Essential Question 18.4: The bulb flickers at 120 Hz. It is dim at the start of the cycle in Figure 18.9, bright when the voltage goes positive, dim when the voltage passes through zero, and bright again when the voltage goes negative. Thus, for each cycle of the voltage, the bulb goes through two cycles, so its frequency is twice that of the voltage. Note that the bulb does not go completely off when the voltage passes through zero because the filament takes time to cool.

18-5 Resistors in Series

In many circuits, resistors are placed in series, as in Figure 18.10(b), so that the charge flows through the resistors in sequence. Let's start with a battery connected to a single 10 Ω resistor, which is made from a wire of a particular resistivity and cross-sectional area. If we break the resistor into two pieces, one 40% as long as the original and one 60% as long, the pieces have resistances of 4 Ω and 6 Ω, respectively, because resistance is directly proportional to length.

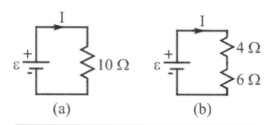

(a)   (b)

Figure 18.10: A single 10 Ω resistor can be thought of as two resistors, whose resistances add to 10 Ω, placed in series with one another.

If the 4 Ω and 6 Ω resistors are connected by a wire of negligible resistance, the battery sees no difference between the single 10 Ω resistor and the 4 Ω and 6 Ω resistors that are connected in series – the battery is still trying to force charge to flow through a total resistance of 10 Ω. We can generalize by splitting the original resistor into more than two pieces, but we always end up with the same total resistance.

Thus, we can say that when we have N resistors connected in series, the equivalent resistance of the set of resistors is:

$$R_{eq} = R_1 + R_2 + \ldots + R_N.$$ (Eq. 18.7: **Equivalent resistance of resistors in series**)

EXPLORATION 18.5A – Finding the current through one resistor

The emf of the battery in Figure 18.10(b) is 20 volts.

Step 1 – *How does the current through the 4 Ω resistor in Figure 18.10(b) compare to that through the 6 Ω resistor? Explain why.* The current is the same at all points in a series circuit. All the charge that flows through the 4 Ω resistor keeps going to flow through the 6 Ω resistor. The rate of flow is also the same, because charge cannot pile up anywhere in the circuit. Because current is the rate of flow of charge this means the current is the same through both resistors.

Step 2 – *What is the current through the 4 Ω resistor in Figure 18.10(b)?* It is tempting to apply Ohm's Law directly to the 4 Ω resistor, with a potential difference of 20 volts. Resist this temptation, because the potential difference across that resistor is something less than 20 volts! The most straightforward way to proceed is to first find the equivalent resistance of the circuit, which is 4 Ω+ 6 Ω = 10 Ω in this case, and apply Ohm's Law to find the current in the circuit:

$$I_{circuit} = \frac{\Delta V}{R_{eq}} = \frac{20 \text{ volts}}{10 \Omega} = 2.0 \text{ A}.$$

The current is the same at all points in a series circuit, so this is the current through each resistor.

Step 3 – *What is the potential difference across each of the two resistors in Figure 18.10(b)?* ***What is the sum of the potential differences? Why do we get this result?*** Now we can apply Ohm's Law to the individual resistors, to find that:

$$\Delta V_{4\Omega} = IR = (2.0\ \text{A})(4.0\ \Omega) = 8.0\ \text{V} \quad \text{and} \quad \Delta V_{6\Omega} = IR = (2.0\ \text{A})(6.0\ \Omega) = 12\ \text{V}.$$

These add together to equal 20 volts, the battery voltage. This is to be expected from Kirchoff's loop rule. In a series circuit, the loop rule tells us that the sum of the potential differences across the resistors equals the battery voltage.

Key ideas for series circuits: Components in a series circuit have equal currents passing through them. The sum of the potential differences across the resistors in a series circuit is equal to the battery voltage. **Related End-of-Chapter Exercises: 25 and 46.**

EXAMPLE 18.5B – Which bulb is brighter?
A standard 40 W light bulb is connected in series with a standard 100 W light bulb and an electrical outlet, which we can treat as a 120 V battery, as shown in Figure 18.11. (a) Provide a qualitative analysis to show which bulb is brighter. (b) Analyze the circuit quantitatively to estimate the power dissipated in each bulb. Hint: we calculated the resistance of these bulbs in Section 18.4.

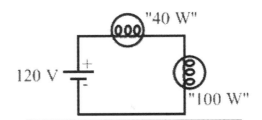

Figure 18.11: A circuit with a 100 W light bulb and a 40 W light bulb connected in series.

SOLUTION
(a) It's easy to think of justifications for all three possible answers. An argument in favor of the 100 W bulb being brighter is that such a bulb is brighter than a 40 W bulb when used at home. We could also argue that the bulbs are equally bright because they are in series, and therefore have the same current. The correct answer, however, is that the bulb marked as 40 W is brighter in this case. The bulbs do have equal currents, but the brightness is determined by the power. In this case let's use the equation $P = I^2R$. Because the current is the same, the bulb with higher resistance has more power dissipated in it and is brighter. We showed in Section 18.4 that the 40 W bulb has a larger resistance. Note that bulbs are designed to be placed in parallel, so it should not be too surprising that they behave in unexpected ways when they are in series.

(b) Let's use the resistances we calculated in Section 18.4, 144 Ω for the 100 W bulb and 360 Ω for the 100 W bulb. The equivalent resistance of the two bulbs in series the sum of these values, about 500 Ω. Knowing the equivalent resistance enables us to find the current in the circuit, from Ohm's Law: $I = \Delta V / R_{eq} = (120\ \text{V})/(500\ \Omega) = 0.24\ \text{A}$. The power dissipated in each bulb is:

$$P_{40W} = I^2R = (0.24\ \text{A})^2(360\Omega) = 21\ \text{W} \quad \text{and} \quad P_{100W} = I^2R = (0.24\ \text{A})^2(144\Omega) = 8.3\ \text{W}.$$

The bulb marked "40 W" dissipates more power, and is thus brighter.

Essential Question 18.5: In the situation above, the bulbs will not actually have the resistances we used in the calculation. Why not? Will the bulbs actually dissipate more power or less power?

Answer to Essential Question 18.5: Each bulb has a potential difference less than the 120 V the bulb is designed for, so the bulbs are dimmer than usual. This reduces the filament temperature, lowering its resistance, which decreases the equivalent resistance of the circuit. This increases the current in the circuit, increasing the power dissipated in each bulb because $P = I^2 R$. The decrease in R is offset by the increase in I, and the extra factor of I gives a net increase.

18-6 Resistors in Parallel

When charge has more than one path to choose from between two points in a circuit, we say that those paths are in parallel with one another. Interestingly, adding a resistor to a circuit can actually decrease the resistance of that circuit if the resistor is placed in parallel with another resistor in the circuit. Such is the case in Figure 18.12(b), where the circuit has a lower net resistance than the circuit in Figure 18.12(a).

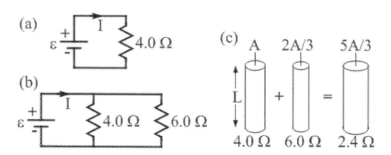

Figure 18.12: In (b), a 6.0 Ω resistor is placed in parallel with the 4.0 Ω resistor. (c) The charge now has a larger effective area to flow through, so the resistance of the circuit has been reduced.

If the resistors in Figure 18.12(b) have the same length and resistivity, then they must have different cross-sectional areas. The two resistors can be replaced by one equivalent resistor with the same length and resistivity, and with a cross-sectional area equal to the sum of the cross-sectional areas of the original resistors, as in Figure 18.12(c). Resistance is inversely proportional to area, so adding the second resistor in parallel actually decreases the resistance of the circuit.

If we have N resistors connected in parallel, their equivalent resistance is given by:

$$\frac{1}{R_{eq}} = \frac{1}{R_1} + \frac{1}{R_2} + \ldots + \frac{1}{R_N}.$$ (Eq. 18.8: **Equivalent resistance of resistors in parallel**)

EXPLORATION 18.6 – Current in a parallel circuit
The emf of the battery in Figure 18.12(b) is 12 volts.

Step 1 – *How does the current through the 4.0 Ω resistor in Figure 18.12(b) compare to that through the 6.0 Ω resistor? Explain why.* When the charge has two paths to choose from, as it does in this circuit, more of the charge passes through the lower resistance path. Thus, the current through the 4.0 Ω resistor is larger than that through the 6.0 Ω resistor.

Step 2 – *Find the equivalent resistance of the circuit in Figure 18.12(b) and use it to find the current through the battery in that circuit.*

Applying equation 18.8 gives: $\dfrac{1}{R_{eq}} = \dfrac{1}{4.0\,\Omega} + \dfrac{1}{6.0\,\Omega} = \dfrac{3}{12\,\Omega} + \dfrac{2}{12\,\Omega} = \dfrac{5}{12\,\Omega}.$

Inverting this to find the equivalent resistance gives $R_{eq} = (12/5)\,\Omega = 2.4\,\Omega$. In other words, the battery acts as if it is connected to a single 2.4 Ω resistor, and thus the current in the circuit, and through the battery, is $I = \varepsilon / R_{eq} = (12\text{ V})/2.4\,\Omega = 5.0\text{ A}.$

Step 3 – *What is the current through each of the two resistors in Figure 18.12(b)? What is the sum of the currents? Why do we get this result?* Each resistor in the circuit is directly connected to the battery, and thus each resistor has 12 V across it. Applying Ohm's Law to each resistor:

$$I_{4\Omega} = \frac{\Delta V}{R} = \frac{12\text{ V}}{4.0\ \Omega} = 3.0\text{ A} \qquad \text{and} \qquad I_{6\Omega} = \frac{\Delta V}{R} = \frac{12\text{ V}}{6.0\ \Omega} = 2.0\text{ A}.$$

These sum to 5.0 A, the current through the battery. This is consistent with Kirchoff's junction rule, which says the current entering a junction equals the current leaving the junction.

> **Key ideas for parallel circuits**: Components in parallel with one another have the same potential difference across them. Current splits between parallel paths, with more current passing through the path with lower resistance. **Related End-of-Chapter Exercises: 21 and 24.**

In a calculation like that in Step 2, a common error is to forget to invert when applying Equation 18.8, stating the answer incorrectly as $R_{eq} = (5/12)\Omega = 0.42\ \Omega$. Checking units can prevent this error. Also, a rule of thumb is that the equivalent resistance of two resistors in parallel is between half the smaller resistance and the smaller resistance. In Step 2 above, in which the smaller resistance is 4.0 Ω, the equivalent resistance must be between 2.0 Ω and 4.0 Ω. The smaller the value of the larger resistance the smaller the equivalent resistance.

EXAMPLE 18.6 – Splitting the current

The current entering a particular section of a circuit is *I*. As shown in Figure 18.13, the current divides between two parallel paths, a current I_1 that passes through a resistor of resistance R_1, and a current I_2 that passes through a resistor of resistance R_2. What fraction of the current passes through each resistor? In other words, express I_1 and I_2 in terms of I, R_1, and R_2.

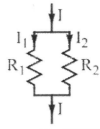

SOLUTION

We have two relationships we can use to find the answer. One is the junction rule, which tells us that $I = I_1 + I_2$, which we can re-arrange to get $I - I_1 = I_2$. The second relationship is the idea that the potential difference across the two resistors is the same, because they are in parallel.

> **Figure 18.13**: A section of a circuit in which two resistors are in parallel.

$$\Delta V = I_1 R_1 = I_2 R_2. \text{ Using } I - I_1 = I_2 \text{ gives } I_1 R_1 = (I - I_1)R_2 = IR_2 - I_1 R_2.$$

Solving for I_1, and using that expression to solve for I_2, gives:

$$I_1 = \frac{R_2}{R_1 + R_2} I \qquad \text{and} \qquad I_2 = \frac{R_1}{R_1 + R_2} I.$$

This is consistent with the idea that current prefers the path of least resistance, and tells us exactly how current splits between two branches that are in parallel with one another.

Essential Question 18.6: Two resistors, with resistances of 10 Ω and 30 Ω, are in parallel with one another (and only one another) in a circuit. What fraction of the current entering this part of the circuit passes through each resistor?

Answer to Essential Question 18.6: The fact that one resistance is three times the other means that three times as much current passes through the smaller resistor as through the larger resistor. Thus ¾ of the current passes through the 10 Ω resistor and ¼ passes through the 30 Ω resistor.

18-7 Series-Parallel Combination Circuits

In many circuits, some resistors are in series while others are in parallel. In such series-parallel combination circuits we often want to know the current through, and/or the potential difference across, each resistor. Let's explore one method for doing this. This method can be used if the circuit has one battery, or when multiple batteries can be replaced by a single battery.

EXPLORATION 18.7 – The contraction/expansion method of circuit analysis

Four resistors are connected in a circuit with a battery with an emf of 18 V, as shown in Figure 18.14. The resistors have resistances $R_1 = 4.0\,\Omega$, $R_2 = 5.0\,\Omega$, $R_3 = 7.0\,\Omega$, and $R_4 = 6.0\,\Omega$.

Our goal is to find the current through each resistor.

Step 1 – *Label the currents at various points in the circuit. This can help determine which resistors are in series and which are in parallel.* This is done in Figure 18.15(a). The current passing through the battery is labeled I. This current splits, with a current I_1 through resistor R_1, and a current I_2 through resistor R_2. The current I_2 goes on to pass through R_3, so R_2 and R_3 are in series with one another. The two currents re-combine at the top right of the circuit, giving a net current of I directed from right to left through resistor R_4 and back to the battery.

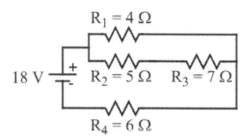

Figure 18.14: A series-parallel combination circuit with one battery and four resistors.

Step 2 – *Identify two resistors that are either in series or in parallel with one another, and replace them by a resistor of equivalent resistance.* Resistors R_2 and R_3 are in series, so they can be replaced by their equivalent resistance of 12.0 Ω (resistances add in series), as shown in Figure 18.15(b). Is this the only place we could start in this circuit? For instance, are resistors R_1 and R_2 in parallel? To be in parallel, both ends of the resistors must be directly connected by a wire, with nothing in between. The left ends of resistors R_1 and R_2 are directly connected, but the right ends are not, with resistor R_3 in between. In fact, the only place to start in this circuit is with R_2 and R_3.

Step 3 – *Continue the process of replacing two resistors by an equivalent resistor until the circuit is reduced to one equivalent resistor.* In the next step, shown in Figure 18.15(c), the two resistors in parallel, R_1 and R_{23}, are replaced by their equivalent resistance of 3.0 Ω. Finally, in Figure 18.15(d), the two resistors in series are replaced by their equivalent resistance, 9.0 Ω.

Step 4 – *Apply Ohm's Law to find the total current in the circuit.* With only one resistor we know both its resistance and the potential difference across it, so we can apply Ohm's Law:

$$I = \frac{\varepsilon}{R_{eq}} = \frac{18\text{ V}}{9.0\,\Omega} = 2.0\text{ A}.$$

Step 5 – *Label the potential at various points in the single-resistor circuit.* Choose a point as a reference. Here we choose the negative terminal of the battery to be $V = 0$. The other side of the battery is therefore +18 volts. Wires have negligible resistance, so $\Delta V = IR = 0$ across each wire.

Thus, all points along the wire leading from the negative terminal of the battery have $V = 0$, while all points along the wire leading from the positive terminal have $V = +18$ V. See Figure 18.15(e).

Figure 18.15: The various steps in the contraction/expansion method.

(a) Labeling the current at various points can help identify which resistors are in series and which are in parallel.

(b) R_2 and R_3 are in series, and can be replaced by one equivalent resistor, R_{23}.

(c) – (d) Resistors are replaced a pair at a time to find the circuit's equivalent resistance.

(e) Apply Ohm's Law to find the total current.

(f - h) Expansion reverses the steps. At each expansion step, we find the current and potential difference for each resistor.

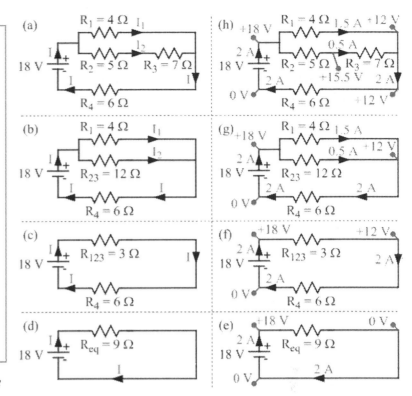

Step 6 – *Expand the circuit back from one resistor to two. Find the current through, and potential difference across, both resistors.* Expansion reverses the steps of the contraction. In Figure 18.15(f), we replace the 9.0 Ω resistor by the 3.0 Ω and 6.0 Ω resistors, in series, it came from. When a resistor is split into two in series, the current (2.0 A in this case) through all three resistors is the same. We can now use Ohm's law to find the potential difference. For the two resistors, we get $\Delta V_{3\Omega} = IR = 6.0$ V and $\Delta V_{6\Omega} = IR = 12$ V. Thus, the wire connecting the 3.0 Ω and 6.0 Ω resistors is at a potential of V = +12 V. This is consistent with the loop rule, and the fact that the direction of the current through a resistor is the direction of decreasing potential.

Step 7 – *Expand the circuit back from two resistors to three.* The 3.0 Ω resistor is replaced by the 4.0 Ω and 12.0 Ω resistors, in parallel, which it came from, as shown in Figure 18.15(g). When one resistor is split into two in parallel, the potential difference across all three resistors is the same. That is 6.0 V in this case. We can then apply Ohm's Law to find the current in each resistor, giving $I_1 = \Delta V / R = 6$ V/4 Ω $= 1.5$ A and $I_2 = \Delta V / R = 6$ V/12 Ω $= 0.5$ A. These add to the 2.0 A through their equivalent resistor, as we expect from the junction rule.

Step 8 – *Continue the expansion process, at each step finding the current through, and potential difference across, each resistor.* In this circuit there is one more step. This is shown in Figure 18.15(h), where the 12.0 Ω resistor is split into the original 5.0 Ω and 7.0 Ω resistors. These resistors have the same current, 0.5 A, as the 12.0 Ω resistor, and their potential differences can be found from Ohm's Law and sum to the 6.0 V across the 12.0 Ω resistor.

Key ideas for the contraction/expansion method: One way to analyze a circuit is to contract the circuit to one equivalent resistor and then expand it back. In each step in the contraction two resistors that are either in series or in parallel are replaced by one resistor of equivalent resistance. In the expansion when one resistor is expanded to two in series all three resistors have the same current, while when one resistor is expanded to two in parallel all three resistors have the same potential difference across them.   **Related End-of-Chapter Exercises: 27, 32, 33, 36.**

Essential Question 18.7: Check the answer above by comparing the power associated with the battery to the total power dissipated in the resistors. Why should these values be the same?

Answer to Essential Question 18.7: The power provided to the circuit by the battery can be found from $P = \varepsilon I = (18 \text{ V}) \times (2.0 \text{ A}) = 36 \text{ W}$. The equation $P = I^2 R$ gives the power dissipated in each resistor: $P_1 = I_1^2 R_1 = (1.5 \text{ A})^2 (4.0\Omega) = 9.0 \text{ W}$; $P_2 = I_2^2 R_2 = (0.5 \text{ A})^2 (5.0\Omega) = 1.25 \text{ W}$; $P_3 = I_2^2 R_3 = (0.5 \text{ A})^2 (7.0\Omega) = 1.75 \text{ W}$; and $P_4 = I^2 R_4 = (2.0 \text{ A})^2 (6.0\Omega) = 24 \text{ W}$, for a total of 36 W. Thus, the power input to the circuit by the battery equals the power dissipated in the resistors, which we expect because of conservation of energy.

18-8 An Example Problem; and Meters

Let's now explore a situation that involves many of the concepts from the last few sections, and allows us to discuss the role of a switch in a circuit.

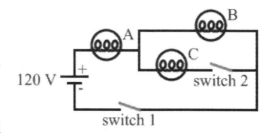

EXPLORATION 18.8 – Three bulbs and two switches

Three identical light bulbs, A, B, and C, are placed in the circuit shown in Figure 18.16 along with two switches, 1 and 2, and a battery with an emf of 120 V (like a standard electrical outlet).

Figure 18.16: A circuit with one battery, two switches, and three identical light bulbs. The switches are initially open.

Step 1 – *Are any bulbs on when the switches are both open? If so, which bulbs are on? If not, explain why not.* For a bulb to glow a current must pass through it. For there to be a current there must be a complete circuit, a conducting path for charges to flow through from one terminal of the battery to the other. With switch 1 open there is not a complete circuit, so all the bulbs are off.

Step 2 – *Kirchoff's loop rule is true even when the switches are open. How is this possible?* This is possible because the potential difference across switch 1 is equal to the battery emf. If we define the wire connecting the negative terminal of the battery to the left side of switch 1 to be at $V = 0$, all other parts of the circuit, including the right side of the switch, are at a potential of $V = +120 \text{ V}$. There are no potential differences across the bulbs because there is no current.

Step 3 – *Complete these sentences. An open switch has a resistance of _____. A closed switch has a resistance of _____.* We generally treat an open switch as having a resistance of infinity. A closed switch acts like a wire, so we assume it has a resistance of zero.

Step 4 – *Rank the bulbs based on their brightness when switch 1 is closed and switch 2 is open. What is the potential difference across each bulb?* Bulb C is off – because switch 2 is open there is no current in that part of the circuit. Thus, the circuit has bulbs A and B in series with one another and the battery. Because bulbs A and B are identical, and have the same current through them, they are equally bright. The ranking is A=B>C. Bulbs A and B share the emf of the battery, with a potential difference of 60 V across each bulb. Bulb C has no potential difference across it.

Step 5 – *What happens to the brightness of each bulb when switch 2 is closed (so both switches are closed)? What is the potential difference across each bulb?* With switch 2 closed, charge flows through bulb C, so C comes on and is brighter than before. Bulbs B and C are in parallel, and have equal resistance, so half the current passes through B and half through C. Bulb B got all the current before switch 2 was closed, so you might think that bulb B is now obviously dimmer. However, closing switch 2 decreases the overall resistance of the circuit, increasing the current. So, bulb B only gets half the current, but the total current increases – which effect dominates?

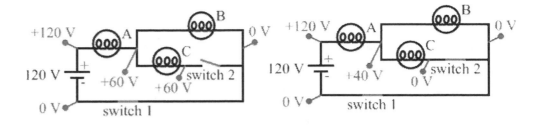

Because all the current passes through bulb A, increasing the current in the circuit increases both the brightness of, and the potential difference across, bulb A. By the loop rule, increasing A's potential difference means B's potential difference decreases, so B's current and brightness must also be less. To summarize, A and C get brighter, while B gets dimmer. B and C are now equally bright, and A is the brightest of all. Assuming the bulbs have the same resistance, A has 80 V across it, while B and C each have 40 V across them, as shown in Figure 18.17.

Figure 18.17: Labeling the potential at various points helps us understand what happens to the bulbs when switch 2 is closed.

Key Ideas for Switches: We can treat an open switch as having infinite resistance, and a closed switch as having no resistance. **Related End-of-Chapter Exercises**: 5 – 7.

Ammeters Measure Current

A meter that measures current is known as an **ammeter**. Should an ammeter be wired in series or parallel? Should the ammeter have a small resistance or a large resistance? Does adding an ammeter to the circuit increase or decrease the current through the resistor of interest?

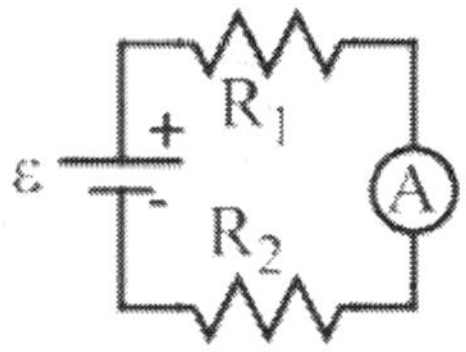

Circuit elements that are in series have the same current passing through them. Thus, to measure the current through a resistor an ammeter should be placed in series with that resistor, as in Figure 18.18. Adding the ammeter, which has some resistance, increases the equivalent resistance of the circuit and thus reduces the current in the circuit. The resistance of the ammeter should be as small as possible to minimize the effect of adding the ammeter to the circuit.

Figure 18.18: An ammeter, represented by an A inside a circle, is used to measure the current through whatever is in series with it. In this case, that's everything in the circuit.

Voltmeters Measure Potential Difference

A meter that measures potential difference is known as a **voltmeter**. Should a voltmeter be wired in series or parallel? Should it have a small or a large resistance? How does adding a voltmeter to a circuit affect the circuit?

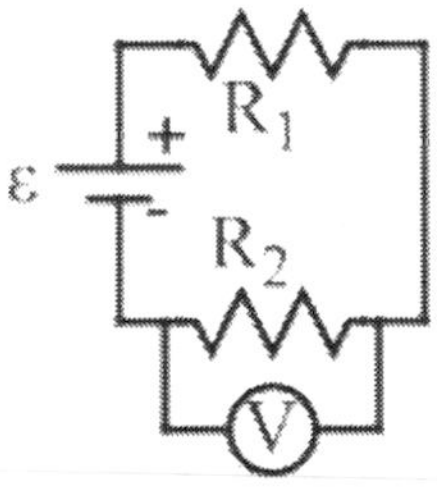

Circuit elements in parallel have the same potential difference across them. Thus, to measure the potential difference across a resistor a voltmeter should be placed in parallel with that resistor, as in Figure 18.19. Connecting the voltmeter, which has some resistance, in parallel decreases the resistance of the circuit, increasing the current. The resistance of the voltmeter should be as large as possible to minimize the effect of adding the voltmeter.

Related End-of-Chapter Exercises: 47, 60.

Figure 18.19: A voltmeter, represented by a V inside a circle, is used to measure the potential difference of whatever is in parallel with it. In this case, that's resistor R_2.

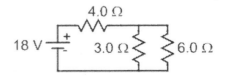

Essential Question 18.8: Can you add a 5.0 Ω resistor to the circuit in Figure 18.20 so that some current passes through it while the current through original resistors is unchanged? Explain.

Figure 18.20: The circuit for Essential Question 18.8.

Answer to Essential Question 18.8: Yes, it can be done by placing the 5.0 Ω resistor in parallel with the original set of resistors. This is how circuits work in your house; turning on a lamp, for instance, does not affect a TV, even when the lamp and the TV are plugged into the same outlet.

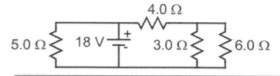

Figure 18.21: Connecting the new resistor across the battery does not change the current elsewhere in the circuit.

18-9 Multi-loop Circuits

In many circuits with more than one battery, the batteries cannot be replaced by a single battery. In such cases we rely on Kirchoff's Rules, the loop rule and the junction rule.

EXPLORATION 18.9 – Analyzing a multi-loop circuit
Solve for the current through each of the resistors in the circuit shown in Figure 18.22. The emf's of the three batteries are: $\varepsilon_1 = 2.0$ V; $\varepsilon_2 = 15$ V; and $\varepsilon_3 = 12$ V. The resistances of the four resistors are $R_1 = 1.0\ \Omega$; $R_2 = 5.0\ \Omega$; $R_3 = 2.0\ \Omega$; $R_4 = 3.0\ \Omega$.

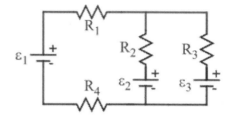

Figure 18.22: A multi-loop circuit with three batteries and four resistors.

Step 1 – Choose a direction for the current, and label the current, in each branch of the circuit. A branch is a path from one junction to another in a circuit. Figure 18.23 shows the two junctions (points at which more than two current paths come together) as circles, with the three different branches running from one junction to the other. Within each branch, everything is in series, so each branch has its own current, labeled $I_1, I_2,$ and I_3. In many cases, such as this one, we do not know for certain which way the current goes in a particular branch. We simply choose a direction for each current, and if we are incorrect we will get a minus sign for that current when we solve for it.

Step 2 – Based on the directions chosen for each current, put a + sign at the higher-potential end of each resistor and a – sign at the lower-potential end. This step is optional, but can make it easier to set up the loop-rule equations correctly in step 3. Current is directed from the high-potential end of a resistor to the low-potential end, giving the signs in Figure 18.23.

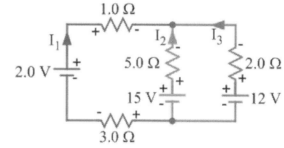

Step 3 – Apply the loop rule to one complete loop in the circuit. This means applying the equation $\sum \Delta V = 0$. ΔV for a battery is the battery emf, which is positive if we go from the – terminal to the + terminal and negative if we go the other way. ΔV for a resistor is IR for that resistor, positive if we go from – to + (with the current) and negative if we go the other way (against the current). Let's start with the inside loop on the left, going clockwise around the loop starting from the lower-left corner. Keeping track of the signs carefully, we get:

Figure 18.23: The three currents in the circuit are labeled, and directions are chosen for them. The resistors are labeled with high-potential (+) and low-potential (–) ends based on the direction chosen for current; the current direction through a resistor is in the direction of decreasing potential.

$$+2.0\ \text{V} - (1.0\ \Omega)I_1 + (5.0\ \Omega)I_2 - 15\ \text{V} - (3.0\ \Omega)I_1 = 0$$

This simplifies to $-13\ \text{V} - (4.0\ \Omega)I_1 + (5.0\ \Omega)I_2 = 0$. [Equation 1]

Step 4 – Keep applying the loop rule to complete loops. Each new equation should involve a branch not involved in any previous equations. Stop writing down loop equations when you have involved every branch at least once. For this particular circuit, we only have one more branch (the one on the right) to involve, so we just need one more loop equation. We can either write a loop equation for the inside loop on the right or for all the way around the outside. Let's try the inside loop on the right. Starting at the lower right and going counter-clockwise, we get:

$$+12\text{ V}-(2.0\ \Omega)I_3+(5.0\ \Omega)I_2-15\text{ V}=0.$$

This simplifies to $-3\text{ V}+(5.0\ \Omega)I_2-(2.0\ \Omega)I_3=0.$ [Equation 2]

We could still go around the outside to obtain a third loop equation; however, because that third equation can be obtained from Equations 1 and 2 above it does not give us any new information, and thus does not give us the third independent equation we need.

Step 5 – *Apply the junction rule to come up with additional equations relating the variables.*
The junction rule says that the sum of the current entering a junction must equal the sum of the current leaving the junction. In this circuit both junctions give the same equation, so no matter which junction we choose we get:

$$I_1+I_2+I_3=0.$$ [Equation 3]

This actually tells us that at least one of the currents must be directed opposite to the way we guessed, because we cannot have three currents coming into a junction and no current directed away (or vice versa), so we expect at least one current to have a minus sign in the end.

Step 6 – *Solve the equations to find the currents in the circuit.* This is now an exercise in algebra, solving three equations in three unknowns. Here is one way to do it. Noting that I_2 appears in both Equation 1 and Equation 2, solve Equation 3 for I_2 and substitute it into both Equations 1 and 2. From Equation 3 we get $I_2=-I_1-I_3$.

Equation 1 becomes: $-13\text{ V}-(9.0\ \Omega)I_1-(5.0\ \Omega)I_3=0.$ [Equation 4]

Equation 2 becomes: $-3\text{ V}-(5.0\ \Omega)I_1-(7.0\ \Omega)I_3=0.$ [Equation 5]

Multiply Equation 4 by a factor of +7: $-91\text{ V}-(63\ \Omega)I_1-(35\ \Omega)I_3=0.$ [Equation 6]

Multiply Equation 5 by a factor of -5: $+15\text{ V}+(25\ \Omega)I_1+(35\ \Omega)I_3=0.$ [Equation 7]

Add Equations 6 and 7: $-76\text{ V}-(38\ \Omega)I_1=0$, which gives $I_1=-2.0\text{ A}$. Substituting this into Equation 4 (or Equation 5) gives $I_3=+1.0\text{ A}$. Using Equation 3 then gives $I_2=+1.0\text{ A}$.

Key idea for a multi-loop circuit: In a circuit with multiple batteries, we can use Kirchoff's loop rule and Kirchoff's junction rule to solve for any unknown parameters. The loop rule is actually a statement of conservation of energy applied to circuits, while the junction rule is a statement of conservation of charge.   **Related End-of-Chapter Exercises: 37 – 41, 61, 62.**

Essential Question 18.9: Re-draw Figure 18.23 with the current in each branch labeled correctly. Also, label the potential at all points in the circuit, if the lower junction is at a potential of $V=0$. What is the potential difference between the two junctions?

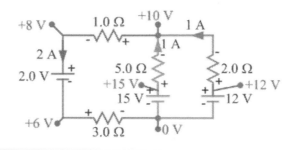

Answer to Essential Question 18.9: Figure 18.24 shows the current in each branch, and the potential at various points relative to $V = 0$ at the lower junction. Labeling potential is like doing the loop rule. Starting at the lower junction and moving up the middle branch, the potential stays at $V = 0$ until we reach the 15-volt battery. Crossing the battery from the – terminal to the + terminal raises the potential by the battery emf to +15 V. The potential difference across the 5.0 Ω resistor is $IR = 5.0 \text{ V}$. As we move through the resistor in the same direction as the current

Figure 18.24: The solution to the circuit in Exploration 18.9, with the correct currents and with the potential labeled at various points.

the potential decreases, reaching +15 V – 5 V = +10 V at the upper end of that resistor, and at the upper junction. Thus, the upper junction has a potential 10 V higher than the lower junction.

Labeling the potential is a way to check the answer for the currents. Going from the lower junction to the upper junction via any branch gives the same answer for the potential of the upper junction. If the answer depended on the path, we would know something was wrong.

18-10 RC Circuits

In some circuits the current changes as time goes by. An example of this is an RC circuit, involving a resistor (R) and a capacitor (C).

EXPLORATION 18.10 – RC Circuits
In the RC circuit in Figure 18.25, the resistor and capacitor are in series with one another. There is also a battery of emf ε, and a switch that is initially in the "discharge" position. The capacitor is initially uncharged, so there is no current in the circuit.

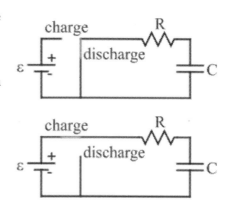

Step 1 – *What are the general equations for the potential difference across a resistor, and the potential difference across a capacitor?* The potential difference across a resistor is given by Ohm's law, $\Delta V_R = IR$, while the potential difference across a capacitor is given by $\Delta V_C = Q/C$.

Figure 18.25: An RC circuit with a battery, resistor, capacitor, and switch.

Step 2 – *Use the loop rule to find the potential difference across the resistor, and across the capacitor, immediately after the switch is moved to the "charge" position.* The capacitor is uncharged, so its potential difference is zero. The closed switch has no potential difference, so by the loop rule the potential difference across the resistor equals the emf of the battery.

Step 3 – *What happens to the potential difference across the capacitor, the potential difference across the resistor, and the current in the circuit as time goes by?* In this circuit, the battery pumps charge from one plate of the capacitor to the other. Because the charge is pumped through the resistor the rate of flow of charge is limited. As time goes by the charge on the capacitor increases, as does the potential difference across the capacitor (being proportional to the charge on the capacitor). By the loop rule, the potential difference across the resistor, and the current in the circuit, decreases as time goes by. Because the current (the rate of flow of charge) decreases, the rate at which the potential difference across the capacitor rises also decreases, slowing the rate at which the current decreases. This gives rise to the exponential relationships reflected in Figure 18.26, and characterized by the product of resistance and capacitance, which has units of time.

$$\tau = RC \ .$$ (Equation 18.9: **Time constant for a series RC circuit**)

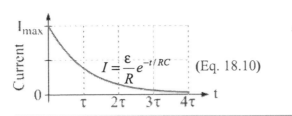 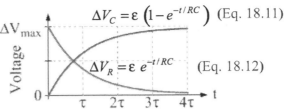

Figure 18.26: Plots of the current, resistor voltage, and capacitor voltage, as a function of time as the capacitor charges.

Step 4 – *If we wanted the capacitor voltage to increase more quickly, could we change the resistance? If so, how? Could we accomplish this by changing the capacitance? If so, how?* To change the capacitor voltage more quickly we could change the resistance or the capacitance. Decreasing the resistance increases the current, so charge flows to the capacitor more quickly. Decreasing the capacitance gives a larger potential difference across the capacitor with the same amount of charge, so that also works. This is consistent with the definition of the time constant, the product RC, which is a measure of how quickly the current, and potential differences, change in the circuit. Decreasing the time constant means that these quantities change more quickly.

Step 5 – *When the switch has been in the "Charge" position for a long time, the circuit approaches a steady state, in which the current and the resistor voltage both approach zero, and the capacitor voltage approaches the battery emf. If the switch is now moved to the "Discharge" position, what happens to the potential difference across the capacitor, the potential difference across the resistor, and the current in the circuit as time goes by?* Now the capacitor discharges through the resistor, so the current is in the opposite direction as it is when the capacitor is charging. The magnitude of the current decreases as time goes by because the potential difference across the resistor, which is the negative of the capacitor voltage by of the loop rule, decreases as time goes by. This gives rise to the relationships shown in Figure 18.27.

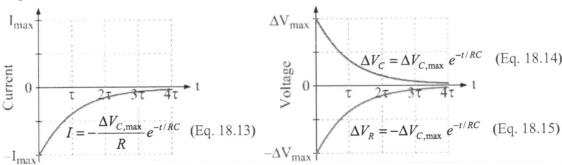

Figure 18.27: Plots of the current, resistor voltage, and capacitor voltage, as a function of time as the capacitor discharges.

Key ideas for RC Circuits: The current in an RC circuit changes as time goes by. In general, the current decreases exponentially with time. The expressions for the potential differences across the resistor and capacitor also involve negative exponentials of a quantity proportional to time, but the loop rule is satisfied at all times. **Related End-of-Chapter Exercises: 11, 12, 63, 64.**

Essential Question 18.10: A particular RC circuit is connected like that in Figure 18.25. The battery emf is 12 V, and the resistor has a resistance of 47 Ω. When the switch is placed in the "Charge" position, it takes 2.5 ms for the capacitor voltage to increase from 0 V to 3.0 V. What is the capacitance?

Answer to Essential Question 18.10: Let's first determine the time constant, using Equation 18.11, and replacing the factor RC by τ. This gives:

$$(3.0 \text{ V}) = (12 \text{ V})\left(1 - e^{-(2.5 \text{ ms})/\tau}\right).$$

Divide both sides by 12 V: $0.25 = 1 - e^{-(2.5 \text{ ms})/\tau}$.

Re-arrange, then take the natural log of both sides: $\ln(0.75) = -\dfrac{2.5 \text{ ms}}{\tau}$.

Solving for the time constant gives: $\tau = -\dfrac{2.5 \text{ ms}}{\ln(0.75)} = 8.7 \text{ ms}$.

Multiplying ohms by farads gives seconds, so $C = \dfrac{\tau}{R} = \dfrac{8.69 \text{ ms}}{47 \, \Omega} = 180 \, \mu C$.

Chapter Summary

Essential Idea: Direct Current Circuits.
Electric circuits are essential to our daily lives, being part of electronic devices like cell phones and iPods. In this chapter we explored the principles of how basic circuits work.

Electric Current
Current is the rate at which charge flows. The symbol we use for current is I:

$$I = \frac{\Delta Q}{\Delta t}$$

(Equation 18.1: **Current, the rate of flow of charge**)

The unit for current is the ampere (A). 1 A = 1 C/s.

The direction of current is the direction positive charges flow.

Ohm's Law and Resistance
$\Delta V = IR$ (Equation 18.2: **Ohm's Law**)

The electrical resistance of a wire of length L and cross-sectional area A is:

$$R = \frac{\rho L}{A} .$$

(Equation 18.3: **Electrical resistance**)

The unit of resistance is the ohm (Ω). 1 Ω = 1 V/A.

The resistivity ρ depends on the material the resistor is made from.

Resistors in Series and Parallel
If N resistors are connected in series, their equivalent resistance is given by:

$$R_{eq} = R_1 + R_2 + \ldots + R_N .$$

(Eq. 18.7: **Equivalent resistance of resistors in series**)

If N resistors connected in parallel, their equivalent resistance is given by:

$$\frac{1}{R_{eq}} = \frac{1}{R_1} + \frac{1}{R_2} + \ldots + \frac{1}{R_N} .$$

(Eq. 18.8: **Equivalent resistance of resistors in parallel**)

Electric Power, and the Cost of Electricity

$$P = I \Delta V = I^2 R = \frac{(\Delta V)^2}{R}.$$ (Equation 18.5: **Electrical power**)

The cost for operating a particular electrical device can be determined from:

$$(\text{Number of hours it is on}) \times (\text{power rating in kW}) \times (\text{cost per kW-hr})$$ (Eq. 18.6: **Electricity cost**)

The contraction/expansion method for analyzing series-parallel combination circuits
When a circuit has one battery with an emf ε, and resistors that have series and parallel connections, the current through, and potential difference across, each resistor can be found by:

1. Identifying two resistors that are either in series or in parallel, and replacing them by their equivalent resistance. Repeat this contraction until one resistor is left.
2. Determining the total current in the circuit by applying Ohm's Law: $I = \varepsilon / R_{eq}$.
3. Gradually reversing the steps in the contraction. When one resistor is expanded to two in series all three resistors have the same current, while when one resistor is expanded to two in parallel all three resistors have the same potential difference across them.

Applying Kirchoff's Rules to analyze Multi-loop Circuits
Kirchoff's Rules can also be applied to analyze a circuit. The loop rule is a statement of conservation of energy, while the junction rule is a statement of conservation of charge.

Loop rule: the sum of all the potential differences around a closed loop is zero.

Junction rule: the total current entering a junction equals the total current leaving the junction.

To analyze a multi-loop circuit (a circuit with multiple batteries connected in a way that they can not be replaced by a single equivalent battery) we use the following steps:
1. Label the currents in the various branches of the circuit.
2. Choose a loop and apply the loop rule to obtain an equation involving one or more of the unknowns (often, these are the currents). Repeat until each branch has been used at least once. Successive equations must involve a branch not involved in previous equations.
3. Apply the junction rule to obtain one or more equations relating the currents.
4. Solve the system of equations for the unknowns.

RC Circuits
In a series RC circuit, the current, and the potential differences across the resistor and the capacitor, change with time. When **the capacitor is being charged** by the battery:

$$I = \frac{\varepsilon}{R} e^{-t/RC}$$ (Eq. 18.10: **Current**) $\Delta V_C = \varepsilon \left(1 - e^{-t/RC}\right)$ (Eq. 18.11: **Capacitor voltage**)

$$\Delta V_R = \varepsilon \, e^{-t/RC}$$ (Eq. 18.12: **Resistor voltage**)

When the battery is removed and **the capacitor is discharging**:

$$I = -\frac{\Delta V_{C,\text{max}}}{R} e^{-t/RC}$$ (Eq. 18.13) $\Delta V_C = \Delta V_{C,\text{max}} \, e^{-t/RC}$ (Eq. 18.14: **Capacitor voltage**)

$$\Delta V_R = -\Delta V_{C,\text{max}} \, e^{-t/RC}$$ (Eq. 18.15: **Resistor voltage**)

End-of-Chapter Exercises

Exercises 1 – 12 are primarily conceptual questions designed to see whether you understand the main concepts of the chapter.

1. You are responsible for running the lift at a ski hill, and you can adjust the speed of the lift so there is a smooth flow of skiers up and down the hill. The ski patrol notifies you that one of the trails down the hill (which is in parallel with other trails) needs to be closed temporarily because of an accident. Should you adjust the speed at which the lift carries skiers up the hill? Justify your answer based on an analogy with a circuit.

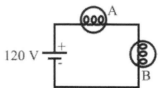

2. Many people think that current gets used up in a circuit, and that in a circuit like that in Figure 18.28, the current through bulb A is different from that through bulb B. (a) Is that true? (b) Use an analogy, such as the fluid analogy or the ski-hill analogy from Section 18.3, to justify your answer to part (a). (c) What gets used up in a circuit?

Figure 18.28: A circuit with two light bulbs, A and B, and a battery, for Exercise 2.

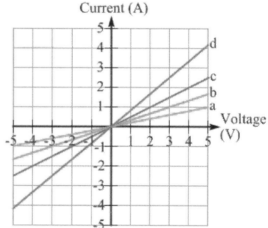

3. You have two resistors. You connect one resistor to a variable power supply (like a battery with a variable emf) and measure the current through it as a function of the potential difference across it. You repeat the process with the second resistor. You then repeat the process with the two resistors connected in parallel, measuring the total current in the circuit as a function of the potential difference across the parallel combination, and do it all a fourth time with the two resistors in series. (a) Which graph in Figure 18.29 goes with which process? (b) What is the resistance of each of the two resistors?

4. Graph c in Figure 18.29 shows the current through a resistor as a function of the voltage across it. (a) Is the resistor an ohmic device? How do you know? (b) What is the resistance of the resistor?

Figure 18.29: Graphs of current vs. potential difference for the situations described in Exercises 3 and 4.

5. Four identical light bulbs are arranged in the circuit in Figure 18.30, connected to a 20-volt battery. There are also two switches in the circuit. Switch *a*, in series with bulb A, is initially closed, while switch *b*, in parallel with bulb B, is initially open. Assume that the resistance of each bulb is the same no matter how much current passes through it at all times, that the switches have negligible resistance when they are closed, and that the battery is ideal (no internal resistance). If switch *b* is then closed (and switch *a* remains closed) what happens to the brightness of (a) bulb A? (b) bulb B? (c) bulb C? (d) bulb D?

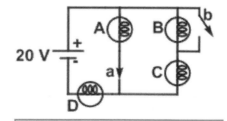

Figure 18.30: A circuit consisting of four identical light bulbs, two switches, and a 20-volt battery. For Exercises 5 – 7.

6. Return to the system described in Exercise 5 and shown in Figure 18.30. In what position should the two switches be to maximize the brightness of (a) bulb A? (b) bulb B? (c) bulb C? (d) bulb D?

7. The four bulbs in Figure 18.30 are identical. Rank the bulbs based on their brightness, from brightest to dimmest, if (a) both switches are closed, (b) switch *a* is closed and switch *b* is open, (c) switch *a* is open and switch *b* is closed, (d) both switches are open.

8. As shown in Figure 18.31, three resistors are connected in a circuit with a 12-volt battery. The resistance of resistor C is neither zero nor infinite, but its exact value is unknown. Resistor A has a resistance of 4.0 Ω. Resistor B has a resistance of 3.0 Ω. (a) Can you say which resistor has the most current through it, or does that depend on the resistance of resistor C? If so, which resistor has the most current through it? (b) Is it possible to rank the resistors based on the potential difference across them? If so, rank them from largest to smallest.

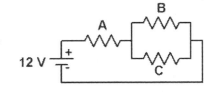

Figure 18.31: A circuit with three resistors and a 12-volt battery, for Exercises 8 and 9.

9. Return to the situation shown in Figure 18.31, and described in Exercise 8. If the resistance of resistor C is increased, what happens to the current through (a) resistor A? (b) resistor B? (c) resistor C? (d) the battery? (e) If the current through the battery is 2.0 A, what is the resistance of resistor C?

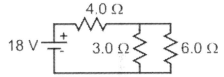

10. Consider the circuit from Essential Question 18.8, which is re-drawn in Figure 18.32. (a) Is it possible to add a 5.0 Ω resistor to the circuit so that the current through at least one of the original resistors increases? (b) If not, explain why not. If so, determine how many ways there are to do this, draw a circuit diagram for each and, in each diagram, circle the resistor(s) with the increased current.

Figure 18.32: A circuit consisting of three resistors and an 18-volt battery, for Exercise 10.

11. Consider the RC circuit shown in Figure 18.33. The battery has an emf of 18 V; and resistor A has a resistance of 3.0 ohms. At *t* = 0 the capacitor is uncharged. At some time later, when the potential difference across the capacitor is 3.0 V, the switch is closed. This brings resistor B, with a resistance of 6.0 ohms, into the circuit. Immediately after the switch is closed, determine: (a) the potential difference across resistor B, (b) the current through resistor A, (c) and the current through resistor B. (d) Describe the effect of closing the switch on the capacitor. In particular, does the rate at which it is charging change? (e) What is going on in the circuit a long time after the switch is closed?

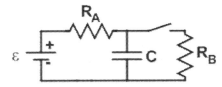

Figure 18.33: An RC circuit consisting of a battery, resistor, and capacitor, along with a second resistor that can be added to the circuit by closing a switch. For Exercise 11.

12. A series RC circuit consists of a resistor, a capacitor that is initially uncharged, and a battery. Two graphs are shown for this situation in Figure 18.34. The lower graph is for the initial situation. The upper graph is for a second trial in which either the resistance or the capacitance has been changed. (a) If the graph shows the potential difference across the resistor as a function of time, as the capacitor charges, what was changed before the second trial, the resistance or the capacitance? Explain, and state whether the resistance or capacitance was increased or decreased. (b) Repeat part (a), but now the graph shows the current in the circuit as a function of time, as the capacitor charges, instead.

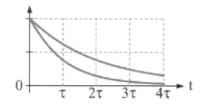

Figure 18.34: The graph shows either the current or the resistor voltage, as a function of time, for a series RC circuit in which the capacitor charges. For Exercise 12.

Exercises 13 – 19 deal with current, batteries, and resistance.

13. At a particular point, there is a current I associated with N protons flowing past the point in a time T. Which of the following changes, done individually (not sequentially) corresponds to a doubling of the current? Justify each answer. (a) The velocity of the protons is doubled, so N protons flow past in a time of $T/2$. (b) The number of protons flowing past in the time T is doubled. (c) Electrons are added to the system so that N protons and N electrons, all flowing the same way, pass by in a time T. (d) The protons are replaced by positive ions that each have a charge of $+2e$, flowing past at the same rate the protons were.

14. A current of 2.0 A is directed to the right past a particular point in a circuit. (a) If the charges flowing are electrons, in what direction are the electrons flowing? (b) Assuming all the electrons flow past in the same direction, how many electrons flow past every second? (c) Do all the electrons really flow past in the same direction? If not, what does the answer to part (b) represent?

15. A particular battery is rated at 2800 milliamp-hours. (a) What kind of unit is the milliamp-hour? (b) If this battery is powering a digital camera, which uses a current of 600 mA, how long will the battery last?

16. The wire connecting a wall switch to an overhead light is 2.4 m long. When the light is turned on there is a net flow of electrons in one direction along the wire; in other words, there is a current in the wire. This net flow is actually rather slow. In this case let's say the average speed of the electrons is 0.20 mm/s (this is known as the drift speed). (a) Based on this, what is the average time it takes an electron at the switch to reach the light bulb filament when the switch is closed (and assuming the electrons flow in that direction!)? (b) In your experience, how long does it take the bulb to come on when you close the switch? (c) How do you reconcile the answers from parts (a) and (b)?

17. A particular wire has a circular cross-section and a resistance R. The wire is then "drawn out" – stretched – until it is four times longer but has the same volume as before. (a) Has its resistance increased, decreased, or stayed the same? (b) What is the resistance of the wire now, in terms of R?

18. A nichrome wire has a length of 2.5 m. How long should a copper wire be to have the same resistance as the nichrome wire? The wires have the same cross-sectional area.

19. We can construct a resistor with a resistance that does not change with temperature by joining a material with a positive temperature coefficient to a material with a negative temperature coefficient, if we choose the parameters correctly. For instance, let's say we want to make a resistor with a total resistance of 100 ohms. We will do this by joining together, end-to-end, a length of copper wire (copper has a temperature coefficient of $+0.0068/°C$) and a length of carbon wire (carbon has a resistivity of $3.0 \times 10^{-4} \, \Omega$ m and a temperature coefficient of $-0.00050/°C$). If each wire has a circular cross-section with a radius of 1.25 mm, and we want the resistance to be 100 ohms even if the temperature fluctuates, how long should the two pieces of wire be?

Exercises 20 – 25 deal with resistors in series or resistors in parallel.

20. **Application: a three-way light bulb.** One way to make a three-way light bulb (a bulb that shines with three different brightnesses, depending on the position of a switch) is to allow connections to the bulb filament so that when the switch is in position 1 the 120 V from the wall socket is connected across the entire filament; in position 2 the 120 V is connected across 40% of the length of the filament; and in position 3 the 120 V is connected across the remaining 60% of the length of the filament. Assume the filament has a uniform cross-section, and that the resistance of the filament is independent of the filament's temperature. (a) Rank the switch positions based on the brightness of the bulb, from brightest to dimmest. (b) If the bulb dissipates 150 W when the switch is in position 2, how much power is dissipated in the other two switch positions?

21. **Application: another three-way light bulb.** A second way to make a three-way light bulb is to have two different filaments. In switch position 1, the wall socket voltage is across one filament, which dissipates 50 W. In position 2, the voltage is across the second filament, which dissipates 100 W. In position 3, the voltage is across both filaments in parallel. (a) What is the resistance of the 50-W filament? (b) What is the resistance of the 100-W filament? (c) What is the equivalent resistance of the two filaments when they are in parallel? (d) What is the power dissipated when the switch is in position 3?

22. You have three identical resistors, one battery, and a number of wires. (a) Show how you can connect the resistors and the battery in a circuit so that the currents through the three resistors are equal (and non-zero!). Is there more than one way to accomplish this? (b) Show how you can connect the resistors and the battery in a circuit so that the current through one of the resistors is twice as large as that through each of the other two resistors. Is there more than one way to accomplish this?

23. You have five identical resistors. You connect one or more of the resistors in some way between point A and point B; one or more in some way between point B and point C; and one or more in some way between point C and point D. There are no other connections between the points, which are all along a straight line in alphabetical order. You then measure the resistance between points A and B to be 5 Ω, between points B and C to be 10 Ω, and between points C and C to be 20 Ω. (a) Find the resistance of each resistor. (b) Sketch a diagram showing how the resistors are connected.

24. You have two identical resistors that are in parallel with one another. When a third identical resistor is wired in parallel with the first two, the equivalent resistance of the set of resistors changes by 6.0 Ω. (a) Does the equivalent resistance increase or decrease? (b) What is the resistance of one of these resistors?

25. You have N identical resistors connected in series with a 12-volt battery. The current through each resistor is 0.50 A. When one more resistor, identical to the others, is added to the circuit, in series, the current through each resistor drops to 0.40 A. (a) What is N? (b) What is the resistance of each resistor?

Exercises 26 – 36 involve series-parallel combination circuits.

26. You have three resistors, with resistances of 23 Ω, 47 Ω, and 100 Ω. By using one, two, or all three, list the different equivalent resistance values can you create with these resistors.

27. As shown in Figure 18.35, four resistors are connected in a circuit with a 20-volt battery. (a) Identify two resistors that are either in series or in parallel, and replace them by a single equivalent resistor. Re-draw the circuit. (b) Repeat part (a) twice more until the circuit has been reduced to a single equivalent resistor. (c) Apply Ohm's Law to find the total circuit current. (d) Reverse the order of the steps in the contraction, and expand the circuit back to its original configuration. At each step in the expansion, draw the circuit diagram and label the current through each resistor. Also, label the potential at various points in the circuit, relative to $V = 0$ at the negative terminal of the battery.

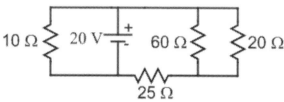

Figure 18.35: Four resistors are connected to a battery that has an emf of 20 V. For Exercise 27.

28. Four resistors are connected to a battery that has an emf of ε , as shown in Figure 18.36. The battery has a current I passing through it, and provides a total power P to the circuit. State whether each of the following statements is true or false, and explain your answer. (a) The sum of the currents through the four resistors is equal to I. (b) The sum of the potential differences across the four resistors is equal to ε . (c) The sum of the power dissipated in the four resistors is equal to P.

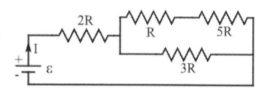

Figure 18.36: Four resistors are connected to a battery that has an emf of ε . For Exercises 28 and 29.

29. Four resistors are connected to a battery that has an emf of ε , as shown in Figure 18.36. (a) What is the equivalent resistance of the circuit, in terms of R? (b) What fraction of I, the current through the battery, passes through the $5R$ resistor? (c) Rank the resistors based on the current through them, from largest to smallest. (d) Rank the resistors based on the potential difference across them, from largest to smallest.

30. Figure 18.37 shows five identical light bulbs in a circuit with a battery. Rank the bulbs based on their brightness, from brightest to dimmest.

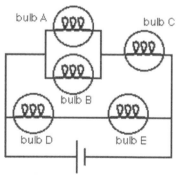

31. Figure 18.37 shows five identical light bulbs in a circuit with a 120-volt battery. Each bulb dissipates 150 W of power if it has a potential difference of 120 V across it. Assume that the resistance of each bulb stays the same no matter what the current is through the bulb. (a) What is the

Figure 18.37: Five identical light bulbs are placed in a circuit with a battery. For Exercises 30 and 31.

resistance of one of these bulbs? (b) Is the total power dissipated in this circuit more than or less than 150 W? Justify your answer without explicitly calculating the power. (c) Find the power dissipated in each bulb, and the total power dissipated in the circuit.

32. Five resistors are wired in a circuit with a single battery, as shown in Figure 18.38. (a) Is resistor A connected in series with resistor B, in parallel with resistor B, or neither? Explain. (b) Rank the resistors based on the current through them, from largest to smallest.

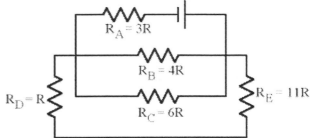

33. Five resistors are wired in a circuit with a single battery, as shown in Figure 18.38. (a) In terms of R, what is the equivalent resistance of the circuit? (b) If the current through resistor D is I_D, find the current through each of the other resistors, in terms of I_D. (c) Rank the resistors based on the potential difference across them, from largest to smallest.

Figure 18.38: Five resistors are wired in a circuit with a single battery. For Exercises 32 and 33.

34. Three resistors are connected in an equilateral triangle, as shown in Figure 18.39. A battery is connected to this set of resistors so that its positive terminal is connected by a wire to one of the lettered points, and the negative terminal is connected to another of the lettered points. Consider the following three cases:

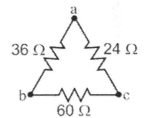

Case 1: Positive terminal connected to a; negative terminal to b.
Case 2: Positive terminal connected to a; negative terminal to c.
Case 3: Positive terminal connected to b; negative terminal to c.
 Find the equivalent resistance of the circuit in (a) case 1, (b) case 2, and (c) case 3.

Figure 18.39: Three resistors connected in an equilateral triangle, for Exercise 34.

35. Four resistors are connected together in the configuration shown in Figure 18.40. A battery is connected to this set of resistors so that its positive terminal is connected by a wire to one of the lettered points, and the negative terminal is connected to another of the lettered points. Consider the following three cases:

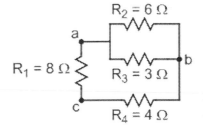

Case 1: Positive terminal connected to a; negative terminal to b.
Case 2: Positive terminal connected to a; negative terminal to c.
Case 3: Positive terminal connected to b; negative terminal to c.
 Rank these cases based on (a) their equivalent resistance; (b) the magnitude of the current through the 4 Ω resistor; (c) the magnitude of the potential difference across the 8 Ω resistor.

Figure 18.40: Four resistors, and three possible locations the terminal of a battery can be connected to. For Exercises 35 and 36.

36. Return to the situation described in Exercise 35, and shown in Figure 18.40. In which of the cases is (a) R_2 in parallel with R_3? (b) R_1 in series with R_4? (c) R_1 in series with the combination of R_2 and R_3? Explain your answer.

Exercises 37 – 45 deal with Kirchoff's Rules and multi-loop circuits.

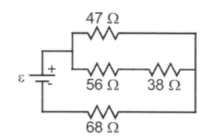

Figure 18.41: Four resistors connected in a circuit with a battery of unknown emf, for Exercise 37.

37. If the current through the 56 Ω resistor in Figure 18.41 is 95 milliamps, what is the emf of the battery?

38. The four resistors from Exercise 37 are connected in a different circuit with a different battery, of unknown emf, as shown in Figure 18.42. If the potential difference across the 56 Ω resistor is 3.2 V, what is the emf of this battery?

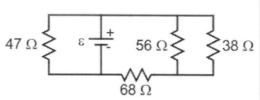

Figure 18.42: Four resistors connected in a circuit with a battery of unknown emf, for Exercise 38.

39. Three batteries are connected in a circuit with three resistors, as shown in Figure 18.43. The currents in two of the branches are known and marked correctly on the diagram. (a) Determine the magnitude and direction of the current through the 2-volt battery. (b) Find the emf of the battery in the middle branch of the circuit. (c) Find the value of the resistance R of the resistor at the bottom left. (d) What is the potential difference, $V_a - V_b$, between points a and b in the circuit?

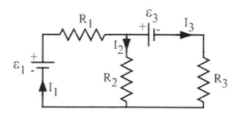

Figure 18.43: A multi-loop circuit, for Exercise 39.

40. The multi-loop circuit in Figure 18.44 is made up of three resistors and two batteries. The various currents in the circuit are labeled. (a) Write out the junction rule at one junction in the circuit. (b) Show that applying the junction rule at the other junction results in the same equation. (c) Apply the loop rule to the inside loop on the left to obtain a loop equation. (d) Apply the loop rule to the inside loop on the right to obtain another loop equation. (e) Apply the loop rule to the outside loop to obtain a third loop equation. (f) Show that one of your loop equations can be obtained by either adding or subtracting the other two loop equations.

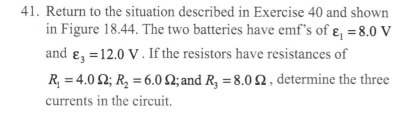

Figure 18.44: A multi-loop circuit consisting of three resistors and two batteries. For Exercises 40 and 41.

41. Return to the situation described in Exercise 40 and shown in Figure 18.44. The two batteries have emf's of $\varepsilon_1 = 8.0$ V and $\varepsilon_3 = 12.0$ V. If the resistors have resistances of $R_1 = 4.0\ \Omega$; $R_2 = 6.0\ \Omega$; and $R_3 = 8.0\ \Omega$, determine the three currents in the circuit.

42. Three 3.00 Ω resistors, three batteries, an ammeter, a voltmeter, and one unknown resistor R are connected in the circuit in Figure 18.45. Assume that the ammeter has negligible resistance and that the voltmeter has infinite resistance. The batteries have emf's of: $\varepsilon_1 =$ 20.0 volts; $\varepsilon_2 = 5.00$ volts; and $\varepsilon_3 = 10.0$ volts. The ammeter reads +0.650 A, the positive

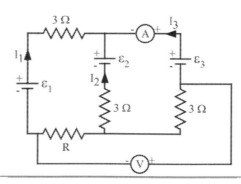

Figure 18.45: A multi-loop circuit consisting of three batteries and four resistors, as well as one ammeter and one voltmeter. For Exercise 42.

value indicating that the current direction shown for I_3 on the diagram is correct. The directions shown for the other two currents may or may not be correct. For parts (a) and (b), include a minus sign if the current is in the direction opposite to that shown on the diagram. (a) What is the value of I_2? (b) What is the value of I_1? (c) What is the resistance of the unknown resistor R? (d) What is the reading on the voltmeter? The voltmeter gives a positive value if the potential at its positive terminal is higher than the potential at its negative terminal.

43. A particular multi-loop circuit consists of three batteries and three resistors, as shown in Figure 18.46. The three currents in the circuit are labeled, but their directions are not necessarily correct. (a) Apply the junction rule to obtain an equation relating the three currents. (b) Based on your junction equation, would you expect one or more of the currents in the circuit to be directed opposite to the direction shown on the diagram? Explain.

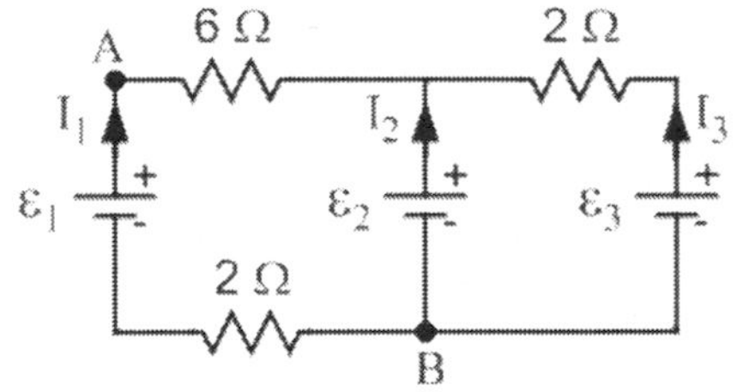

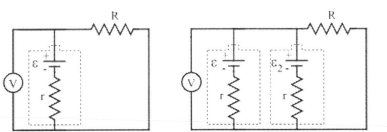

Figure 18.46: A particular multi-loop circuit consists of three batteries and three resistors. For Exercises 43 – 45.

44. A particular multi-loop circuit consists of three batteries and three resistors, as shown in Figure 18.46. The battery emf's are $\varepsilon_1 = 10$ V; $\varepsilon_2 = 2$ V, and $\varepsilon_3 = 6$ V. (a) Solve for the magnitude and direction of the three unknown currents. Note that you should not have to do lots of algebra to solve this problem. (b) What is the potential difference between points A and B in the circuit? Which point is at the higher potential?

45. Return to the situation described in Exercise 44, and shown in Figure 18.46. By adjusting the emf of the second battery (that is, by adjusting ε_2) you can set $I_2 = 0$. (a) What is ε_2 in this situation? (b) What is the potential difference between points A and B in this situation? Which point is at the higher potential?

Exercises 46 and 47 deal with the internal resistance of a battery.

46. A real battery can be modeled as an emf in series with a small internal resistor of resistance r. The circuit on the left in Figure 18.47 shows one such battery connected in a circuit with a voltmeter and a resistor. When R, the resistance of the other resistor in the circuit, is very large the voltmeter reads 6.3 V. When R is 5.0 Ω the voltmeter reads 5.8 V. What is the battery's internal resistance r?

Figure 18.47: The circuit at left shows one battery connected in a circuit with a voltmeter and a resistor, while in the circuit at right a second battery, with a different emf, has been added. For Exercises 46 – 47.

47. As shown in the circuit on the left in Figure 18.47, a resistor R with a resistance of 9.80 Ω is connected to a battery with an emf of $\varepsilon = $ 6.00 volts and an internal resistance of $r = 1.00$ Ω. (a) What is the current through resistor R? (b) The voltmeter measures the terminal voltage on the battery. Assuming the voltmeter has an infinite resistance, what is the terminal voltage? (c) As shown in the figure on the right, a second battery with an unknown emf ε_2 and an internal resistance of r = 1.00 Ω is placed in parallel with the first battery. The terminal voltage measured by the voltmeter is now 5.00 V. What is the current through resistor R now? (d) What is the emf of the second battery?

General problems and conceptual questions

48. A hair dryer is rated at 1850 W. How much does it cost you to use the hair dryer for 10 minutes every day if you are charged 20 cents for every kW-h?

49. A particular 100-W incandescent light bulb costs $0.50, and can be left on for 1000 hours before burning out. A compact fluorescent light bulb of equal brightness, but with a power rating of 20 W, costs $4.00, and lasts for 5000 hours. Over the lifetime of the bulb, what is the total energy used by (a) the incandescent bulb? (b) the fluorescent bulb? For providing 5000 hours of light, compare the total energy used by, and total cost, of (c) five 100-W bulbs, and (d) one fluorescent bulb. Assume you are charged 20 cents for every kW-h.

50. Rank the following based on the daily energy cost associated with them. A: a clock radio rated at 5 W (5 watts) that is on 24 hours a day. B: a 300-W television that is on for 3 hours a day. C: a 1000-W microwave oven that is in use for 20 minutes a day. D: a 90-W computer that is on for 8 hours a day.

51. Four resistors are connected to a battery, as shown in Figure 18.48. If the potential difference across through the 38 Ω resistor is 2.5 V, what is the emf of the battery?

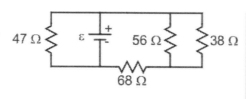

Figure 18.48: Four resistors connected in a circuit with a battery of unknown emf, for Exercise 51.

52. The four resistors from Exercise 51 are connected in a different circuit with a different battery, of unknown emf, as shown in Figure 18.49. If the current through the 38 Ω resistor is 84 milliamps, what is the current through the 47 Ω resistor?

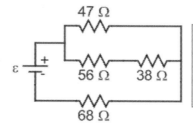

Figure 18.49: Four resistors connected in a circuit with a battery of unknown emf, for Exercise 52.

53. Three resistors are connected in a circuit with a 9-volt battery, as shown in Figure 18.50. Resistor 1 has a resistance of 9 Ω and resistor 3 has a resistance of 6 Ω. The resistance of resistor 2 is some value between 4 Ω and 20 Ω. (a) Can you tell which resistor has the largest current through it, or does that depend on the value of resistor 2? Explain. (b) If the resistance of resistor 2 is increased, what happens to the current through resistor 1? Explain.

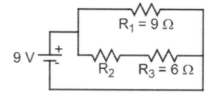

Figure 18.50: A circuit consisting of three resistors and a 9-volt battery. For Exercises 53 – 54.

54. Three resistors are connected in a circuit with a 9-volt battery, as shown in Figure 18.50. Resistor 1 has a resistance of 9 Ω and resistor 3 has a resistance of 6 Ω. The resistance of resistor 2 is 12 Ω. (a) What is the equivalent resistance of the circuit? (b) What is the value of the current through resistor 3?

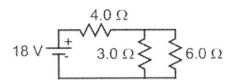

55. Consider the circuit from Essential Question 18.8, which is re-drawn in Figure 18.51. (a) How would you add a 5.0 Ω resistor to the circuit so that the current through the 3.0 Ω resistor is maximized? (b) By how much does the current through the 3.0 Ω resistor increase in that case?

Figure 18.51: A circuit consisting of three resistors and an 18-volt battery, for Exercise 55.

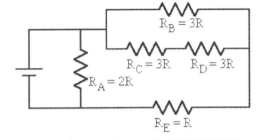

56. Figure 18.52 shows a circuit consisting of five resistors and one battery. (a) Find the equivalent resistance of the circuit, in terms of *R*. (b) Which resistor has the largest potential difference across it? Briefly justify your answer. (c) Rank the resistors based on the current passing through them, from largest to smallest.

57. Figure 18.52 shows a circuit consisting of five resistors and one battery. The battery has an emf of 2.0 V, and R = 2.0 Ω. Find the current through each resistor.

Figure 18.52: A circuit consisting of five resistors and a battery, for Exercises 56 and 57.

58. Return to the Answer to Essential Question 18.9. Note that the current through the 2.0-volt battery is passing through the battery from the positive terminal to the negative terminal. The current through the other two batteries is directed from the negative terminal to the positive terminal. (a) What is happening to the 2.0-volt battery in this situation? (b) Find the total power input to the circuit, and the total power dissipated in the circuit. Should we expect these values to be the same?

59. Five resistors are connected in a circuit with a battery, as shown in Figure 18.53. (a) Which resistor has the largest current passing through it? Explain. Now compare just the 8 Ω resistor and the 10 Ω resistor. Of these two resistors, which has the largest (b) current through it? (c) potential difference across it? Justify your answers.

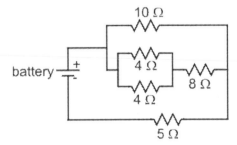

60. Five resistors are connected in a circuit with a battery, as shown in Figure 18.53. (a) Re-draw the circuit diagram, adding an ammeter to measure the current through the 5 Ω resistor, and a voltmeter to measure the potential difference across the 5 Ω resistor. (b) If the battery has an emf of 20 V, what is the reading on the ammeter? (c) What is the reading on the voltmeter?

Figure 18.53: A circuit consisting of five resistors and a battery. For Exercises 59 and 60.

61. Consider the circuit shown in Figure 18.54. (a) Starting at the negative terminal of the 16-V battery, and going clockwise around the loop, write out a loop equation for the inside loop on the left by applying the loop rule. (b) What is the effect of using a different starting point? Write out a loop equation for the same loop, starting at the bottom left corner of the circuit, and going clockwise around the loop. Is this equation equivalent to the equation from (a)? Explain. (c) What is the effect of going counter-clockwise around the loop instead? Write out a loop equation for the same loop, starting at the same place as in (a) but going counter-clockwise. Is this equation equivalent to the equation from (a)? Explain.

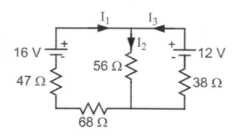

Figure 18.54: A multi-loop circuit consisting of four resistors and two batteries. For Exercises 61 and 62.

62. Consider the circuit shown in Figure 18.54. Find (a) I_1, (b) I_2, and (c) I_3.

63. A series RC circuit consists of a 12-volt battery, a resistor, capacitor, and a switch, all connected in series. The switch is open and the capacitor is initially uncharged. The switch is closed at $t = 0$ and the capacitor begins to charge. At $t = 3.0$ s, the potential difference across the capacitor is 2.0 V. (a) What is the potential difference across the capacitor at $t = 6.0$ s? (b) At what time is the potential difference across the capacitor equal to 6.0 V? (c) At what time is it equal to 12 V?

64. Consider the RC circuit shown in Figure 18.55. The battery has an emf of 12 V; and the two resistors each have a resistance of 3.0 ohms. At $t = 0$, the capacitor is charged, with the top plate positive and the capacitor voltage is equal to 12 V. The capacitance of the capacitor is 1.0 F. (a) Assuming the switch remains open, find the current in the circuit at $t = 5.0$ s. The switch is then closed. Determine the current through each of the two resistors (b) immediately after the switch is closed, (c) when the capacitor voltage is 9.0 V, and (d) a long time after the switch is closed.

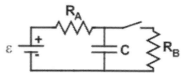

Figure 18.55: An RC circuit consisting of a battery, resistor, and capacitor, along with a second resistor that can be added to the circuit by closing a switch. For Exercise 64.

65. Comment on the statements made by three students who are discussing an issue related to DC circuits. The question that they are discussing is whether it is possible to add a resistor to a circuit and increase the current passing through the single battery in the circuit. There is already a non-zero current in the circuit before the resistor is added.

Terry: I don't think that adding a resistor to the circuit makes any difference to the current. The current comes from the battery, and if we don't change the battery then the current stays the same.

Sarah: I disagree with that. If we connect the resistor in parallel with another resistor in the circuit, then the equivalent resistance of the circuit is going to go down, and the current through the battery is going to go up.

Andy: No way! I disagree with Terry, too, but I think that if you add a resistor to the circuit that you have to be increasing the total resistance, not decreasing it. That's going to decrease the current, not increase it.

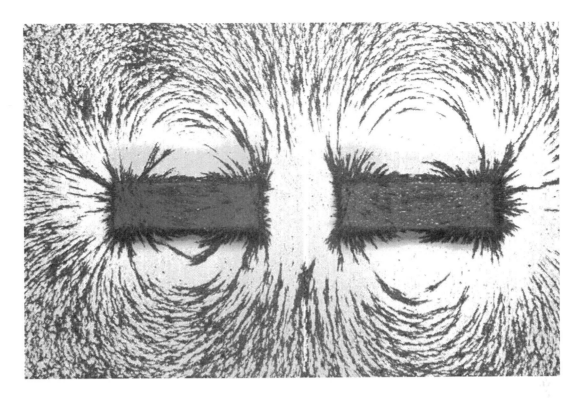

One way to visualize a magnetic field is to use iron filings. Each iron filing aligns with the field. In the photo at right, the iron filings indicate that the magnets repel one another. Also, the field is strongest at the ends (poles) of the magnets, where the density of filings is largest. Photo credit: Thomas Mounsey / iStockphoto.

Chapter 19 – Magnetism

CHAPTER CONTENTS

19-1 **The Magnetic Field**
19-2 **The Magnetic Force on a Charged Object**
19-3 **Using the Right-hand Rule**
19-4 **Mass Spectrometer: An Application of Force on a Charge**
19-5 **The Magnetic Force on a Current-Carrying Wire**
19-6 **The Magnetic Torque on a Current Loop**
19-7 **Magnetic Field from a Long Straight Wire**
19-8 **Magnetic Field from Loops and Coils**

In this chapter, we will explore magnetism. We will discuss how magnetic fields influence charged particles and currents, and we will also discuss where magnetic fields come from. Magnetic fields are very useful things. Among other uses, the magnetic field from a magnet can interact with the steel door of a refrigerator to hold a child's artwork to the fridge; a hiker can use a compass, which has a magnetic needle that interacts with the Earth's magnetic field, to determine which way to walk to get out of the woods; and the magnetic field generated by a magnetic resonance imaging (MRI) machine in a hospital can interact with molecules in your body to investigate what is causing a particular pain in your knee.

The Earth's magnetic field is tremendously important to us, deflecting a dangerous stream of charged particles, coming at Earth from the Sun, around the Earth. The Earth would be a rather different planet were it not for this field.

19-1 The Magnetic Field

Challenge yourself. Write down any similarities you see between magnetic fields and electric fields, as well as any differences you see. Then check your list against the list below. As we will learn later, there are two ways to generate a magnetic field. One way is to use a current, and the similarities and differences below apply to magnetic fields generated by currents. The second way to produce a magnetic field is by changing an electric field, which we will investigate later in the book.

Some Similarities between Electric Fields and Magnetic Fields
- Electric fields are produced by two kinds of charges, positive and negative. Magnetic fields are associated with two magnetic poles, north and south. As we will learn in this chapter, magnetic fields are also produced by charges (but moving charges).
- Like electric charges repel, while unlike charges attract. Like magnetic poles repel, while unlike poles attract.
- The electric field points in the direction of the force experienced by a positive charge. The magnetic field points in the direction of the force experienced by a north pole.

Some Differences between Electric Fields and Magnetic Fields
- Positive and negative charges can exist separately. North and south poles always come together. Single magnetic poles, known as magnetic monopoles, have been proposed theoretically, but a magnetic monopole has never been observed.
- Electric field lines have definite starting and ending points. Magnetic field lines are continuous loops. Outside a magnet, the magnetic field is directed from the north pole to the south pole. Inside a magnet, the magnetic field runs from south to north.

> The magnetic fields we will deal with in this chapter are associated with currents (this includes fields from a refrigerator magnet, the Earth, and an MRI machine). The symbol we use for magnetic field is B.
>
> The SI unit of magnetic field is the tesla (T).

Visualizing Magnetic Fields

As with electric fields, we use field lines and field vectors to visualize magnetic fields. Figure 19.1(a) shows magnetic field lines near a bar magnet, where we can see that the field lines are continuous. Recall that the field is strongest where the field lines are densest, which is inside the magnet itself. The field-vector view, in Figure 19.1(b), emphasizes that the magnetic field exists at all points in space. The field vectors are darkest where the field is strongest.

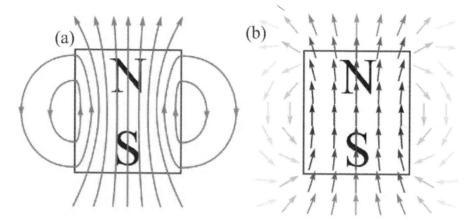

Figure 19.1: Two views of the magnetic field near a bar magnet. Figure (a) shows the magnetic field lines, while figure (b) shows the field vectors. Each of the field vectors can be thought of as compass needles.

The Magnetic Field of the Earth

The Earth's magnetic field is weak compared to a magnet on your refrigerator, but the Earth's field is strong enough to be used by humans, birds, and bacteria for navigation, and to act as a protective shield for the Earth. The Earth's field has a strength of about 5×10^{-5} T at the surface of the Earth, depending on the location. Presently, the Earth's field is gradually decreasing in strength. For comparison, a typical refrigerator magnet has a magnetic field in the millitesla range, while a strong magnet in a research lab has a field of about 10 T. At the surface of the Earth, the magnetic field is stronger near the poles, where the field lines are vertical, and weaker near the equator, where the field lines are horizontal.

The Earth's field is similar in form to that from a bar magnet, and thus resembles the field in Figure 19.1. However, the Earth's magnetic field is not nearly as symmetric as the field from a bar magnet. As shown in Figure 19.2, the form of the Earth's magnetic field is strongly influenced by the solar wind, which consists of charged particles that are emitted by the Sun. Fortunately for us, the Earth's field protects us from much of the effects of the solar wind.

What produces the Earth's magnetic field? This question is one that scientists are working to answer completely, but the basic mechanism is that is associated with electric charge carried by swirling flows of molten iron deep within the Earth's core.

The location of the Earth's magnetic poles changes over time, as the currents within the Earth change. Mostly, the magnetic poles wander around gradually, but roughly every 250 000 years or so, on average, the Earth's magnetic field flips direction (taking a couple of thousand years to flip). The last flip was about 780 000 years ago, so we are overdue for a change. A reversal of the Earth's field direction requires a major change in the flow patterns of the molten iron within the Earth. Understanding such major changes is an area of cutting-edge research.

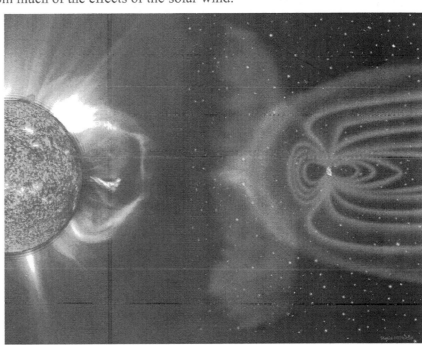

Figure 19.2: In this illustration, the Sun is shown at the left. The solar wind, which streams outward from the Sun, causes the Earth's magnetic field (at the right) to be highly asymmetric, with the field lines extending a significant distance to the right, beyond the extent of this illustration. The Earth's field acts as a rather effective shield against the energetic charged particles in the solar wind. Illustration from SOHO (ESA & NASA).

Related End-of-Chapter Exercises: 10, 33, 36.

Essential Question 19.1: If the north pole of a compass needle points toward the south pole of a magnet, why does the north pole of a compass point north on the Earth?

Answer to Essential Question 19.1: The magnetic south pole of the Earth is actually near the geographic north pole of the Earth (the geographic poles correspond to the Earth's axis of rotation), and the Earth's magnetic north pole is near the geographic south pole.

19-2 The Magnetic Force on a Charged Object

In Chapter 16, we investigated the force experienced by a charged object in an electric field. This force is given by equation 16.3, $\vec{F}_E = q\vec{E}$. The relationship between the magnetic force exerted on a charged particle and the magnetic field is a little more complicated than that between the electric force and the electric field. Let's try to understand the relationship, in the magnetic situation, by making some observations (see Table 19.1) involving various charged particles in a uniform magnetic field (a magnetic field with constant magnitude and direction).

Experiment	Result	Comment or pictorial representation
1. The charged particle is released from rest in the magnetic field.	The particle remains at rest.	The particle experiences no force. (Note: if the particle was in an electric field instead, it would have a constant acceleration.)
2. The particle has an initial velocity parallel to the magnetic field.	The particle moves at constant velocity.	The particle experiences no force! The magnetic field has no influence on a charged particle moving parallel to the field.
3. The particle is given an initial velocity in a direction perpendicular to the magnetic field. The field is directed into the page.	**Figure 19.3**: The particle moves in a circular path at constant speed in a plane perpendicular to the magnetic field. Particles with charges of opposite sign orbit in opposite senses.	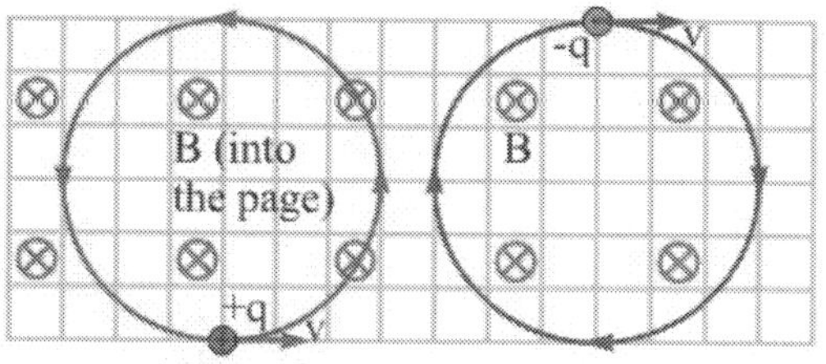
4. The charge of the positively charged particle in experiment 3 is tripled to +3q.	**Figure 19.4**: The radius of the circular path is reduced by a factor of 3, indicating that the force on the particle has tripled.	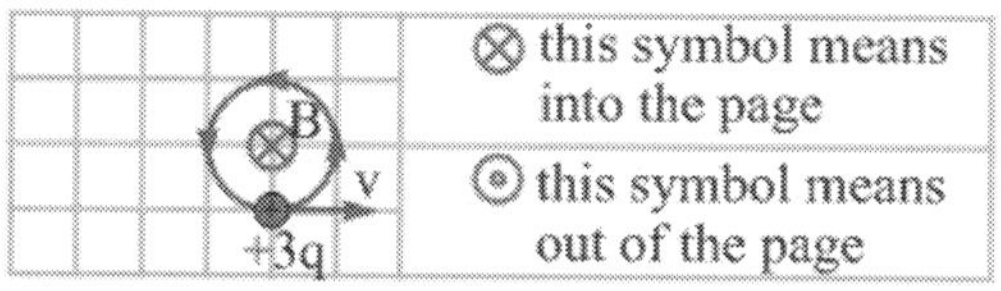 ⊗ this symbol means into the page ⊙ this symbol means out of the page
5. The strength of the magnetic field for the particle in situation 3 is tripled.	**Figure 19.5**: The radius of the circular path is reduced by a factor of 3, indicating that the force on the particle has tripled.	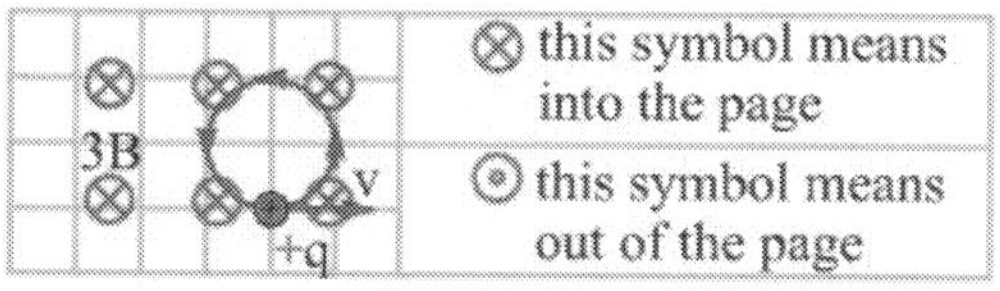 ⊗ this symbol means into the page ⊙ this symbol means out of the page
6. The magnitude of the initial velocity for the particle in situation 3 is tripled.	**Figure 19.6**: The radius of the circular path is increased by a factor of 3, indicating that the force on the particle has tripled. Also, the period of the circular orbit is observed to be independent of the speed of the particle.	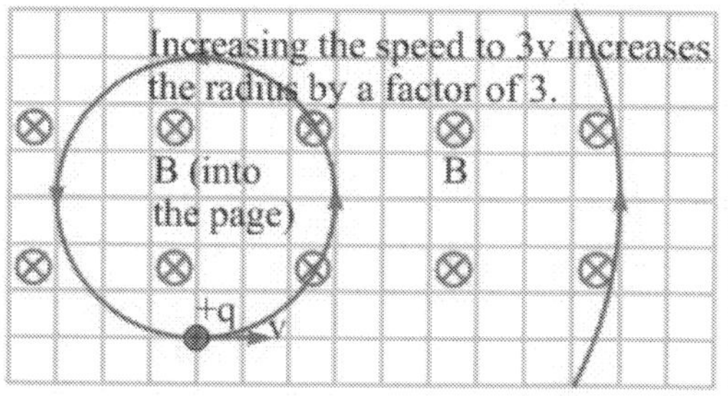 Increasing the speed to 3v increases the radius by a factor of 3.

Table 19.1: The behavior of various charged particles in a uniform magnetic field. All other influences (such as gravity) are neglected.

Related End-of-Chapter Exercises: 6 and 38.

For experiments 3 – 6 in Table 19.1, the particles experience uniform circular motion, so we can apply the analysis methods we used in Chapter 5. The net force on the particle, which is directed toward the center of the circle, comes from the magnetic field. Thus, the magnitude of the magnetic force on a particle of mass m and speed v is: $F_M = mv^2/r$. Solving for the radius, r:

$$r = \frac{mv^2}{F_M} .$$ (Equation 19.1)

In experiments 4 and 5, the mass and speed of the particle are constant. Thus, any change in radius comes from a change in the force. In Experiment 4, tripling the charge decreases the radius by a factor of 3, so the magnetic force must increase by a factor of 3. Thus, we conclude that the magnetic force is proportional to the charge. Analyzing experiment 5 in the same way, we conclude that the magnetic force is proportional to the magnitude of the magnetic field.

The results of experiment 6 are interesting. Based on Equation 19.1, we might expect that tripling the speed would lead to an increase in radius by a factor of 9. Because we observe that the radius increases by a factor of 3, we conclude that there must be a factor of v hidden in F_M . In other words, the magnetic force on a charged particle is proportional to the speed of the particle. Let's write a compact equation for the force exerted by a magnetic field on a charged particle.

$F_M = qvB \sin\theta$,   (Eq. 19.2: **The magnitude of the magnetic force on a charge q**)

where θ is the angle between the velocity, $\bar{v}$, and the magnetic field, $\bar{B}$. The direction of the force is perpendicular to the plane defined by $\bar{v}$ and $\bar{B}$, and given by the right-hand rule.

The right-hand rule for determining the direction of the magnetic force on a moving charge
First, make sure you use your right hand! Also, refer to Figure 19.7.
- Point the fingers on your right hand in the direction of the charge's velocity. *Fingers = velocity*
- While keeping your fingers aligned with the velocity, rotate your hand so that, when you curl your fingers, you can curl them into the direction of the magnetic field.
- Hold out your thumb so it is perpendicular to your fingers. Your thumb points in the direction of the force experienced by a positively charged particle. *thumb = force*
- If the particle has a negative charge, your right hand lies to you. Just reverse the direction of the force. The magnetic force on a negatively charged particle is opposite in direction to that of a positively charged particle if the particles are traveling in the same direction.

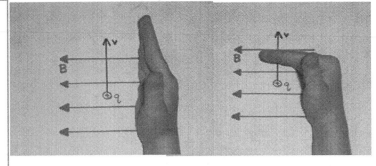

Figure 19.7: To find the magnetic force on a charged particle moving in a magnetic field, point the fingers on your right hand in the direction of the velocity, as in the photo on the left. Orient your hand so you can curl your fingers into the magnetic field, as in the photo on the right. Keep your thumb perpendicular to your fingers, and your thumb points in the direction of the force experienced by a positive charge (out of the page, in this case). A negative charge experiences a force in the opposite direction. Photo courtesy of A. Duffy.

Essential Question 19.2: A charge with an initial velocity directed parallel to a magnetic field experiences no magnetic force, and travels in a straight line. If the initial velocity is perpendicular to the field, the particle travels in a circle. Predict the shape of the path followed by a charge with a velocity component parallel to the field, and a velocity component perpendicular to the field.

Answer to Essential Question 19.2: Try this – walk in a straight line while whirling your hand in a circle around an axis parallel to your velocity. Your hand traces out a spiral, the shape of the path followed by the particle. For the particle, the spiral's axis is parallel to the magnetic field.

19-3 Using the Right-hand Rule

Let's practice using the right-hand rule (described at the end of section 19-2).

EXAMPLE 19.3 – Applying the right-hand rule
Draw a picture of each situation below, and use the picture to help answer the question.
(a) In what direction is the magnetic force on a particle that has a positive charge, and which has a velocity directed to the right in a uniform magnetic field directed up the page?
(b) In what direction is the magnetic force on a negatively charged particle, and which has a velocity directed down the page in a uniform magnetic field that has one component directed out of the page, and the other component directed down the page?
(c) In what direction is the velocity of a particle that has a positive charge, and which experiences a magnetic force directed into the page in a uniform magnetic field directed right?

SOLUTION

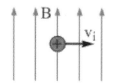

(a) Consider Figure 19.8. The magnetic force is perpendicular to the plane defined by the velocity and the magnetic field, which is the plane of the page. Thus, the force is either directed into or out of the page. Our right hands can distinguish between these directions. Place your right hand on the page, with the fingers pointing right. If the plane of the hand is perpendicular to the page, with the thumb sticking up out of the page, we can curl the fingers into the direction of the field. Thus, the force on the positive charge is out of the page.

Figure 19.8: A positively charged particle is directed right, in a field directed up.

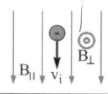

(b) This situation is shown in Figure 19.9. A magnetic field that is parallel to the particle's velocity exerts no force. Thus, we can focus on the perpendicular component of the field. Place your right hand flat on the page with the palm up and the fingers directed to the bottom of the page, so you can curl the fingers up, out of the page. The thumb points to the left, in the direction of the force on a positive charge. Our charge is negative, so the right hand lies to us and the force is directed to the right.

Figure 19.9: A negatively charged particle is moving down, in a field directed both down the page, parallel to the velocity, ($B_{\parallel}$) and out of the page ($B_{\perp}$), perpendicular to the velocity.

(c) Consider the situation shown in Figure 19.10. First, remember that the velocity is perpendicular to the force, so, if the force is perpendicular to the page, the velocity must be in the plane of the page. Let's apply the right-hand rule in reverse order. Curl the fingers of your right hand, and hold your hand with the thumb pointing down, into the page, and with your curled fingers directed right. When you un-curl your fingers 90°, they point up the page in the direction of the velocity component that is perpendicular to the magnetic field. However, there may or may not be a velocity component directed left or right, parallel to the magnetic field lines. This case is ambiguous – we can't say for certain which direction the velocity is in because the angle between $\vec{v}$ and $\vec{B}$ does not have to be 90°.

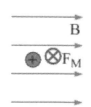

Related End-of-Chapter Exercises: 13 – 16.

Figure 19.10: A moving positively charged particle experiences a force directed into the page, in a field directed right.

A special case: a charged particle with a velocity perpendicular to the magnetic field

When the velocity of a charged particle is perpendicular to the field, the particle follows a circular path. Applying the form of Newton's Second Law for circular motion, $\sum F = mv^2 / r$:

$$qvB \sin\theta = \frac{mv^2}{r}.$$

The velocity and magnetic field are perpendicular, so we have $\sin(90°) = 1$. Solving for r:

$$r = \frac{mv}{qB}.$$ (Equation 19.3: **Radius of the path, when $\vec{v}$ and $\vec{B}$ are perpendicular**)

We can find the time it takes the particle to go around the circle once, which we call the period of the orbit, by dividing the circumference of the orbit by the particle's speed:

$$T = \frac{2\pi r}{v} = \frac{2\pi mv}{vqB} = \frac{2\pi m}{qB}.$$ (Equation 19.4: **The period of the circular orbit**)

Note that *the period is independent of the speed* of the particle.

EXPLORATION 19.3 – Identifying the particles

As shown in Figure 19.11, four particles pass through a square region of uniform magnetic field directed perpendicular to the page.

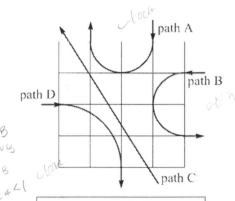

Figure 19.11: The paths followed by the particles.

Particle	Charge	Mass	Speed	Path taken
1	0	2m	6v	C
2	+q	m	4v	D
3	+2q	4m	v	A
4	−q		2v	B

Table 19.2: The charge, mass, and speed for four particles passing through a magnetic field.

Step 1 – *Identify the path taken by particle 1.* Because particle 1 has no charge, the field exerts no force on it. Particle 1 travels through the field in a straight line – so, it follows path C.

Step 2 – *Are the paths clockwise or counterclockwise? Use your answer to determine the direction of the magnetic field.* Positive charges travel in one sense, while negative charges travel in the opposite sense. Paths A and D both show clockwise motion, so these are the paths taken by the two positive charges. Path B shows a counterclockwise motion, so path B is taken by particle 4, the only particle with a negative charge. Applying the right-hand rule the positive particle traveling along path D, we find that the magnetic field is directed out of the page.

Step 3 – *Apply Equation 19.3 to see which particle follows path A and which follows path D.* Equation 19.3 tells us that the radius of the path is proportional to the mass multiplied by speed divided by the charge (we neglect the factor of the magnetic field, which is the same for all the charges). This combination is $4mv/q$ for particle 2, and $2mv/q$ for particle 3. Thus, the path followed by particle 2 has twice the radius (path D) as that followed by particle 3 (path A).

Key idea: The radius of curvature, and the direction of curvature, provides information about the speed, charge, and/or the mass of a particle. **Related End-of-Chapter Exercises: 7, 8, 17, 18.**

Essential Question 19.3: Rank the four particles in Exploration 19.3 in terms of the magnitude of the force applied to them by the magnetic field, from largest to smallest.

Answer to Essential Question 19.3: The direct way to solve this problem is to use Equation 19.2, $F_M = qvB \sin\theta$. The particles are in the same magnetic field, and $\sin\theta = 1$ because each velocity is perpendicular to the magnetic field. Thus, in this situation the force magnitude is proportional to the magnitude of the charge multiplied by the speed. This combination of factors is 0 for particle 1, $4qv$ for particle 2, and $2qv$ for particles 3 and 4, so the ranking is $2 > 3 = 4 > 1$.

19-4 Mass Spectrometer: An Application of Force on a Charge

There are a number of practical devices that exploit the force that a magnetic field applies to a charged particle. Let's investigate one of these devices, the mass spectrometer.

EXPLORATION 19.4 – How to make a mass spectrometer

Mass spectrometers, which separate ions based on mass, are often used by chemists to determine the composition of a sample. Let's explore one type of mass spectrometer, which uses electric and magnetic fields. For each step below, sketch a diagram to help you with the analysis.

Step 1 – The accelerator. *Release a charged particle from rest near one plate of a charged parallel-plate capacitor, so that the particle accelerates toward the other plate of the capacitor. Apply energy conservation to obtain a relation between ΔV, the potential difference across the capacitor, and the speed of the particle when it emerges from a small hole in the second plate.* The particle shown in Figure 19.12 has a positive charge, but the accelerator can also work for negatively charged particles if we reverse the battery attached to the capacitor. Let's apply what we learned in Chapter 17. If we define the particle's electric potential energy to be zero at the negative plate ($U_f = 0$), the potential energy is $U_i = q(\Delta V)$ when the particle is next to

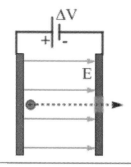

the positive plate. The particle has no initial kinetic energy ($K_i = 0$), but as the particle accelerates across the gap between the plates the electric potential energy is transformed into kinetic energy. There is no work done by non-conservative forces ($W_{nc} = 0$). With three terms being zero, our five-term conservation of energy equation, $K_i + U_i + W_{nc} = K_f + U_f$, becomes $U_i = K_f$. Using our expressions for potential and kinetic energy gives:

Figure 19.12: Charged particles are released from rest near the left-hand plate of a parallel-plate capacitor, accelerate across the gap, and emerge via a hole cut in the right-hand plate.

$$q(\Delta V) = \frac{1}{2}mv^2, \text{ so the particle's speed when it emerges is } v = \sqrt{\frac{2q(\Delta V)}{m}}.$$

Step 2 – The velocity selector. *The particle passes through a second parallel-plate capacitor in which the plates are parallel to the particle's velocity. In addition to the electric field inside the capacitor there is also a magnetic field, directed perpendicular to both the electric field and the velocity of the particle. The combined effect of the two fields is that a particle with just the right speed experiences no net force and passes undeflected through the velocity selector. Determine how this speed relates to the magnitudes of the fields.* In the situation shown in Figure 19.13, the top plate is positively charged, and the bottom plate is negatively charged, so the electric field in the capacitor is directed down. The electric force on the positive charge is also directed down, because $\vec{F}_E = q\vec{E}$. For the particle to experience no net force, the magnetic force must exactly balance the electric force. The magnitude of the magnetic force is given by $F_M = qvB \sin\theta$.

Because the velocity and magnetic field are at right angles, $\theta = 90°$ and $\sin(90°) = 1$.

Setting the magnitudes of the two forces equal, $F_M = F_E$, gives $qvB = qE$. The factors of q cancel. Solving for the speed of the undeflected particles gives:

$$v = \frac{E}{B}.$$ (Eq. 19.5: **The speed of undeflected particles in a velocity selector**)

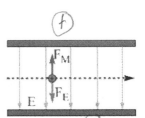

There is a uniform magnetic field, directed perpendicular to the page, between the plates.

Question: Is the magnetic field in Figure 19.13 directed into or out of the page?

Answer: By the right-hand rule, to obtain a magnetic force directed up on a positive charge with a velocity to the right, the magnetic field is into the page.

Figure 19.13: A charged particle with just the right velocity passes undeflected through the velocity selector because the magnetic force balances the electric force.

Question: What happens to particles traveling faster than the undeflected particles? What happens to particles traveling slower than the undeflected particles?

Answer: The magnetic force depends on speed, while the electric force does not. For particles going faster than the selected speed, the magnetic force exceeds the electric force. The net upward force deflects the fast particles up out of the beam (see Figure 19.14). For relatively slow particles, the magnetic force is less than the electric force. The net downward force deflects the slow particles down out of the beam.

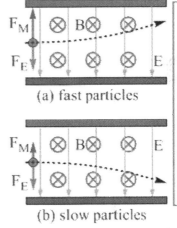

(a) fast particles

(b) slow particles

Figure 19.14: (a) A velocity selector deflects fast particles in one direction (up, in this case) and (b) slow particles in the opposite direction (down, here).

Step 3 – The mass separator. *Particles that pass undeflected through the velocity selector are sent into the mass separator, which consists of a uniform magnetic field that is perpendicular to the velocity of the particles. If the particles are collected after traveling through half-circles in the field, find an expression for the separation between particles of mass m_1 and particles of mass m_2.* Consider equation 19.3 ($r = mv/qB$), which gives the radius of the path followed by a charged particle in a uniform magnetic field. We give the particles the same charge in step 1, and they are in the same magnetic field, so the radius depends on mass and speed. The velocity selector ensures that all the particles reaching the mass separator have the same speed, so the radii differ only because the masses differ. As shown in Figure 19.15, the particles enter the field at the same point. After passing through a half-circle, the separation between the particles is the difference between the diameters of the circular paths. Thus, the separation is given by:

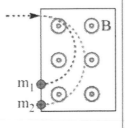

Figure 19.15: After traveling through a half-circle in the mass separator, particles of different mass are separated by a distance equal to the difference between the diameters of their paths.

$$\Delta s = 2r_1 - 2r_2 = 2\left(\frac{m_1 v}{qB} - \frac{m_2 v}{qB}\right) = \frac{2v}{qB}(m_1 - m_2)$$ (Eq. 19.6: **Separation in the mass spec.**)

Key Ideas for the mass spectrometer: The mass spectrometer is an excellent example of how we exploit uniform electric and magnetic fields, both individually and in combination, when working with charged particles. **Related End-of-Chapter Exercises: 45 – 47.**

Essential Question 19.4: If the particles in step 2 of Exploration 19.4 were negatively charged, would we need to reverse the direction of either field (or both fields) for the velocity selector to function properly, or would it work for negatively charged particles without any changes?

Answer to Essential Question 19.4: The velocity selector works for negatively charged particles without any changes. When the particle has a negative charge, the electric force and the magnetic force both reverse direction. For particles traveling at a speed of $v = E/B$, however, the two forces balance and the particles are undeflected. Faster particles would be deflected down, given the orientation of the fields in Figure 19.14, while slower particles would be deflected up.

19-5 The Magnetic Force on a Current-Carrying Wire

Starting with the magnetic force on a moving charge, $F_M = qvB\sin\theta$, let's derive the equation for the magnetic force on a current-carrying wire. In the equation below, we write the velocity as a length L divided by a time interval Δt. We also use the definition of current, as the amount of charge q passing a point in a certain time interval Δt. This gives:

$$F_M = qvB\sin\theta = q\frac{L}{\Delta t}B\sin\theta = \frac{q}{\Delta t}LB\sin\theta = ILB\sin\theta \ .$$

The magnitude of the magnetic force exerted on a wire of length L, carrying a current I, by a magnetic field of magnitude B is:

$$F_M = ILB\sin\theta \ , \qquad \text{(Equation 19.7: \textbf{The magnetic force on a current-carrying wire})}$$

where θ is the angle between the current direction and the magnetic field.

The direction of the force is given by a right-hand rule that follows the rule for charges. Point the fingers on your right hand in the direction of the current. Align your hand so that when you curl your fingers, they point in the direction of the magnetic field. Stick out your thumb, and it points in the direction of the magnetic force. The current direction is defined to be the direction of flow of positive charge, so the direction given by your hand never needs to be reversed, as it does when we find the direction of the magnetic force on a negatively charged particle.

EXPLORATION 19.5A – Three paths from a to b

As shown in Figure 19.16, three wires carry equal currents from point a to point b. The wires are in a uniform magnetic field directed to the right.

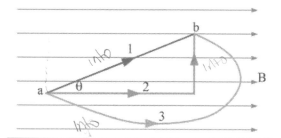

Figure 19.16: Three wires carry currents of the same magnitude from point a to point b in a uniform magnetic field.

Step 1 – *Apply the right-hand rule to find the direction of the magnetic force exerted on each wire by the field.* For wire 1, pointing the fingers of the right hand in the direction of the current and then curling them into the direction of the magnetic field gives a force directed into the page. For wire 2, there is no force on the part that is parallel to the field. Apply the right-hand rule to the part of wire 2 that is perpendicular to the field – the force is directed into the page. For wire 3, we ignore the parts of the wire that are going left or right, parallel to the field, and focus on the sections that are going up or down, perpendicular to the field. Wire 3 goes up more than it goes down. Applying the right-hand rule with the current directed up gives a force into the page.

Step 2 – *Rank the wires based on the magnitude of the force they experience. To help you, re-draw wire 3 as short segments that are either parallel or perpendicular to the magnetic field.* Applying Equation 19.7, we find for wire 1 that $F_{M1} = ILB\sin\theta$, where wire 1 has a length L. For wire 2, only the part directed up, perpendicular to the field, experiences a force. This section of wire 2 has a length of $L\sin\theta$, so applying equation 19.7 for wire 2 gives $F_{M2} = I(L\sin\theta)B\sin(90°)$. The factor of $\sin(90°)$ equals 1, giving: $F_{M1} = F_{M2} = ILB\sin\theta$.

Let's re-draw wire 3 using line segments that are either parallel to or perpendicular to the magnetic field, as in Figure 19.17. We could use shorter segments, so the wire would not look so jagged, but the argument that follows would still apply. The segments that are parallel to the field experience no force. For the segments perpendicular to the field, subtract the total length of segments that carry current down from the total length of segments that carry current up, because their forces are in opposite directions. The net displacement is $L\sin\theta$ up, the same as wire 2. Thus, all three wires experience the same force and the ranking is 1 = 2 = 3. The same argument applies for any wire: all wires carrying equal currents from a to b in a uniform magnetic field experience equal forces.

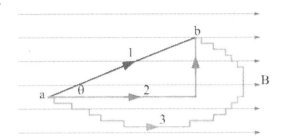

Figure 19.17: Wire 3 is replaced by a joined set of line segments, parallel and perpendicular to the magnetic field.

Key Idea: All wires carrying equal currents from one point to another in a uniform magnetic field experience the same magnetic force.   **Related End-of-Chapter Exercises: 51, 52.**

EXPLORATION 19.5B – The force on a current-carrying loop
Step 1 – *In what direction is the net magnetic force on a rectangular wire loop that carries a clockwise current I in a uniform magnetic field that is directed out of the page? The loop is in the plane of the page.* A diagram is shown in Figure 19.18 (a). To find the net magnetic force on the loop, we add, as vectors, the forces on each side of the loop. As shown in Figure 19.18 (b), the force on the left side cancels the force on the right side, and the force at the top cancels the force at the bottom. Thus, the net force on the loop is zero.

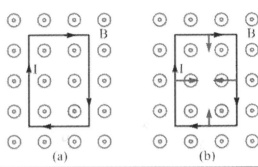

Figure 19.18: (a) A rectangular loop in a uniform magnetic field directed out of the page. (b) The magnetic force on each side of the loop is shown with an inward-directed arrow. The net force on the loop is zero.

Step 2 – *In what direction is the net force on the loop if the field is directed to the left?* In this special case, the forces on the top and the bottom sides are each zero. As Figure 19.19 (b) shows, the remaining forces cancel, and again the net force on the loop is zero. This is always true - as long as the magnetic field is uniform, the net magnetic force on any complete current-carrying loop is zero.

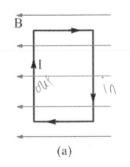

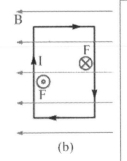

Figure 19.19: (a) A rectangular loop in a uniform magnetic field directed to the left. (b) In this case, the top and bottom sides experience no force because they are parallel to the magnetic field. The force on the left side is out of the page, while the force on the right side is into the page. These forces are of equal magnitude, so the net force is zero.

Key idea for current-carrying loops: In a uniform magnetic field, the net magnetic force acting on a current-carrying loop is zero.   **Related End-of-Chapter Exercise: 53.**

Essential Question 19.5: Consider figures 19.18 and 19.19. The net magnetic force acting on a complete current-carrying loop is zero, but what about the net magnetic torque on the loop?

Answer to Essential Question 19.5: Figure 19.18, with the magnetic field perpendicular to the plane of the loop, is a special case, in which both the net force and the net torque are zero. In Figure 19.19, the forces acting on the loop would cause the loop to rotate, with the left side of the loop coming out of the page and the right side going into the page. In this case there is a net torque acting on the loop, which is generally the case.

19-6 The Magnetic Torque on a Current Loop

In a uniform magnetic field, a wire loop carrying current experiences no net force, but there is generally a net torque acting that tends to make the loop rotate. As we will see, we can exploit the interaction between the loop and the field to make a motor.

Let's begin by drawing a number of views of a rectangular current-carrying loop in a uniform magnetic field. The loop can rotate without friction about an axis parallel to its long sides, passing through the center of the loop. In the front view, imagine viewing the loop as if your eye is at the bottom of the page, looking along the page at the loop. The angles specified in Figure 19.20 are the angles between the loop's area vector, which is perpendicular to the plane of the loop, and the magnetic field.

Specifying area as a vector is something new. The magnitude of the area vector is the loop's length multiplied by its width. The direction of the area vector is perpendicular to the plane of the loop, but there are two directions that are perpendicular to this plane. The orientation of the area vector can be found by applying a right-hand rule. Curl the fingers on your right hand in the direction of the current flow. Your thumb, when you stick it out, gives the direction of the area vector. The angles specified in Figure 19.20 are the angles between the area vector and the magnetic field.

If the loop starts from rest in Figure 19.20(a), it will begin to rotate. The angular velocity increases in a clockwise direction, as observed from the front view. The clockwise torque continues until Figure 19.20(d), at which point there is no torque. In the absence of friction, the loop's angular momentum keeps it moving forward, with the counter-clockwise torque in Figures 19.20(e) – (g) slowing the loop down and bringing it instantaneously to rest in Figure 19.20(g). The loop then reverses direction, and goes through the pictures in reverse order.

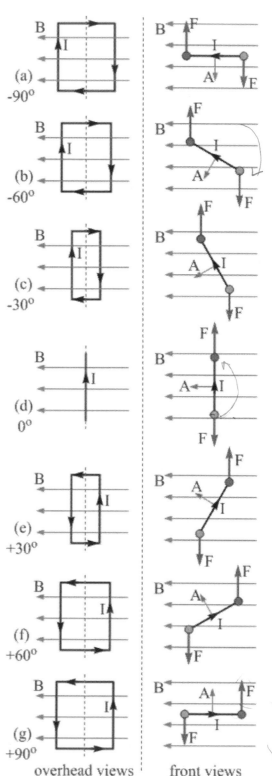

Figure 19.20: The arrow, labeled A in each front view, is the loop's area vector, which is directed perpendicular to the plane of the loop. When the area vector is perpendicular to the magnetic field, the loop experiences maximum torque. If the loop starts from rest in (a), the clockwise torque it experiences in orientations (a) – (d) cause the loop to rotate clockwise. After (d), the torque reverses direction, bringing the loop instantaneously to rest at (g), at which point the motion reverses.

overhead views   front views

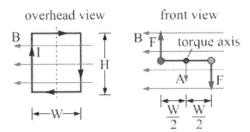

Figure 19.21: The situation from Figure 19.20(a), with dimensions given for the loop.

The situation from Figure 19.20(a) is re-drawn in Figure 19.21, showing the height H and width W of the loop. Apply equation 10.9, $\tau = rF\sin\theta$, to find the torque applied by the field on the loop.

Take torques about the axis through the center of the loop, shown as a dashed red line in Figure 19.21 (overhead view). The distance from each force to this axis is $W/2$, each force has a magnitude of $F = IHB$ (from equation 19.7), and the angle between the line we're measuring the distance along and the line of the force is 90°. A factor of 2 accounts for the two identical torques, which are both clockwise. The magnitude of the net torque on the loop in this orientation is:

$$\tau = 2\frac{W}{2}IHB = IAB\,,\text{ where } A = HW \text{ is the area of the loop.}$$

As the loop rotates, the torque is reduced as the angle between each force and the plane of the loop changes from 90°. This angle is equal to the angle between the area vector and field, so we can write a general expression for the torque, and generalize to coils with more than one loop.

The magnitude of the torque applied by a uniform magnetic field on a coil of N loops carrying a current I, with an angle θ between the area vector $\vec{A}$ and the magnetic field $\vec{B}$, is:

$$\tau = NIAB\sin\theta\,.\quad \text{(Equation 19.8: Torque on a current-carrying coil in a magnetic field)}$$

The DC (Direct Current) Motor – an application of the torque on a current-carrying loop

If the loop starts from rest in the orientation in Figure 19.20(a), then the loop will flop back and forth as the torque alternates between clockwise and counterclockwise. Friction will eventually bring the loop to rest in the zero-torque orientation of Figure 19.20(d). With one change, however, the loop becomes an electric motor – a device for transforming electrical energy into mechanical energy. We simply reverse the direction of the current each time the plane of the loop is perpendicular to the field (as in Figure 19.20(d)). With this change, the torque on the loop, and therefore the rotation of the loop, is always in the same direction.

An easy way to accomplish the current reversal is to use a split-ring commutator, as shown in Figure 19.22. The split-ring commutator is a cylinder divided into two halves that are electrically insulated from one another. The cylinder rotates with the loop, and its left side rubs against a fixed wire connected to the positive terminal of a battery while its right side rubs against a fixed wire connected to the battery's negative terminal. As shown in Figure 19.22, the current reverses direction in the loop every half-rotation. From our perspective, however, the current always goes clockwise around the loop, which is why the torque is always in the same direction.

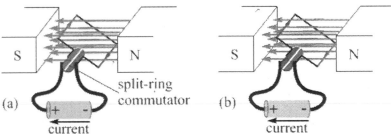

Figure 19.22: The split-ring commutator (the split cylinder that the dark wires from the battery are connect to) reverses the current in the loop every half rotation. From our perspective, the current is always clockwise, but from the loop's perspective the current has reversed between (a) and 180° later in (b).

Related End-of-Chapter Exercises: 54, 59.

Essential Question 19.6: In the situation shown in Figure 19.20, we ignored the magnetic forces acting on the shorter sides of the loop. Why can we do this?

Answer to Essential Question 19.6: In this situation, the forces acting on the two short sides of the loop produce forces that cancel one another. These forces are either zero or are directed along the axis we take torques around, giving no torque about that axis.

19-7 Magnetic Field from a Long Straight Wire

Let's now turn to investigating how to produce a magnetic field. Similar to the way that electric fields can be set up by charged particles and act on charged particles, magnetic fields can be set up by moving charges (or currents) and act on moving charges. The analog of the point charge for magnetism is the long straight current-carrying wire. Figure 19.23 shows the magnetic field from a long straight wire. The magnetic field from a wire decreases with distance from the wire. Instead of the field being proportional to the inverse square of the distance, as is the electric field from a point charge, the magnetic field is inversely proportional to the distance from the wire. Another difference between the electric field situation and the magnetic field situation is that the magnetic field lines are complete loops.

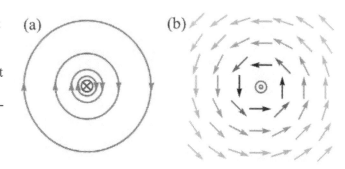

Figure 19.23: The magnetic field lines from a long straight wire wrap around the wire in circular loops. In (a), we see magnetic field lines near a wire that carries current into the page. In (b), field vectors are plotted. The vectors circulate counterclockwise, because the current in (b) is out of the page, opposite to what it is in (a). Both views show the strength of the magnetic field decreasing as the distance from the wire increases.

The magnetic field at a distance r from a long straight wire carrying a current I is:

$$B = \frac{\mu_0 I}{2\pi r}.$$ (Eq. 19.9: **The magnetic field from a long-straight wire**)

The direction of the magnetic field is given by a right-hand rule. In this rule, point the thumb on your right hand in the direction of the current in the wire. When you curl your fingers, they curl the same way that the magnetic field curls around the wire. The constant μ_0 in equation 19.9 is known as the **permeability of free space,** and has a value of $\mu_0 = 4\pi \times 10^{-7}$ T m / A.

In Chapter 8, we analyzed situations involving objects with mass interacting with each other via the force of gravity. In Chapter 16, we investigated situations involving interacting charged particles. Let's investigate analogous magnetic situations involving long straight wires.

EXPLORATION 19.7 – The magnetic force between two parallel wires
A long straight wire (wire 1) carries a current of I_1 into the page. A second long straight wire (wire 2) is located a distance d to the right of wire 1, and carries a current of I_2 into the page. Let's determine the force per unit length experienced by wire 2 because of wire 1.

Step 1 – *Find the magnitude and direction of the magnetic field set up by wire 1 at the location of wire 2.* The magnitude of the field is given by equation 19.9: $B_1 = \mu_0 I_1 /(2\pi d)$. To find the direction of this field at the location of wire 2, recall that the field lines are circular loops centered on wire 1. Applying the right-hand rule (see the previous page), we find that these field lines go clockwise. The field at any point is tangent to the field line, so the field at the location of wire 2 is directed straight down (see Figure 19.24).

Step 2 – Apply equation 19.7 to find the force per unit length that wire 2 experiences because of the magnetic field of wire 1.
Equation 19.7 ($F_M = ILB \sin\theta$) gives us the force a wire of length L experiences in a magnetic field. However, we do not have a length to use for wire 2, so we bring the factor of length to the left side. Substituting the expression for B_1 from step 1 gives:

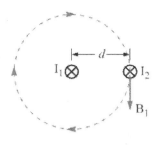

Figure 19.24: The magnetic field set up by wire 1 at the location of wire 2 is directed down the page, tangent to the direction of wire 1's field line at that point.

$$\frac{F_{12}}{L} = I_2 B_1 \sin(90°) = \frac{\mu_0 I_1 I_2}{2\pi d}.$$ (Eq. 19.10: **The force between two parallel wires**)

Applying the right-hand rule associated with equation 19.7, we find that the force experienced by wire 2 is to the left. In other words, when the currents are in the same direction the wires attract. If the currents are in opposite directions, they repel.

Step 3 – Which wire exerts more force on the other, if $I_1 = 3I_2$? No matter how the currents compare, the wires experience forces of equal magnitude in opposite directions – Newton's third law applies. Another way to see this is that equation 19.10 applies equally well to either wire.

Key ideas: Two long straight wires that are parallel to one another exert forces on one another. If the currents are in the same direction, the wires attract one another. If the currents are in the opposite direction, the wires repel one another. **Related End-of-Chapter Exercises: 49 and 57.**

EXAMPLE 19.7 – Finding the net magnetic field
Four long straight parallel wires pass through the x-y plane at a distance of 4.0 cm from the origin. Figure 19.25 shows the location of each wire, and its current. Find the net magnetic field at the origin because of these wires.

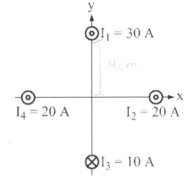

Figure 19.25: The currents carried by four long straight wires.

SOLUTION
Let's apply the principle of superposition to find the net magnetic field at the origin. The four fields, one from each wire, are shown in Figure 19.26. The two fields that are along the y-axis, from the two wires on the x-axis, cancel one another. The two fields along the x-axis, from the wires on the y-axis, add together. Thus, the net magnetic field at the origin is directed along the positive x-axis, and has a magnitude of:

$$B_{net} = B_1 + B_3 = \frac{\mu_0 I_1}{2\pi r} + \frac{\mu_0 I_3}{2\pi r} = \frac{\mu_0}{2\pi r}(I_1 + I_3) = \frac{4\pi \times 10^{-7} \text{ T m/A}}{2\pi (0.040 \text{ m})}(30 \text{ A} + 10 \text{ A}) = 2.0 \times 10^{-4} \text{ T}.$$

Note that magnetic field is a vector. We can add the magnitudes of the two individual fields to find the magnitude of the net field only because the two fields are in the same direction.

Related End-of-Chapter Exercises: 11, 24 – 27.

Essential Question 19.7: For the situation in Exploration 19.7, let's say that the two wires are 40 cm apart and that $I_1 = 3I_2$. Where is the net magnetic field equal to zero near these wires?

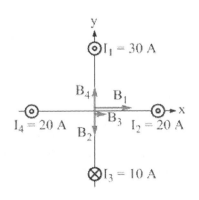

Figure 19.26: The four fields that add as vectors to give the net field.

Answer to Essential Question 19.7: To get a net magnetic field of zero, the two fields must point in opposite directions. This happens only along the straight line that passes through the wires, in between the wires. The current in wire 1 is three times larger than that in wire 2, so the point where the net magnetic field is zero is three times farther from wire 1 than from wire 2. This point is 30 cm to the right of wire 1 and 10 cm to the left of wire 2.

19-8 Magnetic Field from Loops and Coils

The Magnetic Field from a Current Loop

Let's take a straight current-carrying wire and bend it into a complete circle. As shown in Figure 19.27, the field lines pass through the loop in one direction and wrap around outside the loop so the lines are continuous. The field is strongest near the wire. For a loop of radius R and current I, the magnetic field in the exact center of the loop has a magnitude of

$$B = \frac{\mu_0 I}{2R}.$$ (Equation 19.11: **The magnetic field at the center of a current loop**)

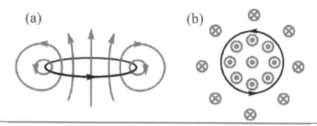

The direction of the loop's magnetic field can be found by the same right-hand rule we used for the long straight wire. Point the thumb of your right hand in the direction of the current flow along a particular segment of the loop. When you curl your fingers, they curl the way the magnetic field lines curl near that segment. The roles of the fingers and thumb can be reversed: if you curl the fingers on your right hand in the way the current goes around the loop, your thumb, when you stick it out, shows the way the field line points inside the loop.

Figure 19.27: (a) A side view of the magnetic field from a current loop. (b) An overhead view of the same loop, showing the field in the plane of the loop.

The magnetic field from a current loop is similar to from a thin disk magnet that you might find on your fridge (as long as the north and south poles of the disk magnet are on opposite faces of the disk, which is generally the case). This similarity between the fields is no coincidence. The disk magnet is made from **ferromagnetic** material – ferromagnetic means having magnetic properties similar to that of iron. In any material, each electron in an atom has an associated angular momentum. In many materials these angular momenta either cancel out or are randomly aligned, giving rise to little or no magnetic field. In ferromagnetic materials, however, the angular momenta of neighboring atoms line up, producing a substantial magnetic field.

A model of a ferromagnetic material is shown in Figure 19.28. Each atom acts like a tiny current loop, with the loops carrying currents that circulate in the same direction. In the inner part of the magnet, nearby currents point in opposite directions and cancel one another out. Around the edge of the magnet, however, there is no cancellation, and the net effect of the currents at the outside is like that of a single current that goes all the way around the outer edge of the disk.

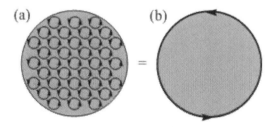

Figure 19.28: A model of a ferromagnetic material. Each atom acts like a tiny current loop, with neighboring current loops aligned with one another. Except around the outer edge of the magnet, nearby currents are directed in opposite directions, and cancel one another. The net effect is that the disk magnet produces a magnetic field similar to that from a current loop of the same radius as the disk.

The Magnetic Field from a Current-Carrying Coil

A cylindrical current-carrying coil, or **solenoid**, is like a stack of current loops. In the ideal case, the solenoid is infinitely long, producing a uniform magnetic field inside the solenoid and negligible field outside the solenoid. The solenoid is the magnetic equivalent of the parallel-plate capacitor – when both devices extend to infinity, the field produced (magnetic field for the solenoid, electric field for the capacitor) is uniform. Just as the electric field from the capacitor depends on the capacitor geometry, the magnetic field produced by the solenoid depends on the solenoid geometry. For a solenoid of length L, with a total of N turns (or $n = N/L$ turns per unit length), and carrying a current I, the magnitude of the magnetic field inside the solenoid is:

$$B = \frac{\mu_0 NI}{L} = \mu_0 nI ,$$ (Equation 19.12: **The magnetic field for an ideal solenoid**)

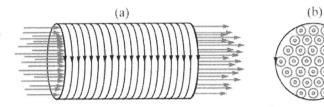

As shown in Figure 19.29, the magnetic field in an ideal solenoid is parallel to the axis of the solenoid. If you curl the fingers of your right hand in the direction of the current, your thumb points in the direction of the field inside the solenoid.

Equation 19.12 applies to an ideal solenoid, of infinite length. Figure 19.30 shows that for a real solenoid, of finite length, the magnetic field is strongest in the center, and reduces in magnitude toward the ends as field lines leak out the sides of the solenoid. The field from a real solenoid has the same form as the field from a typical bar magnet. Like a disk magnet, the bar magnet can be modeled as a number of tiny current loops, all aligned, associated with the angular momentum of electrons in atoms. The net effect of these current loops is a current that circles the outside of the bar magnet, like the current in a solenoid.

Figure 19.29: The magnetic field from an ideal (infinitely long) solenoid. A three-dimensional view is shown in (a), while (b) shows the field from the perspective of someone looking along the axis of the solenoid from the right.

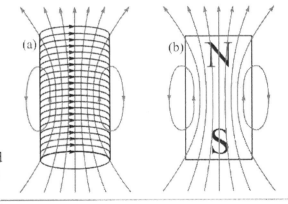

The pictures in Figure 19.30 could be labeled (a) an electromagnet, and (b) a permanent magnet. In a permanent magnet, the magnetic field is always on. An electromagnet can be turned on or off as a current is turned on or off. An electromagnet made by connecting a coil of wire to a battery generally has a weak magnetic field. If a ferromagnetic core, like an iron nail, is placed in

Figure 19.30: The magnetic field from a solenoid of finite length (a), has the same form as the field from a bar magnet (b).

the coil, however, the magnetic field can be increased by a factor of several hundred. The core should consist of "soft" ferromagnetic material, as opposed to the "hard" material that permanent magnets are made from. In hard ferromagnetic materials, neighboring atoms remain aligned when an external magnetic field is removed. In soft materials, the alignment mostly disappears when the external field is removed, so the magnetic field turns off when the current turns off.

Related End-of-Chapter Exercises: 28, 29, 31, 32.

Essential Question 19.8: Starting at a point on the axis of a solenoid, which has 800 turns per meter, an electron is given an initial velocity of 500 m/s in a direction perpendicular to the axis. The electron takes 75 ns to go through a complete circle. What is the current in the solenoid?

Answer to Essential Question 19.8: This situation involves much of what we learned in this chapter. First, what is the connection between all this information? We can connect everything via the magnetic field. With the information about the electron, we can use equation 19.4 to solve for the magnetic field. Equation 19.4 gives the relation between the period of the circular orbit as: $T = 2\pi m/(qB)$. Solving for the magnetic field gives:

$$B = \frac{2\pi m}{qT} = \frac{2\pi\ (9.11\times10^{-31}\ \text{kg})}{(1.60\times10^{-19}\ \text{C})\,(75\times10^{-9}\ \text{s})} = 4.77\times10^{-4}\ \text{T}\ .$$

We can then use equation 19.12, $B = \mu_0 nI$, to solve for the current in the solenoid. This gives: $I = B/(\mu_0 n) = (4.77\times10^{-4}\ \text{T})/(4\pi\times10^{-7}\ \text{T m/A}\times800\ \text{m}^{-1})/ = 0.47\ \text{A}$.

Chapter Summary

Essential Idea: Magnetism and Magnetic Fields.
Magnetism and magnetic fields have many applications, ranging from the Earth's magnetic field, which we use for navigation and which protects us from harmful radiation, to the powerful magnetic field in an MRI machine, which helps doctors diagnose medical problems, to the basic magnets that hold pictures on a refrigerator.

Magnetic Field
- Magnetic fields are associated with two magnetic poles, north and south. Magnetic fields are produced by moving charges.
- Like magnetic poles repel, while unlike poles attract.
- The magnetic field points in the direction of the force experienced by a north pole.
- North and south poles always come in pairs.
- Magnetic field lines are continuous loops. Outside a magnet, the magnetic field is directed from the north pole to the south pole. Inside a magnet, the magnetic field runs from south to north.

The Magnetic Force on a Charged Particle Moving in a Magnetic Field
A charged particle moving through a magnetic field experiences a magnetic force that is perpendicular to both the magnetic field and the velocity of the particle:

$F_M = qvB\ \sin\theta$,   (Eq. 19.2: **The magnitude of the magnetic force on a charge q**)

where θ is the angle between the velocity, $\vec{v}$, and the magnetic field, $\vec{B}$. The direction of the force is perpendicular to the plane defined by $\vec{v}$ and $\vec{B}$, and given by the right-hand rule. The steps in the right-hand rule are:

- Point the fingers on your right hand in the direction of the charge's velocity.
- While keeping your fingers aligned with the velocity, rotate your hand so that, when you curl your fingers, you can curl them into the direction of the magnetic field.
- Hold out your thumb so it is perpendicular to your fingers. Your thumb points in the direction of the force experienced by a positively charged particle.
- If the particle has a negative charge, your right hand lies to you. Just reverse the direction of the force. The magnetic force on a negatively charged particle is opposite in direction to that of a positively charged particle if the particles are traveling in the same direction.

The path followed by a moving charge in a magnetic field

A charged particle moving parallel to a magnetic field experiences no force, and moves in a straight line. A charged particle moving perpendicular to a magnetic field experiences a force that makes it travel in a circle, with a radius and period of:

$$r = \frac{mv}{qB},$$
(Equation 19.3: **Radius of the path, when $\vec{v}$ and $\vec{B}$ are perpendicular**)

$$T = \frac{2\pi m}{qB}.$$
(Equation 19.4: **The period of the circular orbit**)

If the velocity of the particle is neither parallel to, not perpendicular to, the magnetic field, the particle follows a spiral path, spiraling around the magnetic field.

Forces and torques on current-carrying wires and loops

The magnitude of the magnetic force exerted on a wire of length L, carrying a current I, by a magnetic field of magnitude B is:

$$F_M - ILB\sin\theta ,$$
(Equation 19.7: **The magnetic force on a current-carrying wire**)

where θ is the angle between the current direction and the magnetic field.

The direction of the force is given by a right-hand rule that follows the rule for charges. Point the fingers on your right hand in the direction of the current. Align your hand so that when you curl your fingers, they point in the direction of the magnetic field. Stick out your thumb, and it points in the direction of the magnetic force.

A loop of current in a uniform magnetic field experiences no net magnetic force. The magnitude of the magnetic torque on a coil consisting of N loops carrying a current I is:

$$\tau = NIAB\sin\theta ,$$
(Equation 19.8: **Torque on a current-carrying coil in a magnetic field**)

where θ is the angle between the area vector $\vec{A}$ of the loop, which is perpendicular to the plane of the loop, and the magnetic field $\vec{B}$.

Currents produce magnetic fields

Even in permanent magnets, the magnetic field comes from currents, being associated with the angular momentum of electrons in atoms, and the alignment of these angular momenta.

The magnetic field a distance r from a long straight wire carrying a current I is:

$$B = \frac{\mu_0 I}{2\pi r}.$$
(Eq. 19.9: **The magnetic field from a long-straight wire**)

The **permeability of free space** has a value of $\mu_0 = 4\pi \times 10^{-7} \text{ T m / A}$.

$$B = \frac{\mu_0 I}{2R}.$$
(Eq. 19.11: **The magnetic field at the center of a current loop of radius R**)

$$B = \frac{\mu_0 NI}{L} = \mu_0 nI ,$$
(Equation 19.12: **The magnetic field for an ideal solenoid**)

where N is the number of turns, L is the length of the solenoid, and $n = N/L$ is the number of turns per unit length.

End-of-Chapter Exercises

Exercises 1 – 12 are primarily conceptual questions, designed to see whether you understand the main concepts of the chapter.

1. A charged particle is moving with a constant velocity directed up the page when it enters a region of uniform magnetic field. (a) If the particle is deflected into a circular path, but it remains in the plane of the page, what does this tells us about the magnetic field? (b) If the particle has a positive charge and it is deflected to the left when it enters the field, what does this tell us about the magnetic field?

2. What is the direction of the magnetic force experienced by the charged particle in each situation shown in Figure 19.31?

3. A negatively charged particle with an initial velocity directed east experiences a magnetic force directed north. (a) If the initial velocity is perpendicular to a uniform magnetic field, in which direction is the field? (b) If the initial velocity and the magnetic field are not necessarily perpendicular, what can you say about the direction of the field?

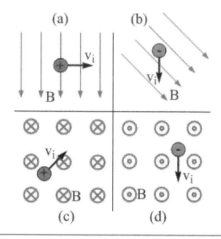

Figure 19.31: Four situations involving a charged particle in a uniform magnetic field, for Exercises 2 and 6.

4. As shown in Figure 19.32, a positively charged particle has an initial velocity directed straight up the page. The initial velocity is perpendicular to a uniform field in the region. (a) If the into-the-page symbol shows the direction of the magnetic force on the particle, in which direction is the magnetic field? (b) If, instead, the into-the-page symbol shows the direction of the magnetic field, in which direction is the force on the particle?

Figure 19.32: A positively charged particle has an initial velocity directed up. Either the magnetic field or the magnetic force is directed into the page. For Exercise 4.

5. Two vectors, which are clearly not perpendicular to one another, are shown in Figure 19.33. (a) Could these two vectors represent the velocity of a negatively charged particle (1) and the direction of a uniform magnetic field (2)? If not, explain why not. If so, in which direction is the magnetic force on the particle? (b) Could the two vectors represent the velocity of a positively charged particle (1), and the direction of the magnetic force acting on the particle (2)? If not, explain why not. If so, in which direction is the magnetic field that exerts the force on the particle?

Figure 19.33: Two vectors, for Exercise 5.

6. Return to the situations shown in Figure 19.31. If the particles have charges of the same magnitude, are in uniform fields of the same magnitude, and have the same initial speed, rank the four situations based on the magnitude of the force experienced by the particle.

7. Figure 19.34 shows the paths followed by three charged particles through a region of uniform magnetic field that is directed perpendicular to the page. If particle 1 has a positive charge, what is the sign of the charge of (a) particle 2? (b) particle 3? (c) In which direction is the magnetic field?

8. Return to the situation described in the previous exercise and shown in Figure 19.34. If the particles have the same mass, and their charges have the same magnitude, rank the particles based on (a) their speed, and (b) the magnitude of the magnetic force they each experience. (c) Which particle spends the most time in this square region? Explain.

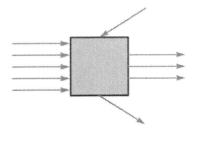

Figure 19.34: The paths followed by three charged particles through a region of uniform magnetic field directed perpendicular to the page, for Exercises 7 and 8.

9. The complete picture of field lines in a particular region is blocked by the square screen in Figure 19.35. (a) Could these field lines be electric field lines? If not, explain. If so, what is behind the screen? (b) Could these field lines be magnetic field lines? If not, explain. If so, what is behind the screen?

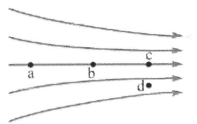

Figure 19.35: Our view of the field lines in a particular region is blocked by a square screen, for Exercise 9.

10. Consider the picture of magnetic field lines in a particular region, shown in Figure 19.36. (a) Points a, b, and c are located on the same field line. Rank these points based on the magnetic field at these points. (b) At which point, a or d, is the magnetic field larger in magnitude? Explain.

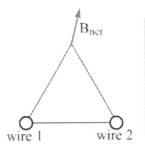

Figure 19.36: The magnetic field in a particular region, for Exercise 10.

11. Two long straight parallel wires are located so that one wire passes through each of the lower corners of an equilateral triangle, as shown in Figure 19.37. The wires carry currents that are directed perpendicular to the page. The net magnetic field at the top vertex of the triangle is also shown in the figure. What is the direction of the current in (a) wire 1? (b) wire 2? (c) Which wire carries more current?

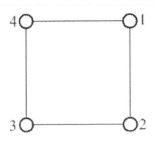

Figure 19.37: Two long straight current-carrying wires pass through the lower corners of an equilateral triangle. For Exercise 11.

12. Four long straight parallel wires pass through the corners of a square, as shown in Figure 19.38. The currents in each wire have the same magnitude, but their directions (either into or out of the page) are unknown. How many possible combinations of current directions are there if the net magnetic field due to these wires at the center of the square is (a) zero? (b) directed to the right? (c) directed toward the top left corner of the square. Draw the various configurations in each case.

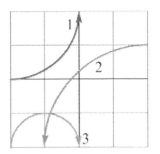

Figure 19.38: Long straight parallel current-carrying wires pass through the corners of a square, for Exercise 12.

Exercises 13 – 16 are designed to give you practice applying the equation for the magnetic force experienced by a charged particle in a magnetic field, $F_M = qvB \sin\theta$.

13. The particle in Figure 19.39 has a charge of 5.0 μC, a speed of 1.4×10^3 m/s, and it is in a uniform magnetic field, directed into the page, of 2.5×10^{-2} T. As the figure shows, the initial velocity of the particle is directed at 30° below the positive x-direction, assuming the x-direction is toward the right. What is the magnitude and direction of the magnetic force acting on the particle?

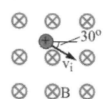

Figure 19.39: A particle with a positive charge has a velocity directed at 30° below the positive x-direction, in a magnetic field directed into the page. For Exercise 13.

14. A charged particle is in a region in which the only force acting on it comes from a uniform magnetic field. When the particle's initial velocity is directed in the positive x-direction (to the right), the particle experiences no magnetic force. When the particle's initial velocity is directed in the positive y-direction (up the page), the particle experiences a magnetic force in the positive z-direction (out of the page). (a) If the particle has a negative charge, in which direction is the magnetic field? In which direction is the magnetic force on the particle if the particle's initial velocity is directed in the (b) +z-direction? (c) –z-direction? (d) –x-direction? (e) –y-direction?

15. In a particular region, there is a uniform magnetic field with a magnitude of $B = 2.0$ T. You take a particle with a charge of +5.0 μC and give it an initial speed of 4.0×10^5 m/s in this field. (a) Under these conditions, what is the magnitude of the maximum magnetic force experienced by the particle? (b) What is the magnitude of the minimum magnetic force experienced by the particle? (c) If the particle experiences a force with a magnitude that is 25% of the magnitude of the maximum force, what is the angle between the particle's velocity and the magnetic field?

16. A particle with a charge of –5.0 μC travels in a circular path of radius 30 m at a speed of 15 m/s. The particle's path is circular because it is traveling in a region of uniform magnetic field. If the particle's mass is 25 μg, what is the magnitude of the magnetic field acting on the particle?

Exercises 17 – 19 involve charged particles moving in circular paths in uniform magnetic fields.

17. Figure 19.40 shows the paths followed by three charged particles through a region of uniform magnetic field that is directed perpendicular to the page. If the particles have the same speed, and their charges have the same magnitude, rank the particles based on (a) their mass, and (b) the magnitude of the magnetic force they each experience.

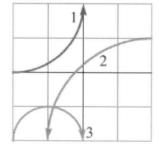

Figure 19.40: The paths followed by three charged particles through a region of uniform magnetic field directed perpendicular to the page, for Exercise 17.

18. Figure 19.41 shows the paths followed by three charged particles through a region of uniform magnetic field that is directed out of the page. The particles have the same magnitude charge. Complete table 19.3, filling in the six pieces of missing data.

Particle	Sign of Charge	Mass	Speed	Path taken
1		2m	4v	
2		4m		
3		m	2v	L

Table 19.3: The charge, mass, and speed for three particles passing through a magnetic field.

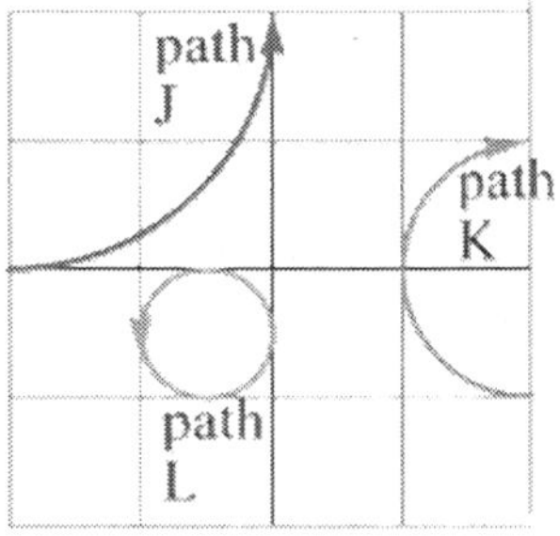

Figure 19.41: The paths of the three particles.

19. A particle with a charge of –8.0 μC and a mass of 9.0×10^{-9} kg is given an initial velocity of 5.0×10^4 m/s in the positive x-direction. The particle is in a uniform magnetic field, with $B = 2.0$ T, that is directed in the negative z-direction, and the particle starts from the origin. Assume the particle is affected only by the magnetic field. (a) How long after leaving the origin does the particle first return to the origin? (b) What is the maximum distance the particle gets from the origin? (c) Where is the particle when it achieves its maximum distance from the origin?

Exercises 20 – 24 deal with long straight wires in situations that are analogous to situations involving masses, from Chapter 8, and charged particles, from Chapter 16.

20. Figure 19.42 shows three long straight parallel wires that are carrying currents perpendicular to the page. Wire 1 experiences a force per unit length of 4 N/m to the left because of the combined effects of wires 2 and 3. Wire 2 experiences a force per unit length of 12 N/m to the right because of the combined effects of wires 1 and 3. Is there enough information given to find the force per unit length experienced by wire 3 because of the combined effects of wires 1 and 2? If so, find it. If not, explain why not.

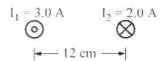

Figure 19.42: Three long straight wires carry currents perpendicular to the page. The net forces per unit length on two of the wires are shown. For Exercise 20.

21. Figure 19.43 shows two long straight parallel wires that are separated by 12 cm. Wire 1, on the left, has a current of 3.0 A directed out of the page. Wire 2, on the right, has a current of 2.0 A directed into the page. (a) What is the magnitude and direction of the net magnetic field from these two wires at the point midway between them? (b) Are there any points a finite distance from the wires at which the net magnetic field from the wires is zero? If not, explain why not. If so, state the location of all such points.

$I_1 = 3.0$ A $I_2 = 2.0$ A

|—— 12 cm ——|

Figure 19.43: Two long straight parallel wires carry currents perpendicular to the page. For Exercise 21.

22. Three long straight parallel wires pass through three corners of a square measuring 20 cm on each side, as shown in Figure 19.44. Find the magnitude and direction of the net magnetic field at (a) the center of the square. (b) the top left corner of the square.

$I_1 = 1.0$ A

23. Three long straight parallel wires pass through three corners of a square measuring 20 cm on each side, as shown in Figure 19.44. Find the magnitude and direction of the magnetic force per unit length experienced by wire 2, in the bottom right corner, due to the other two wires.

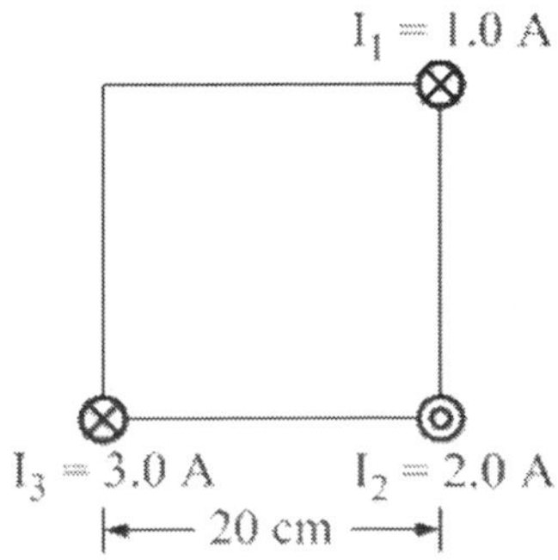

$I_3 = 3.0$ A $I_2 = 2.0$ A

|—— 20 cm ——|

Figure 19.44: Three long straight current-carrying wires pass through three corners of a square. For Exercises 22 and 23.

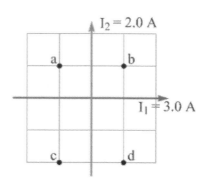

Exercises 21 – 25 involve adding magnetic fields as vectors.

24. Two long straight wires are in the plane of the page, and are perpendicular to one another. As shown in Figure 19.45, wire 2 is placed on top of wire 1, but the wires are electrically insulated from one another. Wire 1 carries a current of 3.0 A to the right, while wire 2 carries a current of 2.0 A up. Rank the four labeled points based on the magnitude of the net magnetic field due to the wires, from largest to smallest.

Figure 19.45: Two long straight perpendicular wires carry currents in the plane of the page. For Exercises 24 and 25.

25. Return to the situation described in Exercise 24, and shown in Figure 19.45. Each of the four points is 20 cm from wire 2. Points a and b are also 20 cm from wire 1, while points c and d are 40 cm from wire 1. Determine the magnitude and direction of the net magnetic field, due to the two wires, at (a) point a, (b) point b, (c) point c, and (d) point d.

26. Two long straight parallel wires pass through two corners of a right-angled triangle, as shown in Figure 19.46. The currents in the wires are directed in opposite directions. Wire 1 produces a magnetic field of 4.0×10^{-5} T at the unoccupied corner of the triangle, which is 20 cm from wire 1. If the current in wire 2 has the same magnitude as that in wire 1, determine the magnitude and direction of the net magnetic field from the two wires at the unoccupied corner of the triangle.

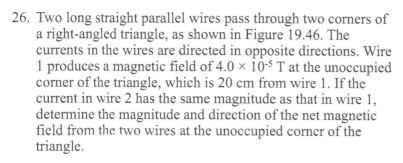

27. Return to the situation described in Exercise 26, and shown in Figure 19.46. Now the magnitude of the current in wire 2 is adjusted so that the net magnetic field at the unoccupied corner of the square, due to the wires, is directed straight up. Find the ratio of the magnitude of the current in wire 2 to that in wire 1.

Figure 19.46: Two long straight parallel wires, carrying currents in opposite directions, pass through two corners of a right-angled triangle, for Exercises 26 and 27.

Exercises 25 – 30 deal with current-carrying loops and solenoids.

28. Figure 19.47 shows three concentric current-carrying loops. The inner loop, with a radius of 20 cm, carries a current of 2.0 A counterclockwise. The middle loop, with a radius of 30 cm, carries a current of 3.0 A clockwise. The outer loop, with a radius of 40 cm, carries a current of 4.0 A counterclockwise. Determine the magnitude and direction of the net magnetic field at point c, the center of each loop, due to the loops.

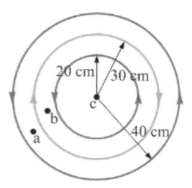

29. Return to the situation described in Exercise 28, but now Figure 19.47 represents a cross-sectional slice through a set of three long current-carrying solenoids. (a) Assuming the solenoids are ideal and have the same number of turns per unit length, and using the currents specified in the previous exercise, rank points a, b, and c based on the magnitude of the net magnetic field at their location. (b) Determine the magnitude and direction of the net magnetic field at point b, if each solenoid has 800 turns/m.

Figure 19.47: Three concentric loops, for Exercise 28, or a cross-section through three concentric solenoids, for Exercise 29.

30. As shown in Figure 19.48, a long straight wire, carrying a current *I* up the page, lies in the plane of a loop. The long straight wire is 40 cm from the center of the loop, which has a radius of 20 cm. Initially, the magnetic field at the center of the loop is due only to the current in the long straight wire. When a current in the loop is turned on, however, the net field at the center of the loop has a magnitude two times larger than the field at that point from the long straight wire. What is the magnitude (in terms of *I*) and direction of the current in the loop? Find all possible solutions.

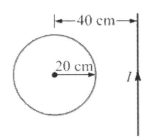

Figure 19.48: A long straight current-carrying wire passes 40 cm from the center of a loop that has a radius of 20 cm, for Exercise 30.

31. A particular ideal solenoid has 1200 turns/m, a radius of 10 cm, and a current of 4.0 A. An electron is fired from the axis of the solenoid in a direction perpendicular to the axis. What is the maximum speed the electron can have if it is not to run into the side of the solenoid?

32. A very strong neodymium-iron-boron (NdFeB) magnet with a magnetic field of around 0.1 T can be purchased for a few dollars. (a) Compare this magnetic field to the field you can get by making your own electromagnet, assuming that you wind some wire at 800 turns/m around an aluminum nail, and the current in the coil is 1.0 A. (b) If you wind the wire around an iron or steel nail instead, you can amplify the magnetic field by a factor of several hundred. How does this field compare to the field from the NdFeB magnet? (c) Name a key advantage that your electromagnet offers over the NdFeB magnet.

General problems and conceptual questions

33. Consider the photograph of two magnets, shown on the opening page of this chapter. (a) Are the magnets attracting one another or repelling? Explain. (b) Can you say which end of the magnet on the right is a north pole and which end is a south pole? Explain.

34. The SI unit of the tesla is named after a very interesting person. Do some research on this individual, and write two or three paragraphs describing his/her contributions to science.

35. We know quite a lot about the historical behavior of the Earth's magnetic field, through a field of study called **paleomagnetism**. Investigate the methods used by researchers in this field, and write 2-3 paragraphs about these methods.

36. What is the predominant direction of the Earth's magnetic field if you are standing on the surface of the Earth (a) at a point on the Earth's equator? (b) in northern Canada, directly above one of the Earth's magnetic poles?

37. The northern lights (*aurora borealis*) and southern lights (*aurora australis*) are colorful displays of light associated with fast-moving charged particles entering the Earth's atmosphere. (a) Recalling that moving charges tend to spiral around magnetic field lines, explain why these light shows are generally confined to Earth's polar regions. (b) In the event that a positively charged particle enters Earth's atmosphere above the equator, with a velocity directed straight down toward the ground, in which direction is the particle initially deflected by the Earth's magnetic field?

38. A charged particle is passing through a particular region of space at constant velocity. One possible explanation for the fact that the particle is undeflected while it is in this region is that there is neither an electric field nor a magnetic field present. Let's consider other possibilities. Explain your answers to the following. (a) Could there be a uniform electric field in the region, but no magnetic field? (b) Could there be a uniform magnetic field in the region, but no electric field? (c) Could there be both a uniform electric field and a uniform magnetic field present?

39. An electron is in a uniform magnetic field. Assume that the electron interacts only with the magnetic field, and that all other interactions, including gravity, can be neglected. 2.0 s after being released, what is the electron's speed if it is (a) released from rest? (b) released with a velocity of 800 m/s in a direction parallel to the magnetic field? (c) released with a velocity of 800 m/s in a direction perpendicular to the magnetic field.

40. Figure 19.49 shows, to scale, the initial velocities of four identical charged particles in a uniform magnetic field. Rank the particles based on the magnitude of the magnetic force they experience from the field if the field is directed (a) up the page (b) to the right (c) out of the page.

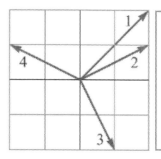

Figure 19.49: The initial velocities of four identical charged particles in a uniform magnetic field, for Exercise 40.

41. When a particle with a charge of +3.0 μC and a mass of 5.0×10^{-8} kg passes through the origin, its velocity components are $\vec{v}_x = +3.0 \times 10^4$ m/s in the x-direction and $\vec{v}_y = +4.0 \times 10^4$ m/s in the y-direction. The particle is traveling through a region of uniform magnetic field which is directed in the negative x-direction, with a magnitude of 2.0×10^{-2} T. The particle follows a spiral path in the field. (a) What is the radius of the spiral path, and in which direction does the axis of the spiral point? (b) How long does it take the particle to make one complete loop on the spiral? (c) How far from the origin is the particle when it has made one complete loop on the spiral?

42. Return to the situation described in Exercise 42, but now we will add a uniform electric field of 1000 N/C in the +x-direction, parallel to the magnetic field. (a) With the addition of the electric field, which of the answers to the previous exercise would change and which would stay the same? Explain. (b) Re-calculate all the answers to the previous exercises that are changed by the addition of the electric field.

43. Four particles pass through a square region of uniform magnetic field, as shown in Figure 19.50. The magnetic field inside the square region is perpendicular to the page, and the field outside the region is zero. The paths followed by particles 1 and 2 are shown; while for particles 3 and 4 the direction of their velocities and the points at which they enter the region are shown. The paths of all particles lie in the plane of the paper. Particle 1 has a mass m, a speed v, and a positive charge $+q$. Assume that the only thing acting on the particles as they move through the magnetic field is the field. (a) In which direction is the uniform magnetic field in the square region? (b) As particle 1 moves through the field, does its kinetic energy increase, decrease, or stay the same? Explain. (c) What is the sign of particle 2's charge? (d) If particle 2's mass is m and its charge has a magnitude of q, what is its speed? (e) Particle 3 has a mass $2m$, a speed $2v$, and a charge $-2q$. Re-draw Figure 19.50, and sketch on this diagram, as precisely as you can, the path followed by particle 3 through the region of magnetic field. (f) Particle 4 has a mass $2m$, a speed $2v$, and no charge. Sketch its path through the region of magnetic field. (g) Which particle feels the largest magnitude force as it passes through the field?

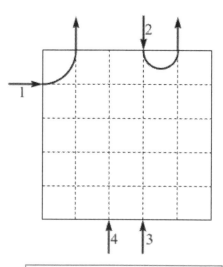

Figure 19.50: Four particles pass through a square region of uniform magnetic field directed perpendicular to the page. Only the paths followed by particles 1 and 2 are shown. For Exercise 43.

44. Two particles are sent into a square region in which there is a uniform magnetic field directed into the page, as shown in Figure 19.51. There is no magnetic field outside of the region. The velocity of the particles is perpendicular to the magnetic field at all times. Particle 1 is sent through the field twice, entering at the same point both times. The first time it has an initial speed v and the second time it has an initial speed of $2v$. (a) Which path, P or Q, corresponds to when the initial speed of particle 1 is $2v$? (b) For which path is the magnitude of the force experienced by particle 1 larger? (c) Is the charge on particle 1 positive or negative? (d) Particle 2 has the same mass and the same magnitude charge as particle 1, but the sign of its charge is opposite to that of particle 1. Particle 2 has an initial speed of v when it enters the field. Re-draw the diagram and sketch the path followed by particle 2.

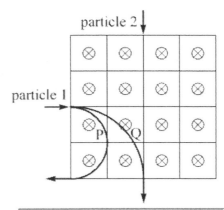

Figure 19.51: Two particles pass through a square region of uniform magnetic field directed into the page. Particle 1 is sent through the field twice, following path P on one occasion and path Q on the other. For Exercise 44.

45. The velocity of the electrons in an electron beam is 1.0×10^5 m/s directed right. The electrons pass through a velocity selector without being deflected in any way, as shown in Figure 19.52. The velocity selector consists of a set of parallel plates with a uniform electric field with $\vec{E} = 2000\, V/m$ directed down, and a

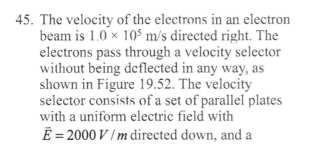
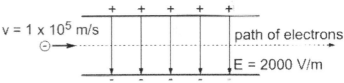

uniform magnetic field directed perpendicular to the velocity of the electrons. (a) In which direction is the magnetic field? (b) What is the magnitude of the magnetic field inside the velocity selector? (c) What will happen to electrons traveling faster than those that are undeflected?

Figure 19.52: Electrons with just the right speed pass undeflected through a velocity selector, for Exercise 45.

46. A charged particle enters a velocity selector with a speed of 2.0×10^4 m/s. A graph of the net force acting on the particle when it enters the velocity selector, as a function of the magnetic field in the velocity selector, is shown in Figure 19.53. A positive net force means that the particle is deflected up, while a negative net force means that it is deflected down. The geometry of the velocity selector is similar to that shown in Figure 19.52, with the initial velocity of the particle directed parallel to the plates of the velocity selector, and perpendicular to both the electric field and the magnetic field, which are perpendicular to one another. Using the graph, determine (a) the magnitude of the electric field in the velocity selector, and (b) the magnitude of the charge on the particle.

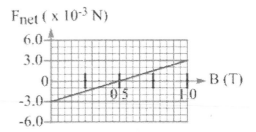

Figure 19.53: A graph of the net force acting on a charged particle when it enters a velocity selector, as a function of the magnetic field in the velocity selector, for Exercise 46.

47. You have a sample of chlorine, and you are trying to determine what fraction of your sample consists of chlorine-35 atoms (with an atomic mass of about 35 atomic mass units) and what fraction is made up of chlorine-37 atoms (with an atomic mass of about 37 atomic mass units). To do this, you use a mass spectrometer, as described in section 19–4, ionizing the atoms so that each atom is singly-ionized, with a charge of $+e$. (a) In

the velocity selector, you use a uniform electric field of 2000 N/C, and a uniform magnetic field of 5.0×10^{-2} T. These two fields are mutually perpendicular, and are each perpendicular to the velocity of the chlorine ions when they enter the velocity selector. What is the speed of chlorine ions that pass undeflected through the velocity selector? (b) The undeflected ions continue to the uniform magnetic field in the mass separator. If you want the ions of different mass to be separated by 1.0 mm after going through a half-circle in the mass separator, to what value should you set the magnitude of the magnetic field in the mass separator?

48. A graph of the magnetic field from a long straight wire, as a function of the inverse of the distance from the wire, is shown in Figure 19.54. What is the magnitude of the current in the wire?

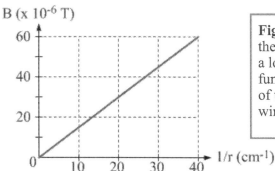

Figure 19.54: A graph of the magnetic field from a long straight wire, as a function of the inverse of the distance from the wire, for Exercise 48.

49. Three long straight parallel wires are placed in a line, with 40 cm between neighboring wires, as shown in Figure 19.53. Wires 1 and 2 carry currents directed into the page, while wire 3 carries a current directed out of the page. The currents all have the same magnitude. (a) Rank the three wires based on the magnitude of the net force per unit length experienced by each wire because of the other two. (b) If the current in each wire is 4.0 A, confirm your ranking from part (a) by determining the magnitude and direction of the force experienced by each wire.

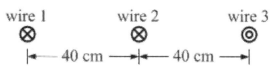

Figure 19.55: Three long straight parallel current-carrying wires, for Exercises 49 and 50.

50. Three long straight parallel wires are placed in a line, with 40 cm between neighboring wires, as shown in Figure 19.55. Wires 1 and 2 carry currents directed into the page, while wire 3 carries a current directed out of the page. Wire 1 experiences no net force due to the other two wires. (a) How does the magnitude of the current in wire 3 compare to that in wire 2? (b) How does the net force per unit length experienced by wire 3 compare to that of wire 2?

51. Two points, a and b, are 5.0 m apart. There is a uniform magnetic field of 2.0 T present that is directed parallel to the line connecting a and b. How much force does the field exert on a wire carrying 3.0 A of current from a to b, if the wire is (a) 5 m long? (b) 10 m long, and follows a circuitous path from a to b?

52. Repeat Exercise 51, but now the magnetic field is directed perpendicular to the line connecting a and b.

53. A rectangular loop with a current of 2.0 A directed clockwise measures 30 cm long by 10 cm wide, as shown in Figure 19.56. The loop is placed 10 cm from a long straight wire that carries a current of 3.0 A to the right. The goal of this exercise is to determine the magnitude and direction of the net magnetic force exerted on the loop by the long straight wire. (a) In section 19-5, we discussed the fact that a current-carrying loop in a uniform magnetic field experiences no net force. Why does the loop in this situation experience a non-zero net force? (b) Draw a diagram showing the force experienced by

each side of the loop. (c) Explain why the forces on the left and right sides of the loop are difficult to calculate, but why we do not need to calculate them to find the net force on the loop. (d) Find the force exerted on the upper side of the loop by the long straight wire. (e) Find the force exerted on the lower side of the loop by the long straight wire. (f) Combine your previous results to find the net force acting on the loop.

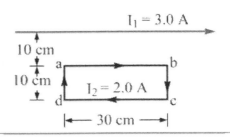

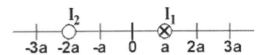

Figure 19.56: A current-carrying loop near a current-carrying long straight wire, for Exercises 53 and 54.

54. Return to the situation described in Exercise 53, and shown in Figure 19.56. Does the rectangular loop experience a net torque, about an axis that passes through the center of the loop, because of the magnetic field from the long straight wire? Explain.

55. Two very long straight wires carry currents perpendicular to the page, as shown in Figure 19.57. The x-axis is in the plane of the page. Wire 1, which carries a current I_1 into the page, passes through the x-axis at $x = +a$. Wire 2, located at $x = -2a$, carries an unknown current. The net field at the origin $(x = 0)$, due to the current-carrying wires, has a magnitude of $B = 2\mu_0 I_1 /(2\pi a)$. What is the magnitude and direction of the current in wire 2? Find all possible solutions.

Figure 19.57: Two long straight parallel wires pass through $x = +a$. and $x = -2a$, respectively. For Exercise 55.

56. Two long straight wires, that are parallel to the y-axis, pass through the x-axis. One wire carries a current of $2I$ in the +y-direction, and passes through the x-axis at the point $x = -d$. The second wire carries an unknown current, and passes through an unknown location on the positive x-axis. The net magnetic field at all points on the y-axis is zero due to the wires, and the force per unit length experienced by each wire has a magnitude of $3\mu_0 I^2 /(2\pi d)$. Find the current in the second wire, and determine at which point it passes through the x-axis.

57. Three equally spaced long straight wires are arranged in a line, as shown in Figure 19.58. The currents in the wires are as follows: wire 1 carries a current I into the page; wire 2 carries a current of $2I$ into the page; wire 3 carries a current of $3I$ out of the page. (a) Rank the three wires based on the magnitude of the net force per unit length they experience, from largest to smallest. (b) In which direction is the net force experienced by wire 1 due to the other two wires?

Figure 19.58: Three long straight parallel wires carry currents perpendicular to the page, for Exercise 57.

58. As shown in Figure 19.59, a long straight wire, supported by light strings tied to the wire at regular intervals, hangs at equilibrium at 30° from the vertical in a uniform magnetic field directed down. The current is directed perpendicular to the page. The wire has a mass per unit length of 0.12 kg/m, and the magnetic field has a magnitude of 0.36 T. What is the magnitude and direction of the current in the wire?

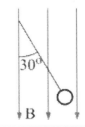

Figure 19.59: A long straight current-carrying wire, supported by strings, is in equilibrium at an angle of 30° from the vertical. For Exercise 58.

59. A rectangular wire loop measures 4.0 cm wide by 8.0 cm long. The loop carries a current of 5.0 A. The loop is in a uniform magnetic field with $B = 2.5 \times 10^{-3}$ T. What is the magnitude of the torque exerted by the field on the loop if the direction of the magnetic field is (a) parallel to the short sides of the loop? (b) parallel to the long sides of the loop? (c) perpendicular to the plane of the loop? In each case, take torques about an axis that maximizes the torque from the field.

60. Return to the situation described in Exercise 59. Rank the three cases, (a), (b), and (c), based on the magnitude of the angular acceleration experienced by the loop, assuming the loop is made from a uniform wire.

61. Return to the situation described in Exercise 59. If each centimeter of the wire loop has a mass of 5.0 grams, determine the angular acceleration of the loop in the situation when the magnetic field is parallel to the short sides of the loop.

62. An electron with a speed of 4.0×10^5 m/s travels in a circular path in the Earth's magnetic field. (a) Assuming that the magnitude of the Earth's field is 5.0×10^{-5} T, what is the radius of the electron's path? (b) Is it reasonable to neglect the force of gravity acting on the electron in this situation? Justify your answer. (c) An electron moving in a circle acts like a current loop, producing a magnetic field of its own. How large is the magnetic field at the center of the circular path, arising from the electron's motion?

63. Comment on each statement made by two students, who are discussing a situation in which a charged object experiences uniform circular motion because it is moving in a uniform magnetic field.

Kailey: The problem says that the object travels at constant speed in a circular path – the motion is confined to a plane. That tells us that the field must be directed perpendicular to that plane, right?

Isaac: I agree with you. Then it says, how does the radius of the path followed by the first object compare to that followed by the second object. Everything about the two objects is the same except that the second one has twice the speed as the first. Well, the force is proportional to the speed, so, if you increase the force, the radius must decrease.

Kailey: I'm not sure about that. Even if the force was the same on the two particles, if you change the speed the faster one is going to travel in a larger circle. That, by itself, would suggest that the radius goes up. Which effect wins, or do they balance?

This photograph shows some large wind turbines. Each turbine transforms energy from the wind into electrical energy, generating electricity. The energy transformation process exploits a phenomenon known as electromagnetic induction, a process that lies at the heart of many electricity generating systems. We will investigate electromagnetic induction in this chapter. Photo credit: Vera Kratochvil, from publicdomainphotos.net.

Chapter 20 – Generating Electricity

CHAPTER CONTENTS

In Chapter 19, we examined how electric current can generate magnetic field. In this chapter we complete the circle, focusing on how magnetic fields can generate electric fields in metal wires and sheets, thereby producing electric currents.

The phenomenon of using magnetic fields to create electric currents has many practical applications. Foremost among these is the generation of electricity, which is crucial to our lives on a daily basis (something we often take for granted until the power goes out!). With global warming being such an important issue, the move now is to use renewable sources of energy, like the wind, to run the generators, instead of non-renewable and polluting sources like coal and oil.

20-1 Magnetic Flux

Let's begin by introducing the concept of flux. Flux means something quite different in physics than it does in everyday conversation. In physics, the flux through an area is simply a measure of the number of field lines passing through an area. In chapter 16, for instance, we could have defined an electric flux, a measure of the number of electric field lines passing through an area, in a way analogous to the following definition of magnetic flux. As we will see in section 20-2, magnetic flux turns out to play a crucial role in the generation of electricity.

> Magnetic flux is a measure of the number of magnetic field lines passing through an area. The symbol we use for flux is the Greek letter capital phi, Φ. The equation for magnetic flux is:
> $$\Phi = BA\cos\theta, \qquad \text{(Equation 20.1: Magnetic flux)}$$
> where θ is the angle between the magnetic field $\vec{B}$ and the area vector $\vec{A}$. The area vector has a magnitude equal to the area of a surface, and a direction perpendicular to the plane of the surface. The SI unit for magnetic flux is the weber (Wb). 1 Wb = 1 T m^2.

EXAMPLE 20.1 – Determining the magnetic flux

A rectangular piece of stiff paper measures 20 cm × 25 cm. You hold the piece of paper in a uniform magnetic field that has a magnitude of 4.0×10^{-3} T. For each situation below, sketch a diagram showing the magnetic field and the paper, and determine the magnitude of the magnetic flux through the paper, when the magnitude of the flux is (a) maximized, (b) minimized, and (c) halfway between its maximum and minimum value.

SOLUTION

(a) How should we hold the paper so that the largest number of field lines pass through it? As shown in Figure 20.1, we hold it so that the plane of the paper is perpendicular to the direction of the magnetic field. We can also understand this orientation by considering equation 20.1. To maximize the flux with an area vector of constant magnitude and a field of constant magnitude, we need to maximize the factor of $\cos\theta$. The factor of $\cos\theta$ reaches its maximum magnitude of 1 when θ, the angle between the area vector and the magnetic field, is either 0° or 180°. In other words, the area vector must be parallel to the magnetic field, which is the case when the plane of the paper is perpendicular to the magnetic field.

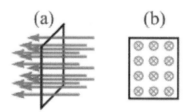

Figure 20.1: To maximize the magnetic flux through a flat area, orient the area so the plane of the area is perpendicular to the direction of the magnetic field. (a) shows a perspective view, while (b) shows the view looking along the field lines. In this case, the area vector is in the same direction as the field lines.

Because $\cos\theta$ has a magnitude of 1, the magnitude of the maximum flux equals the area multiplied by the magnetic field:
$$\Phi_{max} = AB = 0.20 \text{ m} \times 0.25 \text{ m} \times (4.0 \times 10^{-3} \text{ T}) = 2.0 \times 10^{-4} \text{ T m}^2.$$

(b) The factor of $\cos\theta$ in equation 20.1 can be zero. Thus, the minimum magnitude of the magnetic flux is zero ($\Phi_{min} = 0$). How do we hold the paper so that there is no magnetic flux? As shown in Figure 20.2, if the plane of the paper is parallel to the magnetic field, no field lines pass through the paper and the magnetic flux is zero.

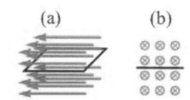

Figure 20.2: There is no flux when the plane of the area is parallel to the field. (a) shows a perspective view, while (b) shows the view looking along the field lines. In this case, the area vector is perpendicular to the field lines.

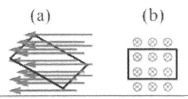

(c) Starting from the situation in Figure 20.1, tilting the loop by 60° (see Figure 20.3) gives a factor of $\cos\theta$ of ½, halfway between its maximum and minimum value. In this case the magnetic flux is $\Phi = 1.0 \times 10^{-4} \text{ T m}^2$, half its value from part (a).

EXPLORATION 20.1 – Ranking situations based on flux

The four areas in Figure 20.4 are in a magnetic field. The field has a constant magnitude, and is directed into the page in the left half of the region and out of the page in the right half. Rank the areas based on the magnitude of the net flux passing through them, from largest to smallest. Note that, when we calculate net flux, field lines passing in one direction through an area cancel an equal number of field lines passing in the opposite direction through an area.

Region 1 is tied with region 4 for the largest area, but all the field lines in region 1 pass through in the same direction, giving region 1 the largest-magnitude flux. In contrast, the net flux through region 4 is zero because the flux through the right half of region 4 cancels the flux through the right side. The field lines in regions 2 and 3 all pass through in the same direction. Region 3 has an area of 3 boxes, while region 2 has an area of πr^2, where the radius is 1 unit, so region 2 has an area of π boxes. Because π is larger than 3, the magnetic flux through region 2 is larger than that through region 3. Thus, ranking by flux magnitude gives $1 > 2 > 3 > 4$.

Figure 20.3: Tilting the loop from the orientation in Figure 20.1 reduces the flux. (a) shows a perspective view, while (b) shows the view along the field lines.

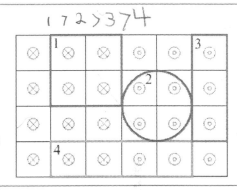

Figure 20.4: Four different regions in a magnetic field. The field has the same magnitude everywhere, but it is directed into the page in the left half of the field and out of the page in the right half.

Key idea for net flux: In calculating net flux, field lines passing one way through an area cancel an equal number of field lines passing in the opposite direction through that area.
Related End-of-Chapter Exercises: 1, 2, 4, 41.

An aside: Electric Flux and Gauss' Law

We did not mention electric flux when we talked about electric field, but we can define electric flux in an analogous way to magnetic field. Electric flux is a measure of the number of electric field lines passing through a surface. The equation for electric flux is:

$$\Phi_E = EA\cos\theta ,$$

where θ is the angle between the electric field $\vec{E}$ and the area vector $\vec{A}$.

There is a law called Gauss' Law, which says that the net electric flux passing through a closed surface is proportional to the net charge enclosed by that surface. Using the correct proportionality constant, Gauss' Law can be used to calculate electric fields in highly symmetric situations. It is interesting to note that the analogous law in a magnetic situation, Gauss' Law for magnetism, is not nearly so useful. Because magnetic field lines are always continuous loops, the net magnetic flux passing through a closed surface is always zero – if a magnetic field line emerges from a surface, it must re-enter the surface at some other location, giving a net flux for that field line of zero, to ensure that the field line is a continuous loop.

Essential Question 20.1: Return to the situation described in Exploration 20.1, and shown in Figure 20.4. If we define out of the page as the positive direction for magnetic flux, rank the four areas by their net flux, from most positive to most negative.

Answer to Essential Question 20.1: In this case, the ranking is 2 > 3 > 4 > 1. Regions 2 and 3 have positive flux, because field lines directed out of the page pass through those regions. Region 4 still has a flux of zero, while region 1 has a negative flux because of the field lines passing through the region into the page.

20-2 Faraday's Law of Induction

Table 20.1 summarizes some experiments we do with a magnet and a loop of wire. The loop has a galvanometer in it, which is a sensitive current meter. When the needle on the meter is in the center there is no current in the loop. The direction and size of the needle's deflection reflects the direction and size of the current in the loop.

Experiment	Initial and final states	Meter reading
1. The north pole of the magnet is brought closer to the loop.	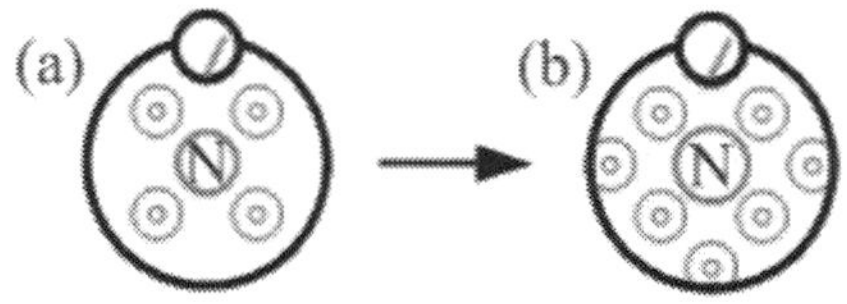 **Figure 20.5**: The loop initially has a small flux (a), which increases as the magnet comes closer (b).	*While the magnet is moving closer*, the meter needle deflects to the right.
2. The north pole is held at rest close to the loop.	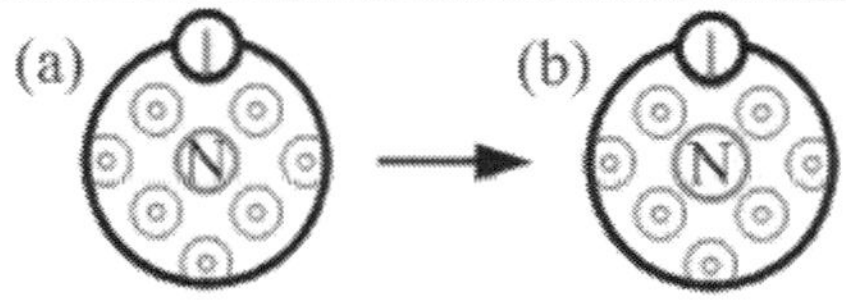 **Figure 20.6**: The magnetic flux is large, but it is also constant the entire time.	The needle does not deflect at all.
3. The north pole of the magnet is moved away from the loop.	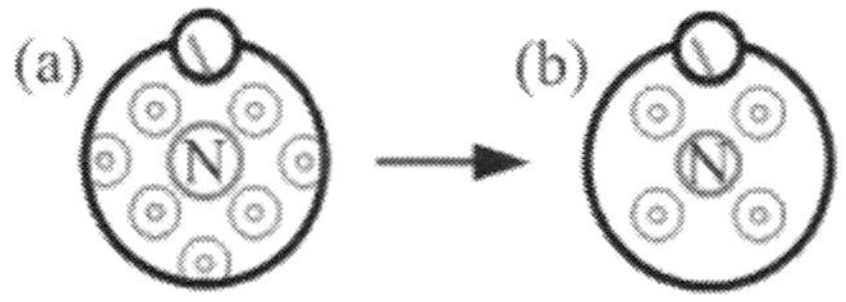 **Figure 20.7**: The loop initially has a large flux (a), which decreases as the magnet is moved away (b).	*While the magnet is moving away*, the meter needle deflects to the left.
4. The north pole of the magnet is brought closer to the loop, but at a faster rate than it was in experiment 1.	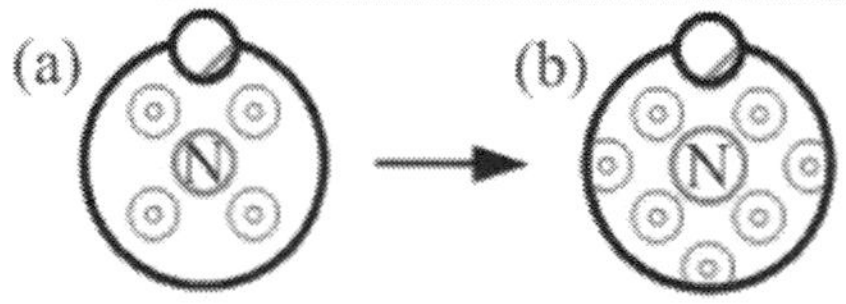 **Figure 20.8**: The same situation as Figure 20.5, but with less time between (a) and (b).	*While the magnet is moving closer*, the needle deflects farther to the right, but for less time, than in experiment 1.
5. The magnet is rotated back and forth in front of the loop.	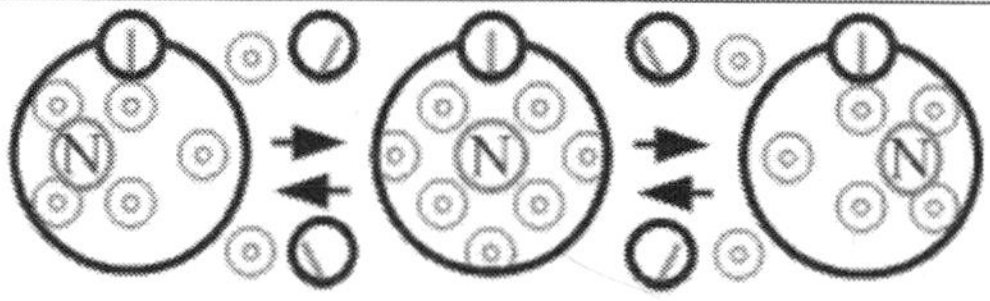 **Figure 20.9**: The flux oscillates from large to small and back again.	As the magnet oscillates, the needle oscillates back and forth at, in this case, double the frequency of the magnet.

Table 20.1: Various experiments involving a magnet and a wire loop connected to a current meter. The views in the figures are looking through the loop at the magnet.

Among the conclusions we can draw from these experiments are the following:

- A magnet interacting with a conducting loop can produce a current in the loop.
- A current arises only when the magnetic flux through the loop is changing. The current is larger when the magnetic flux changes at a faster rate.

Exposing a loop or coil to a *changing* magnetic flux gives rise to a voltage, called an **induced emf.** Following Ohm's law, the induced emf gives rise to an **induced current** in the loop or coil. The emf induced by a changing magnetic flux in each turn of a coil is equal to the time rate of change of that flux. Thus, for a coil with N turns, the net induced emf is given by:

$$\varepsilon = -N\frac{\Delta\Phi}{\Delta t} = -N\frac{\Delta(BA\cos\theta)}{\Delta t}. \qquad \text{(Eq. 20.2: Faraday's Law of Induction)}$$

The minus sign in equation 20.2 will be explained in section 20-3.

EXPLORATION 20.2 – Using graphs with Faraday's Law

A flat square conducting coil, consisting of 5 turns, measures 5.0 cm × 5.0 cm. The coil has a resistance of 3.0 Ω and, as shown in Figure 20.10, moves at a constant velocity of 10 cm/s to the right through a region of space in which a uniform magnetic field is confined to the 20 cm long region shown in the figure. The field is directed out of the page, with a magnitude of 3.0 T.

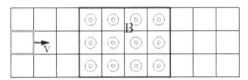

Figure 20.10: A flat conducting coil moves at constant velocity to the right through a region of space in which a uniform magnetic field is confined. The small squares on the diagram are 5.0 cm on each side.

Draw the coil's motion diagram, and a graph of the magnetic flux through the coil as a function of time. Define out of the page as the positive direction for flux. Finally, draw a graph of the emf induced in the coil as a function of time. The diagrams are in Figure 20.11. For the first 1.0 s there is no flux, because no field passes through the coil. The flux grows linearly with time during the half-second the coil moves into the field. The flux is constant, at $B \times A = 7.5 \times 10^{-3}$ T m^2, for the next 1.5 s, and then drops linearly to zero in the half-second the coil takes to leave the field.

Faraday's Law tells us that the induced emf is related to $\Delta\Phi/\Delta t$, which is the slope of the flux versus time graph. While the flux is increasing, the slope of the flux graph is constant with a value of $\Delta\Phi/\Delta t = (7.5 \times 10^{-3}$ T m$^2)/(0.50$ s$) = 1.5 \times 10^{-2}$ V.

Multiplying by the factor of $-N$ from Faraday's Law, where $N = 5$ turns, gives an induced emf of -7.5×10^{-2} V. The induced emf drops to zero while the flux is constant, and then has a constant value of $+7.5 \times 10^{-2}$ V during the half-second period while the coil is leaving the magnetic field and the flux is decreasing.

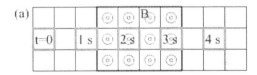

(a)

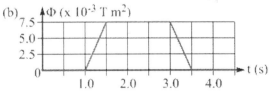

(b)

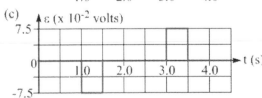

(c)

Figure 20.11: (a) A motion diagram showing the coil's position at 1-second intervals. Graphs, as a function of time, of (b) the magnetic flux through the coil and (c) the emf induced in the coil.

Key idea: The induced emf is proportional to the negative of the slope of the graph of flux as a function of time.
Related End-of-Chapter Exercises: 14 – 17, 23 – 26.

Essential Question 20.2: What is the magnitude of the maximum current induced in the coil in the system described in Exploration 20.2?

Answer to Essential Question 20.2: To find the current of the largest magnitude, we apply Ohm's Law, using the induced emf of the largest magnitude. The resistance was given as 3.0 Ω in Exploration 20.2, so we get $I_{max} = \varepsilon_{max} / R = (0.075 \text{ V})/(3.0 \text{ Ω}) = 0.025 \text{ A}$.

20-3 Lenz's Law and a Pictorial Method for Faraday's Law

Thus far we have discussed the fact that exposing a coil to a changing magnetic flux induces an emf in the coil. If there is a complete circuit, this emf gives rise to an induced current. In what direction is this induced current? The direction of the current also relates to the negative sign in Faraday's law. That negative sign is associated with a whole other law, Lenz's law.

> **Lenz's law:** The emf induced by a changing magnetic flux tends to produce an induced current. The induced current produces a magnetic flux that acts to oppose the original change in flux.

Let's go over a pictorial method for determining the direction of the induced current. As part of this method, recall from chapter 19 that, as shown in Figure 20.12, a current directed clockwise around a loop gives rise to a magnetic field directed into the page inside the loop, while a counterclockwise current produces a field directed out of the page in the loop. This can be confirmed with the right-hand rule. Curl the fingers on your right hand in the direction of the current, and your thumb points in the direction of the field inside the loop.

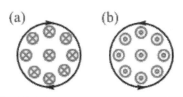

Figure 20.12: (a) A loop with a clockwise current gives rise to a magnetic field directed into the page within the loop. (b) The magnetic field within the loop is directed out of the page if the current is counterclockwise.

EXPLORATION 20.3 – A pictorial method for determining current direction
Step 1 – *The loop in Figure 20.13 is moved from the Before position to the After position, closer to a long straight wire that carries current to the left. Sketch a diagram showing the magnetic field lines, produced by the current in the straight wire, that pass through the loop when the loop is in the Before position.*

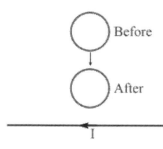

Figure 20.13: A conducting loop, above a long straight current-carrying wire, is moved from the Before position to the After position, closer to the wire.

Let's use the right-hand rule to find the direction of the field from the wire. Point the thumb on your right hand in the direction of the current. Then, curl your fingers on your right hand – they show the direction the field lines circle around the wire. This tells us that, above the wire, where the loop is, the field lines from the wire are directed into the page, as in Figure 20.14.

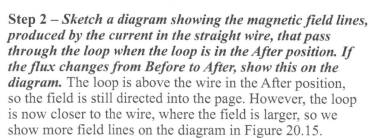

Figure 20.14: When the loop is in the Before position, the field from the long straight wire passes through the loop into the page.

Step 2 – *Sketch a diagram showing the magnetic field lines, produced by the current in the straight wire, that pass through the loop when the loop is in the After position. If the flux changes from Before to After, show this on the diagram.* The loop is above the wire in the After position, so the field is still directed into the page. However, the loop is now closer to the wire, where the field is larger, so we show more field lines on the diagram in Figure 20.15.

Figure 20.15: When the loop moves to the After position, coming closer to the wire where the field from the wire is stronger, we draw more field lines passing through the loop.

Step 3 – Draw a "To Oppose" picture, with one field line representing the direction of the field the induced current in the loop creates to oppose the change in flux the loop experiences in moving from the Before position to the After position. Based on the direction of the field in the To Oppose picture, determine the direction of the induced current. To oppose the increase in magnetic flux that occurs as the loop moves closer to the wire, the induced current in the loop creates a magnetic field in the opposite direction, out of the page. The pictorial method is qualitative, so we only need to draw one field line directed into the page in Figure 20.16.

Figure 20.16: To oppose the change in flux as the loop moves from Before to After, the loop creates a magnetic field out of the page with a counterclockwise induced current.

Using the right-hand rule, when the thumb is directed out of the page, the fingers curl counterclockwise, in the direction of the induced current.

Key ideas for the pictorial method: A pictorial method can be used to determine the direction of the induced current in a loop or coil that experiences a change in magnetic flux. The steps are:
1. Draw a Before picture, showing the field lines passing through before a change is made.
2. Draw an After picture, showing the field lines passing through after a change is made.
3. Draw a To Oppose picture, with a single field line to represent the direction of the field needed to oppose the change from the Before picture to the After picture. Then, apply the right-hand rule to find the direction of the induced current needed to produce this field.

Related End-of-Chapter Exercises: 18 – 20, 22.

EXAMPLE 20.3 – A quantitative analysis

Rank the four single-turn loops in Figure 20.17 based on the magnitude of their induced current, from largest to smallest. The loops are either moving into or out of a region of uniform magnetic field. The field is zero outside the region.

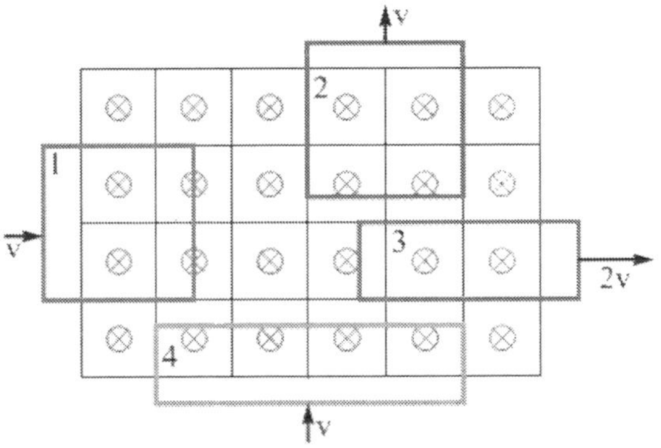

Figure 20.17: Four conducting loops, all of the same resistance, are moving with the velocities indicated either into (loops 1 and 4) or out of (loops 2 and 3) a region of uniform magnetic field.

SOLUTION

Combining Ohm's law with Faraday's law, we find that the magnitude of the current is given by:

$$I_{induced} = \frac{\varepsilon}{R} = \frac{1}{R}\left(\frac{\Delta(BA\cos\theta)}{\Delta t}\right)$$

$\cos\theta = 1$, and B does not change with time so: $I_{induced} = \frac{\varepsilon}{R} = \frac{B}{R}\left(\frac{\Delta A}{\Delta t}\right)$.

Writing the area in terms of the length L and width W: $I_{induced} = \frac{B}{R}\left(\frac{\Delta(LW)}{\Delta t}\right)$.

Defining L as the length of the loop perpendicular to the velocity, which is constant, then the magnitude of $\Delta W / \Delta t$ is the speed. The magnitude of the current is thus: $I_{induced} = BLv / R$.

The field and resistance are the same, so the induced current for the loops is proportional to the product of the speed multiplied by the length of the loop that is perpendicular to the velocity. This gives a ranking by current magnitude of 4 > 1 = 2 = 3. Loop 3 has half the length, perpendicular to its velocity, as loops 1 and 2, but makes up for that factor of two in its speed.

Essential Question 20.3: Find the direction of the induced current in the loops in Figure 20.17.

Answer to Essential Question 20.3: For loops 2 and 3, which are leaving the field, the magnetic flux into the page decreases as the fraction of the loop that is inside the field decreases. The pictorial method for loop 2 is shown in Figure 20.18. The To Oppose picture shows the field directed into the page, opposing the decrease in flux from Before to After, which requires an induced current directed clockwise. Similarly, loop 3 has a clockwise current.

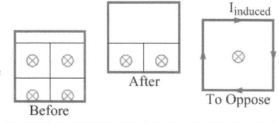

For loops 1 and 4, which are entering the field, the Before and After pictures from Figure 20.18 switch positions. This gives a To Oppose picture with a magnetic field directed out of the page, requiring a counterclockwise current.

Figure 20.18: The pictorial method for loop 2, which tells us that loop 2 has an induced current that is directed clockwise.

20-4 Motional emf

In each of the loops in Figure 20.17, the induced emf is associated with only one side of the rectangle, the side completely in the field, aligned perpendicular to the loop's velocity. Let's address this emf from another perspective.

EXPLORATION 20.4 – A metal rod moving through a magnetic field
As shown in Figure 20.19, a metal rod of length L is moving with a velocity $\vec{v}$ through a uniform magnetic field of magnitude B.

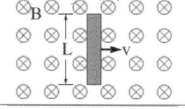

Step 1 – The rod has no net charge, but conduction electrons within the rod are free to move. First, assume that these electrons are moving through the field with the velocity of the rod. Apply the right-hand rule to determine the direction of the force these electrons experience. Hold the right hand above the page with the palm down and the fingers pointing to the right, in the direction of the velocity. When we curl the fingers they curl into the page, in the direction of the magnetic field. The thumb points up the page, showing the direction of the magnetic force on particles with positive charge. Electrons, which are negatively charged, experience a force in the opposite direction, down the page, leaving a net positive charge at the upper end of the rod. As shown in Figure 20.20, the moving rod acts like a battery.

Figure 20.19: A metal rod moving through a magnetic field.

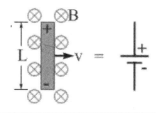

Figure 20.20: A conducting rod moving in a magnetic field acts like a battery, because of the separation of charge from the magnetic force acting on the rod's conduction electrons.

Step 2 – Determine the effective emf of the rod. As the rod becomes polarized, an electric field is set up in the rod. Show that the electric field gives rise to an electric force that is opposite to the magnetic force. Equate these two forces and, by treating the rod as a parallel-plate capacitor, determine the potential difference between the ends of the rod. The upper end of the rod is positive, so the electric field within the rod is directed down. An electron in the rod experiences an electric force that is opposite in direction to this electric field, because $\vec{F}_E = q\vec{E}$, and the charge on the electron is negative. Thus, electrons in the rod experience two forces, an electric force directed up and a magnetic force directed down. An equilibrium charge distribution is reached when these two forces balance, as shown in Figure 20.21.

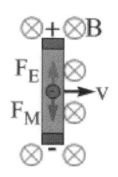

Figure 20.21: Equilibrium is reached when the electric force balances the magnetic force.

The magnetic force balances the electric force, so $qvB = qE$. The factors of q cancel, leaving $vB = E$. Treating the rod as a parallel-plate capacitor, the potential difference between the ends of the rod is $\Delta V = EL$. Solving for this potential difference, which we generally call a motional emf, ε, gives:

$$\Delta V = \varepsilon = -vBL. \qquad \text{(Equation 20.3: Motional emf)}$$

The minus sign indicates that Lenz's law applies, and that the emf tends to produce a current that opposes any change in magnetic flux. Note that equation 20.3 applies when the rod, its velocity, and the direction of the magnetic field are all mutually perpendicular.

Key ideas for motional emf: A conductor moving with a velocity $\vec{v}$ through a magnetic field $\vec{B}$ has an induced emf across it given by $\varepsilon = -vBL$, where L is the length of the conductor that is perpendicular to both $\vec{v}$ and $\vec{B}$. The moving conductor thus acts like a battery.
Related End-of-Chapter Exercises: 9, 10, 28.

EXAMPLE 20.4 – Using a moving rod as a battery

Let's investigate what happens when we use the rod like a battery. We will connect the moving rod from Exploration 20.4 in a simple series circuit, by placing the rod on a pair of parallel conducting rails, as shown in Figure 20.22. The rod moves with negligible friction on the rails, which themselves have negligible resistance but are connected by resistor of resistance R. (a) If the rod moves to the right with a speed v in a field of magnitude B directed into the page, find an expression for the magnitude of the induced current. (b) Use the pictorial method from section 20-3 to find the direction of this current.

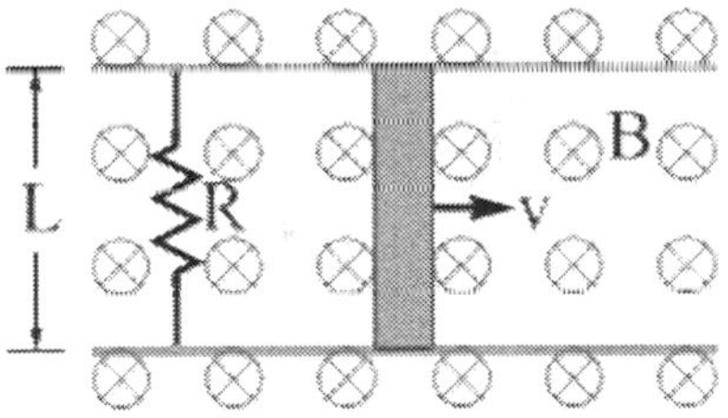

Figure 20.22: The system consisting of a conducting rod on frictionless rails. The rails are connected to one another through a resistor, and the rod moves through a magnetic field that is directed into the page.

SOLUTION

(a) In this situation we combine the equation for motional emf, $\varepsilon = -vBL$, with Ohm's law, $I = \Delta V / R$, to find that the induced current has a magnitude of:

$I = vBL / R$, matching the result from Example 20.3.

(b) As the rod moves to the right, the area of the conducting loop increases. Compared to the Before picture, the After picture shows the rod farther right, with more field lines passing through because the area of the loop has increased. As seen in Figure 20.23, the To Oppose picture shows field directed out of the page, opposing the extra into-the-page field lines in the After picture. To create a field out of the page, the induced current must be directed counterclockwise around the loop.

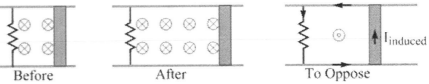

Figure 20.23: Applying the pictorial method to determine the direction of the induced current in the circuit.

Related End-of-Chapter Exercises: 29, 30, 53, and 55.

Essential Question 20.4: The moving rod in Example 20.4 experiences a magnetic force because of the interaction between the external magnetic field and the induced current passing through the rod. In what direction is this magnetic force? How is this direction consistent with Lenz's law?

Answer to Essential Question 20.4: The induced current is directed up through the rod, while the external magnetic field is directed into the page. Make sure you look at the external magnetic field, and not the field from the induced current! Applying the right-hand rule that goes with the expression for magnetic force, $F_M = ILB\sin\theta$, we find that the rod experiences a magnetic force directed to the left. Because the velocity of the rod is directed to the right, the magnetic force tends to slow the rod down, decreasing the time rate of change of flux. The tendency of the induced current to oppose the change in flux is completely consistent with Lenz's law.

20-5 Eddy Currents

Thus far, in this chapter, we have discussed induced currents set up in conducting loops and coils, and in moving rods connected in a circuit, in response to a changing magnetic flux. Something similar happens in solid pieces of conductor, in which a changing magnetic flux gives rise to what are called eddy currents. The meaning of the word "eddy" in this case is the same as its meaning when you refer to an eddy in a river, which is a swirl of water.

> **Eddy currents** are swirling currents that are set up in conductors that are exposed to a changing magnetic flux. Consistent with Lenz's law, these swirling currents create their own magnetic field that tends to oppose the original change in flux.

Eddy currents can arise when there is relative motion between a conductor and a magnet. The relative motion tends to change the flux in the conductor, so by Lenz's law the eddy currents produce a magnetic field that acts to slow the motion. Magnetic interactions in ferromagnetic materials (like iron or steel) are often dominated by the strong attraction between the magnet and the ferromagnetic material. The effects of eddy currents are generally easier to observe in non-ferromagnetic materials like aluminum and copper. Such materials interact weakly with magnets that are at rest with respect to them, but the effects of eddy currents in such materials, which involve a changing magnetic flux, often produced by relative motion, can be dramatic.

Some examples of eddy currents include:
- **Train brakes.** Turning on an electromagnet to create a magnetic field that passes through part of a train wheel causes the wheel, and the train, to slow as different areas of the wheel move into and out of the field (see Exploration 20.5).
- **Magnetic damping on a balance.** Like the system in Figure 20.24, many triple-beam balances have an aluminum plate located near a magnet, to damp out (steadily reduce) the oscillations of the balance so it settles down quickly for a reading.
- **Dropping a magnet down an aluminum or copper tube.** Try this if you can. It is fascinating to look down the tube and see the magnet fall, almost in slow motion, as the eddy currents in the tube wall slow the magnet's fall.

EXPLORATION 20.5 – Using eddy currents to stop a train

Figure 20.25 shows the wheel of a train. When the driver of the train applies the brakes, an electromagnet sets up a magnetic field, directed into the page, which passes through a circular region (in black) at the top of the wheel. Eddy currents are set up in a part of the wheel (in green) that is entering the field, as well as in a part of the wheel (in blue) that is leaving the field.

Figure 20.24: A photograph of the two magnets (dark cylinders) used to damp out the oscillations of the aluminum plate that is part of an apparatus used in a Coulomb's law experiment. Photo credit: A. Duffy.

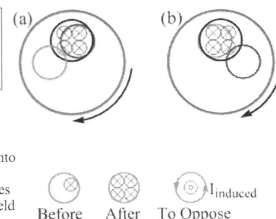

Figure 20.25: Two views of a train wheel exposed to a magnetic field at the top. The wheel is rotating clockwise. View (b) is at a later time than view (a), so the wheel is moving slower in (b).

Step 1 – *Apply the pictorial method to the region of the wheel moving into the field, to determine the direction of the eddy currents induced in this region by the changing flux.* This region is moving into the field, so the After picture shows more field lines passing through the region than the Before picture does (see Figure 20.26). The To Oppose picture shows a field line directed out of the page, opposing the change in flux. By the right-hand rule, the induced eddy currents in this region must swirl counterclockwise to create a field out of the page.

Before   After   To Oppose

Figure 20.26: The pictorial method applied to the region that is entering the magnetic field.

Step 2 – *Consider the interaction between the eddy currents in the region moving into the field and the external magnetic field. Does this interaction give rise to a torque on the wheel? If so, in what direction is the torque?* Figure 20.27 shows a magnified view of the counterclockwise current interacting with the field. The right-hand part of the swirling current, directed up, is in the magnetic field. Applying the right-hand rule shows that there is a force directed to the left on this upward current. The left-hand part of the eddy current is outside the field, so there is no force on that part. The force to the left produces a counterclockwise torque on the wheel, relative to an axis through the wheel's center, acting to slow the wheel's clockwise motion.

Figure 20.27: The magnetic force acting on the eddy current in the region moving into the field produces a torque that acts to slow down the wheel.

Note that we will consider the region moving out of the field in Essential Question 20.5.

Key idea: Eddy currents can give rise to forces or torques that tend to slow relative motion between a conductor and a magnet. **Related End-of-Chapter Exercises: 56 and 57.**

Superconductors and the Meissner effect

One application of eddy currents is the levitation of a magnet above a superconductor, which is a material that has no electrical resistance. Superconductors generally exhibit zero resistance only at very low temperatures. The black disk at the bottom of Figure 20.28 is known as a high-temperature superconductor, because it has no resistance when it is cooled by liquid nitrogen, at a relatively high temperature (at least for superconductors!) of about 77 K (−196°C). When the magnet is brought close to the superconductor, the changing flux gives rise to eddy currents in the superconductor. The eddy currents, which can flow forever because there is no resistance, set up a magnetic field that exactly cancels the field from the magnet (this exclusion of magnetic field from the superconductor is known as the Meissner effect). The magnetic field from the eddy currents in the superconductor can support the magnet without the magnet and the superconductor being in contact.

Figure 20.28: A magnet being levitated by the magnetic field produced by eddy currents in a superconducting disk, which is cooled by liquid nitrogen. Photograph from http://www.lbl.gov

Essential Question 20.5: Repeat steps 1 and 2 of Exploration 20.5 for the region of the wheel (see Figure 20.25) that is leaving the region of magnetic field.

Before   After   To Oppose

Figure 20.29: The pictorial method applied to the region that is leaving the magnetic field.

Answer to Essential Question 20.5: For the region that is moving out of the field, the After picture shows fewer field lines passing through the region than the Before picture does (see Figure 20.29). The To Oppose picture shows a field line directed into the page, opposing the change in flux. By the right-hand rule, the induced eddy currents in this region must swirl clockwise to create a field into the page.

Figure 20.30 shows a magnified view of the clockwise current interacting with the field. The left-hand part of the swirling current, directed up, is in the magnetic field. Applying the right-hand rule shows that there is a force directed to the left on this upward current. The right-hand part of the eddy current is outside the field, so there is no force on that part. The force to the left produces a counterclockwise torque on the wheel, relative to an axis through the wheel's center, acting to slow the wheel's clockwise motion. Note that no matter which region we consider, one entering the field or one leaving the field, the torque always acts to slow the wheel.

Figure 20.30: The magnetic force acting on the eddy current in the region leaving the field produces a torque that acts to slow down the wheel.

20-6 Electric Generators

Faraday's Law, and the induced emf and induced current associated with it, is one of the most practical ideas in physics — it lies at the heart of most devices that generate electricity. The basic components that make up an electric generator are quite simple. All we need is a conducting loop exposed to a changing magnetic flux.

In section 19-6, we spent some time investigating a DC motor, which is a conducting loop in a magnetic field. When current is sent through the loop, the magnetic field exerts a torque on the loop that makes the loop spin, transforming electrical energy into mechanical energy. An electric generator does exactly the opposite, transforming mechanical energy into electrical energy. We can use the same device, a loop in a magnetic field, as a motor or a generator.

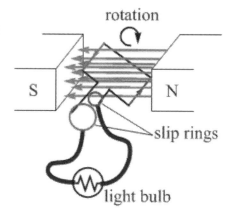

Figure 20.31: The basic components of an electric generator are a conducting loop and a magnetic field. Rotating the loop changes the magnetic flux through the loop, giving rise to an induced emf.

As the loop in Figure 20.31 rotates, the magnetic flux through the loop changes, giving rise to an induced emf. If we completed the circuit by connecting the ends of the loop directly to an electrical device, like a light bulb, the induced emf gives rise to an induced current that lights the bulb, but the wires would twist, either stopping the loop or breaking the wires. One solution to this problem is to have the wires coming from each side of the loop rub against the inside of fixed slip rings, which are then connected to the bulb. Each wire maintains electrical contact with one slip ring as the loop rotates.

Another solution to this problem, which is used in some electricity generating plants, is to keep the loop fixed and spin the magnets around the loop. This has the same effect as spinning the loop through the magnetic field.

EXPLORATION 20.6 – The form of the electric signal from a generator

Step 1 – *The loop in Figure 20.31 rotates at a constant angular speed ω. Plot graphs of the flux through the loop, and the corresponding induced emf, as a function of time. Define the positive direction such that the flux at t = 0 is positive. Plot a second set of graphs to show what happens when the rotation rate doubles.* As the loop rotates, the magnetic flux oscillates. We can understand the cosine dependence of the flux graph, shown at the top in Figure 20.32, by starting with the equation for magnetic flux, $\Phi = BA\cos\theta$. B, the magnetic field, and A, the area of the loop, are constant, but θ changes. Using a relationship from rotational kinematics, the angle between the field and the loop's area vector is given by $\theta = \omega t$. This gives the flux equation a form which shows the time dependence, and the cosine nature, explicitly: $\Phi(t) = BA\cos(\omega t)$.

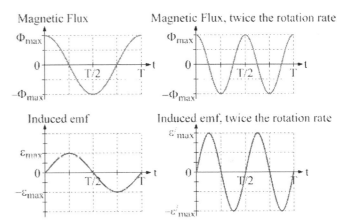

The induced emf is proportional to the negative of the slope of the flux versus time graph. Thus, the emf is zero when the flux graph reaches a positive or negative peak. When the flux is decreasing, the flux graph has a negative slope, so the induced emf is positive. The emf peaks when the flux passes through zero, and the flux graph has a large slope.

If the rotation rate increases, the maximum flux does not change but the flux changes in half the time, doubling the slope on the flux graph. This produces a corresponding doubling in the peak emf. Thus, the peak emf is proportional to the angular speed of the magnets.

Figure 20.32: For the magnetic flux and induced emf graphs on the right, the rotation rate is twice as large as in the graphs on the left.

Step 2 – *Combine Faraday's Law with results from step 1 to find an expression for the emf induced in the loop as a function of time.* If the loop has N turns, applying Faraday's Law gives:

$$\varepsilon = -N\frac{\Delta\Phi}{\Delta t} = -N\frac{\Delta\left[BA\cos(\omega t)\right]}{\Delta t}$$, where we have used the flux expression from step 1.

The field and area are constants, so they can be moved out front: $\varepsilon = -NBA\dfrac{\Delta\left[\cos(\omega t)\right]}{\Delta t}$.

Simplifying, we get: $\varepsilon = \omega NBA \sin(\omega t)$. This last step is easy to show using calculus.

However, we saw both the sine dependence, and the fact that the induced emf is proportional to the angular speed, in step 1, so we can be confident in the result even without using calculus.

Key idea: As a function of time, the emf of an electric generator, in which either the loop or the magnetic field spins at a constant angular velocity ω̄, oscillates sinusoidally. In other words, the electric generator puts out alternating current. Expressed as an equation, the emf is:

$\varepsilon = \omega NBA \sin(\omega t) = \varepsilon_{max}\sin(\omega t)$. (Equation 20.4: **emf from an electric generator**)

The peak voltage, ε_{max}, is the product of ω, N (the number of turns in the loop), B (the strength of the magnetic field), and A (the loop area). **Related End-of-Chapter Exercises: 31 – 34.**

Essential Question 20.6: For generators in North America, which provide 60 Hz AC for a standard wall socket, what is ω, the magnitude of the angular velocity of the rotation?

Answer to Essential Question 20.6: In this case, we can use the relationship $\omega = 2\pi f$. If the frequency is 60 Hz, the angular speed of the magnets in a typical electricity generating facility is 120π, or 377, rad/s. In Europe, the frequency is 50 Hz, with a corresponding decrease in ω.

20-7 Transformers and the Transmission of Electricity

In this day and age, when we rely so heavily on electricity in our daily lives, it is hard to imagine a time when electricity was not so widely available. It was only in the late 1880's, not so long ago when measured against thousands of years of human history, that the "War of Currents" took place. The War of Currents was the battle between Thomas Edison and his General Electric company, who wanted direct current (DC) to be the standard for electricity distribution systems, and Nikola Tesla and George Westinghouse, the proponents of alternating current (AC) as the basis of such systems. The battle even included the electrocution of an elephant, in an Edison-led demonstration of the dangers of AC. The battle lasted a few years before the superiority of AC won the day, not to mention the contract for the large hydroelectric power plant at Niagara Falls.

What makes AC so superior to DC for electricity distribution? As we learned in section 20-6, it is very easy to generate AC. However, the big advantage of AC over DC is that it is extremely easy to change the voltage of AC, with very little loss of energy, while it is much more difficult, and less efficient, to do so for DC. With AC, this transformation of voltage is accomplished with a device called a transformer, which exploits Faraday's Law in its operation.

A transformer is quite a simple device (see Figure 20.33), typically consisting of two coils of wire wrapped around a core made of ferromagnetic material. The role of the core is to ensure that all the magnetic flux, generated by current passing through the primary coil, passes through the secondary coil. If this flux changes, because the current in the primary (p) coil changes, then an emf and a current are induced in the secondary (s). The two coils are exposed to the same changing flux, so we can relate them using Faraday's Law, which we re-arrange to read:

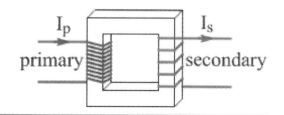

Figure 20.33: A transformer consists of two coils, the primary and the secondary, wrapped around a ferromagnetic core.

$$\frac{\Delta\Phi}{\Delta t} = -\frac{\varepsilon_p}{N_p} = -\frac{\varepsilon_s}{N_s} .$$

Recall from Chapter 18 that electrical power is the product of the potential difference and the current. In an ideal case there is no loss of energy, or power, in the transformer, so the power in the primary, $\varepsilon_p I_p$, is equal to the power in the secondary, $\varepsilon_s I_s$. The power relationship and the relationship from Faraday's Law, above, are combined into one equation below, where potential difference ΔV has replaced the emf ε.

If we assume that no energy is lost in the transformation process, we say that the transformer is ideal. For an ideal transformer, with alternating current in the primary, we have:

$$\frac{\Delta V_p}{\Delta V_s} = \frac{I_s}{I_p} = \frac{N_p}{N_s} .$$ (Equation 20.5: **Relations for an ideal transformer**)

The subscripts p and s stand for primary and secondary, respectively. ΔV represents the potential difference across the coil, I is the current, and N is the number of turns in the coil. Equation 20.5 is generally used to relate peak values of current or voltage in the primary to their corresponding peak values in the secondary, or to relate rms values in one coil to rms values in the other coil.

Note that transformers require a changing magnetic flux to generate a current in the secondary coil. Such a changing magnetic flux is provided by alternating current running through the primary, generating alternating current in the secondary.

What is a transformer good for? In general, transformers are used to change the voltage from a wall socket into a different voltage, which could be higher or lower, for use by a particular device. Some devices, such as microwave ovens and cathode ray tube televisions, require higher voltages than the 120 V rms that is provided by a wall socket in North America. Such devices have a **step-up transformer**, in which the secondary coil has a larger number of turns than the primary, to increase the voltage to the required level. Many devices also require a constant voltage, so the transformers also convert the alternating current to direct current.

Other devices, such as computers and clock radios, require a voltage such as 12 V, much less than the 120 V rms from a wall socket, to operate. The power cord in these devices is thus connected to a **step-down transformer**, in which the secondary coil has a smaller number of turns than the primary, to decrease the voltage. Note that, by energy (or power) conservation, decreasing the voltage is associated with an increase in current. In step-up transformers, on the other hand, the increase in voltage is accompanied by a corresponding decrease in current.

EXPLORATION 20.7 – Transformers in the electricity distribution system

A particular electricity generating facility generates electricity at the rate of 2.4 MW, with an rms current of 1000 A at an rms voltage of 2.4 kV.

Step 1 – If the electricity is sent through a cable (known as a transmission line) that has a resistance of 10 Ω on its way to New York City, determine the power lost during the transmission process. Here we can apply one version of the power equation from chapter 18, $P = I^2 R = (1000 \text{ A})^2 (10 \, \Omega) = 10 \text{ MW}$. This is clearly ridiculous, because this power is more than 4 times larger than the power generated by the plant. The bottom line is that essentially all the electrical energy would be dissipated in the transmission line.

Step 2 – If, instead, the voltage is transformed to 240 kV before the electricity is sent along the transmission line, what is the power lost in the transmission process? Applying equation 20.5, increasing the voltage by a factor of 100 produces a corresponding decrease by a factor of 100, to 10 A, in the current. With 10 A of current in the transmission line, the power lost now is only $P = I^2 R = (10 \text{ A})^2 (10 \, \Omega) = 1000 \text{ W}$. This is why power companies transmit electricity over long distances at high voltages, to minimize the current, thereby minimizing transmission losses.

Step 3 – The voltage is ultimately transformed back down to 120 V in New York City. What is the value of the rms current? Neglecting the power lost in step 2, stepping the voltage down by a factor of 2000, from 240 kV to 120 V, increases the current by this same factor of 2000, from 10 A to 20000 A, enough to meet the needs of a large number of residential customers.

Key ideas: Power companies make extensive use of both step-up and step-down transformers in the process of delivering electricity. **Related End-of-Chapter Exercises: 36 – 40, 60.**

Essential Question 20.7: A particular transformer has a primary coil with 100 turns and a secondary coil with 200 turns. If the primary coil has a constant potential difference of 20 V and a constant current of 5 A, what are the values of the potential difference across, and current in, the secondary? Assume the transformer is ideal.

Answer to Essential Question 20.7: Because the current in the secondary coil is constant, the magnetic flux through the secondary coil is constant. Because the flux through the secondary coil does not change, there is no induced emf in the secondary, and thus there is also no current. Transformers work very well for alternating current, but they do not work for direct current.

Chapter Summary

Essential Idea: Electromagnetic Induction.

One of the most practical applications of physics, the generation of electricity, relies on electromagnetic induction. Exposing a conducting loop to a changing magnetic flux (a change in the number of magnetic field lines passing through the loop) will induce an emf, or voltage, in the loop. In a complete circuit, this induced emf will gives rise to an induced current in the loop.

Magnetic Flux

Magnetic flux is a measure of the number of magnetic field lines passing through an area. The symbol we use for flux is the Greek letter capital phi, Φ. The equation for magnetic flux is:

$$\Phi = BA\cos\theta , \qquad \text{(Equation 20.1: \textbf{Magnetic flux})}$$

where θ is the angle between the magnetic field $\vec{B}$ and the area vector $\vec{A}$.

Faraday's Law of Induction

Exposing a loop or coil to a *changing* magnetic flux gives rise to a voltage, called an **induced emf.** Following Ohm's law, the induced emf gives rise to an **induced current** in the loop or coil. The emf induced by a changing magnetic flux in each turn of a coil is equal to the time rate of change of that flux. Thus, for a coil with N turns, the net induced emf is given by:

$$\varepsilon = -N\frac{\Delta\Phi}{\Delta t} = -N\frac{\Delta(BA\cos\theta)}{\Delta t} . \qquad \text{(Eq. 20.2: \textbf{Faraday's Law of Induction})}$$

In many cases, graphing the magnetic flux as a function of time can be helpful because the induced emf is proportional to the negative of the slope of the graph of flux versus time.

Lenz's Law

Lenz's Law is associated with the minus sign in Faraday's Law. Lenz's Law states that the emf induced by a changing magnetic flux tends to produce an induced current. The induced current produces a magnetic flux that acts to oppose the original change in flux.

A pictorial method for applying Lenz's Law to determine the direction of induced current

A pictorial method can be used to determine the direction of the induced current in a loop or coil that experiences a change in magnetic flux. The steps are:
1. Draw a Before picture, showing the field lines passing through before a change is made.
2. Draw an After picture, showing the field lines passing through after a change is made.
3. Draw a To Oppose picture, with a single field line to represent the direction of the field needed to oppose the change from the Before picture to the After picture. Then, apply the right-hand rule to find the direction of the induced current needed to produce this field.

Applying the pictorial method: an example

The magnetic field, directed into the page through a wire loop, is decreasing in magnitude. In which direction is the current induced in the loop?

Figure 20.29: The situation described here is similar to that of Figure 20.29. In the "To Oppose" picture, the field created by the induced current must also be directed into the page, to oppose the loss in field lines into the page the loop experiences.

Before   After   To Oppose

Motional emf

A conductor moving with a velocity $\vec{v}$ through a magnetic field $\vec{B}$ has an induced emf, generally referred to as motional emf, across it given by:

$$\varepsilon = -vBL .$$ (Equation 20.3: **Motional emf**)

where L is the length of the conductor that is perpendicular to both $\vec{v}$ and $\vec{B}$. The moving conductor thus acts like a battery. Note that in many motional emf situations, and in other induced emf situations, Ohm's Law ($\varepsilon = IR$) is often used to determine the current resulting from the motional or induced emf.

Eddy currents

Eddy currents are swirling currents that are set up in conductors that are exposed to a changing magnetic flux. Consistent with Lenz's law, these swirling currents create their own magnetic field that tends to oppose the original change in flux. A practical application of eddy currents is in train brakes, in which braking forces arise from the interaction of the eddy currents in the train wheel and the magnetic field of an electromagnet. The field is turned on, and the eddy currents are set up, only when the operator of the train applies the brakes.

Electric generators

Alternating current can be generated very easily, simply by spinning a conducting loop at a constant rate in a uniform magnetic field. To avoid wires being twisted, in some cases the magnets producing the field are rotated around a conducting loop that is held fixed.

As a function of time, the emf of an electric generator, in which either the loop or the magnetic field spins at a constant angular velocity $\vec{\omega}$, oscillates sinusoidally. Expressed as an equation, the emf is:

$$\varepsilon = \omega NBA \sin(\omega t) = \varepsilon_{max} \sin(\omega t) .$$ (Equation 20.4: **emf from an electric generator**)

The peak voltage, ε_{max}, is the product of ω, N (the number of turns in the loop), B (the strength of the magnetic field), and A (the area of the loop).

Transformers

A transformer is a device for changing the voltage of an alternating current (AC) signal from one value to a higher or lower value. Transformers usually consist of two coils wrapped around a ferromagnetic core. An emf is induced in the secondary coil by a changing flux produced by the changing current in the primary coil. If no energy is lost in the transformation process, we say that the transformer is ideal.

$$\frac{\Delta V_p}{\Delta V_s} = \frac{I_s}{I_p} = \frac{N_p}{N_s} .$$ (Equation 20.5: **Relations for an ideal transformer**)

The subscripts p and s stand for primary and secondary, respectively. ΔV represents the potential difference across the coil, I is the current, and N is the number of turns in the coil. Equation 20.5 is generally used to relate peak values of current or voltage in the primary to their corresponding peak values in the secondary, or to relate rms values in one coil to rms values in the other coil.

End-of-Chapter Exercises

Exercises 1 – 12 are primarily conceptual questions designed to see whether you understand the main concepts of the chapter.

1. The four areas in Figure 20.34 are in a magnetic field. The field has a constant magnitude, but it is directed into the page on the left half of the figure and out of the page in the right half. (a) Rank the areas based on the magnitude of the net flux passing through them, from largest to smallest. (b) Defining out of the page to be the positive direction for magnetic flux, rank the areas based on the net flux, from most positive to most negative.

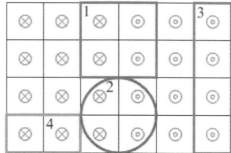

2. Repeat Exercise 1, but now the magnitude of the magnetic field in the left half of the figure is triple the magnitude of the field in the right half.

Figure 20.34: Four different regions in a magnetic field. The field is directed into the page in the left half of the figure and out of the page in the right half. For Exercises 1 and 2.

3. At a particular instant in time, there is no magnetic field passing through a conducting loop. (a) Could the emf induced in the loop be zero at that instant? Explain. (b) Could the emf induced in the loop be non-zero at that instant? Explain.

4. As shown in Figure 20.35, four identical loops are placed near a long straight wire that carries a current to the right. The wire is in the same plane as the loops. (a) Rank the loops based on the magnitude of the net flux passing through them, from largest to smallest. (b) Defining out of the page to be the positive direction for magnetic flux, rank the loops based on the net flux, from most positive to most negative.

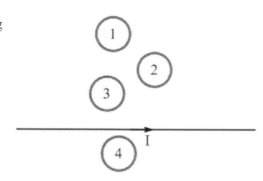

Figure 20.35: Four identical loops near a long straight wire carrying current to the right, for Exercise 4.

5. A long straight wire carrying current out of the page passes through the middle of a circular conducting loop that is in the plane of the page, as shown in Figure 20.36. In what direction is the induced current in the loop when the current in the long straight wire is increasing in magnitude?

I increasing

6. Rank the four loops in Figure 20.37 based on (a) the magnitude of the induced current, from largest to smallest, and (b) the induced current, from largest clockwise to largest counterclockwise. The loops, which all have the same resistance, are moving with the velocities indicated either into (loops 1 and 4) or out of (loops 2 and 3) a region of magnetic field. The field has the same magnitude within the rectangular region, but is directed into the page in the left half of the region and out of the page in the right half.

Figure 20.36: A long straight wire passes through the center of a conducting loop. The current in the long straight wire is directed out of the page, and is increasing in magnitude. For Exercise 5.

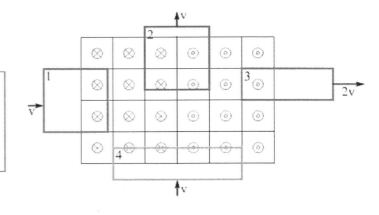

Figure 20.37: Four conducting loops, all of the same resistance, are moving with the velocities indicated either into (loops 1 and 4) or out of (loops 2 and 3) a region of magnetic field, for Exercise 6.

7. A square conducting loop located below a long straight current-carrying wire is moved away from the wire, as shown in Figure 20.38. While the loop is moving, the induced current in the loop is observed to be directed clockwise around the loop. In which direction is the current in the long straight wire?

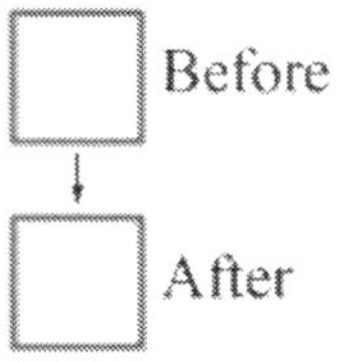

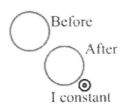

Figure 20.38: A square conducting loop is moved away from a long straight current-carrying wire, for Exercise 7.

8. A conducting loop is moved closer to a long straight wire, as shown in Figure 20.39. The current in the wire is constant, and directed out of the page, while the loop is in the plane of the page. While the loop is moving toward the wire, in what direction is the induced current in the loop? Explain.

Figure 20.39: A conducting loop is moved closer to a long straight wire that carries a constant current directed out of the page, for Exercise 8.

9. Three conducting rods are moving with the velocities shown in Figure 20.40 through a region of uniform magnetic field that is directed into the page. You measure the motional emf of each bar with a voltmeter connected across the long dimension of each rod. For instance, for rods 1 and 3 you measure the potential difference between the right end and the left end of the rod. Rank the rods based on the magnitude of the potential difference you measure.

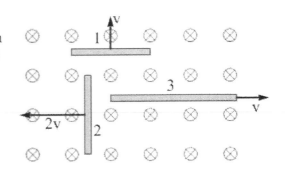

10. Return to the situation shown in Figure 20.40 and described in Exercise 9. Which of the following is at a higher potential, considering the motional emf? (a) The left end of rod 1 or the right end of rod 1? (b) The upper end of rod 2 or the lower end of rod 2? (c) The left end of rod 3 or the right end of rod 3?

Figure 20.40: Three conducting rods are moving through a region of uniform magnetic field. The velocity and orientation of each rods is shown on the diagram. For Exercises 9 and 10.

11. As shown in part (a) of Figure 20.41, a 12-volt battery is connected through a switch to the primary coil of a transformer. The secondary coil, which has fewer turns than the primary, is connected to a resistor. The switch is initially open. When the switch is closed, which of the following statements correctly describes the potential difference across the resistor? Explain your choice.

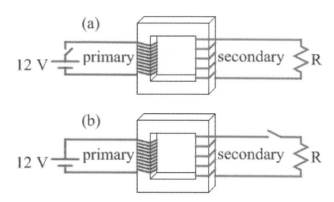

- A constant potential difference of 12 V.
- A non-zero constant potential difference that is smaller than 12 V.
- A constant potential difference that is larger than 12 V.
- A non-zero potential difference when the switch closes that quickly drops to zero.
- The potential difference is zero the whole time.

Figure 20.41: In (a), a 12-volt battery is connected to the primary coil of a transformer through a switch that is initially open. The secondary coil is connected to a resistor. In (b), the switch has been moved to the secondary side of the transformer.

12. Repeat Exercise 11, but use part (b) of Figure 20.41, in which the switch is moved to the secondary side of the transformer.

Exercises 13 – 17 deal with Faraday's Law.

13. A circular wire loop, with a single turn, has a radius of 12 cm. The loop is in a magnetic field that is decreasing at the rate of 0.20 T/s. At a particular instant in time, the field has a magnitude of 1.7 T. At that instant, determine the emf induced in the loop if (a) the field is directed perpendicular to the plane of the loop, (b) the field is directed parallel to the plane of the loop, (c) the angle between the field and the loop's area vector is 30°.

14. A coil of wire is connected to a galvanometer, as shown in Figure 20.42. The galvanometer needle deflects left when current is directed to the left through the coil; is vertical when there is no current, and deflects right when current is directed to the right. When a magnet, with its north pole closer to the magnet, is moved toward the right end of the coil, the galvanometer needle deflects to the left. In which direction will the needle deflect in the following cases? In all cases, the axis of the magnet coincides with the axis of the coil. (a) The magnet remains at rest inside the coil, with the south pole sticking out of the right end of the magnet. (b) The

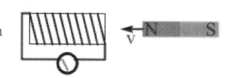

Figure 20.42: A coil, galvanometer, and magnet, for Exercise 14.

magnet, with its south pole closer to the magnet, is moved to the right away from the coil. (c) The magnet, with its south pole closer to the coil, remains at rest, while the coil is moved toward the magnet. The magnet is to the right of the coil at all times.

15. Figure 20.43 shows a square loop traveling at a constant velocity of 10 cm/s to the right through a region of magnetic field. The field is confined to the outlined rectangular region. The field is directed into the page in the left half of this region, and out of the page in the right half of this region. The field has the same magnitude at all points within the region. The small squares on the diagram measure 10 cm × 10 cm, and the loop is shown at $t = 0$. Rank the following times based on the magnitude of the emf induced in the loop as it moves through the field: $t_A = 0.5$ s; $t_B = 1.5$ s; $t_C = 4.5$ s; and $t_D = 6.0$ s.

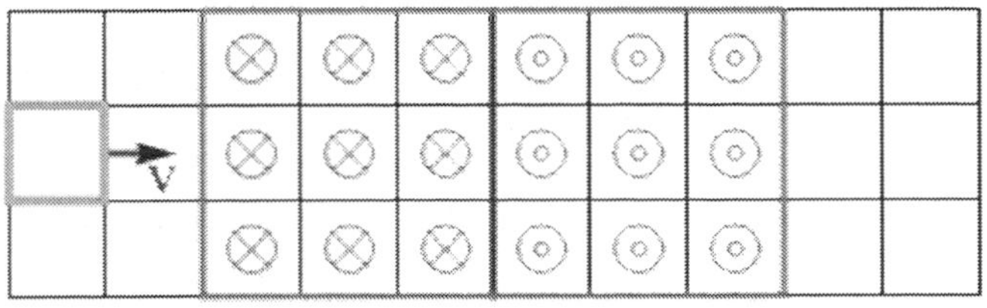

Figure 20.43: A square loop moves through a magnetic field that is confined to the outlined rectangular region outlined. The magnetic field is directed into the page in the left half of this region, and out of the page in the right half. For Exercises 15 – 17.

16. Return to the situation described in Exercise 15, and shown in Figure 20.43. The loop is made from a single turn, and has a resistance of 6.0 Ω. The magnitude of the magnetic field at all locations in the region outlined in red is 3.0 T. (a) Sketch a motion diagram for the loop, showing its location at regular intervals. (b) Defining out of the page to be the positive direction for flux, draw a graph of the magnetic flux through the loop as a function of time. (c) Draw a graph of the emf induced in the loop as a function of time. (d) Find the maximum magnitude of the current induced in the loop as the loop moves through the magnetic field.

17. Return to the situation described in Exercises 15 and 16, and shown in Figure 20.43. If the loop's speed is increased by a factor of 3, do any of the following change as the loop moves through the field? Explain. (a) The maximum magnitude of the loop's magnetic flux. (b) The maximum magnitude of the induced emf. (c) The peak current in the loop.

Exercises 18 – 22 are designed to give you practice applying the pictorial method of solving problems involving Faraday's law.

18. As shown in Figure 20.44, a conducting loop is near a long straight wire that carries current to the right. In case 1, the current is increasing in magnitude. In case 2, the current is decreasing in magnitude. (a) If the magnitude of the induced current in the loop is the same in both cases, what does this tell us about the current in the two cases? (b) What is the direction of the induced current in the two cases? Justify your current directions by drawing the sets of three pictures associated with the pictorial method.

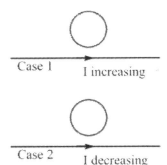

Figure 20.44: Two cases involving a conducting loop near a long straight wire, for Exercise 18.

19. Much like the situation shown in Figure 20.44, a conducting loop is placed above a long straight wire. The loop does not move with respect to the wire, while the current in the wire can be directed either left or right and may be increasing, decreasing, or constant. At a particular instant in time, the induced current in the loop is directed clockwise. (a) Could the current in the wire be directed to the right at this instant? Explain using the pictorial method. (b) Could the current in the wire be directed to the left at this instant? Explain using the pictorial method.

20. A square conducting loop is placed above a long straight wire that carries a constant current to the right. The loop can be moved along one of the three paths shown in Figure 20.45. (a) Rank the paths based on the magnitude of the current induced in the loop, assuming the speed of the loop is the same for each path. (b) Draw the set of three pictures associated with the pictorial method, and use the pictures to determine the direction of the current induced in the loop when it follows path 1. Repeat the process for (c) path 2, and (d) path 3.

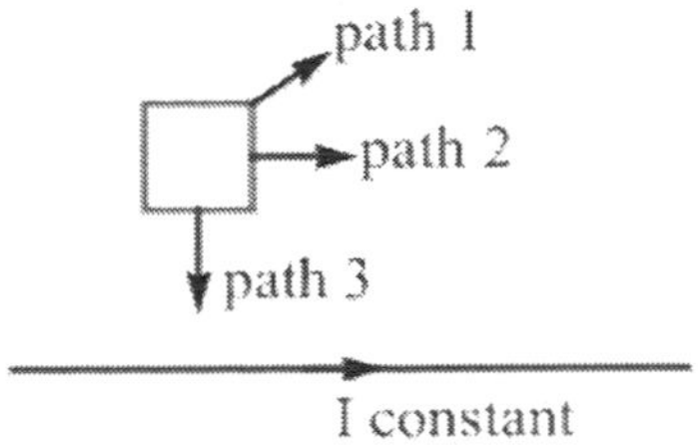

Figure 20.45: A square conducting loop is placed above a long straight wire that carries a constant current to the right. The loop can be moved along one of the three paths shown. For Exercise 20.

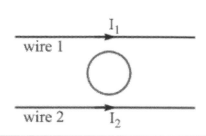

21. A conducting loop is placed exactly halfway between two parallel wires, each carrying a current to the right, as shown in Figure 20.46. At a particular instant in time, the currents in the wires have the same magnitude, but these currents may be increasing or decreasing. At this instant, we observe that there is a clockwise induced current in the loop. (a) Could one of the two currents be neither increasing or decreasing at this instant? Explain. (b) Could both currents be increasing in magnitude (possibly at different rates) at this instant? Explain. (c) Could one current be increasing while the other is decreasing at this instant? Explain.

Figure 20.46: A conducting loop is located halfway between two wires that carry current to the right. At a particular instant, the currents in the two wires have the same magnitude. For Exercise 21.

22. Figure 20.47 shows a particular After picture, showing the field lines passing through a conducting loop After some change is made in the external field, along with six possible Before pictures and two possible To Oppose pictures. (a) For the To Oppose picture P, in which direction is the induced current in the loop? (b) For the To Oppose picture Q, in which direction is the induced current in the loop? (c) Which of the Before pictures could go with the After picture and the To Oppose picture P? Select all that apply. (d) Which of the Before pictures could go with the After picture and the To Oppose picture Q? Select all that apply.

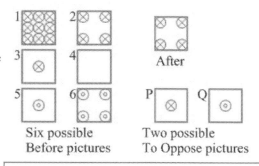

Exercises 23 – 27 are designed to give you practice working with graphs of magnetic flux versus time in induced emf situations.

Figure 20.47: Six possible Before pictures, a particular After picture, and two possible To Oppose pictures, for the situation of a square conducting loop in a magnetic field. For Exercise 22.

23. Consider the graph of magnetic flux through a conducting loop, as a function of time, shown in Figure 20.48. (a) At which time, $t = 10$ s, $t = 20$ s, or $t = 30$ s, does the induced emf have the largest magnitude? Explain. (b) Assuming the loop has a single turn, what is the magnitude of the induced emf at $t = 20$ s?

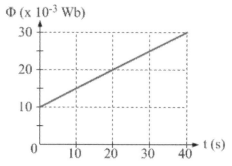

Figure 20.48: A graph of the magnetic flux passing through a particular conducting loop as a function of time, for Exercise 23.

24. The graph in Figure 20.49 shows the magnetic flux through a conducting loop as a function of time. The units of flux are webers, which are equivalent to tesla meters^2. The flux in the loop changes because the magnetic field passing through the loop changes. The loop itself is stationary. Compare the magnitude of the current induced in the loop at the following times. For each pair of times, state at which time the induced current has a larger magnitude, and explain your answer. (a) 5 s and 10 s. (b) 5 s and 18 s. (c) 5 s and 25 s. (d) 5 s and 40 s. (e) 25 s and 40 s.

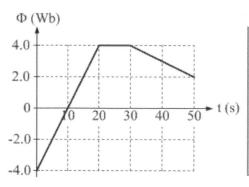

Figure 20.49: A graph showing the magnetic flux passing through a conducting loop as a function of time, for Exercises 24 – 26.

25. Return to the situation described in Exercise 24, and the graph in Figure 20.49. (a) If the conducting loop has a single turn, plot a graph of the emf induced in the loop as a function of time. Use the graph to answer the following questions. If the induced current at $t = 5$ s is directed clockwise, in what direction is the induced current in the loop at (b) $t = 10$ s? (c) $t = 15$ s? (d) $t = 40$ s?

26. Return to the situation described in Exercise 24, and the graph in Figure 20.49. The loop has a resistance of 0.10 Ω, an area of 20 m², and consists of a single turn. (a) What is the magnitude of the magnetic field passing through the loop at $t = 15$ s? Assume that the field is uniform and the direction of the field is perpendicular to the plane of the loop. (b) What is the magnitude of the current in the loop at $t = 15$ s?

27. Figure 20.50 shows a graph of the magnetic flux, as a function of time, through a loop that is at rest. (a) If the induced current in the loop is directed clockwise at the instant corresponding to point 1, at which of the other five points is the induced current directed clockwise? (b) Compare the magnitude of the emf induced in the loop at point 3 and at point 4. (c) At which of the six points does the induced emf have its largest magnitude? Explain. (d) At which of the six points does the induced emf have its smallest magnitude? Explain.

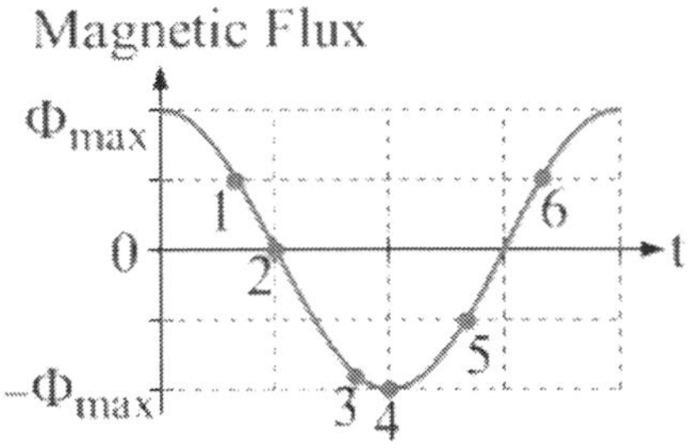

Magnetic Flux

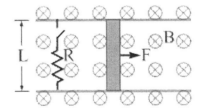

Figure 20.50: The graph shows the magnetic flux through a particular loop, which is at rest, as a function of time. Six points are labeled on the graph. For Exercise 27.

Exercises 28 – 30 are designed to give you practice with motional emf situations.

28. A Boeing 747 has a wingspan of 64 m. (a) If the jet is flying north at a speed of 800 km/h in a region where the vertical component of the Earth's magnetic field is 4.0×10^{-5} T, directed down, what is the motional emf measured from wingtip to wingtip? (b) Which wingtip is at a higher potential?

29. As shown in Figure 20.51, you exert a constant force F to the right on a conducting rod that can move without friction along a pair of conducting rails. The rails are connected at the left end by a resistor and a switch that is initially open. There is a uniform magnetic field directed into the page. You do two experiments, starting with the rod at rest both times. For the first experiment the switch is open, and for the second experiment the switch is closed. If you apply the same force for the same amount of time in each experiment, in which experiment will the rod be moving faster? Explain your answer.

Figure 20.51: The conducting bar can move without friction on the conducting rails. The rails are connected at the left end by a resistor and a switch, which is initially open. For Exercises 29 and 30.

30. As shown in Figure 20.51, you exert a constant force F to the right on a conducting rod of length L that can move without friction along a pair of conducting rails. The rails are connected at the left end by a resistor of resistance R, and we can assume that the resistance of each rail and the rod is negligible in comparison to R. The switch, which is shown open in the diagram, is closed. There is a uniform magnetic field of magnitude B directed into the page, and the rod begins from rest. Both the rails and the magnetic field extend a long way to the right. (a) In terms of the variables specified here, determine the speed of the rod a long time after the rod begins to move. (b) What is the magnitude of the constant force required for the rod to reach a maximum speed of 2.0 m/s, if $L = 20$ cm, $B = 0.20$ T, and $R = 10$ Ω?

Exercises 31 – 35 deal with electric generators.

31. You have a hand-crank generator with a 100-turn coil, of area 0.020 m², which can spin through a uniform magnetic field that has a magnitude of 0.30 T. You can turn the crank at a maximum rate of 3 turns per second, but the hand crank is connected to the coil through a set of gears that makes the coil spin at a rate 8 times larger than the rate at which you turn the crank. What is the maximum emf you can expect to get out of this generator?

32. A particular electric generator consists of a single loop of wire rotating at constant angular velocity in a uniform magnetic field. Figure 20.52 shows three side views of the loop as the loop rotates clockwise, with the loop at various positions. Rank the three positions based on (a) the magnitude of the magnetic flux passing through the loop, and (b) the magnitude of the induced emf in the loop.

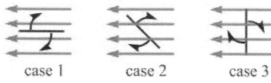

case 1   case 2   case 3

Figure 20.52: Three side views of a loop rotating in a uniform magnetic field. The plane of the loop is parallel to the magnetic field in case 1, at a 45° angle to the field in case 2, and perpendicular to the field in case 3. For Exercise 32.

33. At a particular location, the Earth's magnetic field has a magnitude of 5.0×10^{-5} T. At what angular frequency would you have to spin a 200-turn coil in this field to generate alternating current with a peak emf of 12 V, if each turn of the coil measures 5.0 cm × 5.0 cm?

34. You are designing an electric generator to mimic the alternating current put out by a wall socket in North America, which has a peak voltage of 170 V and a frequency of 60 Hz. The magnets you are using create a uniform magnetic field of 0.10 T. How many turns will you use in your coil, and what area will each of your turns have?

35. **Back emf**. In an electric motor, current flowing through a coil in a magnetic field gives rise to a torque that makes the coil spin. However, a coil spinning in a magnetic field acts as an electric generator, so the coil has an induced emf (known as back emf) that gives rise to an induced current. (a) Based on the principles of physics that we addressed in this chapter, would you expect the induced current to add to the original current in the coil, or subtract from it? Explain. (b) When a motor is first turned on, it takes a few seconds for the coil to reach its operating angular velocity. What happens to the net current in the motor during this start-up period? Is this consistent with a phenomenon you have probably observed, that when a device with a motor, such as an air conditioner, first starts up, the lights in your house dim for a couple of seconds? Note that the compressor in the air conditioner acts as a motor.

Exercises 36 – 40 involve transformers and the distribution of electricity.

36. The power transformer in a typical microwave provides 1000 W at an rms voltage of 2400 V on the secondary side. The primary is connected to a wall socket, which has an rms voltage of 120 V. The primary coil has 100 turns. Assuming the transformer is ideal, determine: (a) the number of turns in the secondary coil, (b) the rms current in the secondary, and (c) the rms current in the primary.

37. The AC adapter for a particular computer is plugged into the wall, so the primary side of the transformer in the adapter is connected to a 60 Hz signal with an rms voltage of 120 V and an rms current of 2.0 A. (a) The secondary of the transformer puts out

alternating current with an rms voltage of 20 V. Assuming the transformer is ideal, what is the rms current in the secondary? (b) After transforming the signal to 20 V AC, the adapter uses capacitors and diodes to turn the signal into direct current electricity at a voltage of 16 V and a current of 5 A. How much power, in this DC electricity, is provided to the computer by the adapter? (c) Use your answers to parts (a) and (b) to explain why, when you hold the adapter in your hand, the adapter feels rather warm.

38. A particular step-down transformer has its primary coil connected to the alternating current from a wall socket. Assuming the transformer is ideal, compare the following: (a) the power in the primary and the power in the secondary, (b) the rms voltage in the primary and the rms voltage in the secondary, (c) the rms current in the primary and the rms current in the secondary, (d) the frequency of the alternating current in the primary and secondary.

39. If you travel from one country to another, such as from the United States to England, you will find that an electronic device designed to plug into a North American wall socket will not plug into a wall socket in England. (a) There is a good reason for this. Explain why it is a good idea that European wall sockets do not directly accept North American plugs, and vice versa. (b) Travelers can purchase a travel adapter, which allows a North American device to be plugged into a European wall socket. In addition to matching the different plugs and sockets, a travel adapter is a transformer. What is the ratio of the number of turns in the two coils in such a travel adapter?

40. A typical pole-mounted transformer, the last step in the power distribution process before electricity is delivered to residential customers, has an rms voltage of 7200 volts on the primary coil. The secondary coil has an rms output voltage of 240 volts, which is delivered to one or more houses. (a) What is the ratio of the number in turns in the primary to the number in the secondary? (b) If the current on the primary side is 6.0 A, and each house requires an rms current of 50 A, how many houses can the transformer supply electricity to?

General problems and conceptual questions

41. A large glass window has an area of 6.0 m^2. The Earth's magnetic field in the region of the window has a magnitude of 5.0×10^{-5} T. (a) What is the maximum possible magnitude of the magnetic flux through the window from the Earth's field? (b) If the flux has a magnitude of 1.0×10^{-4} T m^2, what is the angle between the window's area vector and the Earth's magnetic field?

42. A particular flat conducting loop has an area of 5.0×10^{-3} m^2. The loop is at rest in a uniform magnetic field, directed perpendicular to the plane of the loop, which has a magnitude of 2.0 T. The field has a constant magnitude and direction. (a) Determine the magnitude of the magnetic flux passing through the loop. (b) Determine the magnitude of the emf induced in the loop in a 5.0-second interval.

43. Return to the situation shown in Figure 20.17, which we analyzed qualitatively in Essential Question 20.3. The magnetic field has a magnitude of 2.0 T, the value of v is 6.0 m/s, and the resistance of each single-turn loop is 3.0 Ω. If each small square in the picture measures 10 cm × 10 cm, determine the magnitude and direction of the induced current in (a) loop 1, (b) loop 2, (c) loop 3, and (d) loop 4.

44. A square loop consists of a single turn with a resistance of 4.0 Ω. The loop measures 10 cm × 10 cm, and has a uniform magnetic field passing through it that is directed out of the page. The loop contains a 12-volt battery, connected as shown in Figure 20.53. At the instant shown in the figure, there is no net current in the loop. At what rate is the magnetic field changing, and is the field increasing or decreasing in magnitude?

Figure 20.53: A square loop containing a 12-volt battery has no net current. For Exercise 44.

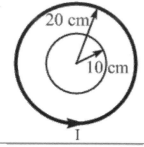

45. Figure 20.54 shows a cross-sectional view of a single-turn conducting loop, with a radius of 10 cm, inside a long current-carrying solenoid. The solenoid has a radius of 20 cm, and has 1000 turns per meter. The current in the solenoid is directed counterclockwise, and is increasing in magnitude. Draw a set of three pictures (Before, After, and To Oppose) to determine the direction of the induced current in the conducting loop.

46. Return to the situation described in Exercise 45, and shown in Figure 20.54. If the current in the solenoid is increasing at the rate of 0.20 A/s, and the loop has a resistance of 4.0 Ω, determine the magnitude of the induced current in the conducting loop.

Figure 20.54: A conducting loop, with a radius of 10 cm, is placed at the center of a long current-carrying solenoid, which has a radius of 20 cm. For Exercises 45 and 46.

47. Figure 20.55 shows a graph of the emf induced in a loop as a function of time. (a) Plot a graph of the corresponding magnetic flux through the loop as a function of time. Your graph should be consistent with Figure 20.54 as well as Faraday's Law. (b) Is there only one possible flux graph for part (a), or are there multiple solutions possible? Explain.

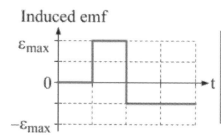

Figure 20.55: A graph of the emf induced in a loop as a function of time. For Exercise 47.

48. You probably have a credit or debit card with a magnetic stripe on the back. The magnetic stripe contains information that is stored in the pattern of magnetic fields on the stripe. When you put your card in a card reader, such as at a checkout counter, you need to slide the card through quickly. Using the principles of physics we have addressed in this chapter, explain how the card reader works, and why sliding the card through the reader slowly does not work.

49. There are many applications of electromagnetic induction, in addition to those we have examined in this chapter. Choose **one** of the following three subjects, and write a couple of paragraphs about how it works, highlighting in particular the role of electromagnetic induction. 1. An electric guitar. 2. A tape recorder. 3. A hard disk in a computer.

50. A biomedical application of electromagnetic induction is magnetoencephalography (MEG), which uses the tiny magnetic fields generated by currents in the brain to create an image of neural activity in the brain. Do some research about MEG, and write a couple of paragraphs describing it.

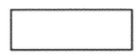

51. As shown in Figure 20.56, a conducting wire loop is placed below a long straight wire that is carrying a current to the right. The current in the wire is increasing in magnitude. (a) In which direction is the induced current in the loop? (b) The loop experiences a net force because of the interaction between the current in the long straight wire and the induced current. In what direction is this force? (c) If the loop moves in response to this net force, explain how this motion is consistent with Lenz's Law.

Figure 20.56: A conducting loop near a current-carrying long straight wire, for Exercise 51. The current in the wire is increasing.

52. There are some flashlights that do not require batteries. Instead, the flashlight has a magnet that moves back and forth through a coil when you shake it. The coil is connected to a capacitor which is in turn connected to a light-emitting diode (LED), which can glow brightly without requiring much current. Explain how such a flashlight works.

53. Figure 20.57 shows a conducting bar of length L that can move without friction on a pair of conducting rails. The rails are joined at the left by a battery of emf ε and a switch that is initially open. The bar, which is initially at rest, has a resistance R, and we will assume the resistance of all other parts of the circuit is negligible. The whole apparatus is in a uniform magnetic field, directed into the page, of magnitude B. Both the rails and the magnetic field extend far to the left and right. In terms of the variables specified above, determine the magnitude and direction of: (a) the current in the circuit immediately after the switch is closed, (b) the net force on the bar immediately after the switch is closed, (c) the current in the circuit a long time after the switch is closed, (d) the net force on the bar a long time after the switch is closed, and (e) the magnitude and direction of the velocity of the bar a long time after the switch is closed.

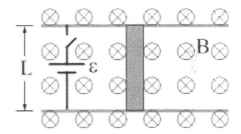

Figure 20.57: The bar, which is initially at rest, can slide without friction on the conducting rails. The rails are connected together on the left by a battery and a switch. For Exercises 53 and 54.

54. Repeat Exercise 53 (illustrated in Figure 20.57), but now use the values $L = 20$ cm, $\varepsilon = 12$ V, $R = 6.0\ \Omega$, and $B = 5.0 \times 10^{-2}$ T.

55. A conducting rod of mass m and length L can slide with no friction down a pair of vertical conducting rails, as shown in Figure 20.58. The rails are joined at the bottom by a light bulb of resistance R. The rails have stops near the bottom to prevent the rod from smashing the bulb. There is a uniform magnetic field of magnitude B directed out of the page. When the rod is released from rest, the force of gravity causes the rod to accelerate down the rails, but the rod soon reaches a terminal velocity (that is, it falls at constant speed). (a) In what direction is the induced current through the light bulb? (b) Sketch two free-body diagrams, one just after the bar is released and one when it has reached terminal velocity. (c) In terms of the variables specified here and g, the magnitude of the acceleration due to gravity, find an expression for the constant speed of the rod. (d) Describe what happens to the brightness of the light bulb as the rod accelerates from rest, and once the rod reaches terminal speed.

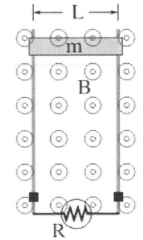

Figure 20.58: When a conducting rod of mass m is released from rest, it slides down the two vertical rails, for Exercise 55.

56. You can demonstrate the effect of eddy currents by dropping a magnet down a pipe made from copper or aluminum, as shown in Figure 20.59. In this case, the magnet is a bar magnet with its north pole at the lower end. Two sections of the tube are outlined, one near the top and one near the bottom. (a) Analyze the lower section of the tube to determine in which direction the eddy currents circulate in that section as the magnet approaches from above. What effect do these eddy currents have on the magnet? (b) Repeat for the upper section of the tube, to determine in which direction the eddy currents circulate in that section as the magnet moves farther away, and the effect of those currents on the magnet.

57. In Exploration 20.5, we explored how eddy currents can be used to bring a train to a stop. In that exploration, the magnetic field from the electromagnets was directed into the page. If the field was reversed, coming out of the page instead, would the train speed up or slow down? Explain.

Figure 20.59: A pipe made of conducting, but non-ferromagnetic, material, through which a magnet has been dropped. For Exercise 56.

58. The secondary of the transformer in a cathode ray tube television provides an rms voltage of 24 kV. The primary is connected to a wall socket, which has an rms voltage of 120 V. The primary coil has 100 turns, and the transformer is ideal. The television has a power rating of 75 W. (a) Find the number of turns in the secondary coil. (b) Find the rms current in the secondary, and (c) the rms current in the primary.

59. As shown in Figure 20.60, the ferromagnetic core of a transformer is generally made from several laminated sheets, which are electrically insulated from one another, rather than being made from a single piece of metal. This method of construction reduces the losses associated with eddy currents in the core itself. Explain how this works.

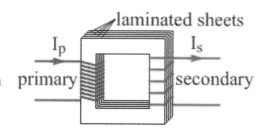

Figure 20.60: The core of a transformer is generally made from several laminated sheets, which are electrically insulated from one another, rather than being made from a single piece of metal. For Exercise 59.

60. A power company experiences losses of 20 kW when it transmits electricity along a particular transmission line at an rms voltage of 250 kV. What would the transmission losses be if the rms voltage was transformed to 500 kV instead?

61. Do some research about the War of Currents, the battle between the advocates of DC and the advocates of AC, which occurred during the late 1800's in the United States, and write a few paragraphs about it.

The photograph shows an ocean wave at Huntington Beach. A water wave is just one example of waves, which is our focus in this chapter. Photo credit: Andrew Schmidt, from publicdomainpictures.net.

Chapter 21 – Waves and Sound

CHAPTER CONTENTS

This chapter marks the beginning of our look at waves, which we will investigate over the next five chapters. Waves are incredibly important to us, and to plants and animals in general. When we talk to a friend, we communicate using sound waves. When we look at a beautiful painting, or simply look at anything, light waves enter our eyes, carrying information and energy. When bats detect moths, or when whales sing songs to one another, waves are involved. When sunlight shines on plants, helping them to grow, it is waves that carry the necessary energy.

In this chapter, we will look at general properties of waves, and we will also discuss sound waves, in particular. An important application of waves, particularly sound waves, is in the creation of music with an instrument. Instruments generally fall into three categories. In wind instruments, the sound waves produced depend on the length of the air column in the instrument. This length can be changed by various methods, such as by opening a particular hole or, in the case of a trombone, changing the length of the instrument itself. In string instruments, waves on the strings excite sound waves in the air. Percussion instruments are similar. On a drum, for instance, striking the drum sets up a particular vibration, which then excites a sound wave. Musical instruments are carefully constructed, exploiting the basic principles of physics that we will investigate in this chapter, to create pleasing tones. Expert musicians know how to play an instrument so as to coax the most pleasing sounds from it, and to put these sounds together in a beautiful way.

21-1 Waves

What is a wave? Put simply, a wave is a disturbance that carries energy from one place to another. Examples include waves on the surface of the ocean, sound waves that carry the sound of chirping birds to your ears on a spring morning, or the waves shown in Figure 21.1.

Figure 21.1: Waves caused by a drop of water hitting the water surface. Photo credit: Jani Ravas, from publicdomainpictures.net.

There are a number of ways to classify waves. One way is the following:

1. *Mechanical waves.* These include water waves, sound waves, and waves on strings, the kind of waves we will investigate in this chapter. Mechanical waves require a medium (such as water, air, or string) through which to travel. There is no net flow of mass through the medium, only energy.

2. *Electromagnetic waves.* Such waves include light, x-rays, microwaves, and radio waves. Electromagnetic waves do not need a medium through which to travel, and thus can travel through a vacuum. We will investigate electromagnetic waves in detail in Chapter 22.

3. *Matter waves.* These waves are associated with objects we often think of as particles, such as electrons and protons. Quantum physics, which we investigate in Chapter 27, tells us that everything, including ourselves, exhibits wave-particle duality, sometimes acting as a wave and sometimes as a particle.

For this chapter, we will confine ourselves to mechanical waves. Another way to classify waves is the following:

1. *Transverse waves.* In these waves, the particles of the medium oscillate in a direction transverse (perpendicular) to the direction the wave travels through the medium. A good example of this is a wave on a string, as shown in Figure 21.2. The various pieces of the string oscillate up and down, while the wave is traveling to the right.

2. *Longitudinal waves.* In these waves, the particles of the medium oscillate along the same direction in which the wave is traveling. A sound wave is a good example, in which air molecules oscillate back and forth along the direction the wave is traveling, as is shown in Figure 21.2. The regions of high density (corresponding to higher then average pressure) and low density (lower pressure) propagate to the right, while the air molecules themselves, on average, oscillate back and forth.

Wavelength and Period

To find the wavelength of a wave, we take a snapshot of the entire wave at one particular instant, as is shown in any of the five images of the string in Figure 21.2. The **wavelength** is the distance from, for instance, one peak to the next peak on the displacement versus position graph. Our symbol for wavelength is λ, the Greek letter lambda. The **period**, T, of the wave is the oscillation period for any particular part of the medium. If we focus on one piece of string, such as the piece colored red in Figure 21.2, and plot its displacement from equilibrium as a function of time, the period is the time between neighboring peaks on the displacement versus time graph.

Figure 21.2: The figure shows fives pictures of a string, separated by equal time intervals. The string has a transverse traveling wave on it. Underneath each picture of the string is a representation of a longitudinal wave, such as a sound wave. The black lines represent the position of molecules of the medium as the wave passes, while the gray lines underneath represent the equilibrium position of these molecules. If time increases down the page, the waves are traveling to the right. If time increases up the page, the motion is to the left.

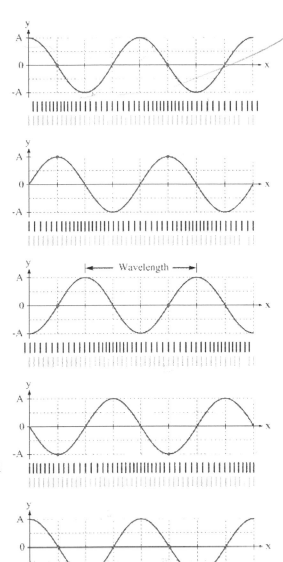

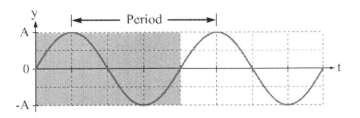

Figure 21.3: A plot of the displacement vs. time for the point on the string that is marked with a dot in Figure 21.2. The shaded region represents the time period covered by the five pictures in Figure 21.2.

Note that we are focusing on a simple kind of wave, a pure sine wave. More complex waves can be built up from sine waves of different wavelength, so our analysis can be generalized to more complicated waveforms.

The wave travels a distance of one wavelength in a time equal to one period. The wave speed is thus the distance over the time, $v = \lambda / T$. Instead of writing the equation in this form, however, we generally use the fact that the frequency, f, of the oscillation is the inverse of the period, $f = 1/T$. This leads to the equation in the box below.

In general, the connection between wave speed, frequency, and wavelength is:

$v = f\lambda$.   (Equation 21.1: **Connecting speed, frequency, and wavelength**)

Equation 21.1, in the form it is presented above, gives the impression that the wave speed is determined by the frequency and wavelength. A better way to write the equation is as:

$$\lambda = \frac{v}{f} ,$$ (alternate form of Equation 21.1)

because the wave speed is set by the properties of the medium (such as the mass and tension of a string), and the frequency of the wave is the frequency of whatever is causing a particular part of the medium to oscillate. The wavelength is then determined by the combination of speed and frequency, as is given above in the alternate form of Equation 21.1.

Related End-of-Chapter Exercise: 42.

Essential Question 21.1: Which representation above, the graph of displacement versus position or the graph of displacement versus time, would you use to find the wave speed?

Answer to Essential Question 21.1: Both representations are needed. The wavelength is found from the graph of displacement versus position, while the period is found from the graph of displacement versus time. Both the wavelength and the period are needed to find the wave speed.

21-2 The Connection with Simple Harmonic Motion

Consider a single frequency transverse wave, like the one shown in Figure 21.4. There is clearly a connection between this wave and simple harmonic motion, because each part of the string experiences simple harmonic motion. Thus, for each part of the string we can use an equation like we used to describe simple harmonic motion, $y = A \cos(\omega t)$ or $y = A \sin(\omega t)$. These equations are good starting points, but they are not sufficient to describe what every point on the string is doing. For instance, at $t = 0$, the equation $y = A \cos(\omega t)$ gives $y = +A$, and only three pieces of the string, marked with dots, have $y = +A$.

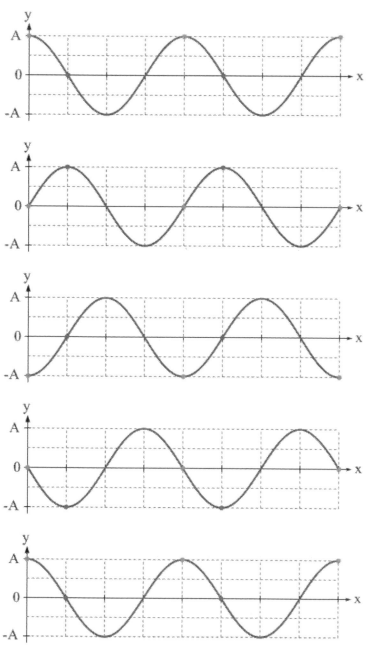

Every point on the string does reach $y = +A$, but not at $t = 0$. For a given point, therefore, we can introduce something called a phase angle, θ, so the equation reflects the position (and the direction of the velocity) of the point at $t = 0$. Thus, each point has an equation of the form $y = A \cos(\omega t + \theta)$, with every point having a unique θ. Having a different equation to describe every point works, but it is cumbersome. Let's see if we can be more efficient in describing the wave mathematically.

First, consider a point on the string just to the right of the left-most point. A point just to the right of the left-most point does exactly what the left-most point does, just at a slightly later time. Thus, θ for that point is a small negative number, reflecting the small delay in the motion compared to the left-most point. Figure 21.5 shows graphs of the displacement versus time for two different sets of points, one set that is at $y = +A$ in the top picture, and the other set which is at $y = 0$, but moving in the positive y-direction, in the top picture. Note that the motion for the second set of points is delayed compared to the first, with a delay proportional to the distance between the points.

Figure 21.4: This figure, like Figure 21.2, shows fives pictures of a string, separated by equal time intervals. Time increases down the page, so the wave is traveling to the right.

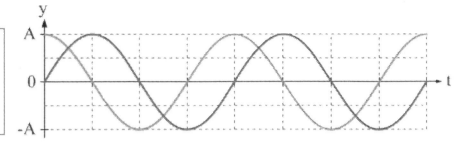

Figure 21.5: Graphs of the displacement versus time for two sets of points, one set initially at $y = +A$, and the other set initially at $y = 0$ and moving in the $+y$ direction.

As x increases, the delay increases, and we find that the phase angle is, in fact, proportional to x, the distance of a point from the left- most point. Thus, we can say that, in the case of a wave traveling in the positive x-direction, $\theta = -kx$, where k is some constant.

If we can identify what the constant k is, we will be finished with our mathematical description. Let's focus now on the point exactly one wavelength to the right of the left-most point. These two points are in phase with one another, which means that whatever one of them does, the other does at the same time. Thus, the equation $y = A \cos(\omega t)$, for the left-most point (at $x = 0$), must agree with the equation for the second point, $y = A \cos(\omega t - k\lambda)$, at $x = \lambda$. Changing the value inside a cosine by a multiple of 2π produces the same result, and because this point is the first point to the right of $x = 0$ that is in phase with the point at $x = 0$, we must have $k\lambda = 2\pi$.

The constant k is known as the **wave number**, and is given by:

$$k = \frac{2\pi}{\lambda}.$$ (Equation 21.2: **the wave number**)

The wave number is, in some sense, the spatial equivalent of the angular frequency. The angular frequency is given by:

$$\omega = \frac{2\pi}{T}.$$ (Equation 21.3: **the angular frequency**)

Note that we now have a single equation that describes the wave. The equation tells us the displacement from equilibrium of each point in the medium at any value of t we might be interested in.

$$y = A\cos(\omega t \pm kx),$$ (Equation 21.4: **Equation of motion for a transverse wave**)

where the plus sign is used when the wave is traveling in the negative x-direction, and the minus sign is used when the wave is traveling in the positive x-direction.

Related End-of-Chapter Exercises: 13, 17, 18, 41.

Essential Question 21.2: In a particular case, the equation of motion of a transverse wave is:
$$y = (8.0 \text{ mm})\cos\left[(3\pi \text{ rad/s})t + (2\pi \text{ rad/m})x\right].$$

Determine the displacement of a point at $x = 2.0$ m at (a) $t = 0$, and (b) $t = 2.5$ s.

Answer to Essential Question 21.2: (a) When $t = 0$, the equation gives, at $x = 2.0$ m,

$y = (8.0 \text{ mm}) \cos [(4.0\pi \text{ rad})] = +8.0 \text{ mm}$, because the cosine of an even multiple of pi is $+1.0$.

Note that if you work this out on a calculator, your calculator needs to be in radians mode.
(b) If $t = 2.5$ s, the equation gives:

$y = (8.0 \text{ mm}) \cos [(7.5\pi \text{ rad}) + (4.0\pi \text{ rad})] = (8.0 \text{ mm}) \cos [(11.5\pi \text{ rad})] = 0$, because the cosine of

$(n + 0.5)\pi$ is always 0. Again, keep your calculator in radians mode to get this answer from your calculator.

21-3 Frequency, Speed, and Wavelength

The speed of a wave depends on the medium the wave is traveling through. If the medium does not change as a wave travels, the wave speed is constant.

As we discussed in section 21-1, in one period, the wave travels one wavelength. Speed is distance over time, so $v = \lambda / T$. The frequency, f, is $1/T$, so the equation relating wave speed, frequency, and wavelength is $v = f\lambda$.

This equation (Equation 21.1), in this form, makes it look like speed is determined by frequency and wavelength, but this is not the case – the speed is determined by the medium. A good example is the speed of a wave on a stretched string. For a string with a tension F_T, a mass m, and a length L, the speed is given by:

$$v = \sqrt{\frac{F_T}{m/L}} = \sqrt{\frac{F_T}{\mu}} , \qquad \text{(Eq. 21.5: } \textbf{The speed of a wave on a string)}$$

where $\mu = m / L$ is the string's mass per unit length.

In general, then, the speed is determined by the medium, the frequency is determined by whatever is producing the wave (such as you, shaking the end of a string back and forth), and the wavelength is determined by Equation 21.1, through the combination of the speed and frequency.

An exception to this rule of thumb is a wave produced by a typical musical instrument. As we will discuss in more detail at the end of the chapter, when you play an instrument you excite a number of frequencies. The size of the instrument (such as the length of a guitar string) then determines the wavelengths of the particular frequencies that are favored. Thus, on a musical instrument, the length of the instrument determines the wavelength, the wave speed is again determined by the properties of the medium, and the combination of the wavelength and speed determines the frequency of the wave.

EXAMPLE 21.3 – Using the equation of motion for a transverse wave

The general equation for a wave traveling on a string is $y = A \cos(\omega t \pm kx)$. In a

particular case, the equation is $y = (8.0 \text{ cm}) \cos [(60 \text{ rad/s})t + (0.50 \text{ rad/m})x]$. Determine:

   (a) the wave's amplitude, wavelength, and frequency.
   (b) the speed of the wave.
   (c) the tension in the string, if the string has a mass per unit length of 0.048 kg/m.
   (d) the direction of propagation of the wave.
   (e) the maximum transverse speed of a point on the string.
   (f) What is the displacement of a point at $x = 2.0$ m when $t = 1.0$ s?

SOLUTION

(a) The amplitude of the wave is the A in the equation, which is whatever is multiplying the cosine. In this particular case, the amplitude is 8.0 cm.

The value of k is whatever is multiplying the x, which is 0.50 rad/m. k is proportional to the inverse of the wavelength, with the wavelength given by $\lambda = 2\pi /(0.50 \text{ m}^{-1}) = 4\pi$ m. Note that we can put in or take out the unit of radians whenever we find it to be convenient.

The value of ω is whatever is multiplying the t, which is 60 rad/s. The frequency, f, is related to the angular frequency, ω, by a factor of 2π:

$$f = \frac{\omega}{2\pi} = \frac{60 \text{ rad/s}}{2\pi} = \frac{30}{\pi} \text{Hz} = 9.5 \text{ Hz} .$$

(b) The speed of the wave can be found from the frequency and wavelength:

$$v = f\lambda = (30/\pi) \text{ Hz} \times (4\pi \text{ m}) = 120 \text{ m/s} .$$

(c) The tension in the string can be found by applying equation 21.5. Solving for tension gives:

$$F_T = v^2\mu = (120 \text{ m/s})^2 \times 0.048 \text{ kg/m} = 690 \text{ N} .$$

(d) In the equation describing the wave, the sign of the x-term is positive. This means that the more positive the x value, the sooner the wave reaches that point, so the wave is traveling in the negative x-direction.

(e) What does "maximum transverse speed" mean? It means the maximum y-direction speed of a point on the string. Any point can be used, because every point experiences the same motion, just at different times. To answer the question, remember that every point on the string is experiencing simple harmonic motion. Thus, this is really a harmonic motion question, not a wave question. Returning to what we learned in chapter 12, the maximum speed of a particle experiencing simple harmonic motion is:

$$v_{\text{max}} = A\omega = (8.0 \text{ cm}) \times 60 \text{ rad/s} = (0.08 \text{ m}) \times 60 \text{ rad/s} = 4.8 \text{ m/s} .$$

(f) We can enter the values of x and t right into the equation, giving:

$$y = (8.0 \text{ cm})\cos\left[(60 \text{ rad/s})t + (0.50 \text{ rad/m})x\right]$$
$$= (8.0 \text{ cm})\cos\left[(60 \text{ rad/s})(1.0 \text{ s}) + (0.50 \text{ rad/m})(2.0 \text{ m})\right]$$
$$= (8.0 \text{ cm})\cos\left[(60 \text{ rad}) + (1.0 \text{ rad})\right] = (8.0 \text{ cm})\cos(61 \text{ rad}) = -2.1 \text{ cm}.$$

Don't forget to put your calculator into radians mode when you do this calculation.

Related End-of-Chapter Exercises: 14, 15, 40.

Essential Question 21.3: Return to Example 21.3. If the wave's equation of motion was unchanged except for a doubling of the angular frequency, which of the answers in Example 21.3 would change, and how would they change?

Answer to Essential Question 21.3: Doubling the angular frequency, ω, causes the frequency to double in part (a). This, in turn, means that that wave speed must double, in part (b). In part (c), the tension is proportional to the square of the speed, so the tension is increased by a factor of 4. In part (e), the maximum transverse speed is proportional to ω, so the maximum transverse speed doubles. Finally, in part (f) we get a completely different value, $y = -0.39$ cm.

21-4 Sound and Sound Intensity

One way to produce a sound wave in air is to use a speaker. The surface of the speaker vibrates back and forth, creating areas of high and low density (corresponding to pressure a little higher than, and a little lower than, standard atmospheric pressure, respectively) in the region of air next to the speaker. These regions of high and low pressure (the sound wave) travel away from the speaker at the speed of sound. The air molecules, on average, just vibrate back and forth as the pressure wave travels through them. In fact, it is through the collisions of air molecules that the sound wave is propagated. Because air molecules are not coupled together, the sound wave travels through gas at a relatively low speed (for sound!) of around 340 m/s. As Table 21.1 shows, the speed of sound in air increases with temperature.

Medium	Speed of sound
Air (0°C)	331 m/s
Air (20°C)	343 m/s
Helium	965 m/s
Water	1400 m/s
Steel	5940 m/s
Aluminum	6420 m/s

Table 21.1: Values of the speed of sound through various media.

For other material, such as liquids or solids, in which there is more coupling between neighboring molecules, vibrations of the atoms and molecules (that is, sound waves) generally travel more quickly than they do in gases. This also is shown in Table 21.1.

Our ears can typically hear sounds with frequencies that lie between 20 Hz and 20 kHz, although the maximum frequency we are sensitive to tends to decrease with age (not to mention with prolonged exposure to high-intensity sound, such as loud music). We are typically most sensitive to sound waves that have frequencies near 2000 Hz, and considerably less sensitive to sounds at the extremes of our frequency range.

Other animals are sensitive to sounds outside of the human range. Elephants, for instance, communicate using sounds below 20 Hz. Because these sounds are not audible to humans, it took scientists quite a while to realize that elephants communicate with one another more than was first thought. Beyond the upper end of the human range, above 20 kHz, we classify sound as **ultrasound**. Dogs, bats, dolphins, and other animals can hear sounds in this range. Ultrasound also has important medical applications, such as in the imaging of a developing fetus in the womb. High-frequency sound waves traveling through the mother's body reflect differently from bone versus tissue, with the pattern of the reflected waves allowing an image to be formed.

Sound intensity

The intensity of a wave is defined to be its power per unit area: $I = P/A$.

For a source broadcasting uniformly in all directions, the wave spreads out like an inflating sphere, so the area in question is the surface area of a sphere.

$$I = \frac{P}{4\pi r^2}.$$ (Eq. 21.6: **Intensity for a source broadcasting uniformly in all directions**)

The sound intensity is proportional to the inverse square of the distance from the source. If the distance is doubled, for instance, the sound intensity decreases by a factor of four. Interestingly, a decrease in sound intensity by a factor of 4 is not perceived as such by the ear-brain system. The ear, in fact, responds logarithmically to sound intensity, and so we use a logarithmic scale for sound that is much like the Richter scale for earthquakes. Just as an earthquake measuring 7.0 on the Richter scale is 10 times more powerful than a quake measuring 6.0, and 100 times more powerful than an earthquake measuring 5.0, a 70 decibel (dB) sound has 10 times the power of a 60 dB sound, and 100 times the power of a 50 dB sound. Every 10 dB represents a change of one order of magnitude in intensity, no matter what the initial intensity is.

For the human ear, the smallest sound intensity that is audible has been determined to correspond to a sound intensity of about $I_0 = 1 \times 10^{-12}$ W/m^2. This value is known as the **threshold of hearing**. On the decibel scale, sounds are viewed in terms of how their intensity compares to the threshold of hearing.

$$\beta = (10 \text{ dB})\log\left(\frac{I}{I_0}\right), \quad \text{(Equation 21.7: \textbf{Absolute sound intensity level, in decibels})}$$

where the equation involves the log in base 10. An interesting reference point on the decibel scale is the **threshold of pain**, the most intense sound an average person can tolerate, which is 120 dB. Substituting 120 dB into equation 21.6, we find that, for the threshold of pain,

$$120 \text{ dB} = (10 \text{ dB})\log\left(\frac{I}{I_0}\right), \text{ so } 12 = \log\left(\frac{I}{I_0}\right).$$

To solve the equation for I, the sound intensity corresponding to the threshold of pain, we do 10 to the power of each side of the equation. 10^x is the inverse function of $\log(x)$, so:

$$10^{12} = 10^{\log\left(\frac{I}{I_0}\right)} = \frac{I}{I_0} = \frac{I}{1 \times 10^{-12} \text{ W/m}^2}.$$

Thus, the intensity of the threshold of pain is 12 orders of magnitude larger than the threshold of hearing, or 1 W/m^2. The most amazing thing about this, however, is what this result tells us about the human ear. The human ear is an incredible instrument, allowing us to hear sounds covering 12 orders of magnitude – that's a factor of 1 trillion.

One convenient feature of the logarithmic scale is that an increase of X decibels corresponds to an increase by a particular factor in intensity, no matter where you start from. This is reflected in the following equation:

$$\Delta\beta = (10 \text{ dB})\log\left(\frac{I_f}{I_i}\right). \quad \text{(Equation 21.8: \textbf{Relative sound intensity level, in decibels})}$$

Related End-of-Chapter Exercises: 18 – 22, 39.

Essential Question 21.4: If a sound intensity level increases by 5 dB, by what factor does the intensity increase?

Answer to Essential Question 21.4: If a sound's intensity level increases by 5 dB, equation 21.7 tells us that: $0.5 = \log\left(\dfrac{I_f}{I_i}\right)$, which gives a ratio of final to initial intensity of $10^{0.5} = 3.2$. In other words, every 5 dB increase corresponds to increasing the sound intensity by a factor of 3.2.

21-5 The Doppler Effect for Sound

We have probably all had the experience of listening to the siren on an emergency vehicle as it approaches us, and hearing a shift in the frequency of the sound when the vehicle passes us. This shift in frequency is known as the Doppler effect, and it occurs whenever the wave source or the detector of the wave (your ear, for instance) is moving relative to the medium the wave is traveling in. Applications of the Doppler effect for sound include Doppler ultrasound, a diagnostic tool used to study blood flow in the heart. There is a related but slightly different Doppler effect for electromagnetic waves, which we will investigate in the next chapter, that has applications in astronomy as well as in police radar systems to measure the speed of a car.

EXPLORATION 21.5 – Understanding the Doppler effect

Let's explore the principles behind the Doppler effect. We will begin by looking at the situation of a stationary source of sound, and a moving observer.

Step 1 – *Construct a diagram showing waves expanding spherically from a stationary source that is broadcasting sound waves of a single frequency. If you, the observer, remain stationary, you hear sound of the same frequency as that emitted by the source. Use your diagram to help you explain whether the frequency you hear when you move toward the source, or away from the source, is higher or lower than the frequency emitted by the source.*

We can represent the expanding waves as a set of concentric circles centered on the stationary source, as in Figure 21.6. This picture shows a snapshot of the waves at one instant in time, but remember that the waves are expanding outward from the source at the speed of sound. If you are stationary at position A, the waves wash over you at the same frequency as they were emitted. If you are at position A but moving toward the source, however, the frequency you observe increases, because you are moving toward the oncoming waves. Conversely, if you move away from the source (and you are traveling at a speed less than the speed of sound), you observe a lower frequency as you try to out-run the waves.

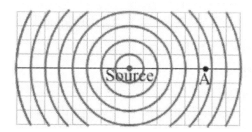

Figure 21.6: Waves emitted by a stationary source expand out away from the source, giving a pattern of concentric circles centered on the source. You, the observer, are at point A. If you are moving, the frequency of the waves you receive depends on both your speed and the direction of your motion.

Step 2 – *Starting with the usual relationship connecting frequency, speed, and wavelength, $f = v / \lambda$, think about whether the observer moving toward or away from a stationary source effectively changes the wave speed or the wavelength. If the speed of sound is v and the observer's speed is v_o, write an equation for the frequency heard by the observer.* As we can see from the pattern in Figure 21.6 above, the wavelength has not changed. What changes, when you move through the pattern of waves, is the speed of the waves with respect to you. When you move toward the source, the effective speed of the waves (the relative speed of the waves with respect to you) is $v + v_o$, while when you move away from the source the wave speed is effectively $v - v_o$. The frequency you observe, f', is thus the effective speed over the wavelength:

$$f' = \frac{v \pm v_o}{\lambda} = \frac{v}{\lambda}\left(\frac{v \pm v_o}{v}\right) = f\left(\frac{v \pm v_o}{v}\right), \quad \text{(Eq. 21.9: Frequency for a moving observer)}$$

where f is the frequency emitted by the source, and where we use the + sign when the observer moves toward the source, and the − sign when the observer moves away from the source.

Step 3 – *Construct a diagram showing waves expanding from a source that is moving to the right at half the speed of sound while broadcasting sound waves of a single frequency. Use your diagram to help you explain whether the frequency you hear when you are stationary is higher or lower than that emitted by the source, when the source is moving toward you and when the source is moving away from you.*

In this situation, the result is quite different from that in Figure 21.6, because each wave is centered on the position of the source at the instant the wave was emitted. Because the waves are emitted at different times, and the source is moving, we get the picture shown in Figure 21.7. To the left of the source, such as at point B, the waves are more spread out. Thus, when the source is moving away from the observer, the observed frequency is less than the emitted frequency. The reverse is true for a point to the right of the source: the waves are closer together than usual, so an observer in this region (such as at point A) observes a greater frequency than the emitted frequency.

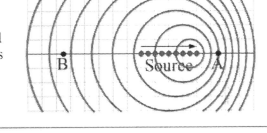

Figure 21.7: When a source of waves is moving relative to the medium, the wave pattern is asymmetric. An observer for which the source moves away observes a lower-frequency wave, while, when the source is moving toward the observer, a higher-frequency wave is observed. In the case shown, the source is moving to the right at half the wave speed.

Step 4 – *Starting with the usual relationship connecting frequency, speed, and wavelength, $f = v / \lambda$, think about whether the source moving toward or away from a stationary observer effectively changes the wave speed or the wavelength. If the speed of sound is v and the source's speed is v_s, write an equation for the frequency heard by the observer.* As we can see from the pattern in Figure 21.7, the movement of the source changes the wavelength. The waves still travel at the speed of sound, however. What changes, when you move through the pattern of waves, is the speed of the waves with respect to you. When the source moves toward the observer, the effective wavelength is $(v − v_s)/f$, while when the source moves away the wavelength is effectively $(v + v_s)/f$. The frequency you observe, f', is thus the speed over the effective wavelength:

$$f' = \frac{v}{\lambda'} = \frac{v}{v \mp v_s} f, \quad \text{(Eq. 21.10: Frequency for a moving source)}$$

where f is the frequency emitted by the source. Use the − sign when the source moves toward the observer, and the + sign when the source moves away from the observer.

Key idea for the Doppler effect: Motion of a source of sound, or motion of an observer, can cause a shift in the observed frequency of a wave. **Related End-of-Chapter Exercises: 23, 24.**

Essential Question 21.5: Is the Doppler effect simply a relative velocity phenomenon? For instance, is the situation of an observer moving at speed v_l toward a stationary source the same as a source moving at speed v_l toward a stationary observer?

Answer to Essential Question 21.5: The Doppler effect for sound (and for all mechanical waves) is not a relative velocity phenomenon. The relative velocity of the source and observer is the same in these two situations, but the observed frequency is different in the two situations. One interesting example is when $v_1 = v$, the wave speed. When the observer moves at speed v toward a stationary source, the observed frequency is twice the emitted frequency. When the source moves at a speed v toward a stationary observer, however, the observed frequency is infinite. We will investigate that situation further in the next section.

21-6 Sonic Booms, and the Doppler Effect in General

Essential Question 21.5 raises the question of what happens when a source of waves travels at the wave speed. We should also consider what happens when the source travels faster than the wave speed.

Let's begin by drawing a diagram like that in Figure 21.7, but with the source traveling to the right at the wave speed. In this special case, in Figure 21.8, because the source keeps up with the waves, the waves pile up at the source, leading to a large amplitude wave that moves with the source. This is known as a **sonic boom**, because a large amplitude corresponds to a loud sound. The observer at position A would hear the sonic boom when the source passed by.

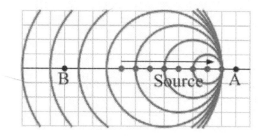

Figure 21.8: When the source moves at the wave speed, the waves pile up on one another at the source, creating a sonic boom.

Let's go further, and see what the picture looks like when the source travels faster than the waves. Figure 21.9 shows what happens when the source travels to the right at twice the wave speed. In this case, the waves pile up along lines that make an angle with the line of travel of the source. This pattern should look familiar to you, given that it looks like the waves left behind by a boat as it travels through water, as in the photograph in Figure 21.10. This tells us that the boat's speed is faster than the speed of the water waves.

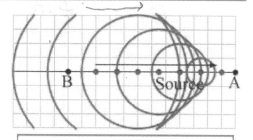

Figure 21.9: When the source moves faster than the waves, the waves create a wake pattern.

In section 21-5, we considered what happens when either the source moves or the observer moves, but not both. Let's now consider what happens in general, when both the source of a wave and the observer are moving with respect to the medium the waves are moving through. The general equation is simply a combination of the equations we derived in section 21-5 for the situations of a moving observer and a moving source.

Figure 21.10: A common example of the situation of a source of waves traveling faster through the medium than the waves themselves is in the wake created when a boat passes through water (here, the Avon Gorge in England). Photo credit: public-domain photo taken by Adrian Pingstone.

The Doppler effect: The Doppler effect describes the shift in frequency of a wave that occurs when the source of the waves, and/or the observer of the waves, moves with respect to the medium the waves are traveling through. The general equation for the observed frequency is:

$$f' = f\left(\frac{v \pm v_o}{v \mp v_s}\right),$$
(Equation 21.11: **The general Doppler equation**)

where f' is the frequency observed by the observer, f is the frequency of the waves emitted by the source, v is the speed of the wave through the medium, v_o is the speed of the observer, and v_s is the speed of the source. In the numerator, use the top (+) sign if the observer moves toward the source, and the bottom (−) sign if the observer moves away from the source. In the denominator, use the top (−) sign if the source moves toward the observer, and the bottom (+) sign if the source moves away from the observer.

EXAMPLE 21.6 – Catching a moth

A particular bat emits ultrasonic waves with a frequency of 56.0 kHz. The bat is flying at 16.00 m/s toward a moth, which is moving at 2.00 m/s away from the bat. The speed of sound is 340.00 m/s. (a) Assuming the moth could detect the waves, what frequency waves would it observe? (b) The waves reflect off the moth and are detected by the bat. What frequency are the waves detected by the bat?

SOLUTION

(a) Here, we use the general Doppler equation, where $f = 56.0$ kHz and $v = 340$ m/s. The observer is the moth, so v_o is 2.00 m/s, and we use the bottom sign (the minus sign) in the numerator because the moth is traveling away from the bat. The bat is the source, so $v_s = 16.00$ m/s, and we use the top sign (the minus sign) in the denominator because the bat is traveling toward the moth. This gives:

$$f' = f\left(\frac{v \pm v_o}{v \mp v_s}\right) = (56.0 \text{ kHz})\left(\frac{340.00 \text{ m/s} - 2.00 \text{ m/s}}{340.00 \text{ m/s} - 16.00 \text{ m/s}}\right) = (56.0 \text{ kHz}) \times 1.0432 = 58.42 \text{ kHz}$$

(b) Again, we use the general Doppler equation. This time, the moth acts as the source (because the moth reflects the waves back to the bat) and the bat is the observer. The frequency emitted by the moth is the frequency we found in part (a). Let's use f'' to denote the frequency of the waves detected by the bat, so $f' = 58.42$ kHz and $v = 340$ m/s. The observer is the bat, so v_o' is 16.00 m/s, and we use the top sign (the plus sign) in the numerator because the bat is traveling toward the moth. The moth is the source, so $v_s' = 2.00$ m/s, and we use the bottom sign (the plus sign) in the denominator because the moth is traveling away from the bat. This gives:

$$f'' = f'\left(\frac{v \pm v_o'}{v \mp v_s'}\right) = (58.42 \text{ kHz})\left(\frac{340.00 \text{ m/s} + 16.00 \text{ m/s}}{340.00 \text{ m/s} + 2.00 \text{ m/s}}\right) = (58.42 \text{ kHz}) \times 1.0409 = 60.8 \text{ kHz}$$

The bat can use the frequency of the detected wave to determine how fast, and in what direction, the moth is flying.

Related End-of-Chapter Exercises: 25 – 27, 46 – 48.

Essential Question 21.6: What happens when a source and observer have identical velocities? Is the observed frequency larger, smaller, or the same as the frequency emitted by the source?

Answer to Essential Question 21.6: When a source and an observer move in the same direction with the same speed, the observed frequency is the same as the frequency emitted by the source.

21-7 Superposition and Interference

What happens when two waves traveling through the same medium encounter one another? In general, we apply the principle of superposition to determine the net displacement of each point in the medium.

> **The principle of superposition:** The net displacement of any point in a medium is the sum of the displacements at that point due to each of the individual waves.

Figure 21.11 shows what happens when two pulses moving in opposite directions along a stretched string meet one another. Both pulses displace the string upward as they travel, so when the peaks of the pulses coincide, the net displacement of the string at that point is equal to the sum of the amplitudes of the pulses. This is known as **constructive interference** - the displacements of the individual waves are in the same direction, and thus add together. An interesting implication of the principle of superposition is that the waves do not change one another's shape as they pass through one another. After passing through, they move away unchanged.

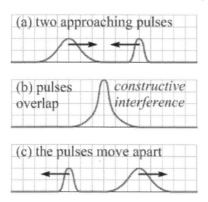

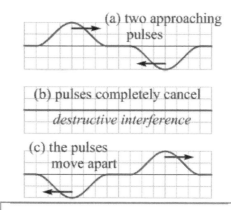

Figure 21.11: The successive images show two pulses moving in opposite directions along a string. At points where the pulses overlap, the net displacement of the string is the sum of the displacements due to the individual waves. In this case, (b) shows **constructive interference**, because the displacements of both pulses are in the same direction.

Figure 21.12: The successive images show two pulses, which are mirror images of one another, moving in opposite directions along a string. In this case, (b) shows completely **destructive interference**, with the pulses exactly canceling one another at the instant they overlap completely.

In Figure 21.12, we see what happens when two pulses that have opposite displacements meet one another while traveling along a stretched string. In this situation, because one pulse is a mirror image of the other, when the pulses coincide the net displacement of the string is zero everywhere, just for an instant. This is known as **destructive** interference, where the displacements of the individual waves are in opposite directions, and thus fully or partly cancel. Once again, after passing through one another, they move away as if they had never met.

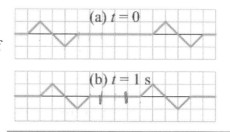

Figure 21.13: Two pulses, one traveling left and one traveling right, on a stretched string. The profile of the string is shown at (a) $t = 0$ and (b) $t = 1.0$ s.

EXPLORATION 21.7 – A process for adding two pulses

Figure 21.13 shows two pulses traveling along a string. The string is shown at two separate times, $t = 0$, and $t = 1.0$ s. We want to know what the string looks like at $t = 4.0$ s, $t = 5.0$ s, and $t = 6.0$ s.

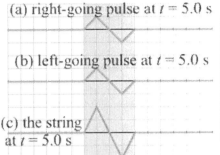

Step 1 – Based on the two pictures in Figure 21.13, determine where each of the two pulses will be at t = 4.0 s. Sketch three diagrams, one above the other. First, sketch a diagram showing the position of the rightward-moving pulse. Second, sketch a diagram of the leftward-moving pulse. Use those two diagrams to determine where the two pulses overlap, and use superposition to draw the string as it looks with both pulses on it. From Figure 21.13, we can see that the pulses travel along the string with a speed of 1 grid unit per second. After three more seconds have passed, the right-going pulse will be three units to the right, and the left-going pulse will be three units to the left, as shown in Figure 21.14. The pulses destructively interfere in the region of overlap, which is shaded in Figure 21.14.

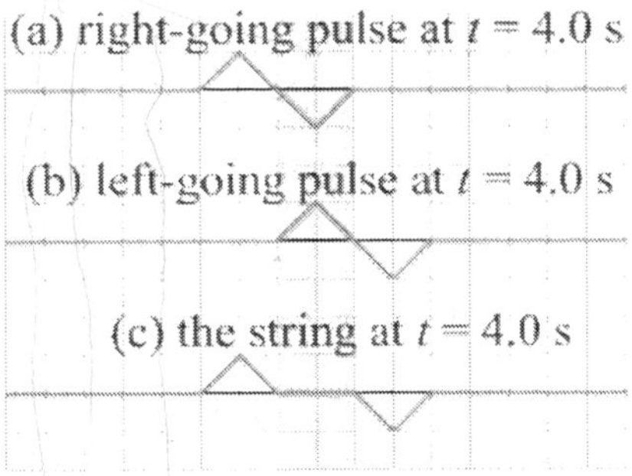

(a) right-going pulse at t = 4.0 s

(b) left-going pulse at t = 4.0 s

(c) the string at t = 4.0 s

Figure 21.14: The two separate pulses, in (a) and (b), and the profile of the string (c), at t = 4.0 s. The region where the pulses overlap is shaded.

Step 2 – Repeat the process outlined in step 1 above, but at t = 5.0 s. Because another second has passed, we slide each pulse over by one unit. At t = 5.0 s, the pulses completely overlap. Because they have identical profiles, the pulses constructively interfere, doubling the displacement at each point in the region of overlap, as shown in Figure 21.15.

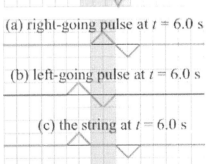

(a) right-going pulse at t = 5.0 s

(b) left-going pulse at t = 5.0 s

(c) the string at t = 5.0 s

Figure 21.15: The two separate pulses, in (a) and (b), and the profile of the string (c), at t = 5.0 s. The region where the pulses overlap is shaded.

Step 3 – Repeat the process outlined in step 1 above, but at t = 6.0 s. An additional second has passed, so we again slide each pulse over by one unit. At t =6.0 s, the net result is the same as at t = 4.0 s, with the two pulses simply swapping positions on the string, as shown in Figure 21.16.

(a) right-going pulse at t = 6.0 s

(b) left-going pulse at t = 6.0 s

(c) the string at t = 6.0 s

Figure 21.16: The two separate pulses, in (a) and (b), and the profile of the string (c), at t = 6.0 s. The region where the pulses overlap is shaded.

Key idea: When two waves meet, apply the principle of superposition. A useful method is to sketch each of the waves separately, and then add the displacements to find the net displacement in the region where the waves overlap. **Related End-of-Chapter Exercises: 1 – 4.**

Interference in Two Dimensions

When two sources emit identical waves, an interesting pattern is created near the sources because of the interference that takes place. The type of interference that takes place at a point depends on the **path-length difference**: the difference between the distance from one source to the point and the distance from the second source to the point. When the sources emit identical waves, any point that is equidistant from the two sources (that is, having a path-length difference of zero), experiences constructive interference. These are not the only places where constructive interference occurs – any point at which the path-length difference is an integral number of wavelengths also experiences constructive interference. Destructive interference, on the other hand, occurs at points that are an integral number of wavelengths, plus half a wavelength, farther from one source than the other. We will discuss these ideas in more detail in chapter 24.

Essential Question 21.7: In the picture of the string in Figure 21.12(b), the string is completely flat. In Figure 21.12(c), the two pulses re-emerge from the flat string. How is this possible? For instance, where is the energy, in (b), necessary to re-form the two pulses?

Answer to Essential Question 21.7: How does the flat string pictured in Figure 21.12(b) differ from a regular flat string, from which no pulses would emerge? What is not obvious from the static image shown in Figure 21.12(b) is that the string, where the pulses overlap, is moving. Some sections of the string are moving down, while others are moving up, with the various parts of the string moving with velocities that are just right to re-create the two pulses properly. Thus, the energy needed to re-form the pulses is in the kinetic energy of various parts of the string.

21-8 Beats; and Reflections

When you listen to two sound waves of similar, but different, frequency, you generally hear the sound rising and falling in intensity, typically at the rate of a few cycles per second. This phenomenon is known as **beats,** and it is caused by interference between the two waves. Let's say the waves are initially in phase, with their peaks coinciding. The waves interfere constructively, producing a large-amplitude sound. Because the waves have different frequencies, however, they gradually drift out of phase. Eventually, the peak from one wave lines up with the trough (negative-displacement peak) in the other wave, leading to destructive interference and, when the interference is completely destructive, no sound. The larger the difference between the two frequencies, the faster the waves drift out of phase with one another. The phase difference continues to grow, but this eventually leads to peaks in the two waves lining up again. This cycle is demonstrated in Figure 21.17.

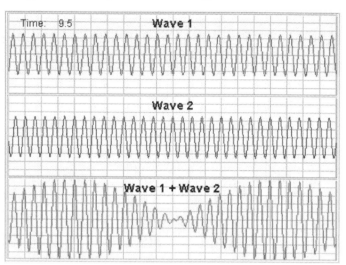

Figure 21.17: The phenomenon of beats is caused by interference between two waves that have different frequencies. The two individual waves are shown at the top and middle, while the superposition of the two waves is shown at the bottom. The rise and fall in the amplitude of the resultant wave is what we hear as beats.

The beat frequency, which is the frequency at which the intensity oscillates, is simply the difference between the frequencies of the two waves.

$$f_{beat} = f_{high} - f_{low}.$$ (Equation 21.12: **the beat frequency**)

String musicians can even tune their instrument using beats, by playing two strings at once and adjusting the tension in one string (which adjusts the frequency of the string). When the beats disappear, the frequencies of the two strings are equal.

Reflections

When a wave traveling along a string encounters the end of the string, the wave reflects. Exactly how the wave reflects depends on whether the end of the string is tied down or loose (or even something in between, such as tied to a spring, but we will consider only the two extremes).

On stringed instruments, for instance, the strings are fixed at the ends. The leading (right-most) edge of an upward going pulse, like that shown in Figure 21.18(a), propagates to the right along the string by each part of the string successively pulling up on the next part of the string. This propagation method works until the pulse reaches the end of the string, which is tied down. The part of the string next to the right end pulls up on the end, but the end does not move. Instead, by Newton's third law, the end exerts a downward force on the piece of the string next to it, leading to an inverted pulse traveling back along the string, as shown in part (e) of Figure 21.18.

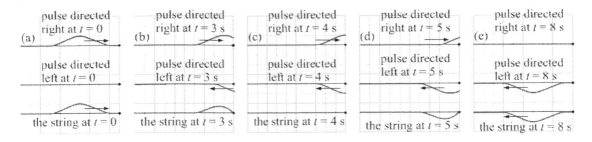

Figure 21.18: When a wave reflects from a fixed end, it reflects upside down.

Note that the string is completely flat in Figure 21.18 (c), halfway through the reflection of the pulse. This is caused by completely destructive interference taking place between the first half of the pulse, which has been inverted and is moving left, and the second half of the pulse, which is still upright and moving right. Figure 21.19 shows a way to visualize the reflection, as if a pulse directed right on the string is interfering with a mirror-image pulse, which is inverted, directed left on the string. Superposition can only work on the string itself, so we don't have to worry about any areas of overlap of the two pulses that are to the right of the end of the string.

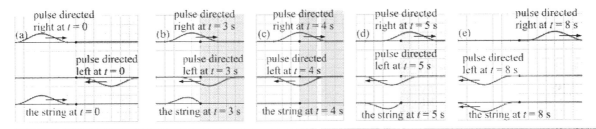

Figure 21.19: Reflection from a fixed end can be visualized as interference between a right-moving pulse, and an inverted copy of the pulse that is moving left. We are imagining the pulses existing to the right of the end of the string (in the shaded region), even though they cannot do so.

If the end of a string is not tied down, but is free to move, it is known as a **free end**. When a wave reflects from a free end, the end responds to the wave by moving, and the wave reflects without being inverted. Figure 21.20 shows the process for a pulse, which we can visualize as if a pulse directed right on the string is interfering with a mirror-image pulse directed left. Note that, in Figure 21.20(c), the end of the string is displaced by twice the amplitude of the pulse, because of constructive interference between the half of the pulse that has been reflected and is moving to the left, and the other half which is still moving to the right.

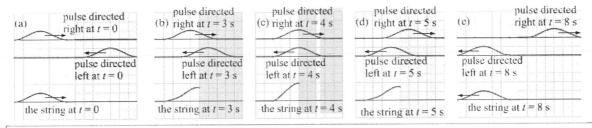

Figure 21.20: Reflection from a free end can be visualized as interference between a right-moving pulse, and an exact copy of the pulse that is moving left.

Reflection of waves: If a wave reflects from a fixed boundary of a medium, the reflected wave is inverted. If, instead, a wave reflects from a free boundary, such as the free end of a string, the reflected wave reflects without being inverted (that is, the reflected wave is upright).
Related End-of-Chapter Exercises: 9, 10, 53 – 55.

Essential Question 21.8: You hear a beat frequency of 6 Hz when you play two guitar strings simultaneously. If one string has a frequency of 330 Hz, what is the frequency of the other string?

Answer to Essential Question 21.8: Because the beat frequency is 6 Hz, we know that the two frequencies differ by 6 Hz. If one string is 330 Hz, the other string is either 336 Hz (6 Hz higher) or 324 Hz (6 Hz lower).

21-9 Standing Waves on Strings

In sections 21-9 and 21-10, we will discuss physics related to musical instruments, focusing on stringed instruments in this section and wind instruments in section 21-10.

Some stringed instruments (such as the harp) have strings of different lengths, while others (such as the guitar) use strings of the same length. We can apply the same principles to understand either kind of instrument. Consider a single string of a particular length that is fixed at both ends. The string is under some tension, so that when you pluck the string it vibrates and you hear a nice sound from the string, dominated by one particular frequency. How does that work?

When you pluck the string, you send waves of many different frequencies along the string, in both directions. Each time a wave reaches an end, the wave reflects so that is inverted. All of these reflected waves interfere with one another. For most waves, after multiple reflections the superposition leads to destructive interference. For certain special frequencies, for which an integral number of half-wavelengths fit exactly into the length of the string, the reflected waves interfere constructively, producing large-amplitude oscillations on the string at those frequencies.

These special frequencies produce **standing waves** on the string. Identical waves travel left and right on the string, and the superposition of such identical waves leads to a situation where the positions of zero displacement (the **nodes**) remain fixed, as do the positions of maximum displacement (the **anti-nodes**), so the wave appears to stand still. Figure 21.21 shows the left and right-moving waves on the string, and their superposition, which is the actual string profile, for the lowest-frequency standing wave on the string at various times.

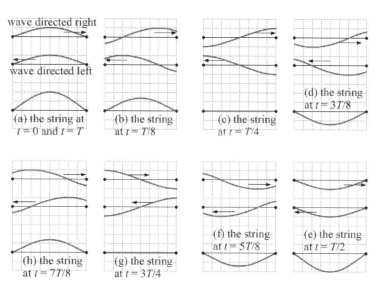

Figure 21.21: The string profile for the lowest-frequency standing wave (the **fundamental**) on the string at $t = 0$, and at regular time intervals after that, showing how the identical left and right-moving waves combine to form a standing wave. Go clockwise around the diagram to see what the string looks like as time goes by.

For a string fixed at both ends: The standing waves have a node (a point of zero displacement) at each end of the string. The various wavelengths that correspond to the special standing-wave frequencies are related to L, the length of the string, by:

$$n\frac{\lambda_n}{2} = L, \qquad \text{so} \qquad \lambda_n = \frac{2L}{n}, \text{ where } n \text{ is an integer.}$$

Using Equation 21.1, $v = f\lambda$, the particular frequencies that tend to be excited on a stretched string are:

$$f_n = \frac{v}{\lambda_n} = \frac{nv}{2L}, \quad \text{(Eq. 21.13: \textbf{Standing-wave frequencies for a string fixed at both ends})}$$

The lowest-frequency standing wave on the string, corresponding to $n = 1$, is known as the **fundamental**. The other frequencies, or **harmonics**, are simply integer multiples of the fundamental. In general, when you pluck a string, the dominant sound is the fundamental, but the harmonics make the sound more pleasing than what a single-frequency note sounds like. Figure 21.22 shows the standing wave patterns for the fundamental and the two lowest harmonics.

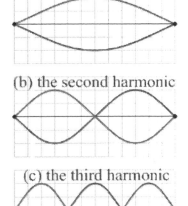

(a) the fundamental

(b) the second harmonic

(c) the third harmonic

Figure 21.22: The standing wave patterns for the fundamental and the second and third harmonics, for a string fixed at both ends.

EXAMPLE 21.9 – Waves on a guitar string

A particular guitar string has a length of 72 cm and a mass of 6.0 grams.
(a) What is the wavelength of the fundamental on this string?
(b) If you want to tune that string so its fundamental frequency is 440 Hz (an A note), what should the speed of the wave be?
(c) When the string is tuned to 440 Hz, what is the string's tension?
(d) Somehow, you excite only the third harmonic, which has a frequency three times that of the fundamental. At $t = 0$, the profile of the string is shown in Figure 21.23, with the middle of the string at its maximum displacement from equilibrium. What is the oscillation period, T?

Figure 21.23: The string profile at $t = 0$ when the third harmonic has been excited on the string, with the middle of the string at its maximum displacement from equilibrium.

SOLUTION

(a) For the fundamental, exactly half a wavelength fits in the length of the string. Thus, the wavelength is twice the length of the string: $\lambda = 144$ cm $= 1.44$ m.

(b) Knowing the frequency and the wavelength, we can determine the wave speed:
$v = f\lambda = (440$ Hz$)(1.44$ m$) = 634$ m/s.

(c) Knowing the speed, we can use Equation 21.5, to find the tension in the string.

$$v = \sqrt{\frac{F_T}{(m/L)}}, \quad \text{so} \quad F_T = v^2 \times \frac{m}{L}.$$

In this case, we get: $F_T = v^2 \times \dfrac{m}{L} = (633.6 \text{ m/s})^2 \times \dfrac{0.006 \text{ kg}}{0.72 \text{ m}} = 3350$ N.

(d) The fundamental frequency is 440 Hz, so the third harmonic has a frequency of 1320 Hz, three times that of the fundamental. The period is the inverse of the frequency, so:

$$T = \frac{1}{f} = \frac{1}{1320 \text{ Hz}} = 760 \text{ } \mu s.$$

Related End-of-Chapter Exercises: 28, 29, 36, 59.

Essential Question 21.9: Return to the situation discussed in Example 21-9. Figure 21.23 shows the string profile at $t = 0$. Show the string profile at times of $t = T/4$, $T/2$, $3T/4$, and T.

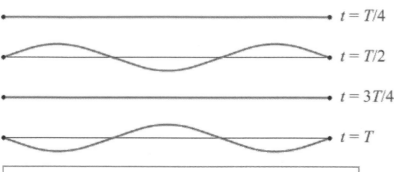

Answer to Essential Question 21.9:
Every half-period, the string profile is inverted, so at $t = T/2$ the string profile is inverted compared to what it is at $t = 0$, and after a full period ($t = T$), the profile is the same as that as $t = 0$. Halfway between these positions, the string profile is flat, as shown in Figure 21.24.

Figure 21.24: The string profile at (a) $t = T/4$, (b) $t = T/2$, (c) $t = 3T/4$, and (d) $t = T$, for the third harmonic.

21-10 Standing Waves in Pipes

Many musical instruments are made from pipes. Such instruments are known as wind instruments. A flute, for instance, is a single pipe in which the effective length can be changed by opening one of several holes in the pipe. In a trombone, the effective length is changed by sliding a tube in or out. In a pipe organ, in contrast, many different pipes, of fixed length, are used, with each pipe having a different fundamental frequency. The connection between all of these instruments is that the effective length of the tube determines the sound the pipe makes.

In contrast with a string instrument, in which a vibrating string sets up a sound wave in the air, the wave in a wind instrument is already a sound wave in a column of air, some of which escapes to make an audible sound. As with strings, however, standing waves produced by reflected waves determine the fundamental frequency of the sound wave produced by a particular pipe. Note that pipes can have both ends open, or have one end open and one end closed. For a sound wave, the open end of a pipe is like a free end, while the closed end of a pipe is like a fixed end. Thus, a pipe with only one end open sounds quite different from a pipe with both ends open, even if the tubes have the same length, because of the different standing waves that are produced by the different reflections in these pipes.

Because an open end acts like a free end for reflection, the standing waves for a pipe that is open at both ends have anti-nodes at each end of the pipe. We can satisfy this condition with standing waves in which an integral number of half-wavelengths fit in the pipe, as shown in parts (a) – (c) of Figure 21.25. This leads to the same equation for standing waves that we had in section 21-9, for the string fixed at both ends.

For a pipe open at both ends: The standing waves produced always have an anti-node at each end of the pipe. The frequencies that produce standing waves in such a pipe are:

$$f_n = \frac{v}{\lambda_n} = \frac{nv}{2L}, \quad \text{(Eq. 21.14: Standing-wave frequencies for a pipe open at both ends)}$$

where n is an integer, and L is the effective length of the pipe.

Because an open end acts like a free end, while a closed end acts like a fixed end, the standing waves for a pipe that is open at only one end have anti-nodes at the open end and nodes at the closed end. We can satisfy this condition with standing waves in which an odd integer number of quarter-wavelengths fit in the pipe, as shown in parts (d) – (f) of Figure 21.25. This leads to new equation for the standing-wave frequencies.

For a pipe open at one end only: The standing waves produced have an anti-node at the open end and a node at the closed end. The frequencies that produce standing waves in such a pipe are:

$$f_n = \frac{nv}{4L},$$ (Eq. 21.15: **Standing-wave frequencies for a pipe open at one end**)

where n is an odd integer, and L is the effective length of the pipe.

Figure 21.25: (a) – (c) A representation of the standing waves in a pipe that is open at both ends, showing the fundamental (a), and the second (b) and third (c) harmonics. (d) – (f) A similar representation for a pipe that is closed at one end only, showing the fundamental (d), and the two lowest harmonics (e) and (f). For a pipe closed at one end only, the harmonics can only be odd integer multiples of the fundamental. Note that the waves in the pipe are sound waves, which are longitudinal waves. This representation shows the maximum displacement from equilibrium for the air molecules as a function of position along each of the pipes. The standing waves oscillate between the profile shown in red and the profile shown in blue.

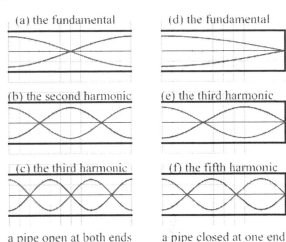

a pipe open at both ends   a pipe closed at one end

EXAMPLE 21.10 – Waves in a pipe

A particular organ pipe has a length of 72 cm, and it is open at both ends. Assume that the speed of sound in air is 340 m/s.

(a) What is the wavelength of the fundamental in this pipe?

(b) What is the corresponding frequency of the fundamental?

(c) If one end of the pipe is now covered, what are the wavelength and frequency of the fundamental?

SOLUTION

(a) For the fundamental, exactly half a wavelength fits in the length of the pipe. Thus, the wavelength is twice the length of the pipe: $\lambda = 2 \times 72$ cm = 144 cm = 1.44 m.

(b) Knowing the speed of sound and the wavelength, we can determine the frequency:

$$f = \frac{v}{\lambda} = \frac{340 \text{ m/s}}{1.44 \text{ m}} = 236 \text{ Hz}.$$

(c) Covering one end of the pipe means the pipe is open at one end only, so now, for the fundamental, only one-quarter of a wavelength fits in the pipe rather than half a wavelength. This doubles the wavelength of the fundamental to 2.88 m. Doubling the wavelength reduces the frequency by a factor of two, so the new fundamental frequency is 118 Hz.

Related End-of-Chapter Exercises: 31, 33, 37, 38.

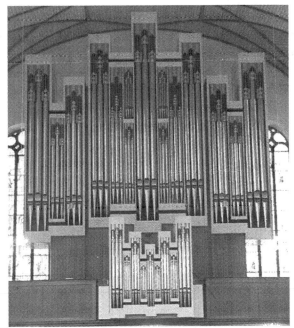

Figure 21.26: The photograph shows a pipe organ in Katharinenkirche, Frankfurt am Main, Germany. Each pipe has a unique frequency.
Photo credit: Wikimedia Commons.

Essential Question 21.10: Musical instruments made from pipes have a variety of pipes or one variable–length pipe. What happens to the fundamental frequency as the pipe length increases?

Answer to Essential Question 21.10: The longer the pipe, the longer the wavelength of the fundamental. Wavelength is inversely proportional to frequency, so the longer the pipe, the smaller the frequency.

Chapter Summary

Essential Idea: Waves.

A wave is a way to transfer energy from one place to another without needing a net flow of material. Waves are in integral part of the way we communicate, whether it be the signals that are picked up by our cell phones, and turned into recognizable speech by the phone's circuitry and speaker, or the light that brings the world to our eyes.

Types of Waves

In this chapter, we dealt with **mechanical waves**, which need a medium through which to travel. Such waves can be **transverse**, in which the particles of the medium oscillate in a direction perpendicular to the direction the wave travels, or **longitudinal**, in which the particles of the medium oscillate along the same direction as the direction the wave travels. A wave on a string is generally transverse, while sound waves are longitudinal.

The wave equation

In general, the relationship between wave speed, v, frequency, f, and wavelength, λ, is:

$$v = f\lambda .$$
(Equation 21.1: **Connecting speed, frequency, and wavelength**)

Equation of motion for a single-frequency transverse wave

In general, the displacement of any point in the medium, at any instant in time, when a single-frequency transverse wave is propagating through the medium in the x-direction, is given by an equation of the form:

$$y = A\cos(\omega t \pm kx),$$
(Equation 21.4: **Equation of motion for a transverse wave**)

where the plus sign is used when the wave is traveling in the negative x-direction, and the minus sign is used when the wave is traveling in the positive x-direction.

The wave number, k, is related to the wavelength, λ, in the same way that the angular frequency, ω, is related to the period, T:

$$k = \frac{2\pi}{\lambda} .$$
(Equation 21.2: **the wave number**)

$$\omega = \frac{2\pi}{T} .$$
(Equation 21.3: **the angular frequency**)

Wave speed

In general, the wave speed is determined not by the frequency and wavelength, but by properties of the medium itself. For example, the speed of a wave on a string is determined by the tension in the string, F_T, and the mass per unit length, μ:

$$v = \sqrt{\frac{F_T}{m/L}} = \sqrt{\frac{F_T}{\mu}} ,$$
(Eq. 21.5: **The speed of a wave on a string**)

Sound Intensity

Intensity is the power per unit area: $I = P/A$. Because of the way the human ear responds to sound, we generally use a logarithmic scale to measure the intensity level of a sound:

$$\beta = (10 \text{ dB}) \log\left(\frac{I}{I_0}\right), \quad \text{(Equation 21.7: \textbf{Absolute sound intensity level, in decibels}})$$

where the reference intensity $I_0 = 1 \times 10^{-12} \text{ W/m}^2$ is known as the threshold of hearing, and the log is in base 10.

The Doppler Effect

The Doppler effect describes the shift in frequency of a wave that occurs when the source of the waves, and/or the observer of the waves, moves with respect to the medium the waves are traveling through. If the source emits a frequency f, the frequency f' received by the observer is:

$$f' = f\left(\frac{v \pm v_o}{v \mp v_s}\right), \quad \text{(Equation 21.11: \textbf{The general Doppler equation}})$$

where v is the speed of the wave through the medium, v_o is the speed of the observer, and v_s is the speed of the source. In the numerator, use the top (+) sign if the observer moves toward the source, and the bottom (−) sign if it moves away. In the denominator, use the top (−) sign if the source moves toward the observer, and the bottom (+) sign if it moves away.

Superposition and interference

When two or more waves overlap, we find the net effect by applying the **principle of superposition**: the net displacement of any point in a medium is the sum of the displacements at that point due to each of the individual waves. If the displacements of the individual waves are in the same direction at a point, we say that the waves experience **constructive interference**, leading to a large net displacement at that point. If the individual displacements are in opposite directions, **destructive interference** occurs, which means that the net displacement is small.

Beats

One example of superposition is when two waves of different frequencies interfere, leading to oscillations in the amplitude of the resultant wave. This is known as beats. The frequency at which the amplitude oscillates is the difference between the two frequencies.

$$f_{beat} = f_{high} - f_{low}. \quad \text{(Equation 21.12: \textbf{the beat frequency}})$$

Standing waves

Standing waves are waves in which the **nodes** (points of zero displacement) and the **antinodes** (points of maximum displacement) remain at rest. Standing waves are generally produced by two identical waves traveling in opposite directions through a medium, and they describe the waves produced by string and wind instruments.

$$f_n = \frac{v}{\lambda_n} = \frac{nv}{2L}, \quad \text{(\textbf{Standing-wave frequencies for strings and for pipes open at both ends}})$$

where L is the length of the string or pipe, v is the wave speed, and n is any integer. The lowest-frequency standing wave (for $n = 1$) is known as the **fundamental**, while the others are known as **harmonics**. Thus, harmonics are integer multiples of the fundamental.

$$f_n = \frac{nv}{4L}, \quad \text{(Eq. 21.15: \textbf{Standing-wave frequencies for a pipe open at one end only}})$$

where n is any odd integer.

End-of-Chapter Exercises

Exercises 1 – 12 are primarily conceptual questions designed to see whether you understand the main concepts of the chapter. For Exercises 1 – 4, the corresponding figure shows the profile of a string at _t_ = 0 and at _t_ = 1.0 s, as two pulses approach one another.

1. Two pulses travel toward one another, as shown in Figure 21.27. Sketch the profile of the string at (a) _t_ = 4.0 s, (b) _t_ = 5.0 s, and (c) _t_ = 6.0 s.

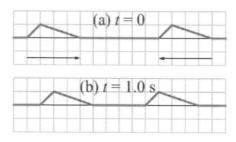

Figure 21.27: Two pulses approach each other along a string, for Exercise 1.

2. Two pulses travel toward one another, as shown in Figure 21.28. Sketch the profile of the string at (a) _t_ = 4.0 s, (b) _t_ = 5.0 s, and (c) _t_ = 6.0 s.

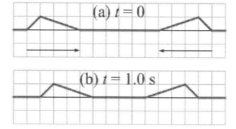

Figure 21.28: Two pulses approach each other along a string, for Exercise 2.

3. Two pulses travel toward one another, as shown in Figure 21.29. Sketch the profile of the string at (a) _t_ = 4.0 s, (b) _t_ = 5.0 s, and (c) _t_ = 6.0 s.

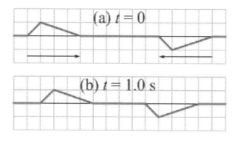

Figure 21.29: Two pulses approach each other along a string, for Exercise 3.

4. Two pulses travel toward one another, as shown in Figure 21.30. Sketch the profile of the string at (a) _t_ = 4.0 s, (b) _t_ = 5.0 s, and (c) _t_ = 6.0 s.

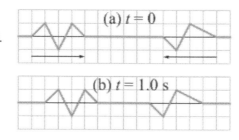

Figure 21.30: Two pulses approach each other along a string, for Exercise 4.

5. Two identical speakers, which are separated by a distance of 7.2 m, are pointed at one another. The speakers, which are in phase with one another, broadcast identical, single-frequency sound waves. There is one point on the line joining the speakers which always experiences constructive interference no matter what the frequency of the identical waves emitted by the speakers is. Where is this point? Explain why the interference is always constructive there.

6. Return to the situation discussed in Exercise 5. If the speed of sound is 340 m/s and the frequency of the waves emitted by each speaker is 170 Hz, find the location of all points along the line between the speakers at which the interference is (a) completely constructive, and (b) completely destructive.

7. Two pulses are traveling along a string, as shown in Figure 21.31. A particular point on the string is marked with a black dot. Plot the displacement of that point as a function of time, over the time interval $t = 0$ to $t = 8.0$ s.

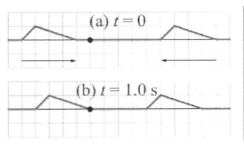

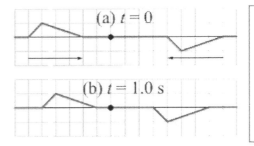

Figure 21.31: Two pulses travel along a string. A particular point on the string is marked with a black dot. For Exercise 7.

8. Two pulses are traveling along a string, as shown in Figure 21.32. A particular point on the string is marked with a black dot. Plot the displacement of that point as a function of time, over the time interval $t = 0$ to $t = 8.0$ s.

Figure 21.32: Two pulses travel along a string. A particular point on the string is marked with a black dot. For Exercise 8.

9. You have four tuning forks, with frequencies of 440 Hz, 445 Hz, 448 Hz, and 452 Hz. By using two tuning forks at a time, how many different beat frequencies can you produce, and what are the numerical values of these frequencies?

10. When you strike two tuning forks and listen to them both at the same time, you hear beats with a beat frequency of 6 Hz. If one tuning fork has a frequency of 512 Hz, what is the frequency of the other?

11. The profile of a string that supports a particular standing wave is shown at $t = 0$ in Figure 21.33. The string is fixed at both ends. At $t = 0$, the standing wave is at its maximum displacement from equilibrium. The standing wave is created by two identical traveling waves on the string, one moving to the right and the other to the left. (a) What is the amplitude of each of these traveling waves? (b) Sketch the profile of the string one-quarter of a period after $t = 0$. (c) Sketch the right-going and left-going waves one-quarter of a period after $t = 0$. Hint: the superposition of these two waves should give the profile in part (b).

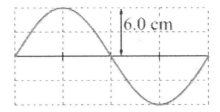

Figure 21.33: The profile of a string, at $t = 0$, that is fixed at both ends. The wave on the string is a standing wave, and the situation shown in the diagram shows the standing wave at its maximum displacement from equilibrium. For Exercise 11.

12. As you are walking along the street, a car blaring loud music passes you. As the car drives away from you, you recognize the music but you realize that it sounds funny. What is the problem?

Exercises 13 – 17 involve applying the equation of motion for a transverse wave.

13. The equation of motion for a particular transverse wave is
$$y = (7.0 \text{ mm}) \sin\left[(4\pi \text{ rad/s})t + (\pi \text{ m}^{-1})x\right].$$
Determine the wave's (a) amplitude, (b) angular frequency, (c) frequency, (d) wavelength, and (e) velocity.

14. For a particular transverse wave that travels along a string that lies on the x-axis, the equation of motion is $y = (6.0 \text{ cm}) \cos\left[(50 \text{ rad/s})t - (0.25 \text{ m}^{-1})x\right]$. Determine (a) the wave's amplitude, wavelength, and frequency, (b) the speed of the wave, (c) the tension in the string, if the string has a mass per unit length of 0.040 kg/m, (d) the direction of propagation of the wave, (e) the maximum transverse speed of a point on the string, (f) the displacement of a point at $x = 1.0$ m when $t = 2.0$ s.

15. The equation of motion for a particular wave traveling along a string along the x-axis is $y = A\cos\left[\omega t + (4.0 \text{ m}^{-1})x\right]$. The tension in the string is 34 N, and the string has a mass per unit length of 0.050 kg/m. The maximum transverse speed of a point on the string is 25 cm/s. Determine (a) the angular frequency, ω, and (b) the amplitude, A, of the wave.

16. At a time of $t = 0$, the profile of part of a string is shown in Figure 21.34. The wave on the string is traveling in the +x direction (to the right) at a speed of 20 cm/s. Write out the equation of motion for the wave.

Figure 21.34: The profile of part of a string at $t = 0$. The wave on the string is traveling to the right at 20 cm/s. For Exercise 16.

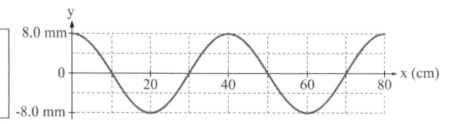

17. A graph of the motion of one point on a string (specifically, the point at $x = 0$), as a function of time is shown in Figure 21.35. The wave is traveling in the negative x-direction on a string that has a tension of 32 N, and with a mass per unit length of 60 grams per meter. Determine (a) the frequency of the wave, (b) the speed of the wave, (c) the wavelength, and (d) the expression for the wave's equation of motion.

Figure 21.35: A graph of the motion of one point on a string, at $x = 0$. The wave on the string is traveling in the negative x-direction. For Exercise 17.

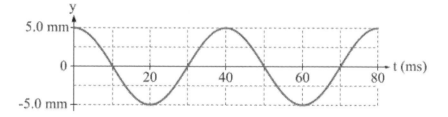

Exercises 18 – 22 involve sound, sound intensity, and the decibel scale.

18. You are listening to the radio when one of your favorite songs comes on, so you turn up the volume. If you managed to increase the sound intensity by 15 dB, by what factor did the intensity of the sound, in W/m², increase?

19. You are working in a room in which the sound intensity is 75 dB. What is the corresponding intensity, in W/m²?

20. When you apply the brakes on your car, they happen to squeak, emitting a 70 dB sound as observed by you sitting in the driver's seat of the car. When you sound the car's horn, however, you observe an 80 dB sound. As you are driving, a dog runs into the road in

front of your car, so you apply the brakes and sound the horn simultaneously. (Fortunately, the dog escapes unharmed.) Do you observe a 150 dB sound while you are stopping, with the brakes squeaking and the horn sounding together? Explain your answer, being as quantitative as possible.

21. When you stand 2.0 m away from a speaker that is emitting sound uniformly in all directions, the sound intensity you observe is 90 dB. What is the sound intensity at a distance of (a) 1.0 m from the speaker, and (b) 4.0 m from the speaker?

22. You are observing fishermen illegally catching fish by using a small explosive device to stun the fish. The explosion takes place near the surface of the water, so the sound of the explosion travels through both the air and the water. You record the sound of the explosion using two separate microphones, one in the air above the water and one below the water surface. (a) Which microphone picks up the sound first? (b) If the time delay between the sounds reaching the two microphones is 0.50 seconds, about how far are you from the fishermen?

Exercises 23 – 27 involve the Doppler effect. Assume the speed of sound is 340 m/s.

23. In a common classroom demonstration, a buzzer is turned on inside a soft football. The buzzer emits a tone of 256 Hz. (a) If the football is thrown directly at you at a speed of 12.0 m/s, what frequency do you hear? (b) Fortunately, you duck in time to have the ball pass over your head. What frequency do you observe as the ball moves away from you?

24. In another common classroom demonstration of the Doppler effect, the instructor whirls a buzzer, on the end of a string or electric cable, in a horizontal circle around their head. If the buzzer has a frequency of 500 Hz, the circle has a radius of 1.0 m, and the period of the buzzer's motion is 0.50 s, what are the maximum and minimum frequencies observed by the students in the classroom as they sit in their seats listening to the buzzer?

25. As you are riding your bicycle at 10.0 m/s north along a road, an ambulance traveling south approaches you. You observe the ambulance's siren to have a frequency of 352 Hz. However, the siren's frequency is actually 325 Hz, when the ambulance is at rest. (a) How fast is the ambulance traveling? (b) After the ambulance has passed you, what frequency do you observe for the siren?

26. Your car horn happens to have the unusual property of emitting a pure tone at a frequency of 440 Hz. You drive at 20 m/s toward a high wall, and sound the horn briefly. After a short time, you hear the echo of the sound, after it was reflected by the wall. What is the frequency of the echo?

27. A particular bat emits ultrasonic waves with a frequency of 68.0 kHz. The bat is flying at 12.00 m/s toward a moth, which is traveling at 3.00 m/s toward from the bat. The speed of sound is 340.00 m/s. (a) Assuming the moth could detect the waves, what frequency waves would it observe? (b) What frequency are the waves that reflect off the moth and are detected by the bat?

Exercises 28 – 32 involve standing waves.

28. A particular guitar string has a length of 75 cm, and a mass per unit length of 80 grams/ meter. You hear a pure tone of 1320 Hz when a particular standing wave, represented by the sequence of images shown in Figure 21.24, is excited on the string. (a) What is the wavelength of this standing wave? (b) What is the speed of waves on this string? (c) What is the tension in the string? (d) What is the fundamental frequency of this string?

29. An Aeolian harp (named after Aeolus, the Greek god of the wind) consists of several strings fixed to a frame or a sounding box. The device is simply placed outside, and the strings are played randomly by the wind. You decide to make such a harp out of strings that all have a mass per unit length of 80 grams per meter, and that all have a tension of 50 N. (a) If you want one of the strings to have a fundamental frequency of 330 Hz, how long should you make it? (b) If you want another of the strings to have a fundamental frequency of 660 Hz (double that of the first string, and therefore exactly one octave higher up the scale), how long should it be?

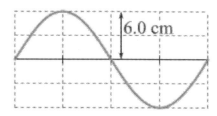

30. The profile of a particular standing wave on a string is shown in Figure 21.36, showing the string at its maximum displacement from equilibrium at $t = 0$. The string has a length of 1.0 m, extending from $x = 0$ to $x = +1.0$ m. Over one period of oscillation for the standing wave, plot a graph of displacement as a function of time for the point at (a) $x = 0.25$ m, (b) $x = 0.50$ m, (c) $x = 0.65$ m.

Figure 21.36: The profile of a particular standing wave on a string at $t = 0$, when the string is in one of its maximum displacement states, for Exercise 30.

31. As shown in Figure 21. 37, the height of an air column in a particular pipe is adjusted by changing the water level in the pipe. In a traditional experiment, a tuning fork is placed over the pipe, and the height of the air column is adjusted, by moving a reservoir of water up and down, until the pipe makes a loud sound, which is when the pipe's fundamental frequency matches the frequency of the tuning fork. If the speed of sound is 340 m/s, and an air column of 22.4 cm produces the loudest sound, what is the frequency of the tuning fork?

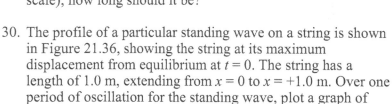
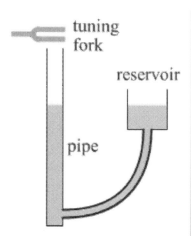

Figure 21.37: The height of the air column in this pipe can be adjusted by changing the water level in the pipe. When a tuning fork, which is emitting sound, is placed over the pipe, the pipe will emit a loud sound when the frequency of the tuning fork matches the fundamental frequency of the pipe. For Exercise 31.

32. A bloogle is a corrugated plastic tube, which is open at both ends, that emits a tone when you whirl it around your head. Generally, if you whirl it faster, the tube will emit a higher-frequency harmonic. You measure the various frequencies of a particular bloogle to be 420 Hz, 560 Hz, 700 Hz, and 840 Hz. (a) What is the fundamental frequency of this bloogle? (b) Estimate the length of the bloogle.

Exercises 33 – 38 involve applications of sound and waves.

33. Some cameras have automatic focusing systems that rely on ultrasonic emitters and detectors. You are trying to take a picture of your friends, who are 4.5 m from your camera. To focus correctly, the camera sends out a short ultrasonic pulse that reflects off your friends. If the speed of sound is 340 m/s, how much time passes between the emission of the pulse and the detection of the pulse by the camera?

34. One useful application of sound waves is a pair of noise canceling headphones. Such headphones have a microphone that picks up ambient noise (such as the noise of the engines inside a jet airplane). The wave representing the sound is then inverted and played through the speakers of the headphones into your ears. Explain, using principles of physics addressed in this chapter, how this works so that you hear a low-amplitude sound in the headphones.

35. One medical application of sound waves is in the use of ultrasound to see inside the womb to create an image of a fetus, as in the photograph shown in Figure 21.38. Do some research about this particular application of sound waves, and write two or three paragraphs describing how it works, and how it exploits the principles of physics discussed in this chapter.

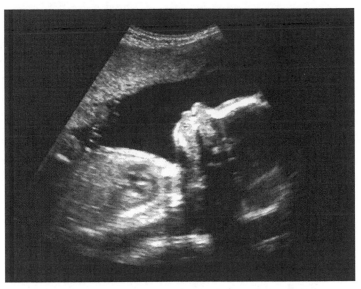

Figure 21.38: An image of a fetus in the womb, at 24 weeks, obtained by ultrasonic imaging. Photo credit: Maciej Korzekwa, via iStockPhoto.com.

36. The frequencies of neighboring notes on a musical scale differ by a factor of $2^{1/12}$. A particular guitar string is tuned to sound an A note, of 440 Hz. The next highest note is $A^{\#}$ (A sharp). (a) What is the frequency of this particular note? (b) By changing the effective length of the string, by pressing the string down onto one of the frets on the guitar, you can get the string to sound $A^{\#}$ instead of A. If the string has a length L when it sounds A, what is the effective length of the string when it is sounding $A^{\#}$? (c) Explain why the spacing between frets on the guitar decreases as the effective length decreases.

37. You want to make a simple set of wind chimes out of metal pipes that are open at both ends. You would like to create a set of three pipes that sound a C-major chord, playing the notes C (264 Hz), E (330 Hz), and G (396 Hz). (a) What is the ratio of the lengths of the three pipes you should use to make your wind chimes? (b) Which pipe is the shortest, and what is its length? Assume the speed of sound is 340 m/s.

38. The human ear can be modeled, to a first approximation, as a pipe that is open at one end only. If the length of the ear canal is 25 mm in a typical person, and the speed of sound in air is 340 m/s, what is the ear's resonance frequency? (This is the frequency of the fundamental frequency of the pipe and, in theory, should correspond to the frequency of sound that a typical person is most sensitive to.)

General problems and conceptual questions

39. A track designed for running 100-meter races is 8 m wide. If the starter fires her starting pistol from one side of the track, near the runners, the runner next to her has an advantage over the runner in the lane on the other side of the track. (a) Approximately how much time passes between when the closest runner hears the sound of the starting gun and when the farthest runner hears the sound of the gun? (b) If the runners run at an average speed of 10 m/s, what distance does this time difference translate to? Note that in serious competitions, the starting gun is electronically connected to speakers attached to the starting blocks for each runner, so that the start is fair.

40. A single-frequency wave, with a wavelength of 25 cm, is traveling in the positive x-direction along a string, causing each particle in the string to oscillate in simple harmonic motion with a period of 0.20 s. If the maximum transverse speed of each particle is 20 cm/s, and the particle at $x = 0$ is at its maximum positive displacement from equilibrium at $t = 0$, determine: (a) the speed of the wave, (b) the amplitude of the wave, and (c) the equation of motion for the wave.

41. Figure 21.39 (a) shows a snapshot of a traveling wave at $t = 0$, while Figure 21.39 (b) is a graph of the displacement versus time for the point at $x = 0$. (a) Is the wave traveling in the positive or negative x-direction? Explain. (b) Write out the equation of motion for the wave. (c) Does the equation of motion change if the graph in Figure 21.39 (b) applies to the point at $x = 20$ cm, instead? If so, what is the equation of motion in that case?

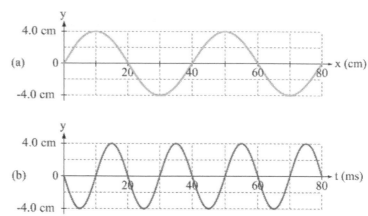

Figure 21.39: (a) A photograph of a wave on a string at $t = 0$. (b) A graph of the displacement versus time for the point at $x = 0$. For Exercises 41 and 42.

42. Figure 21.39 shows two representations of a traveling wave on a string. Figure 21.39 (a) shows a snapshot of a traveling wave at $t = 0$, while Figure 21.39 (b) is a graph of the displacement versus time for the point at $x = 0$. In each part below, state which representation you can use to find the answer, as well as giving the numerical value of the answer. (a) What is the wavelength? (b) What is the period? (c) What is the amplitude? (d) What is the speed of the wave?

43. Two trains are traveling along parallel tracks. Each train has a whistle that emits a tone of 333 Hz when the train is at rest. One train is traveling east at 5.00 m/s. The engineer in that train hears a beat frequency of 4.00 Hz when both train whistles are sounding. What is the velocity of the second train? Assume the speed of sound is 340 m/s. Summarize all the possible solutions to this exercise.

44. As shown in Figure 21.40, a child is swinging back and forth on a swing. The child is near a speaker that is broadcasting a pure (single-frequency) tone. The child is shown in five different positions during a swing. In which position will the child hear (a) the highest-frequency sound, and (b) the lowest-frequency sound? Briefly justify your answers.

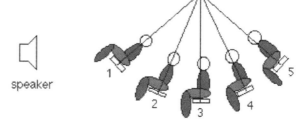

Figure 21.40: A child swinging back and forth on a string near a speaker that is broadcasting a pure tone, for Exercise 44.

45. The flow of blood through the heart can be studied with Doppler ultrasound. Ultrasonic waves are sent toward the heart, and by looking at the frequency of the waves that reflect from a particular spot in the heart, you can determine how fast blood is traveling in that region, and whether the blood is flowing toward or away from the ultrasound probe. An image is usually created from this data, with the colors of the various regions in the image reflecting the velocity of blood in those regions. If the probe sends out ultrasound with a frequency of 3.00 MHz, what is the frequency of waves that reflect back to the probe from an area of the heart (a) that is at rest? (b) where blood is traveling away from the probe with a speed that is 0.5% of the speed of sound in the medium? (c) where blood is traveling toward the probe at a speed of 0.7% of the speed of sound in the medium?

46. An ultrasonic sonar system emits ultrasonic waves that have a frequency of 600 kHz. The waves reflect from a plane that is moving at 50% of the speed of sound, directly toward the sonar system. (a) Find the frequency at which the waves reach the plane. (b) Find the frequency of the waves that are detected by the sonar system, after reflecting from the plane.

47. Repeat Exercise 46, but now have the plane moving directly away from the sonar system.

48. The pattern of sound waves emitted by a source traveling at constant velocity is shown in Figure 21.41. (a) In what direction is the source moving? (b) At what fraction of the speed of sound is the source traveling? If you are at rest, and the source is emitting waves that have a frequency of 480 Hz, what frequency do you observe if you are (c) at point A? (d) at point B?

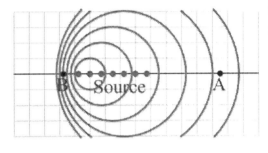

Figure 21.41: The pattern of circular waves emitted by a source that is traveling at a constant velocity. Each of the dots shows the position of the source when it emitted one of the wave peaks. For Exercise 48.

49. Repeat parts (a) – (c) of Exercise 48, but now base your answers on the pattern shown in Figure 21.42.

50. Do some research about what causes the loud sound when someone cracks a whip, and write a couple of paragraphs explaining the physics of whip-cracking.

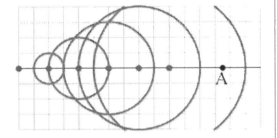

Figure 21.42: The pattern of circular waves emitted by a source that is traveling at a constant velocity. Each of the dots shows the position of the source when it emitted one of the wave peaks. For Exercise 49.

51. Two speakers, which are separated by a distance of 2.4 m, broadcast identical single-frequency sound waves. The speakers are in phase with one another. If you stand at a location that is 1.7 m farther from one speaker than the other, what are the lowest three frequencies at which (a) completely constructive interference occurs at your location, and (b) completely destructive interference occurs at your location?

52. Return to the situation described in Exercise 51. The speed of sound is 340 m/s, and the frequency of the waves emitted by the speakers is 340 Hz. You are initially right next to one of the speakers, and you then walk steadily away from it in a direction that is perpendicular to the line joining the two speakers. (a) At how many locations will you pass through a point at which completely constructive interference occurs? (b) How far are these locations from the speaker that marks your starting point?

53. In Figure 21.43, a pulse is traveling along a string toward the string's right end, which is a fixed end (shown as a dot on the right). Sketch the profile of the string at (a) $t = 4.0$ s, (b) $t = 6.0$ s, and (c) $t = 7.0$ s.

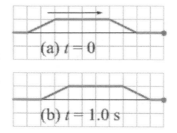

Figure 21.43: A pulse travels along a string toward the right end, which is either a fixed end (Exercise 53) or a free end (Exercise 54).

54. Repeat Exercise 53, but now the right end of the string is a free end instead of a fixed end.

55. In Figure 21.44, a pulse is traveling along a string toward the string's right end, which is a fixed end (shown as a dot on the right). A particular point on the string is shown as a black dot. Plot the displacement as a function of time for this point, over the time interval $t = 0$ to $t = 15.0$ s.

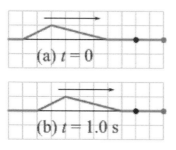

Figure 21.44: A pulse travels along a string toward the right end, which is either a fixed end (Exercise 55) or a free end (Exercises 56 and 57).

56. Repeat Exercise 55, if the right end of the string is a free end instead of a fixed end.

57. Repeat Exercise 56, except now plot the displacement as a function of time for the right end of the string, with the end being a free end.

58. Two stretched strings are placed next to one another, one with a length of 50 cm and the other with a length of 60 cm. The two strings have the same mass per unit length. You pluck the shorter one so that it vibrates at its fundamental frequency. You then adjust the tension in the longer string until it resonates with the first one. To resonate, the two strings must have the same frequency, so that vibrations on one string can cause the second string to vibrate. (a) What is the ratio of the speed of waves on the shorter string to the speed of waves on the longer string? (b) What is the ratio of the tension in the shorter string to the tension in the longer string?

59. A particular guitar string is under a tension of 38.5 N, and has a fundamental frequency of 320 Hz. If you want to tune the string so that it has a fundamental frequency of 330 Hz, to what value should you adjust the tension in the string?

60. A particular pipe that is open at both ends has a fundamental frequency of 442 Hz. When it, and a second pipe, have their fundamental frequencies excited simultaneously, a beat frequency of 8 Hz is observed. What is the ratio of the length of the first pipe to that of the second pipe if the second pipe is (a) also open at both ends, and (b) closed at one end.

61. As shown in Figure 21.45, a string passing over a pulley supports the weight of a 25 N block that hangs from the string. The other end of the string is fixed to a wall. The string has a mass per unit length of 75 grams per meter. The part of the string between the wall and the pulley is observed to oscillate with a fundamental frequency of 44 Hz. (a) What is the speed of waves on the string? (b) What is the distance, L, from the wall to the pulley? (c) If the weight hanging from the string is doubled, what will be the fundamental frequency of the part of the string between the wall and the pulley?

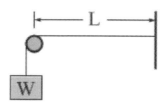

Figure 21.45: One end of a string is tied to a wall. The other end passes over a pulley, and supports a weight tied to the other end of the string. For Exercises 61 and 62.

A photograph of a mobile phone – you probably have one yourself that looks somewhat similar. If you showed this picture to someone in 1980, however, they would probably have difficulty determining what this object is.

Photo credit: Anna Langova, from publicdomainpictures.net.

Chapter 22 – Electromagnetic Waves

CHAPTER CONTENTS

In the 150 or so years since James Clerk Maxwell postulated the existence of electromagnetic waves, the world has been completely transformed by applications of such waves. Among these applications are:

- Radio. The development of radio enabled news and information to be disseminated quickly over wide areas; it allowed wireless communication with ships and planes; and it has filled the airwaves with all sorts of music and talk.
- Television. Television has been transformative, bringing with it instant access, for almost everyone in the world, to news coverage of daily events and natural disasters, coverage of sporting events, game shows, documentaries, and entertainment in various forms. It has only been about 70 years, however, since televisions have been available in the home, and color television was only introduced in the 1950's.
- Medical diagnostic instruments and treatments. It is now commonplace to get diagnostic tests using x-rays or magnetic resonance imaging (MRI), and to be treated using various forms of electromagnetic waves. Thanks, in part, to such technology, humans now enjoy much longer and healthier lives than they did not so long ago.
- Mobile phones and other wireless technology. The world is now unplugged thanks to electromagnetic waves. Phone calls find their way to your mobile phone whether you are at home, at work, or at the beach, both surfing the waves and, while bathed in electromagnetic waves from the Sun, surfing the Internet.

In this chapter, we will explore some of the history, and the principles of physics, that relate to the electromagnetic waves that are such an integral part of our daily lives, but that we do not give a second thought to, most of the time. It would be a very different world without them.

22-1 Maxwell's Equations

In the 19th century, many scientists were making important contributions to our understanding of electricity, magnetism, and optics. For instance, the Danish scientist Hans Christian Ørsted and the French physicist André-Marie Ampère demonstrated that electricity and magnetism were related and could be considered part of one field, electromagnetism. A number of other physicists, including England's Thomas Young and France's Augustin-Jean Fresnel, showed how light behaved as a wave. For the most part, however, electromagnetism and optics were viewed as separate phenomena.

James Clerk Maxwell was a Scottish physicist who lived from 1831 – 1879. Maxwell advanced physics in a number of ways, but his crowning achievement was the manner in which he showed how electricity, magnetism, and optics are inextricably linked. Maxwell did this, in part, by writing out four deceptively simple equations.

Physicists love simplicity and symmetry. To a physicist, it is hard to beat the beauty of Maxwell's equations, shown in Figure 22.1. To us, they might appear to be somewhat imposing, because they require a knowledge of calculus to fully comprehend them. These four equations are immensely powerful, however. Together, they hold the key to understanding much of what is covered in this book in Chapters 16 – 20, as well as Chapters 22 and 25. Seven chapters boiled down to four equations. Think how much work we would have saved if we had just started with Maxwell's equations instead, assuming we understood all their implications immediately.

Equation 1: $\int \vec{E} \bullet d\vec{A} = \dfrac{Q}{\varepsilon_0}$ Equation 3: $\int \vec{E} \bullet d\vec{l} = -\dfrac{d\Phi_B}{dt}$

Equation 2: $\int \vec{B} \bullet d\vec{A} = 0$ Equation 4: $\int \vec{B} \bullet d\vec{l} = \mu_0 I_{enclosed} \boxed{+ \mu_0 \varepsilon_0 \dfrac{d\Phi_E}{dt}}$

term added by Maxwell

Figure 22.1: Maxwell's equations.

Understanding Maxwell's equations.
Equation 1 is known as Gauss' Law for electric fields. It tells us that electric fields are produced by charges. From this equation, Coulomb's Law can be derived. Note that the constant ε_0 in Equation 1, the permittivity of free space, is inversely related to k, the constant in Coulomb's Law: $k = 1/(4\pi \varepsilon_0)$.

Equation 2 tells us that magnetic field lines are continuous loops.

Equation 3 is Faraday's Law in disguise, telling us that electric fields can be generated by a magnetic flux that changes with time.

Consider Equation 4. Setting the left side equal to the first term on the right-hand side, we have an equation known as Ampère's law, which tells us that magnetic fields are produced by currents. Everything up to this point was known before Maxwell. However, when Maxwell examined the equations (the first three, plus Equation 4 with only the first term on the right) he noticed that there was a distinct lack of symmetry. The equations told us that there were two ways to create electric fields (from charges, or from changing magnetic flux), but they only had one way to create magnetic fields (from currents). One of Maxwell's major contributions, then, was to bring in the second term on the right in Equation 4. This gave a second way to generate magnetic fields, by electric flux that changed with time, making the equations much more symmetric.

Maxwell did not stop there. He then asked the interesting question, what do the equations predict if there are no charges and no currents? Equations 3 and 4 say that, even in the absence of charges and currents, electric and magnetic fields can be produced by changing magnetic and electric flux, respectively. Furthermore, Maxwell found that when he solved the equations, the solutions for the electric and magnetic fields had the form $E(x,t) = E_0 \cos(\omega t - kx)$ and $B(x,t) = B_0 \cos(\omega t - kx)$. We recognize these, based on what we learned in Chapter 21, as the equations for traveling waves.

Finally, Maxwell derived an equation for the speed of the traveling electric and magnetic waves,

$$c = \sqrt{\frac{1}{\mu_0 \varepsilon_0}} = \sqrt{\frac{1}{(4\pi \times 10^{-7} \text{ Tm/A})(8.85 \times 10^{-12} \text{ C}^2/(\text{Nm}^2))}} = 3.00 \times 10^8 \text{ m/s} .$$

(**Equation 22.1**: Maxwell's derivation of the speed of light)

Maxwell recognized that this speed was very close to what the French physicists Hippolyte Fizeau and Léon Foucault had measured for the speed of light in 1849. Thus, Maxwell proposed (in 1873) that light consists of oscillating electric and magnetic fields in what is known as an electromagnetic (EM) wave. It was 15 years later, in 1888, that Heinrich Hertz, from Germany, demonstrated the production and detection of such waves, proving that Maxwell was correct.

Despite Hertz's experimental success, he rather famously stated that he saw no application for electromagnetic waves. Only 120 or so years later, the world has been transformed by our use of electromagnetic waves, from the mobile phones (and other communication devices) that almost all of us carry around, to the radio, television, and wireless computer signals that provide us with entertainment and information, and to x-rays used in medical imaging.

Throughout the 19th century, evidence for light behaving as a wave piled up very convincingly. However, all previous known types of waves required a medium through which to travel. Scientists spent considerable effort searching for evidence for the medium that light traveled through, the so-called luminiferous aether, which was thought to fill space. Such an aether could, for instance, explain how light could travel through space from the Sun to Earth. An elegant experiment in 1887, by the American physicists Albert Michelson and Edward Morley, showed no evidence of such a medium, and marked the death knell for the aether idea. Our modern understanding is that light, or any electromagnet wave, does not require a medium through which to travel. This view is consistent with Maxwell's equations. Electromagnetic waves consist of oscillating electric and magnetic fields, and by Maxwell's equations such time-varying fields produce oscillating magnetic and electric fields, respectively. Hence, an electromagnetic wave can be thought of as self-sustaining, with no medium required.

Finally, note how Equation 22.1 reinforces the idea that light, electricity, and magnetism are all linked. The equation brings together three constants, one associated with light (c, the speed of light), one associated with magnetism (μ_0, which appears in equations for magnetic field), and one associated with electricity (ε_0, which appears in equations for electric field).

Related End-of-Chapter Exercises: 36, 37, 38.

Essential Question 22.1 In Chapter 19, we used the equation $E/B = v$ when we discussed the velocity selector. Use this to help show how the units in Equation 22.1 work out.

Answer to Essential Question 22.1 In the denominator under the square root in Equation 22.1, units of $(T \, m/A) \times C^2/(N \, m^2)$ become $T \, C^2/(A \, N \, m)$, canceling a factor of meters. $1 \, A = 1 \, C/s$, so this leads to units of $T \, C \, s/(N \, m)$. The fact that $E/B = v$ tells us that electric field units of N/C, divided by magnetic field units of T, can be replaced by m/s. This leads to units of s^2/m^2. Bringing these up to the numerator, we invert the units to m^2/s^2, and taking the square root gives the required units of the speed of light, m/s.

22-2 Electromagnetic Waves and the Electromagnetic Spectrum

Although we often focus on light, because we rely on light so much as we interact with the world around us, light is just one example of an electromagnetic wave. As shown in Figure 22.2, we classify electromagnetic waves into a variety of categories based on their frequency (or wavelength), as well as on how the waves are produced.

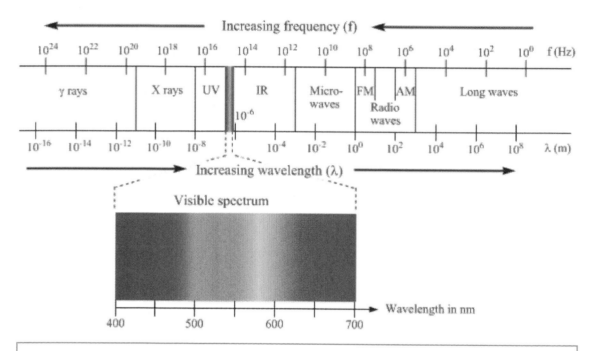

Figure 22.2: An overview of the electromagnetic spectrum. Note that the diagram covers 24 orders of magnitude in both frequency and wavelength.

Our eyes are sensitive to electromagnetic (EM) waves that have wavelengths in the visible spectrum, between 400 and 700 nm, but our bodies can be affected by EM waves in other ways, too. We have some sensors on the backs of our hands, in particular, that are sensitive to infrared radiation, which we can use to tell, without touching it, whether an object is hot. Ultraviolet radiation, in small doses, can produce tanning of the skin, but it is also associated with premature aging of the skin and cataracts. As you move to the left on the diagram in Figure 22.2, the energy associated with packets of EM radiation increases, and thus x-rays and gamma rays have enough energy to pass into and through our bodies. This makes them useful, because they can be used for diagnostic imaging (x-rays) our in cancer treatment (gamma rays), but they also have enough energy to be able to change DNA molecules, which is generally not a good thing. At the other end of the spectrum, we are generally insensitive to EM waves that have a longer wavelength than microwave radiation, such as radio waves.

Electromagnetic waves are produced by accelerating charged particles. Visible light, for instance, can be produced by electrons changing energy levels inside atoms, or by vibrating atoms. X-rays are produced by firing electrons at a metal target, with the x-rays being given off when the electrons are slowed abruptly by the metal atoms. Radio waves are produced by connecting a source of oscillating voltage to one or more metal rods (antennas), causing electrons to oscillate back and forth along the rod. The charge separation is associated with the production of electric fields, while the current associated with the moving electrons generates the magnetic fields.

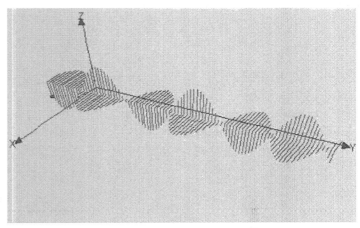

Properties of electromagnetic waves

Figure 22.3 shows a snapshot of a particular kind of electromagnetic wave, known as a plane (or linearly) polarized wave. The vectors parallel to the y-z plane represent the electric field vectors in the wave at the particular instant shown, while the vectors parallel to the x-y plane represent the magnetic field vectors at the same instant. Note that all the electric field vectors are in one plane, while all the magnetic field vectors are aligned in a plane that is perpendicular to the plane defined by the electric field vectors.

Figure 22.3: A linearly-polarized electromagnetic wave. The lines parallel to the y-z plane represent the electric field vectors, while the lines parallel to the x-y plane represent the magnetic field vectors. The wave is shown at a particular instant in time. As time goes by, the wave propagates to the right at the speed of light.

Some general features of electromagnetic waves include:
- the energy carried by an electromagnetic wave is divided equally between the electric fields and the magnetic fields.
- the electric and magnetic fields are in phase with one another.
- both the electric field vectors and the magnetic field vectors are perpendicular to the direction of propagation of the wave. Thus, an EM wave is classified as a transverse wave.
- the direction of propagation can be determined by applying a right-hand rule. Start with the fingers on your right hand pointing in the direction of the electric field at a particular point on the wave. If you align your hand so that you can curl your fingers from the electric field direction to the magnetic field direction, your thumb, when it is stuck out, will point in the propagation direction of the wave.
- at all points on the wave, the ratio of the electric field to the magnetic field is given by

$$c = \frac{E}{B}.$$ (Eq. 22.2: **Ratio of electric and magnetic fields in an EM wave**)

Related End-of-Chapter Exercises: 15 – 17.

Essential Question 22.2: (a) What is the wavelength of an x-ray beam if its frequency is 1×10^{18} Hz? How does it compare to the wavelength of the radio wave emitted by the FM radio station WBUR, which broadcasts at a frequency of 90.9 MHz? Assume both waves travel through the air. (b) If the electric field vectors in a linearly-polarized laser beam oscillate with an amplitude of 75 millivolts / meter, what is the amplitude at which the magnetic field vectors in the beam oscillate?

Answer to Essential Question 22.2: (a) To find the wavelength, we can combine the equation $\lambda = v / f$ with the fact that the speed of light in air is 3.00×10^8 m/s. Thus, a frequency of 1×10^{18} Hz corresponds to a wavelength of 3×10^{-10} m, while a frequency of 90.9 MHz corresponds to a wavelength of 3.30 m. (b) Using Equation 22.2, with $c = 3.00 \times 10^8$ m/s, gives an amplitude of $B_{max} = E_{max} / c = (0.075 \text{ V/m}) / (3.00 \times 10^8 \text{ m/s}) = 2.5 \times 10^{-10}$ T.

22-3 Energy, Momentum and Radiation Pressure

All waves carry energy, and electromagnetic waves are no exception. We often characterize the energy carried by a wave in terms of its intensity, which is the power per unit area. At a particular point in space that the wave is moving past, the intensity varies as the electric and magnetic fields at the point oscillate. It is generally most useful to focus on the average intensity, which is given by:

$$I_{average} = \frac{\text{average power}}{\text{area}} = \frac{E_{max} B_{max}}{2\mu_0} . \qquad \text{(Eq. 22.3: \textbf{The average intensity in an EM wave})}$$

Note that Equations 22.2 and 22.3 can be combined, so the average intensity can be calculated using only the amplitude of the electric field or only the amplitude of the magnetic field.

Momentum and radiation pressure

As we will discuss later in the book, there is no mass associated with light, or with any EM wave. Despite this, an electromagnetic wave carries momentum. The momentum of an EM wave is the energy carried by the wave divided by the speed of light. If an EM wave is absorbed by an object, or it reflects from an object, the wave will transfer momentum to the object. The longer the wave is incident on the object, the more momentum is transferred. This time dependence complicates matters, though, so let's define something about this situation that does not depend on time, which is called radiation pressure, P.

When we looked at an analogous situation for a rubber ball bouncing off an object, in Chapter 7, the ball transfers twice as much momentum to the object when the collision causes the ball's velocity to be equal-and-opposite to what it was before the collision than it does when the ball is stopped completely by the collision. For electromagnetic waves, the pressure is twice as large when the wave reflects from a perfect reflector than when it is 100% absorbed.

$$P = \frac{2I}{c} . \qquad \text{(Equation 22.4: \textbf{Radiation pressure when a wave reflects 100\%})}$$

$$P = \frac{I}{c} . \qquad \text{(Equation 22.5: \textbf{Radiation pressure when a wave is 100\% absorbed})}$$

Simply shining a flashlight onto an object causes a pressure to be exerted on the object. For an ordinary flashlight, however, the pressure is so small that it is negligible. For comparison, atmospheric pressure is approximately 10^5 Pa. To exert that pressure with an electromagnetic wave that reflects 100% requires an electromagnetic wave with an intensity of 1.5×10^{13} W/m², which is about 10 orders of magnitude more intense than bright sunlight!

Related End-of-Chapter Exercises: 18, 20, 35, 45 – 49.

It has been proposed that spacecraft use radiation pressure for propulsion. The idea is that the craft would unfurl a low-mass large-area reflective sail, and sunlight reflecting from the sail would provide a force to accelerate the spacecraft. Such a spacecraft is known as a **solar sailboat**. Radiation pressure associated with sunlight striking solar panels is exploited on some satellites to make minor adjustments in their motions without needing to use the on-board power source.

EXPLORATION 22.3 – Designing a solar sailboat

Let's design a solar sailboat that we can use to explore the solar system, making use of the following data. Mass of the Sun: $M = 2 \times 10^{30}$ kg; mass of the solar sailboat: $m = 1000$ kg; power emitted by the Sun in the form of electromagnetic waves: $power = 4 \times 10^{26}$ W.

Step 1 – *Find an expression for the gravitational force exerted on the satellite by the Sun, if the satellite is a distance r from the Sun.* Applying Newton's Law of universal gravitation, which we covered in Chapter 8, we find

$$F_g = \frac{GmM}{r^2}, \text{ where the constant } G = 6.67 \times 10^{-11} \text{ N m}^2 / \text{kg}^2.$$

Step 2 – *Find an expression for the intensity of sunlight reaching the spacecraft.* For a source like the Sun, which emits waves uniformly in all directions, the intensity at a particular distance is the radiated power divided by the surface area of a sphere with a radius equal to that distance. So,

$$I = \frac{power}{4\pi r^2}.$$

Step 3 – *Assuming the sail deployed by the spacecraft is perfectly reflective, find an expression for the force exerted on the sail by the reflecting sunlight. Assume also that the sails are oriented to reflect the sunlight straight back toward the Sun.* The force associated with the radiation pressure is $F_{rad} = P_{rad}A$, where P_{rad} is the radiation pressure and A is the sail area.

Using Equation 22.4, along with the result from Step 2, to determine the radiation pressure, we get

$$F_{rad} = \frac{2I}{c}A = \frac{2 \times power}{c \times 4\pi r^2}A = \frac{A \times power}{2\pi r^2 c}.$$

Step 4 – *Determine the sail area required to balance the gravitational force exerted on the spacecraft by the Sun.* Setting the gravitational force equal to the force associated with the radiation pressure gives:

$$\frac{GmM}{r^2} = \frac{A \times power}{2\pi r^2 c} \quad \Rightarrow \quad A = \frac{2\pi cGmM}{power}.$$

Interestingly, the area required to balance the forces does not depend on the distance the spacecraft is from the Sun, because both forces are inversely proportional to r^2. Plugging in the values for the various constants, and the values stated above, in this situation the sail area works out to 630000 m², which, if the sail was square, would require a sail almost 800 m × 800 m.

Key ideas for solar sailboats: Radiation pressure from sunlight reflecting from a very light metal sail can be a propulsion mechanism for a spacecraft. Solar sailboats need no fuel, and thus can be much lighter than a conventional spacecraft. **Related End-of-Chapter Exercises: 21, 22, 44.**

Essential Question 22.3: Return to Exploration 22.3. Using a sail area 10% larger than that calculated in Step 4, determine the acceleration of the spacecraft if it is the same distance from the Sun that the Earth is ($r = 1.5 \times 10^{11}$ m). Use this acceleration to approximate the spacecraft's speed one week after it starts from rest.

Answer to Essential Question 22.3: With a sail area 10% larger than that needed to balance the forces, there is a net force on the spacecraft of 10% of F_{rad} from step 3. The acceleration, by Newton's Second Law, is thus

$$a = \frac{0.1 \times F_{rad}}{m} = \frac{0.1 \times A \times power}{c \times 4\pi r^2 m} = 3 \times 10^{-4} \text{ m/s}^2.$$

Using $v = v_i + at$, with $t = 604800$ s in one week, gives a speed of $v = 180$ m/s. Even though the acceleration is very small, after many weeks the spacecraft builds up a large speed. The acceleration decreases as the distance from the Sun increases, but the spacecraft continually picks up speed as it moves away from the Sun.

22-4 The Doppler Effect for EM Waves

In Chapter 21, we spent considerable effort in coming to understand the Doppler effect for sound. We looked at how, for instance, it is not simply a relative-velocity phenomenon. The sound waves travel through a medium, so what matters is how the source of the waves as well as the observer of the waves moves with respect to the medium.

Because electromagnetic waves do not need a medium, the Doppler effect for EM waves is simply a relative-velocity phenomenon. The shift in frequency observed by an observer depends only on the relative velocity, $\vec{v}$, between the source and the observer. If the source emits EM waves that have a frequency f, the observed frequency f' is given by

$$f' = f\left(1 \pm \frac{v}{c}\right),$$ (Equation 22.6: **The Doppler effect for electromagnetic waves**)

where v is the magnitude of the relative velocity between the source and the observer. As with the Doppler effect for sound, we use the top (+) sign when the source and observer are moving toward one another, and the bottom (−) sign when the source and observer are moving farther apart.

Applications of the Doppler effect for EM waves

Figure 22.4 illustrates a common application in which the Doppler effect is exploited by astrophysicists to determine how fast, and in what direction, distant stars or galaxies are moving with respect to us here on Earth. At the bottom is the spectrum received on Earth from the Sun. The dark lines at specific wavelengths correspond to light that is absorbed by hydrogen atoms in the Sun. These same lines are seen in the spectrum received from a distant source, at the top, except all the lines are Doppler-shifted toward the red end (the right end) of the spectrum, indicating that the source is moving away from the Earth. It was data such as this that was used by Edwin Hubble (1889 – 1953) to show that the universe is expanding.

Figure 22.4: The spectrum on the bottom represents light coming to us from the Sun, which is essentially at rest with respect to the Earth. The four dark lines in the spectrum are caused by the absorption of light of particular wavelengths by hydrogen atoms in the Sun. The spectrum at the top represents light coming to us from a distant star, which is moving away from the Earth at 5% of the speed of light. The characteristic hydrogen absorption lines in the top spectrum have been shifted to the right, toward the red end of the spectrum. This is known as redshift.

A second application of the Doppler effect is Doppler radar, which is used in weather forecasting, as it can pick up rotating storm systems. Doppler radar has many sports applications, too, such as detecting the speed of a serve in tennis, or of a pitch in baseball. Doppler radar is also used by police to catch speeding motorists. Let's explore this last application in more detail.

EXPLORATION 22.4 – To catch a speeder

A police officer in a stationary police car aims a radar gun at a truck traveling directly toward the police car. The frequency of the radar gun is 10.525 GHz (10.525×10^9 Hz), and the frequency of the waves reflecting from the truck and returning to the radar gun is shifted from the emitted by frequency by 1600 Hz. If the speed limit on the road is 60 km/h, should the officer pull the truck over to give the driver a ticket? Let's work through the problem to decide.

Step 1 – *Is the frequency of the waves coming back to the radar gun higher or lower than the frequency of the emitted waves?* The relative velocity of the two vehicles brings them closer, which effectively lowers the wavelength of the waves, corresponding to a higher frequency.

Step 2 – *If the truck is traveling at a speed v, write an expression for f′, the frequency of the waves received by the truck.* To do this, we can simply apply Equation 22.6.

$$f' = f\left(1 + \frac{v}{c}\right),$$ where f is the emitted frequency. We use the plus sign in the equation

because the truck is moving toward the police car.

Step 3 – *Write an expression for f″, the frequency of the waves that are picked up by the radar gun after reflecting from the truck. Your expression should be in terms of f, rather than f′.* We apply Equation 22.6 again, and use the result from step 2.

$$f'' = f'\left(1 + \frac{v}{c}\right) = f\left(1 + \frac{v}{c}\right)^2,$$ where f is the emitted frequency. Again, we use the plus

sign in the equation because the truck is moving toward the police car.

Step 4 – *Solve for the speed of the truck, and decide whether the truck driver should get a speeding ticket.*

Expanding the bracket in the expression from step 3 gives $f'' = f\left(1 + \frac{2v}{c} + \frac{v^2}{c^2}\right) \approx f\left(1 + \frac{2v}{c}\right).$

The speed of light is orders of magnitude larger than the speed of the truck, so v^2/c^2 will be so much smaller than v/c that the v^2/c^2 term can be neglected. Writing the expression in terms of the known frequency shift, 1600 Hz, allows us to solve for the speed of the truck.

$$1600 \text{ Hz} = f'' - f = \frac{2fv}{c} \quad \Rightarrow \quad v = \frac{c \times 1600 \text{ Hz}}{2f} = \frac{(3 \times 10^8 \text{ m/s}) \times 1600 \text{ Hz}}{2 \times (10.525 \times 10^9 \text{ Hz})} = 22.8 \text{ m/s}.$$

Converting the speed to km/h gives a speed of 22.8 m/s × 0.001 km/m × 3600 s/h = 82 km/h. This is well over the speed limit, so the driver should certainly get a speeding ticket.

> **Key idea for Doppler radar:** In typical Doppler radar applications, the waves are emitted and detected at the same place, and thus the Doppler effect is applied twice to calculate the frequency shift between the emitted and detected waves. **Related End-of-Chapter Exercises: 3, 4, 50 – 53.**

Essential Question 22.4: Return to Exploration 22.4. If the truck is traveling away from the police car at 82 km/h instead of traveling toward it, would the frequency shift of the waves received by the radar gun still be 1600 Hz?

Answer to Essential Question 22.4: If the truck travels away, the frequency shifts lower instead of higher. The difference when we apply Equation 22.6 is that we use a minus sign instead of a plus sign. None of the numbers change, so the frequency shift still has a magnitude of 1600 Hz.

22-5 Polarized Light

Polarizing film consists of linear molecules aligned with one another. When an electromagnetic wave is incident on the film, electric field components that are parallel to the molecules cause electrons to oscillate back and forth along the molecules. This transfers energy from the wave to the molecules, so that part of the wave is absorbed by the film. Waves with electric field vectors in a direction perpendicular to the molecules do not transfer energy to the molecules, however, so they pass through the polarizing film without being absorbed. An electromagnetic wave emerges from the polarizing film **linearly polarized** – all its electric field vectors are aligned with the transmission axis of the polarizing film (which we call a polarizer), the transmission axis being perpendicular to the long molecules in the film.

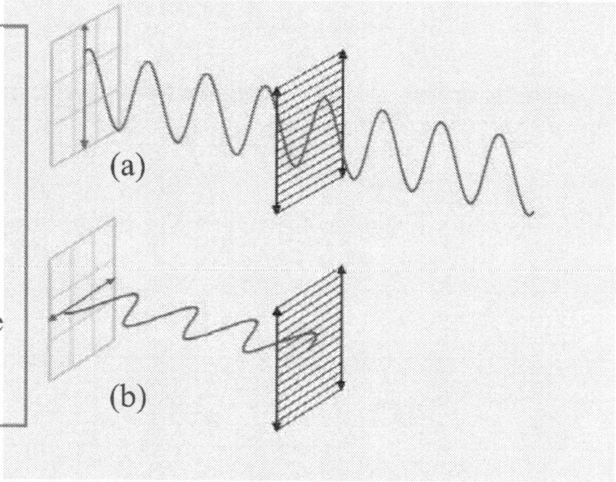

Figure 22.5: A schematic view of what happens to a linearly polarized electromagnetic wave that is incident on a polarizer. The arrow at left shows the polarization direction of the wave, while the vertical arrows on the polarizer indicate the transmission axis for the polarizer. The horizontal lines on the polarizer show the orientation of the long molecules in the polarizer. (a) The wave passes through the polarizer, because the polarization direction matches the transmission axis. (b) The wave is completely blocked by the polarizer, because the polarization direction of the polarizer is perpendicular to the polarizer's transmission axis.

Figure 22.5 shows two special cases, in which the wave is linearly polarized with its electric field vectors either parallel to, or perpendicular to, the polarizer's transmission axis. Figure 22.6 shows the more general case in which the electric field vectors make an angle $\Delta\theta$ with the transmission axis. Splitting the electric field vectors into components parallel to and perpendicular to the transmission axis, the parallel component is transmitted while the perpendicular component is entirely absorbed by the polarizer.

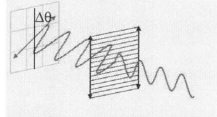

Figure 22.6: The general case of a linearly polarized wave that is incident on a polarizer.

The light emerging from the polarizer has two important features:
- it is linearly polarized, with the polarization direction of the wave matching the transmission axis of the polarizer it just passed through, and
- the intensity of the wave is reduced, because energy is absorbed by the polarizer.

If the magnitude of the electric field in the incident wave is E_0, the magnitude of the electric field in the wave emerging from the polarizer is $E_1 = E_0 \cos(\Delta\theta)$, because it is the cosine component in Figure 22.6 that is transmitted by the polarizer. In general, a wave is characterized by its intensity, which is proportional to the square of the amplitude of the electric fields. Thus,

$I_1 = I_0 \cos^2(\Delta\theta)$. (Eq. 22.7: **Malus' Law for the intensity of light emerging from a polarizer**)

Malus' Law applies when the incident light is linearly polarized. Figure 22.7 shows unpolarized light incident on a polarizer. On average, half the energy is associated with waves in which the electric field vectors are parallel to the polarizer's transmission axis, which are 100% transmitted by the polarizer, while the other half of the energy is associated with waves in which the electric field vectors are perpendicular to the transmission axis, which are 100% absorbed. Thus, the beam emerging from the polarizer is half as intense as that of the incident light. As with a linearly polarized incident wave, the wave emerging from the polarizer is linearly polarized in the direction of the polarizer's transmission axis.

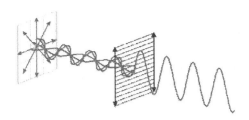

Figure 22.7: When unpolarized light is incident on a polarizer, the emerging beam is (i) half as intense as the incident beam, and (ii) linearly polarized in a direction parallel to the polarizer's transmission axis.

EXAMPLE 22.5 – A sequence of polarizers

As shown in Figure 22.8, light with its polarization direction at 30° to the vertical passes through a sequence of three polarizers. The light has an intensity of 800 W/m². Measured from the vertical, the transmission axes of the polarizers are at angles of 0°, 30°, and 75°, respectively. What is the intensity of the light when it emerges from (a) the first polarizer, (b) the second polarizer, and (c) the third polarizer?

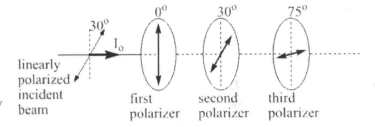

Figure 22.8: Linearly polarized light passes through a sequence of three polarizers, for Example 22.5.

SOLUTION

(a) Because the light is linearly polarized to begin with, we apply Malus' Law. In this case, the angle between the polarization direction of the light and the transmission axis of the polarizer the light is incident on is $\Delta\theta_1 = 30° - 0° = 30°$. Applying Malus' Law gives:

$$I_1 = I_0 \cos^2(\Delta\theta_1) = (800 \text{ W/m}^2)\cos^2(30°) = (800 \text{ W/m}^2) \times \frac{3}{4} = 600 \text{ W/m}^2.$$

(b) When the light emerges from the first polarizer, it is polarized at an angle of 0° to the vertical, matching the orientation of the transmission axis of the polarizer it just passed through. With the second polarizer at 30°, the angle between the light and the transmission axis is $\Delta\theta_2 = 30° - 0° = 30°$. Applying Malus' Law gives:

$$I_2 = I_1 \cos^2(\Delta\theta_2) = (600 \text{ W/m}^2)\cos^2(30°) = (600 \text{ W/m}^2) \times \frac{3}{4} = 450 \text{ W/m}^2.$$

(c) When the light emerges from the second polarizer, it is once again polarized at an angle of 30° to the vertical, matching the orientation of the transmission axis of the polarizer it just passed through. With the third polarizer at 75°, the angle between the light and the transmission axis is $\Delta\theta_3 = 75° - 30° = 45°$. Applying Malus' Law gives:

$$I_3 = I_2 \cos^2(\Delta\theta_3) = (450 \text{ W/m}^2)\cos^2(45°) = (450 \text{ W/m}^2) \times \frac{1}{2} = 225 \text{ W/m}^2.$$

Related End-of-Chapter Exercises: 6 – 12.

Essential Question 22.5: Crossed polarizers are two polarizers with transmission axes that are perpendicular to one another. What is the final intensity (a) if unpolarized light with an intensity of 600 W/m² is incident on crossed polarizers, and (b) if a third polarizer, with a transmission axis at 45° to the first polarizer, is placed between the original two polarizers?

Answer to Essential Question 22.5: (a) For crossed polarizers, no light emerges from the second polarizer. For the second polarizer, applying Malus' Law with an angle of $\Delta\theta = 90°$ leads to zero final intensity. (b) Surprisingly, placing a third polarizer between crossed polarizers leads to some light emerging. With unpolarized light, the first polarizer reduces the intensity by a factor of 2, to 300 W/m². Applying Malus' Law twice, with $\Delta\theta = 45°$ in each case, gives two factors of ½, with an intensity of 150 W/m² after the middle polarizer and 75 W/m² after the last polarizer.

22-6 Applications of Polarized Light

Let's summarize the basic method we applied to the polarizer situation in Example 22.5.

A General Method for Solving a Problem Involving Light Passing Through Polarizers

1. For the first polarizer, what we do depends on whether the incident light is unpolarized or polarized. If the light is unpolarized, the intensity is reduced by a factor of 2. If the light is polarized, we apply Malus' Law, where $\Delta\theta$ is the angle between the polarization direction of the incident light and the transmission axis of the polarizer.

2. The light emerging from the first polarizer is always polarized in the direction of the transmission axis of the first polarizer. For the second polarizer, then, we apply Malus' Law, where $\Delta\theta$ is the angle between the polarization direction of the light that emerges from the first polarizer and the transmission axis of the second polarizer. This angle is the same as the angle between the transmission axes of the first and second polarizers.

3. Repeat step 2 for each of the remaining polarizers in the sequence. Remember that $\Delta\theta$, the angle in Malus' Law, is the angle between the polarization direction of the light (this is the same as the angle of the transmission axis of the polarizer the light just emerged from) and the transmission axis of the next polarizer in the sequence.

Related End-of-Chapter Exercises: 28 – 32.

Applications of polarized light

Light emitted by the Sun is unpolarized, but sunlight can become at least partly polarized when it reflects from a flat surface. In general, the direction of polarization is parallel to the plane of the reflecting surface. Light commonly reflects from horizontal surfaces, such as bodies of water. The lenses in polarized sunglasses have vertical transmission axes, so they block light polarized horizontally. This greatly reduces glare from light reflecting off horizontal surfaces.

You can test whether your sunglasses are polarized by looking at the sky on a sunny day while you are wearing your sunglasses. Sunlight scattered through an angle of 90° in the atmosphere is polarized, so if you look in a direction that is at 90° to the direction of the Sun you will have linearly polarized light incident on your sunglasses. If you tilt your head to one side and then the other while you are looking at this part of the sky, you should see the sky brighten and darken if your sunglasses are polarized. By tilting your head, you change the angle between the light and the transmission axes of your sunglass lenses. By Malus' Law, this changes the intensity of the light passing through the sunglasses into your eyes. If you do not see such a change in intensity, then your sunglasses are not polarized.

Rotating the direction of polarization

As we discussed previously, crossed polarizers (polarizers that have their transmission axes at 90° to one another) generally block all light from passing through them. However, if you add something in between the crossed polarizers that changes the polarization direction of the light, some light can get through the system. In Essential Question 22.5, we looked at how adding a third polarizer between the crossed polarizers can result in light being transmitted. However, other transparent materials, such as Karo syrup, bits of mica, some cellophane tape, or clear plastic objects such as plastic forks, when placed between the crossed polarizers also result in light being transmitted. Interestingly, such materials generally affect different wavelengths of light differently, so by varying the thickness of the transparent material the light passes through, colorful patterns can be created. Such patterns have been used to make art installations, such as the one by Austine Wood Comarow at the Museum of Science in Boston, which has ever-changing pictures when viewed through a rotating polarizer.

Stress in a material can also affect the extent to which the polarization direction of a light wave passing through the material is rotated. Engineers exploit this material property by placing models between crossed polarizers to study the stress patterns.

Liquid-crystal displays

Another common application of polarized light is in liquid-crystal displays (LCDs), such as those on digital watches. You may have noticed that an LCD readout can be unreadable if you look at it through polarized sunglasses. This is because the light coming off the LCD is polarized, and thus can all be absorbed by polarizing sunglasses when the display is at a particular angle. The basis structure of an LCD display is shown in Figure 22.9(b). Key components are the crossed polarizers, separated by liquid crystals, and the mirror surface at the back. Light incident on the display from the right first passes through one polarizer, then through layers of liquid crystals. Successive layers of liquid crystals are rotated with respect to one another, and the net effect is that the polarization direction of the light is rotated by 90°. This aligns the light so that it passes through a second polarizer, with its transmission axis perpendicular to the first. The light then reflects off the mirror and reverses the steps, emerging from the sandwich.

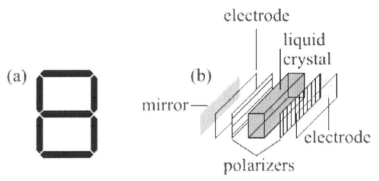

Figure 22.9: (a) Liquid-crystal displays are generally made of seven-segment displays, with different segments being turned on or off to make different numbers. (b) Each segment is formed from a sandwich of crossed polarizers and liquid crystals. Light entering from the right reflects, and the segment appears bright, when no potential difference is applied. Applying a potential difference causes light to be blocked, and the segment to go dark.

By applying a potential difference to the transparent electrodes, however, the liquid crystal layers un-twist, so the light's polarization axis is not rotated. In that case, all the light is blocked by the second polarizer, and that part of the readout looks dark. In a typical LCD readout, numbers are formed using seven-segment displays, in which the appropriate potential difference is applied across a given segment to turn it black, if desired, or no potential difference is applied to make the segment bright, like the background of the display.

Related End-of-Chapter Exercises: 2, 5, 57, 58.

Essential Question 22.6: Light passes through two polarizers. The intensity of the light emerging from the second polarizer is 65% of the intensity of the incident light. The incident light is either unpolarized or linearly polarized. Which is it?

Answer to Essential Question 22.6 If the light is originally unpolarized, then the intensity is reduced to 50% of the incident intensity when the light passes through the first polarizer. The fact that the final intensity is larger than 50% tells us that the incident light could not be unpolarized, and thus it must be linearly polarized.

Chapter Summary

Essential Idea: Electromagnetic Waves.
Electromagnetic waves of various frequencies fill the air around us, including light waves that help us to see, radio waves that can bring us music and news, infrared waves that help to keep us warm, and waves traveling to and from our cell phones. It's hard to imagine life without all the modern conveniences related to electromagnetic waves, but it was only in the 19th century that EM waves were theoretically predicted and then, over 20 years later, detected experimentally.

Maxwell's Equations
    James Clerk Maxwell wrote down four equations that summarized much of electricity and magnetism. With Maxwell's addition to one of the equations, he was able to predict the existence of electromagnetic (EM) waves, show that light was a form of EM wave, and show that electricity, magnetism, and optics were all linked. Maxwell's contributions rank among the most important contributions to physics of all time.

Energy, Momentum, and Radiation Pressure
    Electromagnetic waves consist of oscillating electric and magnetic fields that travel at the speed of light. We characterize the energy carried by an EM wave in terms of its intensity:

$$I_{average} = \frac{\text{average power}}{\text{area}} = \frac{E_{max} B_{max}}{2\mu_0} .$$ (Eq. 22.3: **The average intensity in an EM wave**)

The momentum associated with EM waves is most obvious when a wave, such as that from a flashlight, is absorbed or reflected. The radiation pressure associated with the wave's momentum change is twice as large when the wave completely reflects than when it is completely absorbed.

The Doppler effect for electromagnetic waves
If a source emits EM waves with a frequency f, the observed frequency f' is given by

$$f' = f\left(1 \pm \frac{v}{c}\right),$$ (Equation 22.6: **The Doppler effect for electromagnetic waves**)

where v is the magnitude of the relative velocity between the source and the observer. As with the Doppler effect for sound, we use the top (+) sign when the source and observer are moving toward one another, and the bottom (−) sign when the source and observer are moving apart.

Polarized light
When linearly polarized light (light with all its electric field vectors in one plane) is incident on a polarizer, the intensity of the emerging light (I_1) is related to that of the incident light (I_0) by

$$I_1 = I_0 \cos^2(\Delta\theta),$$ (Eq. 22.7: **Malus' Law for the intensity of light emerging from a polarizer**)

where $\Delta\theta$ is the angle between the polarization direction of the light and the direction of the polarizer's transmission axis. If unpolarized light is incident on the polarizer, the intensity of the transmitted light is half of that of the incident light. In all cases, electromagnetic waves emerging from a polarizer are polarized parallel to the direction of the polarizer's transmission axis.

End-of-Chapter Exercises

Exercises 1 – 12 are primarily conceptual questions, with a particular emphasis on polarized light, designed to test whether you understand the main concepts of the chapter.

1. Many automobile manufacturers are now working on collision avoidance, or collision mitigation, systems. Such systems sense an impending collision, and can warn the driver, start applying the brakes, and/or cause the seat belts to tighten. Using the principles of physics covered in this chapter, describe how such a system might sense an impending collision.

2. Polarizing sunglasses are generally somewhat expensive, so you are a little surprised to find a stall at an indoor market advertising polarizing sunglasses for $10. Using two pairs of the sunglasses, however, you can quickly test to see whether they are really polarized. What can you do?

3. A police officer who is traveling at a constant velocity of 80 km/h north measures a frequency difference of +2800 Hz for waves that reflect from a southbound car, with the frequency difference stated relative to the frequency of the waves emitted by the officer's radar gun. If the officer aims her radar gun at a different car that, like the officer, is traveling north at 80 km/h, what frequency difference will the officer observe between the reflected and emitted waves?

4. A police officer who is traveling at a constant velocity of 80 km/h north measures a frequency difference of +2800 Hz for waves that reflect from a southbound car, with the frequency difference stated relative to the frequency of the waves emitted by the officer's radar gun. After the two vehicles pass one another, and assuming each vehicle maintains its velocity, what will be the frequency difference observed by the officer if she points her radar gun back toward the receding car?

5. Light from the Sun is unpolarized, but when sunlight scatters from molecules in the Earth's atmosphere, it can become at least partly polarized. The most extreme case is when sunlight is scattered through an angle of 90°, in which the scattered light is linearly polarized in a direction that is perpendicular to both the original direction of propagation and the direction the light travels after changing direction. Polarizing sunglasses are designed to block light that is polarized horizontally. Armed with all this information, in which of the following cases would you expect the sky to look darker when you look straight up at a clear sky while wearing your polarizing sunglasses? Explain your answer. Case 1 – the Sun is low in the sky behind you. Case 2 – you turn 90° so the Sun is off to one side.

6. A vertically polarized beam of light with an intensity I_o is incident on a polarizing filter. The transmission axis of the filter is initially vertical, as shown in Figure 22.10. Keeping the polarizing filter perpendicular to the direction of the incident light at all times, the polarizing filter is rotated a full rotation (360°). Which of the six graphs in Figure 22.10 correctly shows the intensity of the light emerging from the polarizer, as a function of the angle of the polarizer's transmission axis? Explain your answer.

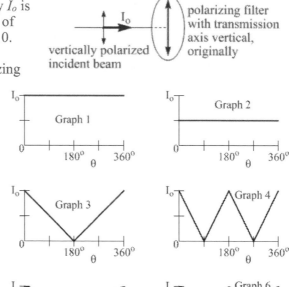

7. Return to the situation described in Exercise 6, and shown in Figure 22.10. Now answer the question when unpolarized light with an intensity of I_o is incident on the polarizer.

8. Return to the situations described in Exercises 6 and 7. The polarizer's transmission axis will now be rotated steadily from 0°, as shown in Figure 22.10, to 180°. Plot a graph of the polarization angle of the light that emerges from the polarizer as a function of the angle of the polarizer's transmission axis if the incident light is (a) polarized vertically, as shown in Figure 22.10, and (b) unpolarized.

Figure 22.10: Light is incident on a polarizer, with the transmission axis of the polarizer initially oriented vertically. Six possible graphs of the intensity of the transmitted light, as a function of the angle of the polarizer's transmission axis, measured from the vertical, are also shown. For Exercises 6 – 8.

9. You have three polarizers. Polarizer A has its transmission axis parallel to the vertical; polarizer B has its transmission axis at 30° to the vertical, and polarizer C has its transmission axis at 90° to the vertical. You are using a light source that emits linearly polarized light that is polarized at 45° to the vertical. Using one, two, or all three of the polarizers, how many different arrangements can you come up with so that no light emerges from the last polarizer? List the various arrangements.

10. Return to the situation described in Exercise 9. Now, your goal is to have the light pass through all three of the polarizers, and emerge from the last polarizer with the maximum possible intensity. (a) In what order do you arrange the polarizers now? Give all possible arrangements that produce maximum intensity. (b) If the intensity of the incident light is 720 W/m², what is the intensity of the light emerging from the third polarizer in the sequence?

11. Return to the situation described in Exercises 10, which involves the three polarizers in Exercise 9. Answer the questions from Exercise 10, if the incident light is unpolarized.

12. Light that is either unpolarized or linearly polarized is incident on a single polarizer. As you rotate the polarizer about an axis parallel to the direction of propagation of the light, you observe that the intensity of the light emerging from the polarizer stays constant. Is the incident light unpolarized or linearly polarized? Explain how you know.

Exercises 13 – 17 deal with the electromagnetic spectrum and EM waves.

13. Do a web search for the "United States Frequency Allocation Chart." The chart shows how different parts of the spectrum are used for various purposes in the United States. Compare the parts of the spectrum reserved for broadcasting AM and FM radio. Which of these parts is wider? Explain your answer.

14. The United States government mandated the replacement of analog television broadcasts with digital television broadcasts. In addition to providing a better TV viewing experience for consumers, eliminating the analog broadcasts freed up a 60 MHz-wide band of the spectrum for other uses. A government-run auction of pieces of this part of the spectrum brought in $19.6 billion. Do some research regarding this issue and write a couple of paragraphs about what you learn.

15. One section of the electromagnetic spectrum is known as "VHF." (a) What does this stand for? (b) What frequency range does the VHF band cover? (c) What are typical uses of waves in the VHF band of the spectrum?

16. Electromagnetic waves associated with hydrogen, with a wavelength of 21.1 cm, have been an important tool in astronomy. Do some research about how a tiny hydrogen atom produces EM waves of such a large wavelength, and about how such waves are used by astronomers. Write a couple of paragraphs regarding what you learn.

17. A typical red helium-neon laser pointer emits light with a wavelength of 632 nm. (a) What is the corresponding frequency of this light? (b) If you shine the beam at a wall 1 meter away from the laser pointer, how many wavelengths fit in this 1 m distance?

Exercises 18 – 22 involve energy, momentum, and radiation pressure.

18. A particular laser pointer emits a beam with an average power of 3.00 mW, and the beam has a circular cross-section with a diameter of 4.00 mm. The beam is linearly polarized. Determine (a) the intensity of the laser beam, (b) the amplitude of the oscillating electric field in the laser beam, and (c) the amplitude of the oscillating magnetic field in the laser beam.

19. The Earth's magnetic field has a strength of about 5×10^{-5} T. If this is the amplitude of the oscillating magnetic field in a linearly polarized EM wave, what is the amplitude of the oscillating electric field in the wave? Comment on the size of your answer.

20. A highly reflective low-density cylinder is being held motionless in mid-air by a laser beam that is directed vertically upward, and which reflects off one of the two circular faces of the cylinder. The beam is large enough that the entire face of the cylinder is uniformly illuminated. The cylinder has a height h, a cross-sectional area A, and a density of 200 kg/m^3. (a) Sketch a free-body diagram of the cylinder. (b) Apply Newton's Second Law, and show that it does not matter what the area A is. (c) Find the height of the cylinder, if the intensity of the laser beam is 1.5×10^7 W/m^2.

21. The sails on a solar sailboat far above the Earth's atmosphere are oriented so that they reflect sunlight through an angle of 90°, with all the reflected light going off in the same direction. (a) In what direction, relative to the direction of the incident light, is the force exerted by the radiation pressure on the sailboat? (b) Sketch a diagram to show the direction of the incident light, the reflected light, and the force.

22. At the location of Earth in the solar system, the intensity of sunlight is about 1300 W/m². The sail on a solar sailboat far above the Earth's atmosphere, but the same distance from the Sun that the Earth is, is circular with a radius of 500 m. The sail is 100% reflective, and is oriented to maximize the force associated with radiation pressure. (a) Calculate the force associated with radiation pressure on the sail. (b) Assuming that the gravitational force exerted on the sailboat by the Sun is the only other force acting on the sailboat, determine the mass the sailboat must have to experience no net force. (c) If the sailboat's mass is actually just 80% of the value you calculated above, determine its acceleration.

Exercises 23 – 27 involve the Doppler effect for electromagnetic waves.

23. How fast would you have to be driving toward a red traffic light for the light to appear green to you? Assume that the wavelength emitted by the red light is 630 nm, and that you will see the light as green if the wavelength you observe is 530 nm.

24. A baseball scout sitting directly behind home plate aims a radar gun at the baseball thrown by the pitcher. The radar gun gives a speed of 160 km/h for the pitch, just slightly less than 100 miles per hour. If the radar gun emits EM waves that have a frequency of exactly 12 GHz, what is the shift in frequency for (a) the waves reaching the moving baseball, and (b) the waves returning to the radar gun after reflecting from the baseball?

25. In the lab, hydrogen gas excited by a high voltage gives off light of particular wavelengths. One of these wavelengths is 435 nm. When an astronomer looks at the spectrum of wavelengths emitted by a distant star, however, she determines that this same line in the spectrum is shifted to a wavelength of 456 nm. (a) Is the star moving toward us or away from us? (b) What is the relative velocity of the star with respect to us?

26. As of the summer of 2010, the women's world-record tennis serve of 130 miles per hour was held jointly by the Dutch player Brenda Schultz-McCarthy, for a serve hit in Cincinnati in 2006, and the American player Venus Williams, for a serve hit in Zurich in 2008. Assuming the speed of each serve was measured by a radar gun that was directly in line with the ball's path, what was the magnitude of the frequency shift between the emitted and reflected waves, if the radar gun emits waves of exactly 22 GHz?

27. You are called to testify in court as an expert witness in a case involving a man who was pulled over for speeding by a police officer who was in a moving police car when she recorded the man's speed on her radar gun. The officer testifies that she was traveling at 50 km/h north on a road that has a speed limit of 100 km/h, and that she observed the man's car traveling south on the road, moving toward her. The officer testifies that her radar gun measured a frequency difference of +1400 Hz for the waves reflecting from the man's car, and that when she uses the radar gun to measure speeds when she is at rest, a frequency difference of +1400 Hz corresponds to a speed of 160 km/h. The police officer had written the man a speeding ticket for traveling at 160 km/h in a 100 km/h zone. How fast was the man traveling in his car, and do you think he deserves the speeding ticket?

Exercises 28 – 32 are designed to give you practice applying the method for solving problems involving polarized light.

28. Light that is linearly polarized in a direction of 20° to the vertical is incident on a polarizer that has its transmission axis at 50° to the vertical. (a) What is the angle between the polarization direction of the light and the transmission axis of the polarizer? (b) If the incident light has an intensity of 400 W/m², what is the intensity of the light emerging from the polarizer? (c) At what angle, measured from the vertical, is the polarization direction of the light emerging from the polarizer?

29. Linearly polarized light, with an intensity of $I_0 = 600$ W/m² is incident on a sequence of two polarizers. The transmission axis of the first polarizer is at 40° from the vertical, while the transmission axis of the second polarizer is at 70° from the vertical, as shown in Figure 22.11. The polarization direction of the light is also at 70° to the vertical. (a) What is the angle between the polarization direction of the light and the transmission axis of the first polarizer? (b) Determine the intensity of the light emerging from the first polarizer. (c) Determine the angle at which the light is polarized when it emerges from the first polarizer. (d) What is the angle between the polarization direction of the light when it emerges from the first polarizer and the transmission axis of the second polarizer? (e) Determine the intensity of the light emerging from the second polarizer.

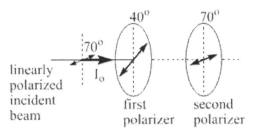

Figure 22.11: A set of two polarizers for light to pass through, for Exercises 29 and 30.

30. Repeat Exercise 29, with the incident light being unpolarized instead. You can skip part (a), which does not make sense when the light is unpolarized.

31. Unpolarized light, with an intensity of $I_0 = 720$ W/m², is incident on a sequence of three polarizers. The transmission axes of the three polarizers with respect to the vertical are, in order, 90° from the vertical, 30° from the vertical, and 60° from the vertical, as shown in Figure 22.12. (a) Determine the intensity of the light emerging from the first polarizer. (b) Determine the angle at which the light is polarized when it emerges from the first polarizer. (c) What is the angle between the polarization direction of the light when it emerges from the first polarizer and the transmission axis of the second polarizer? (d) Determine the intensity of the light emerging from the second polarizer. (e) Determine the angle at which the light is polarized when it emerges from the first polarizer. (f) What is the angle between the polarization direction of the light when it emerges from the second polarizer and the transmission axis of the third polarizer? (g) Determine the intensity of the light emerging from the third polarizer.

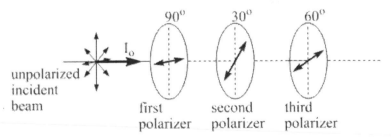

Figure 22.12: A set of three polarizers for light to pass through, for Exercises 31 and 32.

32. Repeat Exercise 31, with the incident light being linearly polarized at 90° to the vertical instead.

Exercises 33 – 35 involve practical applications of electromagnetic waves.

33. A typical Doppler radar system, used in weather forecasting, uses an antenna to emit a pulse of EM waves lasting for 1 microsecond. The antenna then listens for the next millisecond to pick up any reflected signals, such as those from distant thunderclouds. What is the range of such a radar system? In other words, what is the maximum distance a weather system can be from the antenna for the antenna to pick up reflected signals from it within the 1 millisecond window?

34. LASIK (Laser Assisted in Situ Keratomileusis) is now a common eye surgery in which a laser is used to remove a thin layer of cornea, correcting vision in the eye so glasses or contacts do not need to be worn. Typically, a laser known as an excimer laser is used, with a wavelength of 193 nm, which has been shown to produce high-quality incisions in the delicate tissue of the eye. (a) In what section of the electromagnetic spectrum does a wavelength of 193 nm place this laser? (b) If such a laser has a power of 4.0 milliwatts, and is turned on for pulses lasting 20 ns each, how much energy is delivered by the laser to the eye in each pulse?

35. In February of 2007, researchers at the Lawrence Livermore National Laboratory reported that their Solid State Heat Capacity Laser had operated at a record 67 kW average power. (a) How many times larger is this than the power of a typical laser pointer, of approximately 3 mW? The researchers also reported that, with the laser operating at 25 kW and with a beam size of 2.5 cm^2, it took 7 seconds to burn a hole through a 2.5–centimeter thick piece of steel. In this mode of operation, what is (b) the beam's average intensity, and (c) the magnitude of the beam's peak electric field?

General problems and conceptual questions

36. In addition to developing Maxwell's equations, which summarize important features of electricity, magnetism, electromagnetism, and electromagnetic waves, James Clerk Maxwell added much to our understanding of other physical phenomena. Do some research about Maxwell, and write a couple of paragraphs about his other contributions to science.

37. Do some research to create a timeline for the development of Maxwell's equations. In addition to other events and scientists you may want to include, mention specifically the work of Ampère, Coulomb, Faraday, Hertz, and Maxwell, and state where the work of each of these scientists is placed on the timeline.

38. A key part of Maxwell's insights into what the wave solutions to Maxwell's equations meant was Maxwell's knowledge of the speed of light in vacuum, as measured by other scientists. Do some research about our understanding of the value of the speed of light, and the scientists who contributed to this understanding, and then write a couple of paragraphs regarding the evolution of our knowledge about the speed of light.

39. The intensity of sunlight reaching the top of the Earth's atmosphere is about 1300 W/m^2, at a distance of 150 million km from the Sun. Estimate the Sun's power output.

40. How long does light take to reach Earth after leaving the Sun?

41. You are watching a lightning storm from the safety of your home. When you observe a particular flash of lightning, you slowly count the seconds until you hear the thunder associated with the flash. The two are 9 seconds apart. (a) How far away was the lightning, assuming a speed of sound of 340 m/s? (b) What is a general rule to convert seconds to kilometers for lightning? (c) Does the same rule apply to fireworks?

42. You see your friend 100 m away down the street. If you yell back and forth, your conversation will be affected by the speed of sound, which is 340 m/s. If, instead, you use your cell phone to call her on her cell phone, the conversation is relayed via a communications satellite that is 36000 km above you. What is the delay between a sound leaving your mouth and its arrival at your friend's ear if (a) you yell, or (b) you use your cell phone? (c) How far would you have to be from your friend for the delays to be equal?

43. The intensity of sunlight reaching the top of the Earth's atmosphere is about 1300 W/m^2. Use that number to estimate the intensity of sunlight at the location of (a) Mercury, (b) Venus, (c) Mars, and (d) Jupiter.

44. In Essential Question 22.3, we determined that the acceleration of a particular solar sailboat was 3×10^{-4} m/s^2 when the sailboat was the same distance the Earth is from the Sun. Assume the sailboat was released from rest from that location. (a) If this sailboat then traveled directly away from the Sun, what would its acceleration be when it got to the same distance from the Sun that Mars is? (b) Estimate the sailboat's average acceleration during the period while it travels from Earth's orbit to Mar's orbit. Then use your value of the average acceleration to estimate (c) the time the sailboat takes to cover this distance, and (d) the sailboat's speed when it arrives at Mar's orbit.

45. While giving a lecture, your professor absent-mindedly shines his laser pointer out into the audience. When the students point this out, the professor claims that the laser's intensity is less than that of sunlight. Typically, bright sunlight has an intensity of about 1000 W/m^2. Your professor's laser pointer has a power of 4.0 mW, and a beam diameter of 3.0 mm. What is the intensity of light emitted by the laser pointer? Use your answer to comment on the professor's assertion about the brightness of the pointer.

46. The power emitted by the Sun in the form of electromagnetic waves is about 4×10^{26} W. The distance from the Earth to the Sun is about 150 million km. Estimate (a) the intensity of sunlight reaching Earth, and (b) the total energy every second in the sunlight that is incident on the Earth.

47. A particular laser beam has a constant cross-sectional area of 5.0 mm. What is the ratio of the intensity of the beam 1.0 m from the laser to that 4.0 m from the laser?

48. Return to Exercise 47, but now replace the laser beam with a light bulb that emits light uniformly in all directions. What is the ratio of the intensity of the light 1.0 m from the bulb to that 4.0 m from the bulb?

49. A typical 100–W incandescent light bulb emits only about 5% of its energy in the form of visible light. How close would your eye have to be to the filament of a 100–W bulb for the intensity associated with the visible light to be equal to that emitted by a 3.0 mW laser pointer with a beam diameter of 4.0 mm?

50. When a baseball scout aims his radar gun at a stationary car, the radar gun displays a frequency shift of 0 for the difference in frequency between the emitted waves and the reflected waves. (a) Is that what the scout should expect to see for the frequency shift? (b) When the scout aims the radar gun at a car that is moving directly away from him at 60 km/h, the radar gun displays a frequency shift of −720 Hz between the emitted waves and reflected waves. If the gun displays a frequency shift of +1728 Hz when the scout aims the radar gun at a baseball thrown by a pitcher in a high school game, what is the speed of the baseball? The baseball is traveling directly toward the radar gun.

51. A police officer uses a radar gun that emits electromagnetic waves with a frequency of 10.525 GHz (10.525×10^9 Hz). Assume the police officer is at rest, and is aiming the radar gun at vehicles traveling either directly toward, or directly away from, the radar gun. What is the magnitude of the frequency shift between reflected and emitted waves for a vehicle traveling at (a) 60 km/h, (b) 100 km/h, and (c) 150 km/h?

52. Return to the situation described in Exercise 51, and assume the speed limit is 110 km/h, as it is on many Canadian highways. If the police officer observes reflected waves that have a frequency of 2440 Hz below the frequency of the emitted waves, (a) in which direction is the vehicle that reflected the waves moving, (b) what is the vehicle's speed, and (c) should the officer issue a speeding ticket?

53. A police officer, who is at rest, records reflected waves that have a frequency 4200 Hz larger than that of the waves emitted by the officer's radar gun for waves reflecting from a vehicle that is moving at a constant speed of 100 km/h in a direction that is either directly toward or directly away from the officer. (a) In what direction is the vehicle moving? (b) What is the frequency of the officer's radar gun?

54. Light that is either unpolarized or linearly polarized is incident on a single polarizer. The intensity of the light emerging from the polarizer is half that of the incident light. Is the incident light unpolarized or linearly polarized? Explain your answer.

55. Linearly polarized light with an intensity of 240 W/m² is incident on a sequence of two polarizers. The angle between the transmission axes of the two polarizers is 30°. When the incident light is replaced by unpolarized light of the same intensity, the intensity of the light emerging from the second polarizer is the same as with the polarized light. (a) What is the intensity of light emerging from the second polarizer? (b) In the original situation, when the incident light is linearly polarized, what is the angle between the polarization direction of the incident light and the transmission axis of the first polarizer?

56. Repeat part (b) of Exercise 55, except now when the incident polarized light is replaced by unpolarized light of the same intensity, the intensity of the light emerging from the second polarizer is half of the value it is when the polarized light is used.

57. Light passes through two polarizers. The intensity of the light emerging from the second polarizer is 35% of the intensity of the incident light. The incident light is either unpolarized or linearly polarized. Which is it? Explain your answer.

58. Return to the situation described in Exercise 57. (a) If the incident light is unpolarized, determine the angle between the transmission axes of the two polarizers. (b) If the incident light is linearly polarized, determine the range of possible values for the angle between the transmission axes of the two polarizers.

Light can be described by several different models. In Chapters 23 and 24, we will make use of a relatively simple model of light known as the ray model. The ray model will help us to understand how light interacts with mirrors and lenses, and help us understand how the reflection of light can produce beautiful photographs, such as the photograph of Mt. Hood, shown here twice! (Photo credit: public-domain image from Oregon's Mt. Hood Territory)

Chapter 23 – Reflection and Mirrors

CHAPTER CONTENTS

In Chapter 22, we looked at how light is an electromagnetic wave, made up of oscillating electric and magnetic fields propagating through space. In this chapter, we look at the reflection of light, how light interacts with mirrors, and how mirrors can be used to form images. To understand reflection and image formation, we will use a model of light based on rays and wave fronts – this is a much simpler model than the electromagnetic wave perspective. In this chapter and in Chapter 24, we are investigating geometrical optics, in which the optics of mirrors and lenses can be understood in terms of the geometry of similar triangles.

Light plays an integral role in the lives of almost all living things. It is certainly the primary way that most of us interact with the world around us. Light provides us with information about the current state of the traffic light as we wait at an intersection; with the state of mind of someone we are talking to, by communicating facial expressions and body language; with the pictures on a movie screen; and with information about the entire universe via the light of stars, emitted many years ago, as we gaze up at the sky on a clear night.

The reflection of light also plays an important role in our lives. Almost all of us take at least one look in a mirror before leaving the house, to make sure we are presentable, with our hair neatly brushed, etc. When driving, mirrors are also incredibly useful to us, providing information about what is going on beside us and behind us, without our needing to look behind. Reflect on this (pun intended): how many times do you think you use a mirror every day? Try to keep a count – the total will probably surprise you.

23-1 The Ray Model of Light

We will start our investigation of geometrical optics (optics based on the geometry of similar triangles) by learning the basics of the ray model of light. We will then apply this model to understand reflection and mirrors, in this chapter, and refraction and lenses, in chapter 24. Using the triangles that result from applying the ray model, we will derive equations we can apply to predict where the image created by a mirror or lens will be formed.

A **ray** is as a narrow beam of light that tends to travel in a straight line. An example of a ray is the beam of light from a laser or laser pointer. In the ray model of light, a ray travels in a straight line until it hits something, like a mirror, or an interface between two different materials. The interaction between the light ray and the mirror or interface generally causes the ray to change direction, at which point the ray again travels in a straight line until it encounters something else that causes a change in direction. An example in which the ray model of light applies is shown in Figure 23.1, in which the beams of sunlight travel in straight lines.

Figure 23.1: The photograph shows rays of light passing through openings in clouds above the Washington Monument. Each ray follows a straight line. (Photo credit: public-domain photo from Wikimedia Commons)

A laser emits a single ray of light, but we can also apply the ray model in situations in which a light source sends out many rays, in many directions. Examples of such sources include the filaments of light bulbs, and the Sun. If we are far away from such a source, in relation to the size of the source itself, we often treat the source as a **point source**, and assume that the source emits light, usually in all directions, from a single point. Light bulbs, and the Sun, are often treated as point sources. In other situations, such as when we are close to a light bulb that has a long filament, we treat the source as a **distributed source**. Each point on the source can be treated as a point source, so a distributed source is like a collection of point sources, as shown in Figure 23.2.

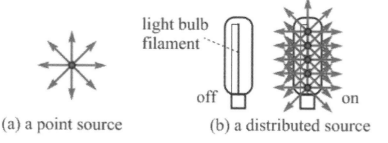

Figure 23.2: (a) A point source of light emits light uniformly in all directions. (b) A distributed source of light, such as a light-bulb filament in the shape of a line, can generally be treated as a collection of point sources.

Wave fronts

In addition to rays of light, we will also mention wave fronts. A wave front is a surface connecting light that was emitted by the light source at the same time. As shown in Figure 23.3, the wave fronts for a point source are spherical shells centered on the source, which propagate away from the source at the speed of light. For a beam of light, like that from a flashlight, in which the rays are parallel, the wave fronts are parallel lines that are perpendicular to the beam.

Figure 23.3: (a) For a point source, the wave fronts are spherical shells that are centered on the source. The larger the radius of the wave front, the more time has passes since the light was emitted. (b) When the rays are part of a beam of light that is traveling in a particular direction, the wave fronts are parallel lines that are perpendicular to the beam.

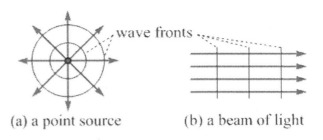

(a) a point source   (b) a beam of light

Shadows

The ray model of light can also be used to understand shadows. Figure 23.4 shows how the shadow cast by a point source can be larger than the object creating the shadow, while that from parallel rays is the same size as the object, as long as the surface on which the shadow is cast is perpendicular to the direction of the rays. Distributed sources create more complicated shadows, but they can be understood as the superposition of the shadows from multiple point sources.

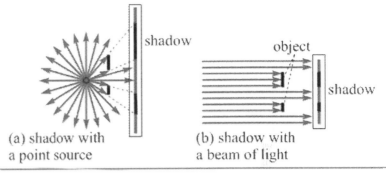

(a) shadow with a point source   (b) shadow with a beam of light

Figure 23.4: We can use the ray model of light, in which travels in straight lines, to explain shadows. When the light source is a point source (a), the shadow is generally larger than the object casting the shadow. When the light rays are all going in the same direction, however, the shadow is the same size of the object when the shadow is cast on a surface that is perpendicular to the light rays.

Treating sources that do not themselves emit light as light sources

In some cases, we will use objects that actually emit light, such as a light bulb or the Sun, as the objects that send light toward a mirror or lens. In other cases we will use objects, such as you, that do not emit light themselves. How can we do this? In general, objects that do not emit light themselves are illuminated by other light sources. As shown in Figure 23.5, such sources can be treated as if they emit light, because they scatter much of the light incident on them in many different directions.

Figure 23.5: An illuminated object can itself be treated as a source of light, because much of the light shining on it is scattered off the object in all directions.

How we see objects

To see an object, rays of light need either to be emitted by, or reflected from, the object, and then pass into our eyes. Our brains assume that the rays of light travel in straight lines, so we trace the rays of light back until they meet at the location of the object, as shown in Figure 23.6.

Related End-of-Chapter Exercises: 3, 4, 30.

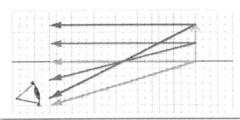

Figure 23.6: Objects send out light in many directions. We see the object if enough of this light enters our eye. Only a small number of the rays are shown for this object, color-coded red for rays from the top, blues for rays from the middle, and green for rays from the bottom of the object.

Essential Question 23.1: The Sun is a very large object, much larger than the Earth. Give an example in which we can treat the Sun as a point source when applying the ray model. Give an example in which the Sun must be treated as a distributed source of light.

Answer to Essential Question 23.1: To explain the formation of your own shadow on a sunny day, we can treat the Sun as a point source located 150 million km away. On the other hand, the shadow that the Earth casts on the Moon during a lunar eclipse has a very dark region (the umbra) and a semi-dark region (the penumbra). This more complex shadow pattern can be partly explained by treating the Sun as a distributed source.

23-2 The Law of Reflection; Plane Mirrors

A ray of light that reflects from a surface obeys a very simple rule, known as the law of reflection. See, also, the illustrations in Figure 23.7.

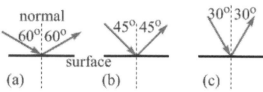

The Law of Reflection: for a ray of light reflecting from a surface, the angle of incidence is equal to the angle of reflection. These angles are generally measured from the normal (perpendicular) to the surface.

Figure 23.7: In each case, the ray obeys the Law of Reflection, in that the angle of incidence, measured from the normal, is equal to the angle of reflection. In the specific examples shown, both the incident ray and the reflected ray are at an angle, measured from the normal, of (a) 60°, (b) 45°, and (c) 30°.

A surface acts as a mirror when the law of reflection is followed on a large scale, as shown in Figure 23.8 (a). In that case, the whole beam of light, with many parallel rays, reflects as expected according to the law. This is known as **specular reflection**: mirror-like reflection that preserves the wave-front structure. In Figure 23.8 (b), however, the surface does not seem to obey the law of reflection. If we look at the magnified view, in (c), however, we see that the surface is irregular. The law of reflection is obeyed for each ray individually, but the irregularities in the surface cause the rays to move off in many different directions after being reflected. This is known as **diffuse reflection**: reflection in which the wave fronts are not preserved. Diffuse reflection explains why some surfaces that appear to be flat, such as a table or a road, do not act as mirrors. As far as light is concerned, these surfaces are far from flat.

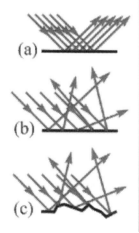

Figure 23.8: (a) Specular reflection from a flat mirror, in which all rays reflect at the same angle. (b) Many flat surfaces exhibit diffuse reflection, in which rays reflect at different angles. (c) A magnified view of the situation in (b). Even though the surface may appear flat to us, the surface is actually quite irregular as far as light is concerned.

The surface in Figure 23.8(a) is known as a plane mirror. Common examples are the mirrors in every bathroom. When we look at ourselves in such a mirror, where do we see our image? How large is the image? To answer such questions, we can use a ray diagram to determine where an image is formed and what its characteristics are.

EXPLORATION 23.2 – Using a ray diagram to find the location of an image

Step 1 – *An arrow is placed in front of a vertical plane mirror, as shown in Figure 23.9. Sketch two rays of light, which travel in different directions, that leave the tip of the arrow and reflect off the mirror. Show the direction of these rays after they reflect from the mirror.*

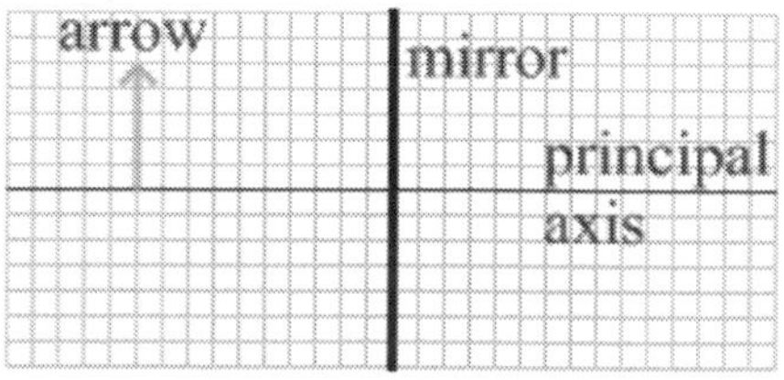

Figure 23.9: An arrow located some distance in front of a plane mirror.

Figure 23.10 shows a number of rays leaving the arrow and reflecting from the mirror, obeying the law of reflection. One of these, the horizontal ray, in red, that strikes the mirror at a 0° angle of incidence (measured from the normal to the mirror) is special, in that it reflects back along the path the ray came in on, and is thus easy to draw. However, you do not need to use this particular ray in the ray diagram – any two rays of light that reflect from the mirror can be used.

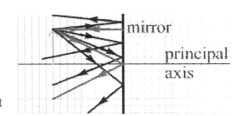

Figure 23.10: A selection of reflected rays. In each case, the reflected rays are drawn in the direction that is consistent with the law of reflection.

Step 2 – *The point where the reflected rays meet is where the tip of the image is located. Sketch the image of the arrow.* The reflected rays diverge as they travel away from the mirror to the left, but if we extend the reflected rays back through the mirror to the right, we find a point where they intersect. This is where the image of the tip of the arrow is located. Note that all the reflected rays, when they are extended back, pass through this point, which is why we can use any two reflected rays to create the ray diagram. Because the base of the arrow, which we call the object, is located on the principal axis (the horizontal line bisecting the mirror), we know that the base of the image will also be located there, so we draw an image of the arrow between the point where the image of the tip is and the principal axis.

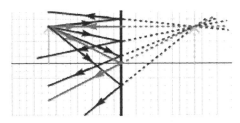

Figure 23.11: For a plane mirror, the reflected rays must be extended back through the mirror to find the location where they meet. Because the rays left the tip of the arrow, the point where the reflected rays meet is the location of the tip of the image of the arrow.

Step 3 – *Prove that the image is located as shown in Figure 23.11 by drawing two more ray diagrams, one showing the location of the image of the midpoint of the arrow, and one showing the location of the image of the base (bottom) of the arrow.* Figure 23.12 combines the ray diagram for the tip with those of the arrow's base and midpoint, showing that the image really is at the location shown in Figure 23.11.

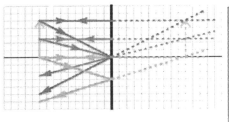

Figure 23.12: No matter which point on the object the rays start from, the reflected rays can be extended back to meet to the right of the mirror.

Step 4 – *Compare the reflected rays in Figure 23.12 to the rays in Figure 23.6.* Our brains cannot tell the difference between the two situations, which is why we see an image of the arrow formed at the location shown in Figure 23.13.

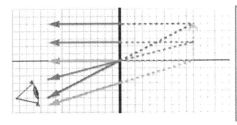

Figure 23.13: Our brains trace the reflected rays back along straight lines until they meet, and we see an image at that location.

Key idea for ray diagrams: The location of the image of any point on an object, when the image is created by a mirror, can be found by drawing rays of light that leave that point on the object and reflect from the mirror. The direction of the reflected rays must be consistent with the law of reflection. The point where the reflected rays meet is where the image of that point is.
Related End-of-Chapter Exercises: 1, 2, 5, 14.

Essential Question 23.2: First, make a prediction. When the arrow in Exploration 23.2 is moved closer to the mirror, will its image be larger, smaller, or the same size as the image we found in Step 2 above? Sketch a new ray diagram to check your prediction.

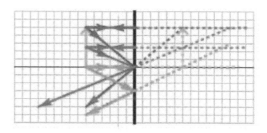

Answer to Essential Question 23.2: Images from plane mirrors are always the same size as the original object. We can see this in the ray diagram in Figure 23.14.

23-3 Spherical Mirrors: Ray Diagrams

Let us move now from plane mirrors to spherical mirrors, which curve like the surface of a sphere. Spherical mirrors can be convex, such as the mirrors on the passenger side of cars, or concave, such as shaving or makeup mirrors. Unlike plane mirrors, which always produce an image that is the same size as the object, the image in a convex mirror is always smaller than the object, while the image in a concave mirror can be larger, smaller, or the same size as the object.

Figure 23.14: Compare this figure to Figure 23.12. Moving the object closer to the mirror moves the image closer to the mirror, but the height of the image is still equal to the height of the object.

The focal point of a spherical mirror is defined by what the mirror does to a set of rays of light that are parallel to one another and to the principal axis of the mirror. As shown in Figure 23.15, a concave mirror reflects the rays so they converge to pass through the focal point, F. A convex mirror, in contrast, reflects parallel rays so that they diverge away from the focal point. Note that each ray obeys the law of reflection when it reflects from the mirror. The location of the focal point depends on the curvature of the mirror. The smaller the radius of curvature of the mirror, the closer the focal point is to the mirror's surface.

Smaller R = *closer focal point* (handwritten marginal note)

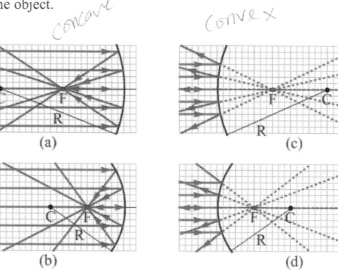

concave   convex (handwritten labels)

(a)   (c)

(b)   (d)

What we show here for spherical mirrors is an approximation, valid for rays that are not too far from the principal axis, in relation to the magnitude of the mirror's focal length. A mirror actually needs to have a parabolic shape to reflect all parallel rays through one point (or away from one point, for a diverging mirror). The fact that spherical mirrors do not really bring all such rays to a point (or diverge them away from a point) is a defect called **spherical aberration**.

Figure 23.15: Focal points are shown for four different mirrors. In (a) and (b), concave mirrors reflect parallel rays so that they converge to a single point called the focal point. In (c) and (d), convex mirrors reflect parallel rays so that they diverge away from the mirror's focal point. The location of the focal point depends on the mirror's radius of curvature. In each case, C is the mirror's center of curvature, R is the radius of curvature, and F is the focal point.

Focal length of a spherical mirror: The focal point of a spherical mirror is located halfway between the surface of the mirror and the mirror's center of curvature. Thus, the focal length of a spherical mirror has a magnitude of $R/2$, where R is the radius of curvature of the mirror. By convention, an object that diverges parallel light rays has a negative focal length (f), while an object that converges parallel light rays has a positive focal length. Thus:

For a concave mirror: $f = +\dfrac{R}{2}$.   (Eq. 23.1)   For a convex mirror: $f = -\dfrac{R}{2}$.   (Eq. 23.2)

In the limit that the radius of curvature approaches infinity, the mirror becomes a plane mirror and the focal length is either + infinity or – infinity (it does not matter which sign is used).

Figure 23.16: (a) For light rays that are parallel to the principal axis, a converging mirror re-unites the wave front at the focal point. From there, the wave front diverges as if the focal point is a point source. (b) A diverging mirror deflects the parallel rays so the wave fronts appear to diverge from the mirror's focal point.

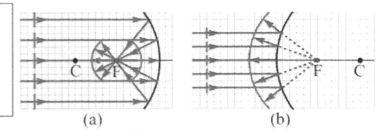

(a)   (b)

Following the wave fronts

Figure 23.16 shows what spherical mirrors do to wave fronts. For the converging mirror, the waves take the same time to get from the left to the focal point. For the diverging mirror, once the waves reflect from the mirror, it is as if they left the focal point at the same time.

EXPLORATION 23.3 – Ray diagram for a convex mirror

We will follow a process similar to that for plane mirrors to draw a ray diagram for a convex (diverging) mirror.

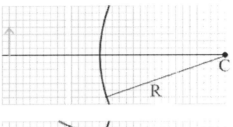

Figure 23.17: An object, represented by an arrow, is placed in front of a convex (diverging) mirror. The mirror's center of curvature is shown.

Step 1 – *First, locate the mirror's focal point. Then, draw a light ray that leaves the tip of the object (its top) and goes parallel to the principal axis. Show how this parallel ray reflects from the mirror.* The focal point is halfway between the point where the principal axis intersects the mirror, and the center of curvature. For a convex mirror, all parallel rays appear to diverge from the focal point, so we draw the reflected ray reflecting along a line that takes it directly away from the focal point, as in Figure 23.18.

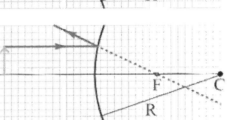

Figure 23.18: The parallel ray reflects from the mirror in such a way that it travels directly away from the focal point.

Step 2 – *Sketch a second ray that leaves the tip of the object and reflects from the mirror. Using your reflected rays, draw the image.* One useful ray, the lower ray in Figure 23.19, reflects from the point on the mirror that the principal axis passes through. At that location, the reflection is like that from a vertical plane mirror. Another useful ray goes straight toward the mirror's center of curvature. This ray has a 0° angle of incidence, and thus reflects back along the same line. The reflected rays diverge on the left of the mirror, but we can extend them back to meet on the right side of the mirror, showing us where the tip of the image is. The image, smaller than the object, is drawn from the tip down to the principal axis.

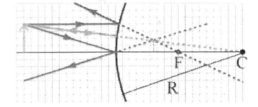

Figure 23.19: In addition to the parallel ray, two other rays are easy to draw the reflection for. The ray that strikes the mirror at the principal axis reflects as if the mirror was a vertical plane mirror. The ray that travels directly toward the center of curvature strikes the mirror at 90° to the surface, and thus reflects straight back.

Key idea: As with a plane mirror, when a number of rays leave the same point on an object and reflect from a spherical mirror, the corresponding point on the image is located at the intersection of the reflected rays.   **Related End-of-Chapter Exercises: 6, 9, 22, 60, 61.**

Essential Question 23.3: (a) Modify the ray diagram in Figure 23.19 to show what happens to the image when the object is moved closer to the mirror. (b) Add several more rays (leaving the tip of the object) to your modified ray diagram, showing what the rays do when they reflect from the mirror. How do you know how to draw the reflected rays?

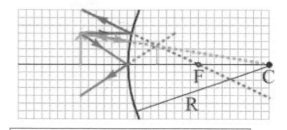

Answer to Essential Question 23.3: (a) The parallel ray is nice to work with, because its reflection does not change direction when the object is moved left or right. The ray that strikes the mirror at the principal axis, however, comes in at a larger angle of incidence, and thus reflects at a larger angle. As shown in Figure 23.20, when the object is closer to the mirror, the image increases in size and moves closer to the mirror.

Figure 23.20: The new ray diagram shows that moving the object closer to the mirror results in the image increasing in size (but remaining smaller than the object) and moving closer to the surface of the mirror.

(b) We have two ways of knowing how to draw the reflected rays properly. First, the law of reflection must be obeyed when the rays reflect from the mirror. Second, we know that when we extend the reflected rays back, they will pass through the tip of the image, which we located in Figure 23.20. Three additional rays are shown in Figure 23.21.

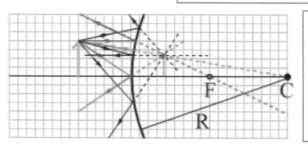

Figure 23.21: For all rays of light that leave the tip of the object and reflect from the mirror, the reflected rays can be extended back to pass through the tip of the image.

23-4 A Qualitative Approach: Image Characteristics

So far, we have looked at ray diagrams in two cases, the case of a plane mirror (section 23-2) and that of a convex mirror (section 23-3). In both cases, we had to extend the reflected rays back through the mirror to get the rays to intersect, giving us an image behind the mirror. We see this all the time with plane mirrors. If you stand 1.0 m in front of a plane mirror, you see an image of yourself 1.0 m behind the mirror. Is it always true that the image created by a mirror is located behind the mirror? We will first investigate the concave mirror, the last case we will deal with, and we will then summarize various image characteristics.

EXAMPLE 23.4 – A ray diagram for a concave mirror
An object, represented by an arrow, is located 15.0 cm in front of a concave mirror that has a focal length of +10.0 cm, as shown in Figure 23.22. Sketch a ray diagram to find the location of the tip of the image of the arrow, and sketch the image on the diagram.

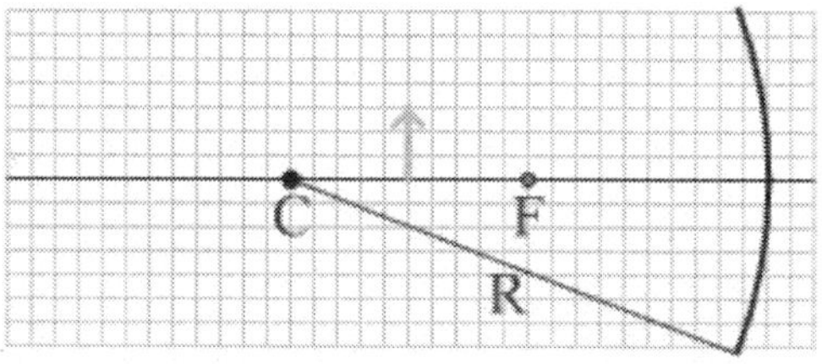

Figure 23.22: An object in front of a concave mirror. The squares in the grid measure 1.0 cm × 1.0 cm.

SOLUTION
Let's use the same procedure we have used previously, starting by drawing rays of light that leave the tip of the object. Again, one useful ray, shown in red, is the ray that travels parallel to the principal axis. The concave mirror reflects this ray so that it passes through the mirror's focal point. A second ray that we know how to draw is the one, in blue, that reflects from the mirror at the principal axis, reflecting at that point as if the mirror were a vertical plane mirror. Note that these two rays meet to the left of the mirror, giving us the location of the tip of the image. As usual, we draw the image of the arrow from that point to the principal axis, because the base of the object is also on the principal axis.

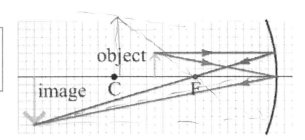

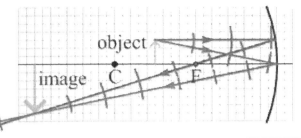

Figure 23.23: A ray diagram showing two of the several possible rays we can draw to locate the tip of the image.

Wave fronts

Figure 23.24 shows how the converging mirror affects the wave fronts (in purple). The light leaving the tip of the object, reflecting from the mirror, and arriving at the tip of the image, takes the same time no matter which path it takes.

Image characteristics

The image in Figure 23.23 is quite different from images we have seen earlier in the chapter. First, the image is inverted (upside down) compared to the object. Second, the image is larger than the object. Third, the image is formed from light rays that actually pass through the image. Note that concave mirrors do not always form images with these characteristics, as we will investigate in more detail in section 23-6. For now, however, let's discuss some general issues related to image characteristics.

Figure 23.24: The wave fronts that leave the tip of the object converge on the tip of the image – the rays take the same time to reach the image. The wave fronts then diverge away from the image. To your brain, the image looks like an object.

Upright or inverted?

As we have seen, plane mirrors and convex mirrors produce an upright image. This is an image that is in the same orientation as the object. An inverted image, like that in Example 23.4, is one in which the image is upside down in relation to the object.

Real or virtual?

virtual = upright

Most of the images we see on a daily basis in mirrors are virtual. A virtual image is one that the light does not actually pass through. Instead, our brains see an image there because, when we look in the mirror at the object, our brains are so used to light traveling in straight lines that we trace all the reflected rays back to their apparent source, the point behind the mirror where the light appears to come from. For a single mirror, when the image is virtual it is also upright.

real = inverted

In Example 23.4, we saw a situation in which the light rays passed through the mirror, creating a real image. Real images, from concave mirrors, have a three-dimensional quality that virtual images do not have, and it is worth going out of your way to see one. For a single mirror, when the image is real it is also inverted.

Larger or smaller?

A plane mirror, as we investigated earlier, produces an image that is always the same size as the object. Convex (diverging) mirrors, on the other hand, always produce an image that is smaller than the object. Concave (converging) mirrors, as we will investigate further in section 23-6, can produce an image that is larger, smaller, or the same size as the object. We will discuss these ideas in a more quantitative way when we define the magnification of a mirror, in section 23-5.

Related End-of-Chapter Exercises: 7, 11, 12.

Essential Question 23.4: Consider the ray diagram in Figure 23.23. If an object of the same size of the image was placed at the image's position, where would its image be located?

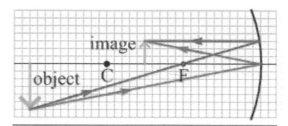

Figure 23.25: When the image is a real image, the ray diagram is reversible.

Answer to Essential Question 23.4: The image would be located where the object in Figure 23.23 is located. This demonstrates an important fact about light rays – they are reversible. As Figure 23.25 shows, we can simply reverse the direction of the rays from Figure 23.23 to obtain the appropriate ray diagram. Note that we can do this only when the image is a real image.

23-5 A Quantitative Approach: The Mirror Equation

The branch of optics that involves mirrors and lenses is generally called geometrical optics, because it is based on the geometry of similar triangles. Let's investigate this geometry, and use it to derive an important relationship between the image distance, object distance, and the focal length.

Let's look again at the ray diagram we drew in Figure 23.23 of section 23-4, shown again here in Figure 23.26.

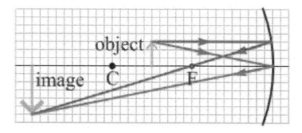

Figure 23.26: The ray diagram we constructed in section 23-4, for an object in front of a concave mirror.

Remove the parallel ray (and its reflection), and examine the two shaded triangles in Figure 23.27, bounded by the other ray and its reflection, the principal axis, and the object and image. The two triangles are similar, because the three angles in one triangle are the same as the three angles in the other triangle. We can now define the following variables: d_o is the object distance, the distance of the object from the center of the mirror; d_i is the image distance, the distance of the image from the center of the mirror; h_o is the height of the object; h_i is the height of the image.

Using the fact that the ratios of the lengths of corresponding sides in similar triangles are equal, we find that:

$$-\frac{h_i}{h_o} = \frac{d_i}{d_o}. \qquad \text{(Equation 23.3)}$$

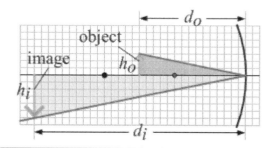

Figure 23.27: Similar triangles, bounded by the principal axis, the object and image, and the ray that reflects from the center of the mirror.

The image height is negative because the image is inverted, which is why we need the minus sign in the equation. Let's now return to Figure 23.26, and use the parallel ray instead. This gives us the shaded similar triangles shown in Figure 23.28.

We use an approximation, which is valid as long as the object height is relatively small, that the length of the smaller triangle is f, the focal length. Again, using the fact that the ratios of the lengths of corresponding sides in similar triangles are equal, we find that:

$$\frac{d_i - f}{f} = -\frac{h_i}{h_o}.$$

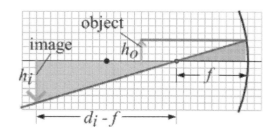

Figure 23.28: Similar triangles, with two sides bounded by the principal axis and the reflection of the parallel ray, and a third side that is equal to the object height (small triangle) or the image height (large triangle).

Simplifying the left side, and bringing in equation 23.3, we get: $\dfrac{d_i}{f} - 1 = \dfrac{d_i}{d_o}$.

Dividing both sides by d_i gives: $\dfrac{1}{f} - \dfrac{1}{d_i} = \dfrac{1}{d_o}$, which is generally written as:

$$\frac{1}{f} = \frac{1}{d_o} + \frac{1}{d_i}.$$ (Equation 23.4: **The mirror equation**)

The mnemonic "If I do I di" can help you to remember the mirror equation.

Often, we know the focal length f and the object distance d_o, so equation 23.4 can be solved for d_i, the image distance:

$$d_i = \frac{d_o \times f}{d_o - f}$$ (Equation 23.5: **The mirror equation, solved for the image distance**)

Sign conventions

We derived the mirror equation above by using a specific case involving a concave mirror. The equation can also be applied to a plane mirror, a convex mirror, and all situations involving a concave mirror if we use the following sign conventions.

The focal length is positive for a concave mirror, and negative for a convex mirror.

The image distance is positive if the image is on the reflective side of the mirror (a real image), and negative if the image is behind the mirror (a virtual image).

The image height is positive when the image is above the principal axis, and negative when the image is below the principal axis. A similar rule applies to the object height.

$f \Rightarrow$ (+) Concave
(−) Convex

$d \Rightarrow$

Magnification

The magnification, m, is defined as the ratio of the height of the image (h_i) to the height of the object (h_o). Making use of Equation 23.3, we can write the magnification as:

$$m = \frac{h_i}{h_o} = -\frac{d_i}{d_o}.$$ (Equation 23.6: **Magnification**)

The relative sizes of the image and object are as follows:

- The image is larger than the object if $|m| > 1$.

- The image and object have the same size if $|m| = 1$.

- The image is smaller than the object if $|m| < 1$.

The sign of the magnification tells us whether the image is upright (+) or inverted (−) compared to the object.

Related End-of-Chapter Exercises: 15 – 19.

Essential Question 23.5: As you are analyzing a spherical mirror situation, you write an equation that states: $\dfrac{1}{f} = \dfrac{1}{+12 \text{ cm}} + \dfrac{1}{+24 \text{ cm}}$. What is the value of $1/f$ in this situation? What is f?

To add fractions you need to find a common denominator.

$$\frac{1}{f} = \frac{1}{+12 \text{ cm}} + \frac{1}{+24 \text{ cm}} = \frac{2}{+24 \text{ cm}} + \frac{1}{+24 \text{ cm}} = \frac{3}{+24 \text{ cm}} . \text{ This gives } f = \frac{+24 \text{ cm}}{3} = 8.0 \text{ cm}.$$

23-6 Analyzing the Concave Mirror

In section 23-4, we drew one ray diagram for a concave mirror. Let's investigate the range of ray diagrams we can draw for such a mirror.

EXPLORATION 23.6 – Ray diagrams for a concave mirror

Step 1 – Draw a ray diagram for an object located 40 cm from a concave mirror that has a radius of curvature of 20 cm. Verify the image location on your diagram with the mirror equation. In drawing a ray diagram, it is helpful to know where the mirror's focal point is. For a spherical mirror, the focal point is halfway between the mirror's center of curvature and the point at which the principal axis intersects the mirror. Thus, the focal length in this case is +10 cm.

In Figure 23.29, two rays are shown. One is the parallel ray, which leaves the tip of the object, travels parallel to the principal axis, and reflects from the mirror so that it passes through the focal point. The second ray reflects off the mirror at the point at which the principal axis meets the mirror, reflecting as if the mirror was a vertical plane mirror.

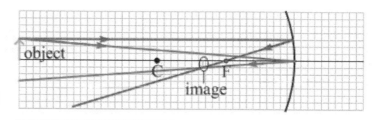

Figure 23.29: A ray diagram for the situation in which the object is far from the mirror. The squares in the grid measure 1.0 cm × 1.0 cm.

Applying the mirror equation, in the form of equation 23.5, to find the image distance:

$$d_i = \frac{d_o \times f}{d_o - f} = \frac{(40 \text{ cm}) \times (+10 \text{ cm})}{(40 \text{ cm}) - (+10 \text{ cm})} = \frac{+400 \text{ cm}^2}{30 \text{ cm}} = +13.3 \text{ cm} .$$

This image distance is consistent with the ray diagram in Figure 23.29.

Step 2 – Repeat step 1, with the object now moved to the center of curvature. The parallel ray follows the same path as it did Figure 23.29. As shown in Figure 23.30, the ray that reflects from the center of the mirror follows a different path, because shifting the object changes the angle of incidence for that ray. This situation is a special case. When the object is located at the center of curvature, the image is inverted, also at the center of curvature, and the same size as the object because the object and image are the same distance from the mirror.

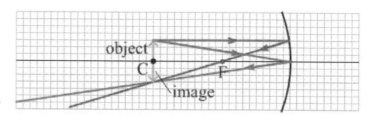

Figure 23.30: When the object is at the mirror's center of curvature, so is the image.

Applying the mirror equation to find the image distance, we get:

$$d_i = \frac{d_o \times f}{d_o - f} = \frac{(20 \text{ cm}) \times (+10 \text{ cm})}{(20 \text{ cm}) - (+10 \text{ cm})} = \frac{+200 \text{ cm}^2}{10 \text{ cm}} = +20 \text{ cm} , \text{ matching the ray diagram.}$$

Step 3 – *Repeat step 1, with the object 15 cm from the mirror.* No matter where the object is, the parallel ray follows the same path. The path of the second ray, in blue, depends on the object's position. The ray diagram (Figure 23.31) shows that the image is real, inverted, larger than the object, and about twice as far from the mirror as the object.

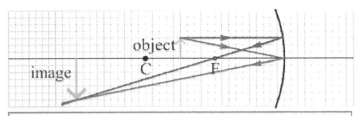

Figure 23.31: A ray diagram for a situation in which the object is between the mirror's center of curvature and its focal point.

Applying the mirror equation gives:

$$d_i = \frac{d_o \times f}{d_o - f} = \frac{(15 \text{ cm}) \times (+10 \text{ cm})}{(15 \text{ cm}) - (+10 \text{ cm})} = \frac{+150 \text{ cm}^2}{5.0 \text{ cm}} = +30 \text{ cm}, \text{ matching the ray diagram.}$$

Step 4 – *Repeat step 1, with the object at the mirror's focal point.* As shown in Figure 23.32, the two reflected rays are parallel to one another, and never meet. In such a case the image is formed at infinity.

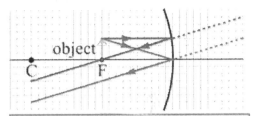

Figure 23.32: A ray diagram for a situation in which the object is at the focal point.

Applying the mirror equation gives:

$$d_i = \frac{d_o \times f}{d_o - f} = \frac{(10 \text{ cm}) \times (+10 \text{ cm})}{(10 \text{ cm}) - (+10 \text{ cm})} = \frac{+100 \text{ cm}^2}{0 \text{ cm}} = +\infty,$$

which agrees with the ray diagram.

Step 5 – *Repeat step 1, with the object 5.0 cm from the mirror.* When the object is closer to the mirror than the focal point, the two reflected rays diverge to the left of the mirror, and they must be extended back to meet on the right of the mirror. The result is a virtual, upright image that is larger than the object, as shown in Figure 23.33.

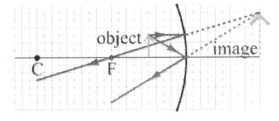

Figure 23.33: A ray diagram for a situation in when the object is between the mirror's surface and its focal point.

Applying the mirror equation to find the image distance gives:

$$d_i = \frac{d_o \times f}{d_o - f} = \frac{(5.0 \text{ cm}) \times (+10 \text{ cm})}{(5.0 \text{ cm}) - (+10 \text{ cm})} = \frac{+50 \text{ cm}^2}{-5.0 \text{ cm}} = -10 \text{ cm}.$$

Recalling the sign convention that a negative image distance is consistent with a virtual image, the result from the mirror equation is consistent with the ray diagram.

Key idea for concave mirrors: Depending on where the object is placed relative to a concave mirror's focal point, the mirror can form an image of the object that is real or virtual. If the image is real, it can be larger than, smaller than, or the same size as the object. If the image is virtual, the image is larger than the object.   **Related End-of-Chapter Exercises: 27 and 47 – 49.**

Essential Question 23.6: When an object is placed 20 cm from a spherical mirror, the image formed by the mirror is larger than the object. What kind of mirror is it? What, if anything, can you say about the mirror's focal length?

Answer to Essential Question 23.6: The mirror must be concave, because a convex mirror cannot produce an image that is larger than the object. A concave mirror produces an image larger than the object only when the object is between the mirror and twice the focal length. So, twice the focal length must be at least +20 cm, and the focal length must be at least +10 cm. All we can say is that the focal length is greater than or equal to +10 cm.

23-7 An Example Problem

Let's begin by discussing a general approach we can use to solve problems involving mirrors. We will then apply the method to a particular situation.

A general method for solving problems involving mirrors
1. Sketch a ray diagram, showing rays leaving the tip of the object and reflecting from the mirror. Where the reflected rays meet is where the tip of the image is located. The ray diagram gives us qualitative information about the location and size of the image and about the characteristics of the image.
2. Apply the mirror equation and/or the magnification equation. Make sure that the signs you use match those listed in the sign convention in section 23-5. The equations provide quantitative information about the location and size of the image and about the image characteristics.
3. Check the results of applying the equations with your ray diagram, to see if the equations and the ray diagram give consistent results.

Rays that are easy to draw the reflections for

To locate an image on a ray diagram, you need a minimum of two rays. If you draw more than two rays, however, you can check the image location you find with the first two rays. Remember, too, that you can draw any number of rays reflecting from the mirror, and that all the rays should obey the law of reflection. There are at least four rays that are easy to draw the reflections for. These rays are shown on Figure 23.34, and include:

1. The ray that goes parallel to the principal axis, and reflects so that it passes through the focal point (concave mirror), or away from the focal point (convex mirror).
2. The ray that reflects from the point on the principal axis that intersects the surface of the mirror. The principal axis is perpendicular to the surface of the mirror, so the angle between the incident ray and the principal axis is the same as the angle between the reflected ray and the principal axis.
3. The ray that travels along the straight line connecting the tip of the object and the mirror's center of curvature. This ray is incident on the mirror along the normal to the mirror's surface, and thus reflects straight back along the same line.
4. The ray that travels along the straight line connecting the tip of the object and the focal point. This ray reflects to go parallel to the principal axis.

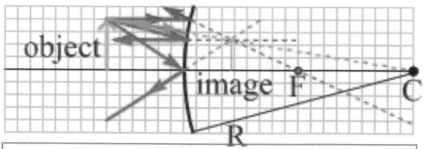

Figure 23.34: An example of the four rays that are easy to draw the reflections for. All the reflected rays meet at the tip of the image.

EXAMPLE 23.7 – Applying the general method

When you stand in front of a mirror that has a radius of curvature of 40 cm, you see an image that is half your size. What kind of mirror is it? How far from the mirror are you? Sketch a ray diagram to check your calculations.

SOLUTION

In this case, let's first apply the equations and then draw the ray diagram. The mirror could be convex, because convex mirrors always produce images that are smaller than the object. A convex mirror produces a virtual, upright image, so the sign of the magnification is positive. Applying the magnification equation, we get:

$$m = +\frac{1}{2} = -\frac{d_i}{d_o}, \text{ which tells us that } \frac{1}{d_i} = -\frac{2}{d_o}.$$

For a convex mirror, the focal length is $-R/2$, which in this case is -20 cm. Applying the mirror equation:

$$\frac{1}{f} = \frac{1}{d_o} + \frac{1}{d_i} = \frac{1}{d_o} - \frac{2}{d_o} = -\frac{1}{d_o}.$$

Thus, we find that $d_o = -f = +20$ cm, and we can show that $d_i = -10$ cm.

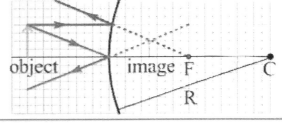

The ray diagram for this situation is shown in Figure 23.35, confirming the calculations.

Figure 23.35: A ray diagram for the solution involving a convex mirror. Each box on the grid measures 2 cm × 2 cm.

The solution above is only one of the possible answers. The mirror could also be concave, because a concave mirror can produce a real, inverted image, so the sign of the magnification is negative. Applying the magnification equation, we get:

$$m = -\frac{1}{2} = -\frac{d_i}{d_o}, \text{ which tells us that } \frac{1}{d_i} = \frac{2}{d_o}.$$

For a concave mirror the focal length is $+R/2$, which in this case is $+20$ cm. Applying the mirror equation:

$$\frac{1}{f} = \frac{1}{d_o} + \frac{1}{d_i} = \frac{1}{d_o} + \frac{2}{d_o} = \frac{3}{d_o}.$$

Thus, we find that $d_o = 3f = +60$ cm, and we can show that $d_i = +30$ cm.

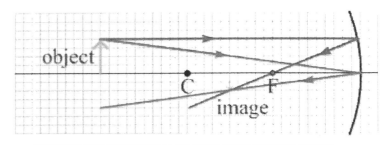

The ray diagram for this situation is shown in Figure 23.36, again confirming the calculations above.

Related End-of-Chapter Exercises: 20, 21, 23, 24, 43.

Figure 23.36: A ray diagram for the situation involving a concave mirror. Each box on the grid measures 2 cm × 2 cm.

Essential Question 23.7: Return to the situation described in Example 23.7. Would there still be two solutions if the image was larger than the object? Explain.

Answer to Essential Question 23.7: Yes, there would still be two solutions, but we would not have a solution associated with a convex mirror, because the convex mirror cannot produce an image that is larger than the object. Instead, both solutions would be associated with a concave mirror. One solution would be for a real image, and the other solution would be for a virtual image.

Chapter Summary

Essential Idea: Reflection and Mirrors.
To understand the image that is created by a mirror, we make use of a simple model of light called the ray model. In the ray model, a ray of light travels in a straight line until it encounters an object. When a ray of light reflects from an object, the light obeys the Law of Reflection.

The Law of Reflection
The angle of incidence is equal to the angle of reflection. These angles are generally measured from the normal to the surface.

Image Formation
For a mirror to form an image of an object, light rays must leave the object and be reflected by the mirror. If the rays leaving a single point on the object are reflected so that they pass through a single point, a real image is formed. If, instead, such reflected rays appear to diverge from a single point behind the mirror (as is the case for a typical bathroom mirror), a virtual image is formed.

Ray Diagrams
When drawing a ray diagram, we generally show rays leaving the tip of the object and reflecting from the mirror. Where the reflected rays meet is where the tip of the image is located. The ray diagram gives us qualitative information about the location and size of the image and about the image characteristics. All rays obey the Law of Reflection when they reflect from the mirror, but some reflected rays are particularly easy to draw. A summary of four such rays is given in section 23-7.

Plane and Spherical Mirrors

Type of mirror	Focal length	Image characteristics
Plane	∞	The image is virtual, upright, the same size as the object, and the same distance behind the mirror that the object is in front of the mirror.
Convex (diverging)	$-R/2$, R being the mirror's radius of curvature	The image is virtual, upright, smaller than the object, and located between the mirror and the mirror's focal point.
Concave (converging)	$+R/2$	The image can be real or virtual, and larger than, smaller than, or the same size as the object. See the table below for details.

Table 23.1: A summary of the mirrors we investigated in this chapter.

Images formed by a Concave (Converging) Mirror

Object position	Image position	Image characteristics
∞	At the focal point.	Real image with height of zero.
Moving from ∞ toward the center of curvature.	Moving from the focal point toward the center of curvature.	The image is real, inverted, and smaller than the object. The image moves closer to the center of curvature, and increases in height, as the object is moved closer to the center of curvature.
At the center of curvature.	At the center of curvature.	The image is real, inverted, and the same size as the object.
Moving from the center of curvature toward the focal point.	Moving from the center of curvature toward infinity.	The image is real, inverted, and larger than the object. The image moves farther from the mirror, and increases in height, as the object is moved closer to the focal point.
At the focal point.	At infinity.	The image is at infinity, and is infinitely tall.
Closer to the mirror than the focal point.	Behind the mirror	The image is virtual, upright, and larger than the object. The image moves closer to the mirror, and decreases in height, as the object is moved closer to the mirror.

Table 23.2: A summary of the image positions and characteristics for various object positions with a concave mirror.

The mirror equation

The mirror equation relates the object distance, d_o, the image distance, d_i, and the mirror's focal length, f. The mnemonic "If I do I di" can help you to remember the mirror equation.

$$\frac{1}{f} = \frac{1}{d_o} + \frac{1}{d_i} .$$ (Equation 23.4: **The mirror equation**)

$$d_i = \frac{d_o \times f}{d_o - f}$$ (Equation 23.5: **The mirror equation, solved for the image distance**)

Sign conventions

The focal length is positive for a concave mirror, and negative for a convex mirror.

The image distance is positive if the image is on the reflective side of the mirror (a real image), and negative if the image is behind the mirror (a virtual image).

The image height is positive when the image is above the principal axis, and negative when the image is below the principal axis. A similar rule applies to the object height.

The image height is positive when the image is upright, and negative when the image is inverted. A similar rule applies to the object height.

Magnification

The magnification, m, is the ratio of the image height (h_i) to the object height (h_o).

$$m = \frac{h_i}{h_o} = -\frac{d_i}{d_o} .$$ (Equation 23.6: **Magnification**)

- The image is larger than the object if $|m| > 1$.

- The image and object have the same size if $|m| = 1$.

- The image is smaller than the object if $|m| < 1$.

The magnification is positive if the image is upright, and negative if the image is inverted.

End-of-Chapter Exercises

Exercises 1 – 12 are conceptual questions designed to see whether you understand the main concepts of the chapter.

1. You and your friend Leigh can both see Leigh's reflection in the same plane mirror. Leigh sees her image at a particular location. You are observing the situation from a different position than Leigh. Do you observe Leigh's image in the same position that Leigh does, or in a different position? Explain.

2. Figure 23.37 shows a red ball that is near a small plane mirror. Is it possible to see a reflection of the ball in this mirror? If so, sketch a diagram showing the location of the ball's image, and showing where your eye could be located so as to see the ball's image. If not, explain why not.

ball

Figure 23.37: A ball near a small plane mirror, for Exercise 2.

3. As shown in Figure 23.38, a point source of light shines on a card that has a narrow vertical slit cut in it. The card is halfway between the source and a screen. Sketch the resulting pattern on the screen.

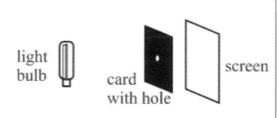

Figure 23.38: A point source of light illuminates a card with a narrow slit. For Exercise 3.

4. As shown in Figure 23.39, a light bulb with a long narrow vertical filament illuminates a card that has a small hole in it. The card is two-thirds of the way from the light bulb to a screen. Sketch the resulting pattern on the screen.

Figure 23.39: A light bulb with a vertical filament (in red) illuminates a card that has a small hole in it. For Exercise 4.

5. Two plane mirrors are placed so they are perpendicular to one another, and you stand in front of this pair of mirrors. You are represented by the red dot in Figure 23.40. (a) How many images of yourself do you observe? (b) Draw a diagram to show where the image(s) is/are located. (c) For the image farthest from you, sketch a ray diagram to show how the image is formed.

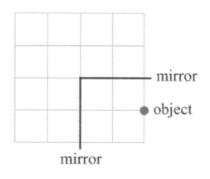

Figure 23.40: You, the object in this figure, stand in the position shown in front of a right-angled mirror. For Exercise 5.

6. Figure 23.41 shows a single ray on a ray diagram. Duplicate the diagram, and add a second ray to show the position of the image.

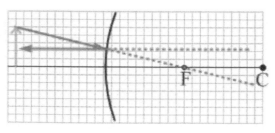

Figure 23.41: A single ray, directed toward the focal point of the mirror, is shown on the figure. For Exercise 6.

7. Two rays of light are shown on the ray diagram in Figure 23.42, along with two arrows representing the object and the image. Which arrow represents the object, and which represents the image?

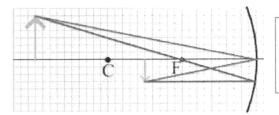

Figure 23.42: A ray diagram, for Exercise 7.

8. Figure 23.43 shows an object in front of a concave mirror. First, re-draw the diagram, preferably on a piece of graph paper. For parts (a) – (d), start the ray from the tip of the object and show how the ray reflects from the mirror. (a) On your diagram, draw a ray that travels parallel to the principal axis toward the mirror. (b) Draw a ray that reflects from the mirror at the point the principal axis intersects the mirror. (c) Draw a ray that travels toward the mirror along the line connecting the tip of the object and the focal point. (d) Draw a ray that travels toward the mirror along the line connecting the tip of the object and the center of curvature. (e) Use the rays to locate the image on the diagram.

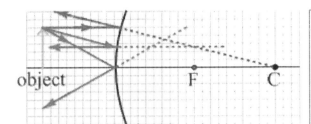

Figure 23.43: An object in front of a concave mirror, for Exercise 8.

9. Three rays of light are shown on the ray diagram in Figure 23.44. One of them is drawn incorrectly. (a) Identify the ray that is drawn incorrectly. (b) Draw a corrected diagram and show the location of the image.

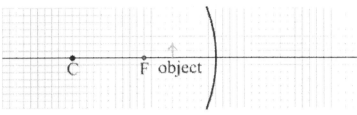

Figure 23.44: An incorrect ray diagram for an object in front of a convex mirror, for Exercise 9.

10. Figure 23.45 shows a concave mirror, and the virtual image formed by this mirror. Draw a ray diagram to show the location of the object in this situation.

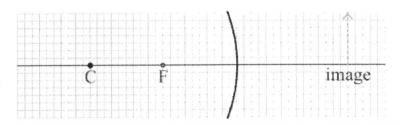

Figure 23.45: The virtual image created by a concave mirror. Draw the corresponding ray diagram to find the object. For Exercise 10.

11. Identify whether the mirror in this situation is a plane mirror, a convex mirror, or a concave mirror. For each part below, determine which of the possibilities you can rule out, if any. (a) First, when an object is placed 10 cm from a mirror, a virtual image is observed. (b) Then, when the object is moved a little closer to the mirror, the image is observed to move closer to the mirror. (c) As the object is moved closer to the mirror, the image is observed to decrease in size.

12. Identify whether the mirror in this situation is plane, convex, or concave. For each part below, determine which of the possibilities you can rule out, if any. (a) First, when an object is placed 10 cm from a mirror, a virtual image is observed. (b) Then, when the object is moved a little closer to the mirror, the image is observed to increase in size. (c) What, if anything, can you conclude about the mirror's focal length?

Exercises 13 – 14 involve plane mirrors.

13. A laser beam reflects from a plane mirror, as shown in Figure 23.46. The incident beam is at 40°, as measured from the horizontal. Initially, the reflected beam is horizontal. (a) What is the angle, initially, between the mirror and the horizontal? (b) When the mirror is tilted by 10°, what is the angle between the reflected beam and the horizontal?

Figure 23.46: In (a), a laser beam shining on a mirror reflects so the reflected beam is horizontal. In (b), the mirror is tilted by 10° from its orientation in (a). For Exercise 13.

14. Maria stands 50 cm away from a small plane mirror that is mounted on the wall in front of her. Maria's face is about 20 cm tall. (a) If Maria is to be able to see the image of her entire face without moving, what is the minimum height the mirror can be? (b) Describe how this minimum-height mirror should be mounted on the wall. (c) If Maria is 1.0 m from the mirror, what is the minimum height the mirror can be? (d) Sketch one or more ray diagrams to support your answers to parts (a) – (c).

Exercises 15 – 19 are designed to give you practice applying the mirror equation.

15. As you are analyzing a spherical mirror situation, you write an equation that states:
$$\frac{1}{f} = \frac{1}{+20 \text{ cm}} + \frac{1}{+30 \text{ cm}}.$$ (a) What is the value of $1/f$ in this situation? (b) What is the mirror's focal length? (c) What kind of mirror is this?

16. Return to Exercise 15. What is the object distance in this situation, and what is the image distance?

17. As you are analyzing a spherical mirror situation, you write an equation that states:
$$\frac{1}{f} = \frac{1}{+20 \text{ cm}} + \frac{1}{-60 \text{ cm}}.$$ (a) What is the value of $1/f$ in this situation? (b) What is the mirror's focal length? (c) What kind of mirror is this?

18. As you are analyzing a spherical mirror situation, you write an equation that states:
$$\frac{1}{f} = \frac{1}{+20 \text{ cm}} + \frac{1}{-10 \text{ cm}}.$$ (a) What is the value of $1/f$ in this situation? (b) What is the mirror's focal length? (c) What kind of mirror is this?

19. As you are analyzing a spherical mirror situation, you write an equation that states:
$$\frac{1}{+20 \text{ cm}} = \frac{1}{+20 \text{ cm}} + \frac{1}{d_i}.$$ (a) What is the value of $1/d_i$ in this situation? (b) What is the image distance?

Exercises 20 – 24 are designed to give you practice applying the general method for analyzing a problem involving mirrors.

20. An object is placed 25 cm away from a mirror that has a focal length of 10 cm. (a) Sketch a ray diagram, to show the position of the image and the image characteristics. (b) Determine the image distance. (c) Determine the magnification.

21. An object is placed 25 cm away from a mirror that has a focal length of –10 cm. (a) Sketch a ray diagram, to show the position of the image and the image characteristics. (b) Determine the image distance. (c) Determine the magnification.

22. Some people like to place large reflective balls in their gardens. Let's say that such a ball has a radius of curvature of 20 cm, and a bird perched on a branch is 1.0 m from the ball. (a) Determine the image distance for the bird's image. (b) Sketch a ray diagram to verify your calculation in part (a).

23. A shiny spoon can be approximated as a spherical mirror with a radius of curvature of 8.0 cm. You hold your finger 12.0 cm from the spoon's concave side. (a) Sketch a ray diagram to determine the image location and the image characteristics. (b) Apply the mirror equation and the magnification equation to determine the location of the image of your finger, and the size of the image compared to the size of your finger.

24. Repeat Exercise 23 but, this time, flip the spoon over so your finger is 12.0 cm from the spoon's convex side.

Exercises 25 – 29 involve applications of reflection and mirrors.

25. The Federal Motor Carrier Safety Administration, in the United States, mandates that all cars have a convex passenger-side rear-view mirror. A particular rear-view mirror on a car has a radius of curvature of 1.2 m. If a large truck is 8.0 m away from this mirror, determine (a) the image distance, and (b) the magnification. (c) Which of these results explains why "Objects in mirror are closer than they appear," which is the warning stamped on the mirror.

26. Figure 23.47 shows a picture of a periscope, such as that used by submarine captains to see what is at the surface of the water, or by spectators at golf tournaments to see over the heads of people standing in front. Two parallel plane mirrors are used to make the periscope. (a) Re-draw the diagram, and show the location of the image, created by mirror 1, of the object. (b) The image created by mirror 1 is the object for mirror 2. Show where the image created by mirror 2 is located. This is the image you see when you look in the periscope.

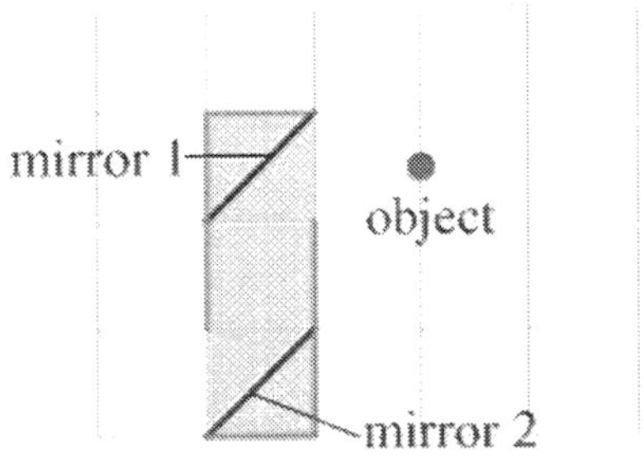

Figure 23.47: A periscope, which has two parallel plane mirrors, is an example of a practical device that involves reflection, for Exercise 26.

27. As part of a show about optical illusions, the illusionist sets up a large concave mirror, with a focal length of 2.0 m, at the back of the stage. This arrangement is shown in Figure 23.48. Hidden underneath the stage is a model of a lion, only ¼ as tall as a real lion. (a) How far should the model be placed from the mirror so that the audience sees a life-size lion hovering over the stage? (b) If the audience is to see an upright lion, should the model be upright or inverted?

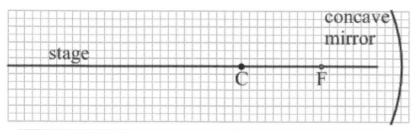

Figure 23.48: A large concave mirror, used by an illusionist to create an optical illusion, for Exercise 27.

28. A concave mirror used by a person who is shaving or applying makeup creates a virtual image of the face that is 1.5 times larger than the person's actual face when the face is 25 cm from the mirror. Determine (a) the image distance, and (b) the mirror's focal length.

29. A particular convex mirror has a radius of curvature of 1.5 m. The mirror is mounted in the corner of a store, near the ceiling, to help prevent shoplifting. (a) Let's say that you are 1.8 m tall, and that you are standing 2.0 m away from the mirror. How tall is your image in this mirror? (b) If, when you look in that mirror, you can see the face of the clerk at the checkout counter, does that mean that the clerk can see you in the mirror? Briefly justify your answer.

General problems and conceptual questions

30. If you hold your hand up in the beam of a film projector or slide projector, you can cast a shadow (in the shape of a rabbit's head, for instance) on a screen. Is this shadow consistent with the ray model of light? Explain.

31. When you are on a train, you can often see clear reflections of some of the other passengers when you look at the train windows. Explain why it is generally easier to see these reflected images when the train is passing through a dark tunnel than when the train is traveling outside on a bright sunny day.

32. A well-known philosophical question about sound is "If a tree falls in a forest and there is no-one there to hear it, does it make a sound?" A similar question regarding light could be "If a mirror is placed near a tree and there is no-one there to look into the mirror, is there an image?" Based on the principles of physics covered in this chapter, what do you think?

33. When you are looking in a plane mirror at the image of an object that remains at rest, does the image of the object move when you shift position? Draw one or two ray diagrams to help explain your answer. Also, in your explanation, make a comparison between a window and a mirror. When you look through a window at an object that remains at rest, the object's position stays fixed as you change your vantage point. Does the same thing happen with the image of a stationary object when viewed in a mirror, or not?

34. You are walking toward a plane mirror with a speed of 2.5 m/s. What is the velocity of your image, relative to you?

35. You are standing between two plane mirrors that are parallel to one another, and separated by 4.0 m. This situation, as you have probably observed, creates multiple images. You are 1.5 m from one of the mirrors, and 2.5 m from the other (neglecting your own width). (a) How far away is your image that is closest to you? (b) How far away are the three next-closest images?

36. You are 2.0 m from a plane mirror, holding a camera, and you want to take a photograph of your image in the mirror. To what distance should the camera be focused so you get the sharpest image of yourself?

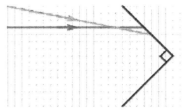

37. Two rays of light are incident on a pair of plane mirrors that are mounted at 90° to one another, as shown in Figure 23.49. (a) After experiencing two reflections, one from each mirror, through what angle has the red ray been deflected? Sketch the path of the red ray. (b) Repeat for the green ray. Note that the grid can help you to sketch the path for the green ray.

Figure 23.49: Two rays of light are incident on a pair of plane mirrors that are mounted at right angles to one another. For Exercise 37.

38. A small plane mirror is mounted on a wall. A laser beam is incident on the mirror at a 45° angle, as shown in Figure 23.50. The reflected beam makes a spot on a wall 5.0 m from, and parallel to, the wall with the mirror. When you push on the wall with the mirror, you deform the wall a little, shifting the angle of the mirror slightly and moving the spot on the other wall. If the spot moves by 12.0 cm (down, in the figure), through what angle has the mirror been rotated?

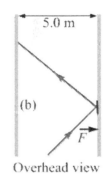

Figure 23.50: Mounting a mirror on a wall and reflecting a laser beam off it can help show how the wall is deformed when you push on it. (a) The initial situation. (b) The situation when you are pushing on the wall with a force F, as shown. For Exercise 38.

39. A common application of mirrors is in astronomical telescopes. Choose one of the following, do some research about how it works, and write a couple of paragraphs describing it: 1. The Keck 1 and Keck 2 telescopes. 2. The Large Zenith Telescope. 3. The Hubble Space Telescope.

40. Exercise 39 involves research-grade optical telescopes. Mirrors are also used in many telescopes used by amateur astronomers. Choose one of the following, do some research about how it works, and write a couple of paragraphs describing it: 1. A Newtonian telescope. 2. A Cassegrain telescope. 3. A Dobsonian telescope.

41. In a particular case of an object in front of a spherical mirror, the object distance is 12 cm and the magnification is +4.0. Find (a) the image distance, (b) the mirror's focal length.

42. In a particular case of an object in front of a spherical mirror with a focal length of 12 cm, the magnification is +4.0. Find (a) the object distance, (b) the image distance.

43. An object is placed 60 cm from a spherical mirror. When you look in the mirror, you see an image of the object that is 3.0 times larger than the object. (a) What kind of mirror is it? (b) How far from the mirror is the object? (c) Sketch a ray diagram to check your calculations. Make sure you find all possible solutions.

44. Figure 23.51 shows an object and a real image created by a spherical mirror. Assume the boxes on the grid measure 10 cm × 10 cm. Find the position of the mirror, and the mirror's focal length.

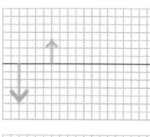

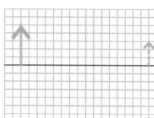

Figure 23.51: This figure shows an object and a real image created by a spherical mirror. For Exercise 44.

45. Figure 23.52 shows an object (the larger arrow) and the virtual image of that object, created by a spherical mirror. Assume the boxes on the grid measure 10 cm × 10 cm. Find the position of the mirror, and the mirror's focal length.

46. Sketch a ray diagram for the situation shown in (a) Figure 23.51, (b) Figure 23.52.

Figure 23.52: The larger arrow represents an object, while the smaller arrow represents the virtual image of that object, created by a spherical mirror. For Exercise 45.

47. In the situation shown in Figure 23.53, a small red LED (light-emitting diode) is placed on the principal axis 7.0 cm from a concave mirror that has a radius of curvature of 14 cm. The LED can be considered to be a point source. Draw a ray diagram to show what happens to rays of light that are emitted by the LED and reflect from the mirror.

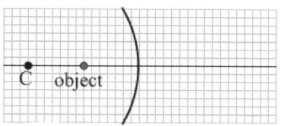

Figure 23.53: A small red LED (light-emitting diode) is placed in front of a concave mirror, for Exercise 47.

48. In the situation shown in Figure 23.54, a small red LED (light-emitting diode) is placed on the principal axis 4.0 cm from a concave mirror that has a radius of curvature of 14 cm. The LED can be considered to be a point source. Find the location of the image of the LED.

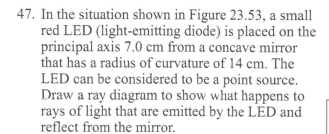

Figure 23.54: A small red LED (light-emitting diode) is placed in front of a concave mirror, for Exercises 48 and 50.

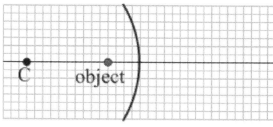

49. The LED from the previous exercise is moved to a location 2.0 cm above the principal axis, and 3.0 cm horizontally from the center of the mirror, as shown in Figure 23.55. Find the location of the image of the LED.

50. Draw several rays showing how the image of the LED is formed in (a) Figure 23.54, and (b) Figure 23.55.

Figure 23.55: A small red LED (light-emitting diode) is placed in front of a concave mirror, for Exercises 49 and 50. The LED is 2.0 cm above the principal axis.

51. A model of a horse is placed 32 cm away from a mirror that has a focal length of 20 cm. The model is 5.0 cm tall. Determine (a) the location of the image, (b) the height of the image, (c) whether the image is real or virtual, and (d) whether the image is upright or inverted.

52. Repeat Exercise 51, with the model of the horse 12 cm from the mirror instead.

53. A model of a horse is placed 32 cm away from a mirror that has a focal length of –20 cm. The model is 5.0 cm tall. Determine (a) the location of the image, (b) the height of the image, (c) whether the image is real or virtual, and (d) whether the image is upright or inverted.

54. Return to the situation described in Exercise 53. Describe what happens to the position and size of the image if the model is moved a little bit closer to the mirror.

55. Return to the situation described in Exercise 51. Describe what happens to the position and size of the image if the model is moved a little bit closer to the mirror.

56. Figure 23.56(a) shows an object placed in front of a vertical plane mirror. A convex mirror is then placed behind the plane mirror, in the location shown in Figure 23.56(b). Does adding the convex mirror cause the object's image to shift to the left, the right, or does it have no effect? Explain.

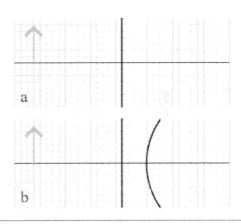

Figure 23.56: (a) An object in front of a plane mirror. (b) A convex mirror is placed behind the plane mirror, at the location shown. For Exercise 56.

57. A particular mirror has a focal length of +20 cm. (a) For this mirror, plot a graph of $1/d_i$ as a function of $1/d_o$ for object distances between +10 cm and +40 cm. (b) How can you read the focal length directly from the graph?

58. Repeat the previous exercise, but now plot a graph of d_i as a function of d_o.

59. Figure 23.57 shows a graph of $1/d_i$ as a function of $1/d_o$ for a particular mirror. What kind of mirror is it, and what is the mirror's focal length?

60. Refer to Figure 23.21, just above the start of section 23-4. (a) What, if anything, happens to the image if you cover up the bottom half of the mirror, preventing any light from reaching that part of the mirror? (b) Does your answer change if you cover up the top half of the mirror instead? Explain, and refer to Figure 23.21 in your explanations.

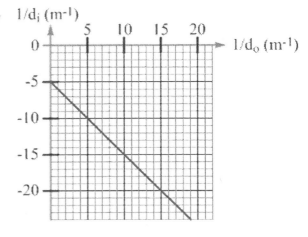

Figure 23.57: A graph of $1/d_i$ as a function of $1/d_o$ for a particular mirror. For Exercise 59.

61. Figure 23.58 shows a small rectangular box, placed so its front surface is 10 cm from a convex mirror, while its back surface is 16 cm from the mirror. The mirror has a radius of curvature of 40 cm, while the box has a height of 8.0 cm. (a) Sketch a ray diagram to show the location of the image of the front surface of the box. (b) On the same diagram, show the location of the rear of the box. (c) How tall is the image of the front of the box? (d) How tall is the image of the rear of the box, when you look at the box in the mirror?

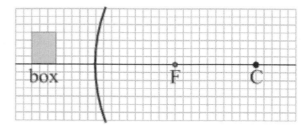

Figure 23.58: A box in front of a convex mirror. The boxes on the grid measure 2 cm × 2 cm. For Exercise 61.

62. As shown in Figure 23.59, an object is placed halfway between two identical concave mirrors, at a point that corresponds to the center of curvature of both mirrors. (a) How many images of the object could you see if you looked in either one of the mirrors? (b) How many of these images would be larger than the object, and how many would be smaller than the object? (c) Where would these images be located, relative to the object?

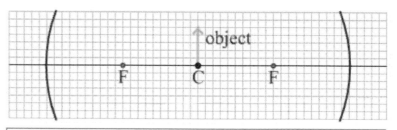

Figure 23.59: An object located at the common center of curvature of two identical concave mirrors. For Exercise 62.

63. Two students are having a conversation about a problem involving mirrors. Comment on each of their statements below. The problem is the following. We have an object in front of a spherical mirror. The image created by the mirror is smaller than the object, and real. We're asked what happens to the image when the object is moved a little closer to the mirror – does the image get larger or smaller, and does it move toward the mirror or away from the mirror?

Heather: Doesn't this depend on what kind of mirror we have? The problem just says it's a spherical mirror. It must matter whether the mirror is concave or convex, don't you think?

Mike: We could try both and see, I guess. How about we draw a ray diagram? Here, look at this. The ray that goes from the tip of the object, parallel to the principal axis, and then through the focal point, that ray stays the same even if we slide the object toward the mirror or away from the mirror. I think this tells us that, if the image gets closer to the mirror, it gets smaller, and if it gets farther away, it gets taller.

Heather: You drew that for a concave mirror. Does the same thing work for a convex mirror? I still think we need to know what kind of mirror we have.

Mike: Maybe there's something in the problem that gives that away? What does it say about the image? It's smaller, and real. Does that tell us what kind of mirror we have?

The photograph shows a rainbow, which is produced by refraction, dispersion, and total internal reflection, all concepts that we will address in this chapter. Note the second, inverted, rainbow higher in the photograph – this is known as a double rainbow.

Photo credit: Petr Kratochvil, from publicdomainpictures.net.

Chapter 24 – Refraction and Lenses

CHAPTER CONTENTS

In this chapter, we will again apply the ray model of light to understand what happens when light (or any EM wave) is transmitted from one medium to another. In general, this change of medium is associated with a change in speed, which generally causes the light to change direction. This phenomenon, known as refraction, has several important applications.

One of these applications is the optical fibers that carry telephone conversations and data. Even with a lot of technology going wireless these days, fiber-optic cables are a key link in the international communications network. In late 2006, for instance, Japan and Russia announced a $40 million joint venture to lay two 900 km under-sea fiber optic cables to link the countries. Each cable is capable of sending data at the rate of 640 Gbps (gigabits per second), a factor of 1000 increase over the cable previously used. Similar under-sea cables are used in many other parts of the world.

A second important application of refraction is in the lenses that we use to make microscopes and telescopes, as well as the glasses and contact lenses that many of us rely on to see properly. For most of us, by carefully grinding corrective lenses to just the right shape, the inherent imperfections in our own eyes can be compensated for, giving us the improved vision that can be so important when dealing with today's fast-paced, and highly visual world.

24-1 Refraction

To understand what happens when light passes from one medium to another, we again use a model that involves rays and wave fronts, as we did with reflection. Let's begin by creating a short pulse of light – say we have a laser pointer and we hold the switch down for just an instant. We shine this light onto the interface between two transparent media so that the beam of light is incident along the normal to the interface. For instance, Figure 24.1 shows a beam of light in air incident on a block of glass. Part of the beam is reflected straight back, and the rest is transmitted along the normal into the glass.

Figure 24.1 shows a sequence of images, showing the pulse of light at regular time intervals. Note that the light in the glass travels with an average speed that is about 2/3 of the average speed of the light in air. The average speed at which light travels through a medium depends on the medium. We can quantify this dependence of average speed on the medium by defining a unitless parameter known as the index of refraction, n.

$$n = \frac{c}{v}, \qquad \text{(Equation 24.1: Index of refraction)}$$

where $c = 3.00 \times 10^8$ m/s is the speed of light in vacuum. Table 24.1 gives indices of refraction for several media, and the corresponding speed of light in the media.

We can make sense of what happens in Figure 24.1 by looking at the wave fronts associated with the beam of light (see Figure 24.2). Because the wave slows down when it is transmitted into the glass, the wave fronts get closer together, shortening the distance from the front of the pulse to the end of the pulse. Going further, we can apply the wave equation (Equation 21.1, $v = f\lambda$) to re-write equation 24.1. The frequency of the wave in the two media is the same, so:

$$n = \frac{c}{v} = \frac{f\lambda}{f\lambda'} = \frac{\lambda}{\lambda'},$$

(Eq. 24.2: Index of refraction, in terms of wavelength)

where λ is the wavelength in vacuum, and λ' is the wavelength in the medium.

What happens when the beam of light is not incident along the normal? In this case, as shown in Figure 24.3, we observe that part of the wave is reflected from the interface, obeying the law of reflection, while the rest is transmitted into the second medium. However, the transmitted light travels in a different direction from the direction the

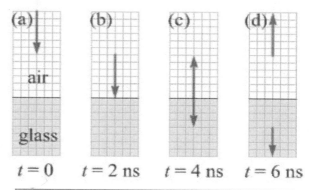

Figure 24.1: When a beam of light is incident along the normal to the interface between two transparent media, part of the beam is reflected straight back, while the rest is transmitted along the normal into the second medium. The squares on the grid measure 10 cm × 10 cm, and the images are shown at 2 ns intervals.

Medium	Index of refraction	Speed of light ($\times 10^8$ m/s)
Vacuum	1.000	3.00
Water	1.33	2.25
Glass	1.5	2.0
Diamond	2.4	1.25

Table 24.1: Typical indices of refraction for various media, and their corresponding speeds of light.

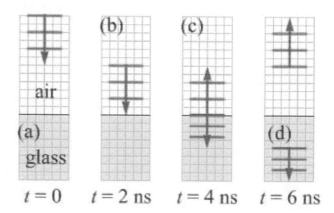

Figure 24.2: Looking at the wave fronts associated with the light as it passes from one medium to another, we can see that when the wave slows down, the wavelength decreases.

incident light was traveling in the first medium. This change in direction experienced by light transmitted from one medium to another is known as **refraction**.

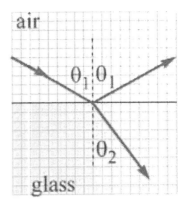

Figure 24.3: When light is not incident along the normal, the light changes direction when it is transmitted into the second medium, as long as the two media have different indices of refraction.

We can understand refraction by looking at the wave fronts as the beam of light passes from one medium to another, as shown in Figure 24.4. The part of the wave front that enters the glass first slows down, while the part still traveling in air maintains its speed. As can be seen, this causes the wave to change direction.

What happens if the light is traveling in the glass instead, and is then incident on the glass-air interface? The reflected ray is quite different from what we had above, but if the angle between the normal and the ray of light in the glass is the same in Figure 24.2 and Figure 24.4, the angle between the normal and the ray in air is also the same in the two cases. Snell's Law applies, no matter which direction the light is going.

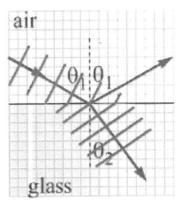

Figure 24.4: The part of the wave front that enters the slower medium (the glass) first slows down before the rest of the wave front, causing the ray to change direction. Wave fronts associated with the reflected wave have been omitted for clarity.

When light is transmitted from one medium to another, the angle of incidence, θ_1, in the first medium is related to the angle of refraction, θ_2, in the second medium by:

$$n_1 \sin\theta_1 = n_2 \sin\theta_2$$

(Equation 24.3: **Snell's Law**)

These angles are measured from the normal (perpendicular) to the surface. The n's represent the indices of refraction of the two media.

In general, when light is transmitted from one medium to a medium with a higher index of refraction, the light bends toward the normal (that is, the angle for the refracted ray is less than that for the incident ray). Conversely, the light bends away from the normal when light is transmitted from a higher-n medium to a lower-n medium.

Related End-of-Chapter Exercises: 1, 2, 3 – 5.

Figure 24.5: The photograph shows a picture of a pencil in a glass of water. Refraction of the light leaving the water makes the pencil appear to bend. Photo credit: A. Duffy.

Essential Question 24.1: Figure 24.6 shows a beam of light as it is transmitted from medium 1 to medium 3. Rank these media based on their index of refraction.

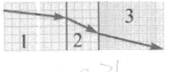

Figure 24.6: A beam of light passing through three media.

3 > 2 > 1

Answer to Essential Question 24.1: Because all of these interfaces are parallel to one another, we can apply a rule-of-thumb that the smaller the angle between the normal and the beam, within a medium, the larger is that medium's index of refraction. Based on this rule-of-thumb, the ranking by index of refraction is 1 > 3 > 2.

24-2 Total Internal Reflection

Let's now explore an implication of Snell's Law that has practical applications. Under the right conditions, light incident on an interface between two media is entirely reflected back into the first medium. This phenomenon, that none of the light is transmitted into the second medium, is known as **total internal reflection**. Let's begin by examining light being transmitted from air to glass, a situation in which total internal reflection does not occur.

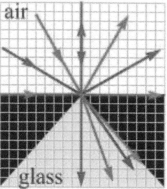

Figure 24.7 shows various beams of light traveling in air and incident on the same point on an air-glass interface. The beams are shown in different colors to make it clear what path the reflected and transmitted beams follow. Four beams are shown, corresponding to angles of incidence of 0°, 30°, 60°, and almost 90°. As the angle of incidence increases, so does the refracted angle (the angle between the beam refracted into the glass and the normal). By choosing an appropriate angle of incidence, any point in the region shaded in blue in the glass can be illuminated by the light refracted into the glass from the air. Note, however, that no point in the region shaded in black in the glass can be illuminated by the light coming from the air, no matter what angle of incidence is used, if the beam of light is always incident on the same point on the interface.

Figure 24.7: Four rays of light, incident on the same point on an air-glass interface. When the light strikes the interface, part of the beam is reflected back into the air, and part is refracted into the glass. The glass is rectangular, but the part of the glass that can not be reached by the light incident on this point has been shaded black.

Table 24.2 shows the angles of the refracted rays in the glass. These angles are determined by applying Snell's Law, using indices of refraction of 1.00 for air and 1.50 for glass.

Incident angle	Refracted angle
0°	0°
30.0°	19.5°
60.0°	35.3°
90.0°	41.8°

Table 24.2: The angles of the incident rays and the refracted rays in Figure 24.7.

EXPLORATION 24.2 – Total internal reflection

Let us now draw a similar picture to that in Figure 24.7, but the light will now travel through the glass before it is incident on the air-glass interface.

Step 1 – First, draw the rays in glass incident at angles of 0°, 19.5°, 35.3°, and 41.8°, the same values as the refracted angles in Table 24.2 . Show both the reflected rays and the rays that refract into the air.

This figure is very similar to Figure 24.7, with the incident and refracted rays reversing directions. The values in Table 24.2 apply again, with the refracted angles in Table 24.2 now the incident angles, and vice versa.

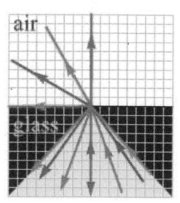

Figure 24.8: Four rays of light, incident on the same point on a glass-air interface. When the light strikes the interface, part of the beam is reflected back into the glass, and part is refracted into the air.

Step 2 – *Now draw two additional rays, with angles of incidence of 60° and 75°, incident on the same point as the incident rays in Figure 24.8, and originally traveling in glass. What happens when you apply Snell's Law for these rays? What do these rays do when they encounter the interface?* Let's apply Snell's Law to the situation of the 60° angle of incidence:

$$n_1 \sin(60°) = n_2 \sin\theta_2 .$$

$$\sin\theta_2 = \frac{n_{glass} \sin(60°)}{n_{air}} = \frac{1.5 \times 0.866}{1.0} = 1.3 .$$

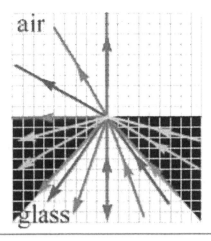

This equation cannot be solved, because the largest value that $\sin\theta$ can be is 1. So, according to Snell's Law, there is no angle of refraction in the air for an angle of incidence of 60° in glass. A similar argument holds for all incident angles greater than 41.8° for rays traveling in glass and striking a glass-air interface. This is consistent with the forbidden region shaded in black in Figure 24.7. No light can refract into the glass from air at angles of refraction that exceed 41.8°, and so no refraction can occur in the opposite direction if the angle of incidence exceeds this critical angle of 41.8°. If the light does not refract into the air, it is entirely reflected back into the glass, obeying the law of reflection. This is known as **total internal reflection**.

Figure 24.9: When the incident angle for light rays incident on the glass-air interface from the glass side exceed a particular critical angle (in the region colored black), 100% of the light reflects back into the glass. This is known as total internal reflection.

Step 3 – *In general, the critical angle of incidence beyond which total internal reflection occurs is the angle of incidence, in the higher-n medium, that results in a 90° angle of refraction in the lower-n medium. Apply Snell's Law to this situation to derive an equation for the critical angle, θ_c.* Applying Snell's Law, and recognizing that $\sin(90°) = 1$, gives:

$$n_1 \sin\theta_c = n_2 \sin(90°) = n_2 .$$

$$\theta_c = \sin^{-1}\left(\frac{n_2}{n_1}\right).$$ (Eq. 24.4: **The critical angle beyond which total internal reflection occurs**)

Key ideas: Total internal reflection can only occur when light in one medium strikes an interface separating that medium from a medium with a lower index of refraction. If the angle of incidence exceeds a particular critical angle, no light is transmitted into the second medium. Instead, all the light is reflected back into the first medium.   **Related End-of-Chapter Exercises: 16 – 20.**

An important application of total internal reflection

Fiber optic cables, which are very important for communications, exploit total internal reflection. Data, as well as voice signals from telephone conversations, are sent through optical fibers by means of laser signals encoded to carry the information. The fibers can even bend around corners and, as long as the bend is not too abrupt, total internal reflection keeps the light inside the fiber.

Essential Question 24.2: Light traveling in medium 1 reaches the surface of the medium, which is surrounded by air ($n = 1.00$). The angle of incidence is 39°, and the light experiences total internal reflection. What can you say about the index of refraction of medium 1?

Answer to Essential Question 24.2: We know that 39° exceeds the critical angle for total internal reflection. If the critical angle is 39°, equation 24.4 gives $n_1 = 1.59$. (n_2 in the equation is 1.00, because medium 2 is air.) For the critical angle to be less than 39°, the index of refraction must be larger than 1.59. Thus, we know that the index of refraction of the medium is at least 1.59.

24-3 Dispersion

When a beam of white light shines onto a prism, a rainbow of colors emerges from the prism. Such a situation is shown in Figure 24.11. What does this tell us about white light? What does this tell us about the glass that the prism is made from?

Figure 24.11: The triangular prisms in the picture on the left arc suspended beneath the skylights at the Albuquerque Convention Center. Sunlight passing through these prisms produces beautiful rainbows on the floor, as in the photo on the right. Photo credits: A. Duffy.

The fact that a prism splits white light up into various colors of the rainbow tells us that white light consists of all the colors. For the prism to be able to separate out these individual colors, however, the index of refraction of the glass must depend on the wavelength of the light passing through the prism. This phenomenon is known as **dispersion**. Because white light contains all wavelengths of light in the visible spectrum, the different wavelengths (corresponding to different colors) refract at different angles, spreading out the colors.

In general, what we observe for a prism is that violet light, with a wavelength of 400 nm, experiences the largest change in direction, while red light, with a wavelength of 700 nm, experiences the smallest change in direction. As the wavelength increases from 400 to 700 nm, the change in direction decreases, and thus the index of refraction of the glass must also decrease. A graph of the index of refraction as a function of wavelength is shown in Figure 24.12.

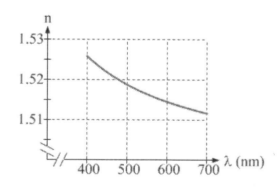

Figure 24.12: A graph of the index of refraction, as a function of wavelength, for a typical sample of glass. The graph is confined to the visible spectrum, from 400 nm (violet) to 700 nm (red).

Refraction, dispersion, and total internal reflection are all important in the formation of a rainbow, such as that in the photograph on the opening page of this chapter. To understand how a rainbow is formed, remember that you can generally only see a rainbow after it has rained, when the Sun is fairly low in the sky, and when you are looking away from the Sun. The rain is important because it produces a lot of water droplets in the sky. Let's begin by looking at how the red light, which is part of the white light coming from the Sun, interacts with a spherical water droplet.

Figure 24.13 shows how red light refracts into a spherical water droplet. Some of the red light reflects from the back of the droplet, and refracts again as it exits the droplet.

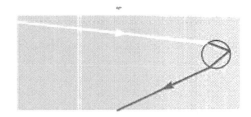

Figure 24.13: The path followed by red light inside a water droplet.

Let us now consider the path followed by violet light through the droplet. Water exhibits some dispersion, so when the violet light refracts into the water droplet the angle of refraction for the violet light is a little different from that for red light. Like the red light, some of the violet light reflects from the back of the water droplet, and experiences more refraction as it exits the droplet. As shown in Figure 24.14, the end result is that the beam of violet light coming from that droplet is higher in the sky than the beam of red light from the same droplet. All of the other colors of the visible spectrum lie between those of red and violet, because red and violet are at the two extremes of the visible spectrum.

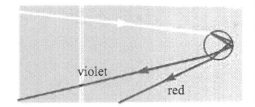

Figure 24.14: The path followed by violet light inside a water droplet, compared to that followed by the red light.

Compare Figure 24.14 to the photograph of the rainbow shown on the first page of this chapter. In Figure 24.14, the beam of violet light that leaves the droplet is higher in the sky than the red light that leaves the droplet. When you look at a rainbow, however, you see the red on the upper arc of the rainbow, and the violet on the lower arc. Is Figure 24.14 at odds with the photo?

Let's try adding the eye of an observer, as well as a second water droplet below the first, to the diagram. These additions are shown in Figure 24.15. When the observer looks at the upper water droplet, what color does it appear to be? It appears to be red, because red light from that droplet enters the observer's eye, while the violet light from that same droplet passes well over the observer's head. The lower droplet appears violet, however, because only violet light from that droplet enters the observer's eye.

Figure 24.15: Adding an observer, and a second water droplet, to the diagram in Figure 24.14 shows why the top arc of the rainbow is red, while the bottom arc is violet. The other colors come from droplets in between these two. The size of the water droplets is exaggerated so that the path taken by the beams is visible.

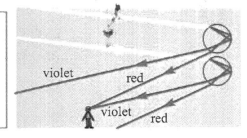

Other issues related to rainbows

Under the right conditions, you can see a double rainbow (or even a triple rainbow). As can be seen in the chapter-opening photograph, the second rainbow is dimmer than the primary rainbow, higher in the sky, and the order of colors is reversed. The secondary rainbow comes from light that reflects twice inside each water droplet, rather than once for the primary rainbow. Another issue is that rainbows are generally half-circles. Is it possible to see a rainbow that is a complete circle, or at least more than a half circle? Yes, it is. If you view a rainbow from a high vantage point, such as out the window of a plane, you may be able to see the entire circle.

Related End-of-Chapter Exercises: 7, 40.

Essential Question 24.3: After a light rain shower, you look out your window and see a beautiful rainbow in the sky. Your neighbor happens to look out her window at the same time, and also observes a rainbow. Does your neighbor see exactly the same rainbow that you do?

Answer to Essential Question 24.3: Everyone has their own rainbow! The light you see enters your eye only. In addition, the light making up the colors of the rainbow that you see comes from droplets that are at just the right position in the sky to refract and reflect sunlight back toward you. Your neighbor would see a very similar rainbow, but at least some of the droplets responsible for her rainbow are different from the water droplets that create your rainbow.

24-4 Image Formation by Thin Lenses

Lenses, which are important for correcting vision, for microscopes, and for many telescopes, rely on the refraction of light to form images. As with mirrors, we draw ray diagrams to help us to understand how such images are formed. Let's first begin by looking at what a lens does to a set of parallel rays of light, such as the five rays in Figure 24.16.

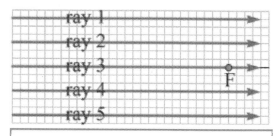

Figure 24.16: Five parallel rays of light.

How can we change the direction of the five rays of light in Figure 24.16 so that they all pass through the point labeled F? Ray 3 already passes through point F, so we don't need to change its direction at all. We can deflect ray 2 with a triangular piece of glass, as shown in Figure 24.17. Passing from air into the glass prism, the ray deflects toward the normal at that surface, while when it emerges back into the air the deflection is away from the normal at the second surface. We can follow a similar process for ray 1, except that we need to produce a larger change in direction for ray 1 compared to ray 2. The glass prism we use for ray 1 thus has its sides at a greater angle from the vertical. For ray 4, we use an identical prism to that used for ray 2, except that we invert it, and for ray 5 the prism is identical to the prism for ray 1, but inverted.

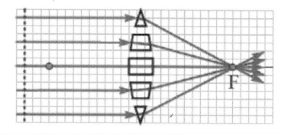

Figure 24.17: We can deflect four of the five rays, so that all five rays meet at point F, by using prisms of the appropriate shapes to produce the deflection required for each beam.

Now, not only do we want all five rays to converge on point F, but we want them to take equal times to travel from the vertical dashed line in Figure 24.17 to point F. Remember that the light travels more slowly in glass than in air. Because rays 1 and 5 travel the greatest distance, they need to pass through the least amount of glass. Ray 3 travels the shortest distance, so we need to delay ray 3 by having it pass through the thickest piece of glass. We can do this without deflecting ray 3 by using a piece of glass with vertical sides, so that ray 3 is incident along the normal. Rather than using various individual glass rectangles and prisms to do the job, we can use a single piece of glass that is thickest in the middle. This piece of glass gets thinner, and its surfaces curve farther away from the vertical, as you move away from ray 3 (that is, as you move away from the principal axis). This is shown in Figure 24.18 – we use a lens. Point F is a focal point of the lens. Lenses allow light to pass through from left-to-right or from right-to-left, so a lens has two focal points, one on each side of the lens.

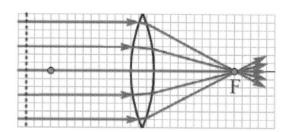

Figure 24.18: A convex lens with surfaces that are spherical arcs brings all parallel rays to one of the focal points, and ensures that the parallel rays take the same time to travel from the vertical line to this focal point.

How does a concave lens, which is thinner in the center than at the edges, affect parallel rays? As shown in Figure 24.19, such lenses generally diverge parallel rays away from the focal point that is on the same side of the lens that the light comes from.

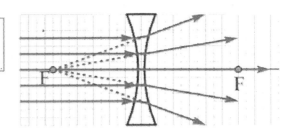

Figure 24.19: The influence of a diverging lens on a set of parallel rays.

EXPLORATION 24.4 – Using a ray diagram to find the location of an image

When drawing ray diagrams for lenses, we follow a process similar to that for mirrors.

Step 1 – *Figure 24.20 shows an object in front of a converging lens. Sketch two rays of light, which travel in different directions, that leave the tip (the top) of the object and pass through the lens. Show the direction of these rays after they are refracted by the lens.* Figure 24.21 shows three rays that leave the tip of the object and which are incident on the lens. Because the ray in red is parallel to the principal axis, it is refracted by the lens to pass through the focal point that is to the right of the lens. The ray in green travels along the line connecting the tip of the object to the focal point on the left of the lens. This ray is refracted so that it is parallel to the principal axis. We know this because of the reversibility of light rays – if we reversed the direction of the ray, it would come from the right and be refracted by the lens to pass through the focal point on the right. Finally, the ray in blue passes through the center of the lens without changing direction. This is an approximation, which is valid as long as the lens is thin so the ray enters and exits the lens very close to the principal axis.

Step 2 – *The point where the refracted rays meet is where the tip of the image is located. Use this information to sketch the image of the object.* In this situation, the refracted rays meet at a point to the right of the lens, below the principal axis. Thus, we draw an inverted image between the point where the image of the tip is (where the rays meet) and the principal axis. This is a real image, because the rays pass through the image. All the rays take the same time to travel from the tip of the object to the tip of the image.

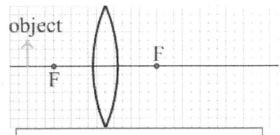

Figure 24.20: An object located some distance in front of a converging lens.

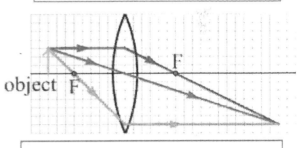

Figure 24.21: Three of the rays that are refracted by the lens. In general, we draw a ray changing direction once, inside the lens. Each ray really changes direction twice, once at each air-glass interface. We are drawing the ray's path incorrectly within the lens, but it is correct outside of the lens, which is where it really matters.

Figure 24.22: Because the rays originate at the tip of the object, the point where the refracted rays meet is the location of the tip of the image of the object. The base of the object is on the principal axis, so the base of the image is on the principal axis, too.

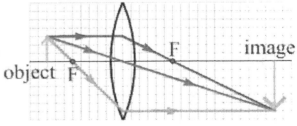

Key idea for ray diagrams: The location of the image of any point on an object, when the image is created by a lens, can be found by drawing rays of light that leave that point on the object and are refracted by the lens. The point where the refracted rays meet (or where they appear to diverge from) is where the image of that point is. **Related End-of-Chapter Exercises: 5, 49**

Essential Question 24.4: Are the three rays in Figure 24.22 the only rays that pass through the tip of the image? Explain.

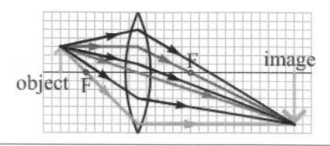

Answer to Essential Question 24.4: The three rays in Figure 24.21 are easy to draw, because we know what they do after passing through the lens. However, as shown in Figure 24.23, all rays that leave the tip of the object and which are refracted by the lens will converge at the tip of the image.

Figure 24.23: All rays that leave the tip of the object and pass through the lens will converge at the tip of the image.

24-5 Lens Concepts

In general, a lens has a larger index of refraction than the medium that surrounds it. In that case, a lens that is thicker in the center than the ends is a converging lens (it converges parallel rays toward a focal point), while a lens that is thinner in the middle than at the ends is a diverging lens (it diverges parallel rays away from a focal point). Like a concave mirror, converging lenses can produce a real image or a virtual image, and the image can be larger, smaller, or the same size as the object. Like a convex mirror, diverging lenses can only produce a virtual image that is smaller than the object.

As with mirrors, the focal point of a lens is defined by what the lens does to a set of rays of light that are parallel to one another and to the principal axis of the mirror. As we discussed in Section 24-4, a converging lens generally refracts the rays so they converge to pass through a focal point, F. A diverging lens, in contrast, refracts parallel rays so that they diverge away from a focal point.

Because of dispersion (the fact that the index of refraction of the lens material depends on wavelength), a lens generally has slightly different focal points for different colors of light. This range of focal points is a defect called **chromatic aberration**.

Focal length of a lens with spherical surfaces: The focal length of a lens depends on the curvature of the two surfaces of the lens, the index of refraction of the lens material, and on the index of refraction of the surrounding medium. The focal length is given by:

$$\frac{1}{f} = \left(\frac{n_{lens}}{n_{medium}} - 1 \right)\left(\frac{1}{R_1} + \frac{1}{R_2} \right), \qquad \text{(Equation 24.5: \textbf{The lensmaker's equation})}$$

where the two R's represent the radii of curvature of the two lens surfaces. A radius is positive if the surface is convex, and negative if the surface is concave.

Companies that make eyeglasses exploit the lensmaker's equation in creating lenses of the desired focal length. By choosing a lens material that has a high index of refraction, a smaller radius of curvature (and thus a thinner and lighter lens) can be used, compared to a lens made from glass with a smaller index of refraction.

The factor of $[(n_{lens}/n_{medium}) -1]$ in Equation 24.5 has an interesting implication. First, consider the diagram in Figure 24.24, which shows a familiar situation of a lens, made from material with an index of refraction larger than that of air, surrounded by air. This lens causes the parallel rays to change direction so that they pass through the focal point on the right, and $[(n_{lens}/n_{medium}) -1]$ is positive, so the focal length of the lens is positive.

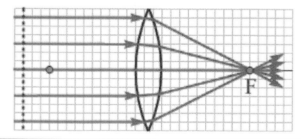

Figure 24.24: A ray diagram for a set of parallel rays encountering a convex lens, made of plastic, surrounded by air.

What happens to the light if the medium around the lens has the same index of refraction as the lens material (perhaps we immerse the lens in some kind of oil)? In this case, the factor of $[(n_{lens}/n_{medium}) -1]$ is zero - this means that the lens does no focusing at all, which we would expect if the lens and the medium have the same index of refraction. Going further, if the surrounding medium has a larger index of refraction than the lens material, $[(n_{lens}/n_{medium}) -1]$ is negative and so is the focal length: parallel rays would be *diverged* in this situation.

Thus, depending on the situation, a lens with a convex shape can be a converging lens, a diverging lens, or neither. To minimize any ambiguity, for the rest of this chapter we will refer to lenses by their function (converging or diverging) rather than their shape (convex or concave).

EXPLORATION 24.5 – Ray diagram for a diverging lens

We will follow a process similar to that of mirrors to draw a ray diagram for a diverging lens, starting with the situation in Figure 24.25. The ray diagram will show us where the image of an object is.

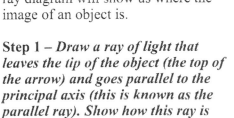

Figure 24.25: An object in front of a diverging lens.

Step 1 – Draw a ray of light that leaves the tip of the object (the top of the arrow) and goes parallel to the principal axis (this is known as the parallel ray). Show how this ray is refracted by the lens. For a diverging lens, all parallel rays appear to diverge from the focal point on the side of the lens that the light comes from, so we draw the refracted ray (see Figure 24.26) refracting along a line that takes it directly away from that focal point.

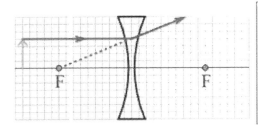

Figure 24.26: The parallel ray refracts to travel directly away from the focal point on the side of the lens the light came from.

Step 2 – Sketch a second ray that leaves the tip of the object and is refracted by the lens. Using the refracted rays, draw the image. One useful ray, shown in Figure 24.27, passes through the center of the lens without changing direction. This is something of an approximation, but the thinner the lens, the more accurate this is. Another useful ray goes straight toward the focal point on the right of the lens. This ray refracts so as to emerge from the lens going parallel to the principal axis. The refracted rays diverge to the right of the lens, but we can extend them back to meet on the left side of the lens, showing us where the tip of the image is.

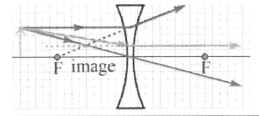

Figure 24.27: In addition to the parallel ray, two other rays are easy to draw the refracted rays for. The ray that passes through the center of the lens is undeflected, approximately. The ray that travels directly toward the focal point on the far side of the lens is refracted so it emerges from the lens traveling parallel to the principal axis. If you look at the object through the lens, your brain interprets the light as coming from the image.

Key idea: When a number of rays leave the same point on an object and are refracted by a lens, the corresponding point on the image is located at the intersection of the refracted rays.
Related End-of-Chapter Exercises: 3, 4, 52.

Essential Question 24.5: Starting with Figure 24.27, show a few more rays of light leaving the tip of the object and being refracted by the lens. How do you know how to draw the refracted rays?

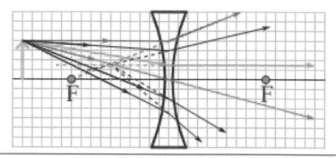

Answer to Essential Question 24.5: To draw the refracted rays properly, we know that when we extend the refracted rays back, they will pass through the tip of the image, which we located in Figure 24.27. Three additional rays are shown in Figure 24.28.

Figure 24.28: For all rays of light that leave the tip of the object and reflect from the mirror, the refracted rays can be extended back to pass through the tip of the image.

24-6 A Quantitative Approach: The Thin-Lens Equation

Even though mirrors and lenses form images using completely different principles (the law of reflection versus Snell's law), we use the same equation to relate focal length, object distance, and image distance, for both mirrors and lenses. This surprising result comes from the fact that the formation of images with both mirrors and lenses can be understood using the geometry of similar triangles. Let's look at how that works for lenses.

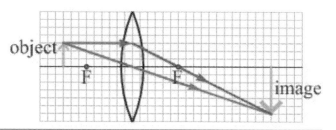

Figure 24.29: The ray diagram we constructed in section 24-4, for an object in front of a converging lens.

Let's look at the ray diagram we drew in Figure 24.22 of section 24-4, shown again here in Figure 24.29.

Remove the red rays, and examine the two triangles in Figure 24.30, one shaded green and one shaded yellow, bounded by the blue rays, the principal axis, and the object and image. The two triangles are similar, because the three angles in one triangle are the same as the three angles in the other triangle. We can now define the following variables: d_o is the object distance, the distance of the object from the center of the mirror; d_i is the image distance, the distance of the image from the center of the mirror; h_o is the height of the object; h_i is the height of the image.

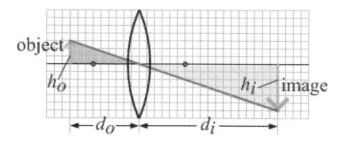

Figure 24.30: Similar triangles, bounded by the principal axis, the object and image, and the ray that passes through the center of the lens.

Using the fact that the ratios of the lengths of corresponding sides in similar triangles are equal, we find that:

$$-\frac{h_i}{h_o} = \frac{d_i}{d_o}.$$ (Equation 24.6)

The image height is negative because the image is inverted, which is why we need the minus sign in the equation. Let's now return to Figure 24.29, and remove the ray that passes through the center. This gives us the shaded similar triangles shown in Figure 24.31.

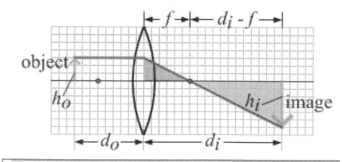

Figure 24.31: Similar triangles, with two sides bounded by the principal axis and the parallel ray, and a third side that is equal to the object height (smaller triangle) or the image height (larger triangle).

Again, using the fact that the ratios of the lengths of corresponding sides in similar triangles are equal, we find that: $\dfrac{d_i - f}{f} = -\dfrac{h_i}{h_o}$.

Simplifying the left side, and bringing in equation 24.3, we get: $\dfrac{d_i}{f} - 1 = \dfrac{d_i}{d_o}$.

Dividing both sides by d_i gives: $\dfrac{1}{f} - \dfrac{1}{d_i} = \dfrac{1}{d_o}$, which is generally written as:

$$\frac{1}{f} = \frac{1}{d_o} + \frac{1}{d_i}.$$ (Equation 24.7: **The thin-lens equation**)

The mnemonic "If I do I di" can help you to remember the thin-lens equation.

Often, we know the focal length f and the object distance d_o, so equation 24.4 can be solved for d_i, the image distance:

$$d_i = \frac{d_o \times f}{d_o - f}$$ (Equation 24.8: **The thin-lens equation, solved for the image distance**)

Sign conventions

We derived the lens equation above by using a specific case involving a convex lens. The equation can be applied to all situations involving a convex lens or a concave lens if we use the following sign conventions.

The focal length is positive for a converging lens, and negative for a diverging lens.

The image distance is positive, and the image is real, if the image is on the side of the lens the light passes through to, and negative, and the image is virtual, if the image is on the side the light comes from.

The image height is positive when the image is above the principal axis, and negative when the image is below the principal axis. A similar rule applies to the object height.

Magnification

The magnification, m, is defined as the ratio of the height of the image (h_i) to the height of the object (h_o). Making use of Equation 24.6, we can write the magnification as:

$$m = \frac{h_i}{h_o} = -\frac{d_i}{d_o}.$$ (Equation 24.9: **Magnification**)

The relative sizes of the image and object are as follows:

- The image is larger than the object if $|m| > 1$.

- The image and object have the same size if $|m| = 1$.

- The image is smaller than the object if $|m| < 1$.

The sign of the magnification tells us whether the image is upright (+) or inverted (–) compared to the object.

Related End-of-Chapter Exercises: 21 – 24.

Essential Question 24.6: As you are analyzing a thin-lens situation, you write an equation that states: $\dfrac{1}{f} = \dfrac{1}{+12 \text{ cm}} + \dfrac{1}{+24 \text{ cm}}$. What is the value of $1/f$ in this situation? What is f?

Answer to Essential Question 24.6: To add fractions, you need to find a common denominator.
$$\frac{1}{f} = \frac{1}{+12 \text{ cm}} + \frac{1}{+24 \text{ cm}} = \frac{2}{+24 \text{ cm}} + \frac{1}{+24 \text{ cm}} = \frac{3}{+24 \text{ cm}}. \text{ This gives } f = \frac{+24 \text{ cm}}{3} = 8.0 \text{ cm}.$$

24-7 Analyzing a Converging Lens

In section 24-4, we drew one ray diagram for a converging lens. Let's investigate the range of ray diagrams we can draw for such a lens. Note the similarities between a converging lens and a concave mirror. This section is very much a parallel of section 23-6, in which we analyzed the range of images formed by a concave mirror.

EXPLORATION 24.7 – Ray diagrams for a converging lens

Step 1 – Draw a ray diagram for an object located 40 cm from a converging lens that has a focal length of +10 cm. Verify the image location on your diagram with the thin-lens equation. Two rays are shown in Figure 24.32. One is the parallel ray, which leaves the tip of the object, travels parallel to the principal axis, and is refracted by the lens to pass through the focal point on the far side of the lens. The second ray passes straight through the center of the lens, undeflected.

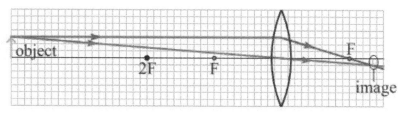

Figure 24.32: A ray diagram for the situation in which the object is far from the mirror. The squares in the grid measure 1.0 cm × 1.0 cm.

Applying the thin-lens equation, in the form of equation 24.5, to find the image distance:
$$d_i = \frac{d_o \times f}{d_o - f} = \frac{(40 \text{ cm}) \times (+10 \text{ cm})}{(40 \text{ cm}) - (+10 \text{ cm})} = \frac{+400 \text{ cm}^2}{30 \text{ cm}} = +13.3 \text{ cm}.$$

This image distance is consistent with the ray diagram in Figure 24.32.

Step 2 – Repeat step 1, with the object now twice the focal length from the lens. We draw the same two rays again, with the parallel ray (in red) being refracted so that it passes through the focal point on the far side of the lens, and the second ray (in blue) passing undeflected (approximately) through the center of the lens. As shown in Figure 24.33, this situation is a special case. When the object is located at twice the focal length from the lens, the image is inverted, also at twice the

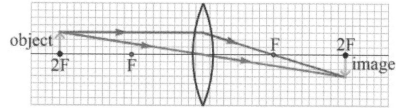

Figure 24.33: When the object is at twice the focal length from the lens, so is the image.

focal length from the lens (on the other side of the lens), and the same size as the object because the object and image are the same distance from the lens.

Applying the thin-lens equation to find the image distance, we get:
$$d_i = \frac{d_o \times f}{d_o - f} = \frac{(20 \text{ cm}) \times (+10 \text{ cm})}{(20 \text{ cm}) - (+10 \text{ cm})} = \frac{+200 \text{ cm}^2}{10 \text{ cm}} = +20 \text{ cm}, \text{ matching the ray diagram.}$$

Step 3 – Repeat step 1, with the object 15 cm from the mirror. No matter what the object distance is, the parallel ray always does the same thing, being refracted by the lens to pass through the focal point on the far side. The path of the second ray, in blue, depends on the object's position. The ray diagram (Figure 24.34) shows that the image is real, inverted, larger than the object, and about twice as far from the lens as the object.

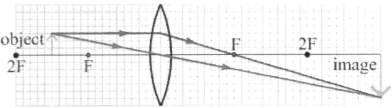

Figure 24.34: A ray diagram for a situation in which the object is between twice the focal length from the lens and the focal point.

Applying the thin-lens equation gives:

$$d_i = \frac{d_o \times f}{d_o - f} = \frac{(15 \text{ cm}) \times (+10 \text{ cm})}{(15 \text{ cm}) - (+10 \text{ cm})} = \frac{+150 \text{ cm}^2}{5.0 \text{ cm}} = +30 \text{ cm}, \text{ matching the ray diagram.}$$

Step 4 – Repeat step 1, with the object at a focal point. As shown in Figure 24.35, the two refracted rays are parallel to one another, and never meet. In such a case the image is formed at infinity.

Applying the thin-lens equation gives:

$$d_i = \frac{d_o \times f}{d_o - f} = \frac{(10 \text{ cm}) \times (+10 \text{ cm})}{(10 \text{ cm}) - (+10 \text{ cm})} = \frac{+100 \text{ cm}^2}{0 \text{ cm}} = +\infty,$$

which agrees with the ray diagram.

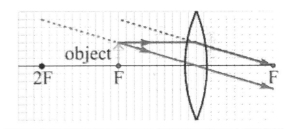

Figure 24.35: A ray diagram for a situation in which the object is at the focal point.

Step 5 – Repeat step 1, with the object 5.0 cm from the lens. When the object is closer to the lens than the focal point, the refracted rays diverge to the right of the lens, and they must be extended back to meet on the left of the lens. The result is a virtual, upright image that is larger than the object, as shown in Figure 24.36. If you look at the object through the lens, your brain interprets the light as coming from the image.

Applying the thin-lens equation gives:

$$d_i = \frac{d_o \times f}{d_o - f} = \frac{(5.0 \text{ cm}) \times (+10 \text{ cm})}{(5.0 \text{ cm}) - (+10 \text{ cm})} = \frac{+50 \text{ cm}^2}{-5.0 \text{ cm}} = -10 \text{ cm}.$$

Recalling the sign convention that a negative image distance is consistent with a virtual image, the result from the thin-lens equation is consistent with the ray diagram.

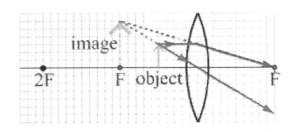

Figure 24.36: A ray diagram for a situation in when the object is between the lens and its focal point.

Key idea for converging lenses: Depending on where the object is relative to the focal point of a converging lens, the lens can form an image of the object that is real or virtual. If the image is real, it can be larger than, smaller than, or the same size as the object. If the image is virtual, the image is larger than the object. **Related End-of-Chapter Exercises: 44, 50, 53.**

Essential Question 24.7: When an object is placed 20 cm from a lens, the image formed by the lens is real. What kind of lens is it? What, if anything, can you say about the lens' focal length?

Answer to Essential Question 24.7: The lens must be converging, because a diverging lens cannot produce an image that is larger than the object. A converging lens produces a real image only when the object distance is larger than the focal length, so the focal length in this case must be positive but less than 20 cm.

24-8 An Example Problem Involving a Lens

Let's begin by discussing a general approach we can use to solve problems involving a lens. We will then apply the method to a particular situation.

A general method for solving problems involving a lens

1. Sketch a ray diagram, showing rays leaving the tip of the object and being refracted by the lens. Where the refracted rays meet is where the tip of the image is located. The ray diagram gives us qualitative information about the location and size of the image and about the characteristics of the image.

2. Apply the thin-lens equation and/or the magnification equation. Make sure that the signs you use match those listed in the sign convention in section 24-5. The equations provide quantitative information about the location and size of the image and about the image characteristics.

3. Check the results of applying the equations with your ray diagram, to see if the equations and the ray diagram give consistent results.

Rays that are easy to draw

To locate an image on a ray diagram, you need a minimum of two rays. If you draw more than two rays, however, you can check the image location you find with the first two rays. You can draw any number of rays being refracted by the lens, but some are easier to draw than others because we know exactly where the refracted rays go for these rays. Such rays are shown on Figure 24.37, and include:

1. The ray that goes parallel to the principal axis, and refracts to pass through the focal point on the far side of the lens (converging lens), or away from the focal point on the near side of the lens (diverging lens).

2. The ray that passes straight through the center of the lens without being deflected.

3. The ray that travels along the straight line connecting the tip of the object and the focal point not associated with the first ray. This ray is refracted by the lens to go parallel to the principal axis.

Figure 24.37: An example of the three rays that are easy to draw the refracted rays for. When you look at the object through the lens, your brain interprets the light as traveling in straight lines, so you see the image, and not the object.

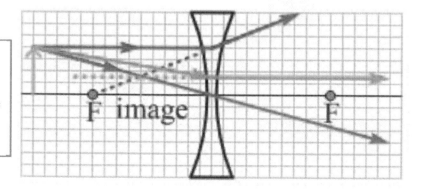

EXAMPLE 24.8 – Applying the general method

When you look at a cat through a lens that has its focal points at distances of 24 cm on either side of the lens, you see an image of the cat that is 1.5 times as large as the cat. How far is the cat from the lens? Sketch a ray diagram to check your calculations.

SOLUTION

In this case, let's first apply the equations and then draw the ray diagram. The lens is clearly a converging lens, because only converging lenses produce images that are larger than the object. One possibility is that the lens produces a virtual, upright image, so the sign of the magnification is positive. Applying the magnification equation, we get:

$$m = +1.5 = -\frac{d_i}{d_o}, \text{ which tells us that } \frac{1}{d_i} = -\frac{1}{1.5d_o}.$$

Applying the thin-lens equation:

$$\frac{1}{f} = \frac{1}{d_o} + \frac{1}{d_i} = \frac{1}{d_o} - \frac{1}{1.5d_o} = \frac{3}{3d_o} - \frac{2}{3d_o} = \frac{1}{3d_o}.$$

Thus, we find that $3d_o = f = +24$ cm, so

$d_o = +8.0$ cm and we can show that $d_i = -12$ cm.

The ray diagram for this situation is shown in Figure 24.38, confirming the calculations.

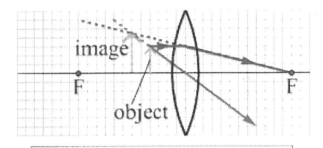

Figure 24.38: A ray diagram for the solution involving a virtual image. Each box on the grid measures 2 cm × 2 cm.

The solution above is only one of the possible answers. The image could also be real and inverted, so the sign of the magnification is negative. Applying the magnification equation, we get:

$$m = -1.5 = -\frac{d_i}{d_o}, \text{ which tells us that } \frac{1}{d_i} = +\frac{1}{1.5d_o}.$$

Applying the mirror equation:

$$\frac{1}{f} = \frac{1}{d_o} + \frac{1}{d_i} = \frac{1}{d_o} + \frac{1}{1.5d_o} = \frac{3}{3d_o} + \frac{2}{3d_o} = \frac{5}{3d_o}.$$

Thus, we find that $d_o = \frac{5}{3}f = +40$ cm, and we can show that $d_i = +60$ cm.

The ray diagram for this situation is shown in Figure 24.39, again confirming the calculations above. These two rays, and all rays that travel from the tip of the object to the tip of the image, take the same time to get there.

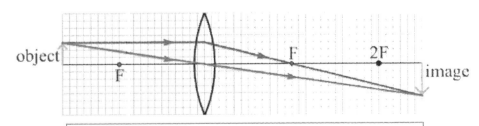

Figure 24.39: A ray diagram for the situation involving a real image. Each box on the grid measures 2 cm × 2 cm.

Related End-of-Chapter Exercises: 25 – 28.

Essential Question 24.8: Return to the situation described in Example 24.8. Would there still be two solutions if the image was smaller than the object? Explain.

Answer to Essential Question 24.8: Yes, there would still be two solutions. One solution involves a real image produced by a converging lens, while the other involves a virtual image produced by a diverging lens.

24-9 The Human Eye and the Camera

There are some interesting parallels between the human eye and a camera. Before continuing, stop and make a list of some of the similarities between the two systems, as well as some of the differences.

An important component of both the eye and a camera is the lens that is used to create a real image. Interestingly, in both systems the image is inverted.

In each case, there is a diaphragm that controls the amount of light that gets through to the lens. In the eye, the pupil plays this role, while the camera shutter does this job in the camera.

Another important component of both systems is the system for recording the focused image. In the eye, this system is the set of rods and cones that cover the retina at the back of the eye. In older cameras, the film does the job of recording the image, while in a digital camera the system is quite similar to that of the eye, with a large number (four million being typical) of tiny light sensors sensing the image.

A key difference between the human eye and the camera is in how the focusing is done. In your own eyes, when you look from one object to another that is a different distance away from you, your eyes adjust to bring the second object into focus. To understand why this adjustment is necessary, considering the thin-lens equation:

$$\frac{1}{f} = \frac{1}{d_o} + \frac{1}{d_i}.$$
(Equation 24.7: **The thin-lens equation**)

In the human eye, the distance between the lens and retina, which is the image distance d_i, is fixed. Changing the object distance thus requires a change in the focal length of the eye to produce a focused image on the retina. This adjustment is done via the ciliary muscles in the eye, which actually change the shape of the lens to change the focal length. This process, known as accommodation, occurs so quickly that we don't even notice it.

In the camera, the lens is generally a solid piece of glass with a fixed focal length. Adjusting the object distance, therefore, requires an adjustment in the image distance (the distance from the lens to the film or light sensors). This is accomplished by moving the lens. The farther the object is from the camera, the closer the lens must be to the film or light sensors.

EXPLORATION 24.9 – Correcting human vision

Consider the ray diagram shown in Figure 24.40, in which the object is quite close to the eye.

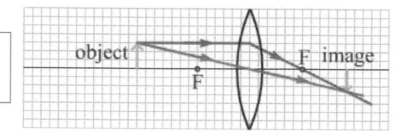

Figure 24.40: A ray diagram for a situation in which you are focusing on an object quite close to you.

Step 1 – *Where is the retina located in Figure 24.40?* The retina must be where the image is for you to see the focused image. Note that the image is upside down, but when you look at objects you do not see them upside down. The flipping of the image back to upright is accomplished by the image-processing system in your brain.

Step 2 – *If the object in Figure 24.40 is moved farther away, and the eye's focal length is unchanged, where is the focused image? Sketch a new diagram to show this.* When the object distance increases, and the focal length remains the same, the image distance decreases, as shown in Figure 24.41.

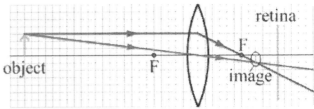

Figure 24.41: If the eye's focal length is fixed, moving the object farther from the eye produces a focused image at a point within the eye, and a blurred image on the retina.

Step 3 – *In a normal eye, the shape of the lens is changed so that the image is formed not before the retina, as in Figure 24.41, but on the retina. To accomplish this for the situation shown in Figure 24.41, should the lens be flatter or rounder than the lens that produces the correctly focused image in Figure 24.40, when the object is closer to the eye?* The lens in Figure 24.41 is doing too much focusing, because it is too round. The ciliary muscles flatten the lens to shift the image position farther from the lens, onto the retina, as shown in Figure 24.42.

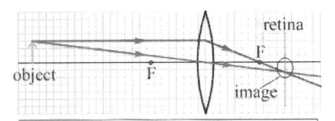

Figure 24.42: In a normal eye, when the object is farther from the eye, the curvature of the lens is reduced to focus the image onto the retina.

Step 4 – *In the eye of a near-sighted person (someone who can only focus properly on objects close to them), the lens can not be made flat enough to focus the image of a far-off object on the retina, so a corrective lens is placed in front of the eye. Does a near-sighted person need a diverging lens or a converging lens? Support your answer with a diagram.* A near-sighted person requires a diverging lens. Before the light from the object reaches the eye, the lens diverges it enough that the eye then deflects the rays to create a focused image on the retina, as shown in Figure 24.43. Another way to understand this is that the diverging lens creates a virtual image of the object close enough to the eye that the near-sighted person can focus on it.

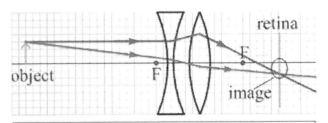

Figure 24.43: Placing a diverging lens, of the appropriate shape, in front of the eye of a nearsighted person allows the person to see far-away objects correctly.

Key ideas for corrective lenses: In a normal eye, the ciliary muscles adjust the shape of each eye's lens to form images on the retina. In a near-sighted person, the lens cannot be made flat enough to view far-away objects correctly, so a diverging lens is used to correct this deficiency. In a far-sighted person, the lens cannot be made round enough to properly focus on close objects, so the corrective lens is a converging lens. **Related End-of-Chapter Exercises: 29, 63.**

Essential Question 24.9: What if Figure 24.40 represents a camera in which the image is focused on the film? When the object is moved to the position shown in Figure 24.41, what changes in the camera to produce a focused image again? Draw a ray diagram reflecting this change.

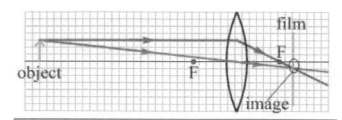

Figure 24.44: In a camera, the lens has a fixed shape, so the lens is moved to the right, closer to the film, when the object is moved farther from the camera. This produces a focused image on the film.

Answer to Essential Question 24.9: If Figure 24.40 shows a focused image for the camera, the image location is where the film is in the camera. When the object is moved farther away, as in Figure 24.41, the lens must be moved to the right, closer to the film, so that the image is again focused on the film. This is shown in Figure 24.44.

24-10 Multi-lens Systems

How do we handle systems in which there is more than one lens (or more than one mirror, or combinations of mirrors and lenses)? The standard approach is to do the analysis one lens (or mirror) at a time. Starting from the object, follow the light until it reaches the first lens or mirror. Apply the methods we learned earlier to find the image created by the first lens or mirror. That image is then the object for the next lens or mirror in the sequence. Continue the process, one lens or mirror at a time, until we have followed the light through every lens or mirror.

EXAMPLE 24.10 – Analyzing a two-lens system

As shown in Figure 24.45, a toy train with a height of 4.0 cm is placed 24 cm from a converging lens that has a focal length of 8.0 cm. A second converging lens, identical to the first, is placed 18 cm from the first lens, and on the opposite side of the lens from the train.

(a) Calculate the position of the image created by the first lens, and sketch a ray diagram to support your calculations.

(b) Repeat part (a), but for the second lens, to find the final image.

(c) Determine the overall magnification of this two-lens system.

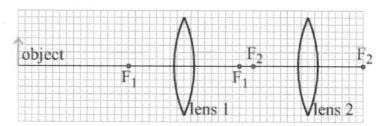

Figure 24.45: An object 24 cm in front of one lens, with a second lens 18 cm behind the first lens. Each box on the grid measures 1 cm × 1 cm.

SOLUTION

(a) To find the first image, let's apply the thin-lens equation. In this case, we get:

$$d_i = \frac{d_o \times f}{d_o - f} = \frac{(+24 \text{ cm}) \times (+8.0 \text{ cm})}{24 \text{ cm} - 8.0 \text{ cm}} = \frac{192 \text{ cm}^2}{16 \text{ cm}} = 12 \text{ cm}.$$

A ray diagram for this situation is shown in Figure 24.46, confirming the calculation above and showing that the image is real, inverted, and smaller than the object.

(b) We use the image produced by the first lens as the object for the second lens. The first thing we need to determine is the object distance for the second lens. If the image is 12 cm from the first lens, and the second lens is 18 cm from the first lens, then the image is only 6 cm (18 cm minus 12 cm)

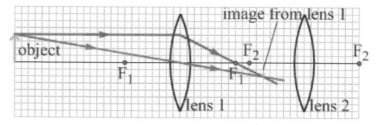

Figure 24.46: A ray diagram showing the real inverted image produced by the first lens. Each box on the grid measures 1 cm × 1 cm.

from the second lens (and is now the object for that lens). With an object distance of 6 cm, applying the thin-lens equation gives:

$$d_i = \frac{d_o \times f}{d_o - f} = \frac{(+6 \text{ cm}) \times (+8.0 \text{ cm})}{6 \text{ cm} - 8.0 \text{ cm}} = \frac{48 \text{ cm}^2}{-2 \text{ cm}} = -24 \text{ cm} .$$

The ray diagram for this situation is shown in Figure 24.47, confirming the calculations.

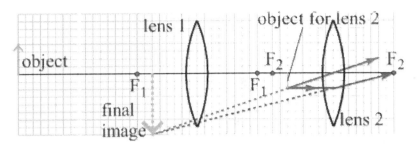

(c) We can determine the magnification in two different ways (which should give the same answer). First, the overall magnification is the ratio of the height of the final image to the height of the original object. We can find the heights of the two images by applying the magnification equation twice. For the first lens,

Figure 24.47: A ray diagram showing the image produced by the second lens. Each box on the grid measures 1 cm × 1 cm.

$$m_1 = \frac{h_{i1}}{h_{o1}} = -\frac{d_{i1}}{d_{o1}} = -\frac{12 \text{ cm}}{24 \text{ cm}} = -0.5 \text{, which, with an object height of 4.0 cm, gives an}$$

image height of −2.0 cm.

For the second lens,

$$m_2 = \frac{h_{i2}}{h_{o2}} = -\frac{d_{i2}}{d_{o2}} = -\frac{(-24 \text{ cm})}{6 \text{ cm}} = +4.0 \text{, which, with an object height of −2.0 cm, gives an}$$

image height of −8.0 cm.

Thus, the overall magnification is: $M = \frac{h_{i2}}{h_{o1}} = \frac{-8.0 \text{ cm}}{4.0 \text{ cm}} = -2.0$.

The final image is twice as large as the original object, with the minus sign telling us that the image is inverted compared to the original object.

The second way to find the overall magnification is to combine the magnifications of the individual lenses. The first lens gives an image that is inverted and half as large as the original object, while the second lens increases the size by a factor of four while maintaining the orientation. The overall magnification is the product of the individual magnifications:

$$M = m_1 \times m_2 = -0.5 \times (+4.0) = -2.0 .$$

The method we applied here can be applied to any number of lenses and/or mirrors. Simply follow the light through the system, using the image created by one lens or mirror as the object for the next lens or mirror in the sequence.

Related End-of-Chapter Exercises: 64, 65.

Essential Question 24.10: In an astronomical telescope, which uses two converging lenses, the distance between the lenses is the sum of the two focal lengths. Explain why this is the case.

Answer to Essential Question 24.10: In an astronomical telescope, the object (such as a distant galaxy) is very far away, so the first lens (the objective) creates an image at the focal point of that lens. To produce as large a final image as possible, the second lens (the eyepiece) should create an image at infinity, so the first image needs to be located at the eyepiece's focal point. Because the focal points coincide, the distance between the lenses is the sum of the focal lengths.

Chapter Summary

Essential Idea: Refraction and Lenses.
When light passes from one medium to a second medium in which the speed of light is different, the change in speed is generally associated with refraction (a change in direction of the light). The phenomenon of refraction is exploited in optical fibers and corrective lenses.

Index of Refraction
The index of refraction of a medium is a unitless parameter that is equal to the ratio of the speed of light in vacuum to the speed of light in the medium. In general, the speed of light in vacuum represents the maximum speed of light, so we expect $n \geq 1$.

$$n = \frac{c}{v}.$$ (Equation 24.1: **Index of refraction**)

The index of refraction is also equal to the ratio of the wavelength of light in vacuum to the wavelength of light in the medium.

$$n = \frac{\lambda_{vacuum}}{\lambda_{medium}}.$$ (Equation 24.2: **Index of refraction, in terms of wavelength**)

Snell's Law
When light is transmitted from one medium to another, the angle of incidence, θ_1, in the first medium is related to the angle of refraction, θ_2, in the second medium by:

$$n_1 \sin\theta_1 = n_2 \sin\theta_2$$ (Equation 24.3: **Snell's Law**)

These angles are measured from the normal (perpendicular) to the surface. The n's represent the indices of refraction of the two media.

Total Internal Reflection
Optical fibers are an important application of total internal reflection. When light traveling in a medium encounters an interface separating that medium from a medium with a lower index of refraction, 100% of the light will be reflected back into the first medium if the angle of incidence exceeds a particular critical angle given by:

$$\theta_c = \sin^{-1}\left(\frac{n_2}{n_1}\right).$$ (Eq. 24.4: **The critical angle beyond which total internal reflection occurs**)

Ray Diagrams
To draw a ray diagram, we show rays leaving the tip of an object and being refracted by a lens. The tip of the image is where the refracted rays meet. If the rays leaving a single point on the object are refracted so they pass through a single point, a real image is formed. If, instead, such refracted rays appear to diverge from a single point behind the lens, a virtual image is formed. A summary of three rays that are particularly easy to draw is given in section 24-7.

Thin Lenses

Type of lens	Focal length	Image characteristics
Diverging (usually concave)	Negative	The image is virtual, upright, smaller than the object, and between the lens and the focal point on the side of the lens the object is on.
Converging (usually convex)	Positive	The image can be real or virtual, and larger than, smaller than, or the same size as the object. See the table below for details.

Table 24.2: A summary of the lenses we investigated in this chapter.

Images formed by a Converging Lens

Object position	Image position	Image characteristics
∞	At the focal point.	Real image with height of zero.
Moving from ∞ toward twice the focal length.	Moving from the focal point toward twice the focal length.	The image is real, inverted, and smaller than the object. The image moves closer to twice the focal length, and increases in height, as the object is moved closer to twice the focal length from the lens.
At twice the focal length.	At twice the focal length.	The image is real, inverted, and the same size as the object.
Moving from twice the focal length toward the focal point.	Moving from twice the focal length toward infinity.	The image is real, inverted, and larger than the object. The image moves farther from the lens, and increases in height, as the object is moved closer to the focal point.
At the focal point.	At infinity.	The image is at infinity, and is infinitely tall.
Closer to the lens than the focal point.	On the same side of the lens as the object.	The image is virtual, upright, and larger than the object. The image moves closer to the lens, and decreases in height, as the object is moved closer to the lens.

Table 24.3: A summary of the image positions and characteristics for a converging lens.

The thin-lens equation (same as the mirror equation from Chapter 23)

The thin-lens equation relates the object distance, d_o, the image distance, d_i, and the focal length, f. The mnemonic "If I do I di" can help you to remember the equation.

$$\frac{1}{f} = \frac{1}{d_o} + \frac{1}{d_i} .$$
(Equation 24.7: **The thin-lens equation**)

$$d_i = \frac{d_o \times f}{d_o - f}$$
(Equation 24.8: **The thin-lens equation, solved for the image distance**)

Sign conventions

The image distance is positive if the image is on the reflective side of the mirror (a real image), and negative if the image is behind the mirror (a virtual image).

The image height is positive when the image is upright, and negative when the image is inverted. A similar rule applies to the object height.

Magnification

The magnification, m, is the ratio of the image height (h_i) to the object height (h_o).

$$m = \frac{h_i}{h_o} = -\frac{d_i}{d_o} .$$
(Equation 24.9: **Magnification**)

The image is larger than the object if $|m| > 1$, smaller if $|m| < 1$, and of the same size if $|m| = 1$. The magnification is positive if the image is upright, and negative if the image is inverted.

End-of-Chapter Exercises

Exercises 1 – 12 are conceptual questions designed to assess whether you understand the main concepts of the chapter.

1. Figure 24.48 shows a beam of light in air (medium 1) incident on an interface along the normal to the interface, with some of the light refracting into medium 2 and some reflecting straight back. What, if anything, is wrong? If nothing is wrong, determine the index of refraction of medium 2. If something is wrong, explain.

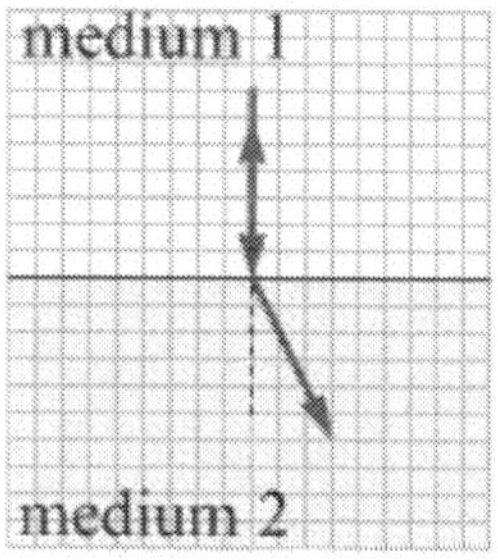

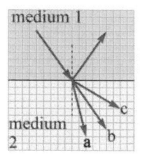

Figure 24.48: A beam of light incident along the normal to an interface, refracting into medium 2. For Exercise 1.

2. As shown in Figure 24.49, a beam of light in medium 1 is incident on an interface separating medium 1 from medium 2. Part of the beam reflects from the interface, and part refracts into medium 2. Describe, qualitatively, how the indices of refraction of the two media compare if the refracted beam follows (a) path a, (b) path b, or (c) path c.

Figure 24.49: Three possible paths for the refracted beam to follow when light in medium 1 encounters the interface between two media, for Exercise 2.

3. Figure 24.50 shows an object near a diverging lens. A single ray is drawn on the ray diagram. Duplicate the diagram, and add a second ray to show the position of the image.

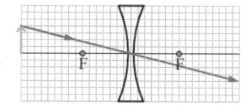

Figure 24.50: A single ray, passing undeflected through the center of a thin lens, is shown on the figure. For Exercise 3.

4. Two rays of light are shown on the ray diagram in Figure 24.51, along with two arrows representing the object and the image. (a) Could this diagram be incorrect? If so, how could it be corrected? (b) Could this diagram be correct? If so, explain how this could be possible.

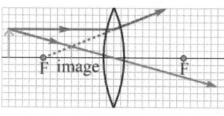

Figure 24.51: A ray diagram, for Exercise 4. Could it be correct, or is it incorrect?

5. Figure 24.52 shows an object in front of a converging lens. First, re-draw the diagram, preferably on a piece of graph paper. For parts (a) – (c), start the ray from the tip of the object and show how the ray is refracted by the lens. (a) On your diagram, draw a ray that travels parallel to the principal axis toward the lens. (b) Draw a ray that travels directly toward the center of the lens. (c) Draw a ray that travels through the focal point between the object and the lens. (d) Use the rays to locate the image on the diagram.

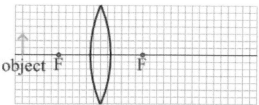

Figure 24.52: An object in front of a converging lens, for Exercise 5.

6. Figure 24.53 shows the virtual image formed by a thin-lens, as well as the locations of the two focal points of the lens. The small squares on the grid measure 1.0 cm × 1.0 cm. The goal of this exercise is to determine the type of lens creating the image, as well as the position and size of the object. (a) How many different solutions are there to this exercise? (b) For each of the solutions, describe the kind of lens that produces the image, the object distance, and the object height. (c) For each solution, sketch a ray diagram.

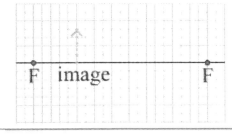

Figure 24.53: The virtual image created by a thin lens, and the locations of the two focal points of the lens. The small squares on the grid measure 1.0 cm × 1.0 cm. For Exercise 6.

7. Assume that the index of refraction of a particular piece of glass varies as shown in the graph in Figure 24.54. If light traveling in this piece of glass encounters a glass-air interface, is the critical angle for total internal reflection larger for violet light, red light, or is it equal for both? Explain.

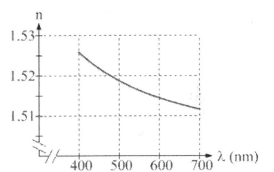

Figure 24.54: A graph of the index of refraction, as a function of wavelength, for a typical sample of glass. The graph is confined to the visible spectrum, from 400 nm (violet) to 700 nm (red). For Exercise 7.

8. In a common demonstration, a small glass beaker virtually disappears when it is placed in a larger beaker full of a particular type of oil. Using the principles of physics addressed in this chapter, explain how this could work.

9. If you wear your eyeglasses under water, can you still see clearly? Explain.

10. Identify whether the lens in this situation is a converging lens or a diverging lens. For each part below, determine which possibility you can rule out, if either. (a) First, when an object is placed 15 cm from a lens, a virtual image is observed. (b) Then, when the object is moved a little closer to the lens, the image is observed to move closer to the lens. (c) As the object is moved closer to the lens, the image is observed to decrease in size.

11. Identify whether the lens in this situation is a converging lens or a diverging lens. For each part below, determine which possibility you can rule out, if either. (a) First, when an object is placed 15 cm from a lens, a virtual image is observed. (b) Then, when the object is moved a little closer to the lens, the image is observed to increase in size. (c) What, if anything, can you conclude about the focal length of the lens?

12. You have an unknown optical device that you are trying to identify. The device could be any one of five things, a plane mirror, convex mirror, concave mirror, converging lens, or diverging lens. To identify the device you make the following observations, in order. For each part, state what, if anything, the observation tells you about what kind of mirror or lens the device is. (a) You observe that when you place an object in front of the device that the device creates an image of the object that is larger than the object. What could the device be? (b) You then observe that the image is inverted compared to the object. Based on this and the previous observation, what could the device be? (c) You then observe that the larger inverted image is on the opposite side of the device as the object. Based on this and the previous observations, what is the device? (d) If the device has a focal length of 10 cm and the object distance is 15 cm, what is the image distance?

Exercises 13 – 15 involve refraction.

13. A beam of light is incident on an interface separating two media. When the angle of incidence, measured from the normal, is 10.0°, the angle of refraction is 18.0°. What is the angle of refraction when the angle of incidence is 30.0°?

14. Return to the situation described in Exercise 13. If the speed of light in one of the media is 2.80×10^8 m/s, what is the speed of light in the other medium? Is there more than one possible answer? Explain.

15. A beam of light is incident on an interface separating two media, as shown in Figure 24.55. The squares in the grid on the diagram measure 10 cm × 10 cm. (a) Which medium has a larger index of refraction? Explain your answer. (b) If the index of refraction of one of the media is 1.10, what is the index of refraction of the other medium? Is there more than one possible answer? Explain.

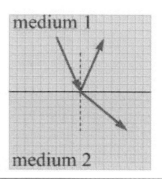

Figure 24.55: When a beam of light is incident on the interface separating two media, part of the beam is reflected back into medium 1 and part refracts into medium 2, for Exercise 15.

Exercises 16 – 20 involve total internal reflection.

16. As shown in Figure 24.56, a beam of light traveling in medium 2 experiences total internal reflection when it encounters the interface separating medium 2 from medium 1. The angle of incidence is 45°, and medium 1 is air, with an index of refraction of 1.00. What, if anything, can you say about the index of refraction of medium 2?

17. As shown in Figure 24.56, a beam of light traveling in medium 2 experiences total internal reflection when it encounters the interface separating medium 2 from medium 1. The angle of incidence is 45°, and the speed of the light in medium 2 is 1.50×10^8 m/s. What, if anything, can you say about the index of refraction of medium 1?

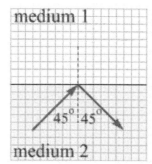

Figure 24.56: A beam of light in medium 2 experiences total internal reflection when it encounters the interface separating medium 2 from medium 1. For Exercises 16 and 17.

18. As shown in Figure 24.57, a beam of light in medium 1 is incident on an interface separating medium 1 from medium 2. Part of the beam reflects from the interface, and part refracts into medium 2, traveling along the interface between the media. Using the diagram, calculate the critical angle for total internal reflection for this situation.

19. A beam of light is incident on an interface separating two media. When the angle of incidence, measured from the normal, is 18.0°, the angle of refraction is 10.0°. What is the critical angle for this particular interface?

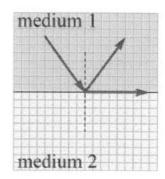

Figure 24.57: When a beam of light in one medium encounters an interface, part of it reflects and part refracts into medium 2, traveling along the interface. For Exercise 18.

20. A small red light-emitting diode (LED) is placed 12.0 cm below the surface of the water in a bathtub. A circle of red light is observed at the water surface. What is the diameter of this circle? Assume that the air above the water has an index of refraction of 1.00, and that the water has an index of refraction of 1.33.

Exercises 21 – 24 are designed to give you practice applying the thin-lens equation.

21. As you are analyzing a thin-lens situation, you write an equation that states:
$$\frac{1}{f} = \frac{1}{+40 \text{ cm}} + \frac{1}{+30 \text{ cm}}.$$ (a) What is the value of $1/f$ in this situation? (b) What is the focal length of the lens? (c) What kind of lens is this?

22. Return to Exercise 21. What is the object distance in this situation, and what is the image distance?

23. As you are analyzing a thin-lens situation, you write an equation that states:
$$\frac{1}{f} = \frac{1}{+15 \text{ cm}} + \frac{1}{-30 \text{ cm}}.$$ (a) What is the value of $1/f$ in this situation? (b) What is the focal length of the lens? (c) What kind of lens is this?

24. As you are analyzing a thin-lens situation, you write an equation that states:
$$\frac{1}{+24 \text{ cm}} = \frac{1}{+24 \text{ cm}} + \frac{1}{d_i}.$$ (a) What is the value of $1/d_i$ in this situation? (b) What is the image distance?

Exercises 25 – 28 are designed to give you practice applying the general method for analyzing a problem involving lenses.

25. An object is placed 30 cm away from a lens that has a focal length of +10 cm. (a) Sketch a ray diagram, to show the position of the image and the image characteristics. (b) Determine the image distance. (c) Determine the magnification.

26. An object is placed 30 cm away from a lens that has a focal length of –10 cm. (a) Sketch a ray diagram, to show the position of the image and the image characteristics. (b) Determine the image distance. (c) Determine the magnification.

27. You are examining an ant through a magnifying glass, which is simply a converging lens. When the ant is 10 cm from the lens and you look through the lens, you see an upright image of the ant that is 3.0 times larger than the ant itself. (a) Determine the image distance in this situation. (b) Determine the focal length of the magnifying glass. (c) Sketch a ray diagram for this situation.

28. An object is placed 48 cm from a lens. When you look through the lens, you see an image of the object that is 3.0 times larger than the object. (a) What kind of lens is it? (b) Sketch a ray diagram to check your calculations. Make sure you find all possible solutions.

Exercises 29 – 33 involve applications of refraction and lenses.

29. The lens in your digital camera has a focal length of 5.0 cm. You are using the camera to take a close-up picture of a flower that is 12.0 cm from the lens. (a) Determine how far the lens should be from the image-sensing system inside the camera. (b) Determine the magnification in this situation. (c) If you then use your camera to take a photograph of your friend, who is 3.0 m from the lens of the camera, how far should the lens be from the image-sensing system now? (d) Determine the magnification in this new situation.

30. Binoculars generally use pairs of prisms in which the light experiences total internal reflection. Each prism (in blue on the diagram) is right-angled, with the other two angles being 45°. A diagram of the path followed by light as it travels through the prisms to your eyes is shown in Figure 24.58. If the prisms are surrounded by air, determine the minimum index of refraction of the prism material.

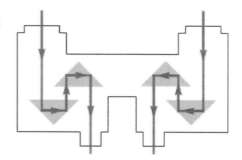

Figure 24.58: A pair of binoculars, with two right-angled prisms on each side to shift the light from the path it is following when it enters the binoculars to a path that takes it right into your eye. For Exercise 30.

31. Diamonds are particularly colorful and sparkly. One reason for this is the relatively large index of refraction of diamond of around 2.4. Another reason is that diamond does exhibit dispersion. Use these facts to explain why light that enters a diamond often experiences a number of reflections within the diamond before emerging, and why this would help spread white light out into its constituent colors.

32. When you are driving along a highway, or walking in the desert, on a hot day, you often see a mirage in the distance, where it looks like there is water on the road, or sand, ahead of you. Do a little research about this phenomenon and write a couple of paragraphs describing it, and the physics relevant to this chapter that are responsible for producing a mirage.

33. As a participant on the reality show *Survivor*, you are stranded on a sunny island with several other individuals. You get the bright idea of trying to start a fire by using your eyeglasses to focus the Sun's rays onto a piece of dry wood. (a) How far from your glasses should the wood be? (b) Do you need to be nearsighted or farsighted for this method to have a chance of working?

General problems and conceptual questions

34. A pulse of light takes 3.00 ns to travel through air from an emitter to a detector. When a piece of transparent material with a length of 40 cm is introduced into the light's path, the pulse takes 3.40 ns. What is the index of refraction of the transparent material?

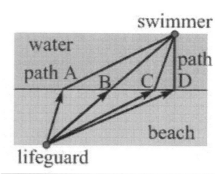

Figure 24.59: A situation involving a lifeguard trying to reach a swimmer in the shortest time, which is analogous to light as it travels from one medium to another. For Exercise 35.

35. Answer this problem by analogy with optics and refraction. A lifeguard can run at 5.0 m/s along the sandy beach, and can swim at 2.0 m/s through the water. The initial positions of the lifeguard and a swimmer who needs the lifeguard's help are shown in Figure 24.59. Four possible paths for the lifeguard to take are shown on the diagram. Which path should the lifeguard take to minimize the time it takes to reach the swimmer? Explain.

36. A particular converging lens has a focal length of +20 cm. A second lens of exactly the same shape as the first lens has a focal length of +25 cm. Is this possible? Explain.

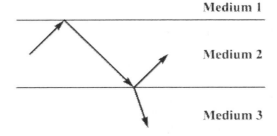

37. As shown in Figure 24.60, light traveling in medium 2 experiences total internal reflection at the boundary with medium 1, then experiences reflection and refraction at the boundary with medium 3. Rank the media based on their indices of refraction, from largest to smallest.

Figure 24.60: The path followed by light that is initially in medium 2. For Exercise 37.

38. A horizontal beam of monochromatic (single wavelength) laser light is incident on a block of glass, as shown in Figure 24.61. The faces of the block of glass are all either horizontal or vertical except for the face at the top right, which is inclined at 45°. The glass has an index of refraction of 1.50, while the air surrounding the glass has an index of refraction of 1.00. Copy the diagram, and sketch the path the light takes through the block, accounting for both refraction and reflection at each air-glass interface the light encounters. Label every point where light emerges from the glass back into the air and, at each of these points, determine the angle at which the light emerges from the glass.

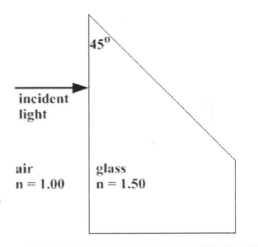

Figure 24.61: Light is incident on a piece of glass that is surrounded by air. For Exercises 38 and 39.

39. Repeat Exercise 38, but now the medium surrounding the glass is water, with an index of refraction of 1.33, instead of air.

40. As shown in Figure 24.62, a beam of red and violet light is incident along the normal to one surface of a right-angled triangular glass prism. The glass has an index of refraction of 1.52 for red light, and 1.54 for violet light. (a) Draw a sketch showing how the red and violet beams travel from the point at which they enter the prism to the side *ab* of the prism. (b) If the angle at vertex *a* of the prism is $\theta = 30.0°$, determine the angles of refraction for the red and violet beams that emerge from the prism from the side *ab*. Show these refracted beams on your sketch. (c) On your sketch, show how the red and violet beams reflect from side *ab* of the prism.

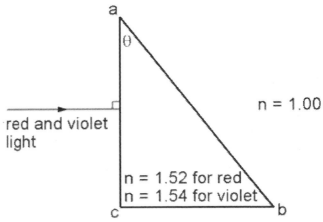

Figure 24.62: A beam of red and violet light entering a right-angled glass prism. For Exercise 40.

41. As shown in Figure 24.63, when an object is placed 12 cm in front of a particular optical device (either a single mirror or a single lens) a **virtual image** is formed 24 cm from the device on the same side of the device as the object. (a) What kind of mirror or lens could this optical device be? (b) Find the focal length of the device.

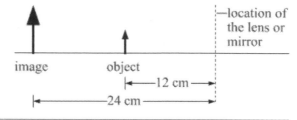

Figure 24.63: When an object is placed 12 cm in front of an optical device, which is either a lens or a mirror, a virtual image is formed 24 cm from the device. For Exercise 41.

42. Galileo Galilei used a telescope to carry out detailed observations of the moons of a particular planet. Do some research about Galileo's telescope, and about the observations he made with it, and write a couple of paragraphs describing the telescope and the observations.

43. In a particular situation, involving an object in front of a lens, the object distance is 20 cm and the magnification is +4.0. Find (a) the image distance, (b) the focal length of the lens.

44. In a particular situation, involving an object in front of a lens with a focal length of +20 cm, the magnification is +4.0. Find (a) the object distance, (b) the image distance.

45. An object is placed a certain distance from a lens. The image created by the lens is exactly half as large as the object. If the two focal points of the lens are 20 cm from the lens, where is the object? Where is the image? (a) Find one solution to this problem. (b) Find a second solution. (c) Sketch ray diagrams for your two solutions.

46. Figure 24.64 shows an object and a real image created by a lens. Assume the boxes on the grid measure 10 cm × 10 cm. Find the position of the lens, and its focal length.

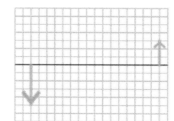

Figure 24.64: This figure shows an object and a real image created by a lens. For Exercises 46 and 48.

47. Figure 24.65 shows an object (the larger arrow) and the virtual image of that object, created by a lens. Assume the boxes on the grid measure 10 cm × 10 cm. Find the position of the lens, and its focal length.

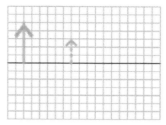

Figure 24.65: The larger arrow represents an object, while the smaller arrow represents the virtual image of that object, created by a lens. For Exercise 47.

48. Sketch a ray diagram for the situation shown in (a) Figure 24.64, (b) Figure 24.65.

49. In the situation shown in Figure 24.66, a small red LED (light-emitting diode) is placed on the principal axis at one of the focal points of a particular converging lens. The LED can be considered to be a point source. Draw a ray diagram to show what happens to rays of light that are emitted by the LED and are refracted by the lens.

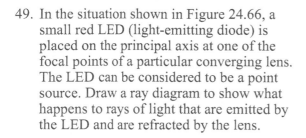

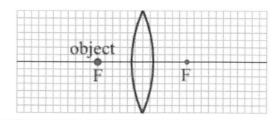

Figure 24.66: A small red LED (light-emitting diode) is placed at one focal point of a converging lens, for Exercise 49.

50. In the situation shown in Figure 24.67, a small red LED (light-emitting diode) is placed on the principal axis, 7.0 cm from a converging lens that has a focal length of 14 cm. The LED can be considered to be a point source. Find the location of the image of the LED.

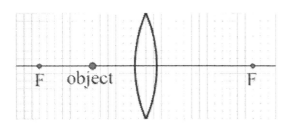

Figure 24.67: A small red LED (light-emitting diode) is placed in front of a converging lens, for Exercises 50 and 52.

51. In the situation shown in Figure 24.68, a small red LED (light-emitting diode) is placed 4.0 cm above the principal axis, and 8.0 cm away from a diverging lens that has a focal length of −12 cm. The LED can be considered to be a point source. Find the location of the image of the LED.

52. Draw several rays showing how the image of the LED is formed in (a) Figure 24.67, and (b) Figure 24.68.

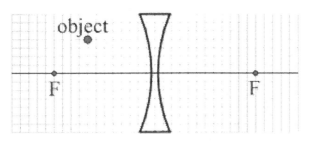

Figure 24.68: A small red LED (light-emitting diode) is placed in front of a diverging lens, for Exercises 51 and 52. The LED is 4.0 cm above the principal axis.

53. A model of a dinosaur is placed 36 cm away from a converging lens that has a focal length of +20 cm. The model is 8.0 cm tall. Determine (a) the location of the image, (b) the height of the image, (c) whether the image is real or virtual, and (d) whether the image is upright or inverted.

54. Repeat Exercise 54, with the model of the dinosaur 15 cm from the lens instead.

55. A model of a horse is placed 36 cm away from a lens that has a focal length of −20 cm. The model is 8.0 cm tall. Determine (a) the image location, (b) the height of the image, (c) whether the image is real or virtual, and (d) whether the image is upright or inverted.

56. Return to the situation described in Exercise 55. Describe what happens to the position and size of the image if the model is moved a little bit farther from the lens.

57. Return to the situation described in Exercise 53. Describe what happens to the position and size of the image if the model is moved a little bit farther from the lens.

58. A particular lens has a focal length of +40 cm. (a) For this lens, plot a graph of $1/d_i$ as a function of $1/d_o$ for object distances between +20 cm and +80 cm. (b) How can you read the focal length directly from the graph?

59. Repeat Exercise 58, but now plot a graph of d_i as a function of d_o.

60. Figure 24.69 shows a graph of $1/d_i$ as a function of $1/d_o$ for a particular lens. What kind of lens is it, and what is the focal length of the lens?

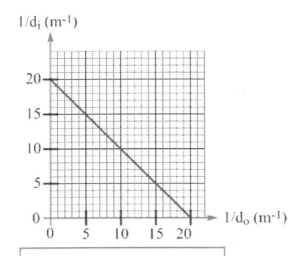

Figure 24.69: A graph of $1/d_i$ as a function of $1/d_o$ for a particular lens. For Exercise 60.

61. A particular lens has a focal length of +25 cm. (a) For this lens, plot a graph of the magnification as a function of the object distance, for object distances between +10 cm and +40 cm. (b) How can you read the focal length directly from the graph?

62. Refer to Figure 24.23, just above the start of Section 24-5. (a) What, if anything, happens to the image if you cover up the bottom half of the lens, preventing any light from reaching that part of the lens? (b) Does your answer change if you cover up the top half of the lens instead? Explain, and refer to Figure 24.23 in your explanations.

63. Refer to Figure 24.42, in Section 24-9, which shows how light from a distant object is focused on to the retina in your eye. (a) Sketch a ray diagram showing where the image is located when the object is only half the distance from the lens. Assume that neither the focal length of the lens, nor the distance from the lens to the retina, changes. (b) Sketch a second ray diagram showing a corrective lens placed in front of the eye, to correctly focus the image of the object onto the retina.

64. Figure 24.70 shows five parallel rays that are incident on a pair of lenses. The first lens is a converging lens with a focal length of +10 cm, while the second lens is a diverging lens with a focal length of -5 cm. One of the focal points of the converging lens is at the same location as one of the focal points of the diverging lens. (a) Sketch a ray diagram to show what this combination of lenses does to the parallel rays. (b) Describe the function of arranging two lenses in this way.

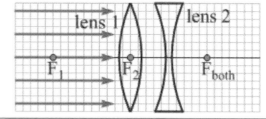

Figure 24.70: A set of parallel rays incident on a pair of lenses, for Exercise 64.

65. As shown in Figure 24.71, an object that is 4.0 cm tall is placed 10 cm to the left of a converging lens that has a focal length of 5.0 cm. A plane mirror is located 9.0 cm to the right of the lens. (a) Find the height and location of the image created by the lens. (b) After passing through the lens, the light encounters the mirror, and a second image is formed. Find the location and height of the image created by the mirror. (c) The mirror sends the light back toward the lens, and the lens creates another image. Find the location and height of this final image. (d) Determine which of these images are real and which are virtual. Justify your answer.

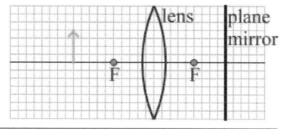

Figure 24.71: An object placed in front of a converging lens, with a plane mirror on the far side of the lens. For Exercise 65. Each square on the grid measures 1.0 cm × 1.0 cm.

66. Comment on following conversation, between two students discussing a situation in which they are trying to determine whether a particular lens is converging or diverging.

Jeremy: The first thing they tell us is that the image produced in the situation is virtual. Doesn't that mean the lens must be a diverging lens? That always gives virtual images.

Bridget: I don't think this tells us much, actually. Converging lenses can also give virtual images, if the object is farther from the lens than the focal point.

Jeremy: OK, so, then they say the image is smaller in size than the object. That doesn't tell us much either, right? Both lenses can produce images smaller than the object.

This set of pictures shows snapshots of the interference of identical waves emitted by two sources. From top to bottom, the wavelength steadily increases, while from left to right, the distance between the two sources increases (being zero in each picture on the far left).

Chapter 25 – Interference and Diffraction

CHAPTER CONTENTS

In this chapter, we will get more sophisticated with our model of light, and examine the wave nature of light. It turns out that the way light behaves when it encounters openings and obstacles is the same way water waves and sound waves behave when they encounter obstacles and openings. The main factor in determining the resulting pattern is the ratio of the size of the obstacle or opening to the wavelength of the particular wave. Thus, the pattern that results when a sound wave passes through an opening that is three wavelengths wide is similar to the pattern resulting from a light wave passing through an opening three wavelengths wide. The wave model of light can be used to explain things ranging from the pretty colors of a soap bubble to why two objects (like two stars) can look like one object when viewed with the naked eye, but are clearly two separate objects when viewed through a telescope.

25-1 Interference from Two Sources

In this chapter, our focus will be on the wave behavior of light, and on how two or more light waves interfere. However, the same concepts apply to sound waves, and other mechanical waves. We will begin by considering two sources, separated by some distance, which are broadcasting identical single-frequency waves in phase with one another. The sources could be two speakers of sound, or could be two sources of light waves. We briefly discussed this situation at the end of section 21-7, and we will now investigate this quantitatively and in detail.

To begin with, we can represent the waves emitted by each source by a set of concentric circles, with a dark region corresponding to a trough in the wave, and a white region corresponding to a peak in the wave. If we then overlap the circles, as shown in Figure 25.1, we get an interesting pattern that is the result of interference between the two sets of the waves. The dark lines radiating out from the center of the pattern correspond to destructive interference, while the bright areas correspond to constructive interference. If you set up two speakers broadcasting identical single-frequency sounds, you can create an interference pattern like this, and you can walk through it to hear areas of constructive and destructive interference.

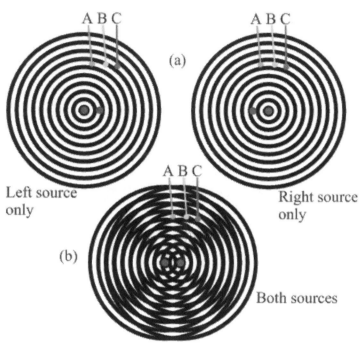

The interference pattern in Figure 25.1 (b) looks complicated, but we can understand it using interference ideas. First, let's define the path-length difference as the difference between the distance a point is from one source and the distance the point is from the second source. For point *A* in Figure 25.1 (b), which is on the perpendicular bisector of the line connecting the sources, the path-length difference is zero because point *A* is equidistant from both sources. Because the sources are in phase with one another, at the instant a peak in the wave is emitted by the left source, a peak is also emitted by the right source. These peaks travel the same distance to point *A* at equal speeds, and thus they arrive at *A* simultaneously. Two peaks arriving at the same time produce constructive interference. This argument holds for any point on the perpendicular bisector to the line connecting the speakers, because all those points have a path-length difference of zero.

Figure 25.1: (a) The waves emitted by two sources can be represented by a pattern of concentric circles (in three dimensions they are a set of spherical shells) that expand out from the source at the wave speed. (b) The pattern of constructive and destructive interference that results when both sources emit waves simultaneously can be seen when the two sets of concentric circles are overlapped.

Point B in Figure 25.1(b) is closer to the right source than it is to the left source, and thus the path-length difference is not zero. Point B happens to be exactly half a wavelength farther from the left source than it is from the right source. When a peak emitted by the right source reaches point B, the peak that was emitted at the same time from the left source is still half a wavelength from point B. Half a wavelength from a peak is a trough, so a trough arrives at point B from the left source at the same time a peak arrives from the right source (and vice versa), leading to destructive interference. All such points that are half a wavelength farther from one source experience destructive interference.

Point C in Figure 25.1(b) is one wavelength closer to source 2 than it is to source 1, so when a peak emitted by source 2 reaches point C, the peak that was emitted at the same time from source 1 is still a wavelength from point C. A full wavelength from a peak is another peak, so peaks arrive at C simultaneously from the two sources, leading to constructive interference. All such points that are a full wavelength farther from one source than the other experience constructive interference.

The trend continues. The bottom line is that all locations that are an integer number of wavelengths farther from one source than the other experience constructive interference, and all locations that are an integer number of wavelengths plus half a wavelength farther from one source than the other experience destructive interference. These general conditions for interference are summarized in the box below.

For two sources, which are in phase with one another, that broadcast identical waves in all directions, the interference can be understood in terms of the path-length difference.

$$L_1 - L_2 = \Delta L = m\lambda,$$ (Equation 25.1: **condition for constructive interference**)

where m is an integer.

$$L_1 - L_2 = \Delta L = (m + 1/2)\lambda,$$ (Equation 25.2: **condition for destructive interference**)

where m is an integer.

For locations that are far from the sources, in comparison to d, the distance between the sources, the waves from the two sources are essentially parallel to one another. As illustrated in Figure 25.2, the path-length difference in this case is given by $\Delta L = d\sin\theta$, where the angle θ is shown in Figure 25.2. Thus, in this situation the angles at which constructive or destructive interference occur are:

$$d\sin\theta = m\lambda,$$ (Equation 25.3: **constructive interference, for two sources in phase**)

where m is an integer, and

$$d\sin\theta = (m + 1/2)\lambda,$$ (Equation 25.4: **destructive interference, for two sources in phase**)

where m is an integer.

Figure 25.2: When a point is a long way from both sources, the geometry of the situation allows us to approximate the path-length difference in terms of d, the distance between the sources, and θ, the angle between the perpendicular bisector of the line joining the sources and the straight line going from the midpoint between the sources and the point.

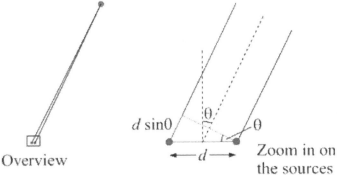

Overview

$d\sin\theta$

Zoom in on the sources

Related End-of-Chapter Exercises: 4, 13 – 15.

Essential Question 25.1 (a) A particular point experiences constructive interference no matter what the wavelength is of the waves sent out by the sources. Where is the point?
(b) What happens to the angles at which destructive interference occurs when (i) the wavelength of the waves is decreased, and (ii) d, the distance between the sources, is decreased?

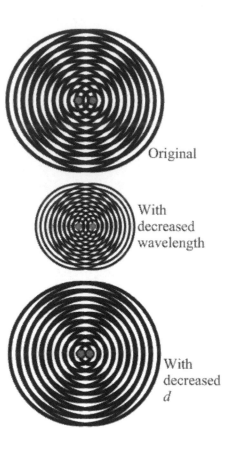

Answer to Essential Question 25.1 (a) The only points for which changing the wavelength has no impact on the interference of the waves are points for which the path-length difference is zero. Thus, the point in question must lie on the perpendicular bisector of the line joining the sources. (b) If we re-arrange Equation 25.3 to solve for the sine of the angle, we get

$$\sin\theta = \frac{m\lambda}{d}.$$

Thus, decreasing the wavelength decreases $\sin\theta$, so the pattern gets tighter. Decreasing d, the distance between the sources, has the opposite effect, with the pattern spreading out. We can understand the wavelength effect conceptually in that, when the wavelength decreases, we don't have to go as far from the perpendicular bisector to locate points that are half a wavelength (or a full wavelength) farther from one source than the other. Figure 25.3 shows the effect of decreasing the wavelength, or of decreasing the distance between the sources.

Original

With decreased wavelength

With decreased d

> **Figure 25.3**: The top diagram shows the interference pattern produced by two sources. The middle diagram shows the effect of decreasing the wavelength of the waves produced by the sources, while the bottom diagram shows the effect of decreasing the distance between the sources.

25-2 The Diffraction Grating

Now that we understand what happens when we have two sources emitting waves that interfere, let's see if we can understand what happens when we add additional sources. The distance d between neighboring sources is the same as the distance between the original two sources, and the sources are arranged in a line. All the sources emit identical waves that are in phase.

EXPLORATION 25.2 – Adding sources
Step 1 – Consider a point a long way from two sources. The sources are a distance d apart. The point is one wavelength farther from one source than the other, so constructive interference occurs at the point. When we add a third source, so that we have three sources equally spaced in a line, separated by d, do we still get constructive interference taking place at the point?
Yes. As the diagram in Figure 25.4 shows, the path-length difference for the third source and the source it was placed closest to will also be one wavelength. Now we get constructive interference for three waves at once, not just two, so the amplitude of the resultant wave is larger than it was with only two sources.

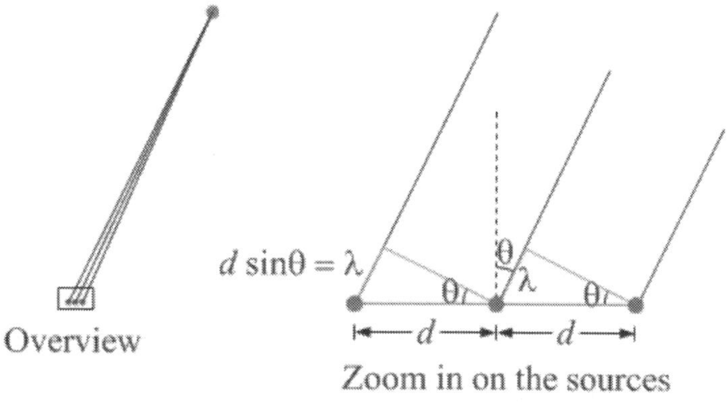

Overview

$d\sin\theta = \lambda$

Zoom in on the sources

> **Figure 25.4**: For a point that is one wavelength farther from one source than another, adding a third source results in even larger amplitude because of constructive interference.

Step 2 – If we consider a different point that is half a wavelength farther from one of two sources than the other, destructive interference occurs at the point. When we add a third source, so that we have three sources equally spaced in a line, separated by d, do we still get destructive interference taking place at the point?

No. The destructive interference at the point was caused by the cancellation between the waves from the first two sources. Adding a third source does not change the fact that the first two waves cancel one another, so there is nothing to cancel the third wave.

Step 3 – For three sources, what path-length difference (between zero and one wavelength) between neighboring sources results in completely destructive interference? With three sources, it turns out that there are two path-length differences between 0 and one wavelength that result in completely destructive interference, these being one-third and two-thirds of a wavelength.

Step 4 – What if we have N sources, where N is any integer greater than 1. Is there a general rule for predicting the angles at which constructive interference occurs? What about destructive interference? Constructive interference occurs at the same points for N sources that it does for 2 sources, so the equation $d \sin\theta = n\lambda$ still applies for situations with $N > 1$ sources.

There are $N - 1$ places where destructive interference happens in between each interference maximum, so we generally dispense with an equation for destructive interference when $N > 2$.

Key idea: The equation $d \sin\theta = m\lambda$ applies to any number of sources > 1, as long as the sources are equally spaced. With multiple sources, it is much easier to produce destructive interference than it is to produce completely constructive interference, so there is no simple equation for destructive interference.
Related End-of-Chapter Exercises: 7, 16 – 18, 38, 39, 48.

The Diffraction Grating
A diffraction grating is essentially a large number of equally spaced sources, and thus the $d \sin\theta = m\lambda$ equation applies.

One application of diffraction gratings is in spectroscopy, which involves separating light into its different wavelengths, a process that astronomers, or chemists, can use to determine the chemical makeup of the source producing the light. In actuality, a diffraction grating is typically a glass or plastic slide with a large number of slits (long thin openings between long thin lines). A diffraction grating (which should probably have been named an interference grating) offers two main advantages over a double slit. First, the more openings the light passes through, the brighter the interference maxima are. Second, the more openings there are, the narrower the bright lines are in the interference pattern, which is important when trying to resolve two similar wavelengths. Figure 25.5 shows the increased sharpness that results from adding slits.

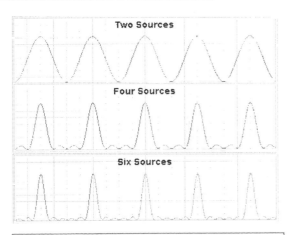

Figure 25.5: Adding sources (or slits that light goes through) results in sharper interference maxima. Each case shows the relative intensity at various points. The amplitude of the peaks also grows as sources are added.

Essential Question 25.2: A beam of light made up of three wavelengths, 660 nm (red light), 530 nm (green light), and 400 nm (violet light) is incident on a diffraction grating that has a spacing of $d = 1300$ nm. The first order spectrum, consisting of a violet line, a green line, and a red line, produced by the grating is shown in Figure 25.6. What are the colors of the other three beams (1 – 3) that come from the grating?

Figure 25.6: The first-order ($m = 1$) spectrum for the situation of Essential Question 25.2.

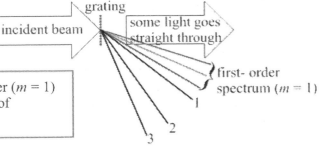

Answer to Essential Question 25.2: You might think that beams 1, 2, and 3 would also be violet, green, and red, respectively. However, in the equation $\sin\theta = m\lambda / d$, the right-hand side cannot exceed 1, because that is the limit on $\sin\theta$. If we use the three wavelengths with $m = 2$ or $m = 3$, we get the values shown in Table 25.1. It is possible to see the second-order violet and green lines, and the third-order violet lines, but none of the others because they correspond to values of $\sin\theta$ that are greater than 1, and are thus not possible. The beams are violet, green, and violet.

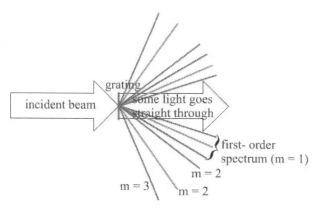

Figure 25.7: The entire spectrum for the situation of Essential Question 25.2, showing how the light splits because of passing through the diffraction grating.

Order	Violet (400 nm)	Green (530 nm)	Red (660 nm)
$m = 2$	$\sin\theta = 800/1300$	$\sin\theta = 1060/1300$	$\sin\theta = 1320/1300$
$m = 3$	$\sin\theta = 1200/1300$	$\sin\theta = 1590/1300$	$\sin\theta = 1980/1300$

Table 25.1: Values of $\sin\theta$ for m > 1 for the situation of Essential Question 25.2.

25-3 Diffraction from a Single Slit

In Section 25-1, we considered what happens when two sources broadcasting identical waves interfere. With light, we typically shine a laser beam through two closely-spaced slits (a double slit, in other words). Each slit acts as a source of waves, but it turns out that each slit does not send out light uniformly in all directions. Instead, a wave passing through a slit (or striking an object) experiences **diffraction**. Each point on the opening, or on the object, acts as a source of waves, and the resulting diffraction pattern is the result of the interference between all these waves. As Figure 25.8 shows, the waves interfere constructively in the forward direction, with more destructive interference in most other directions.

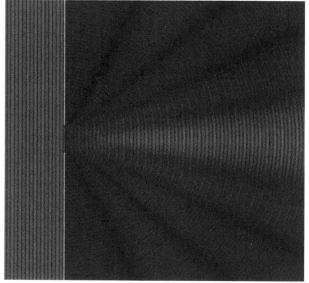

Figure 25.8: The diffraction pattern that results from a plane wave striking an opening that is four wavelengths wide. In color, the picture has alternating red (representing peaks), and blue (represents troughs) regions, separated by black regions denotes zero, or small, amplitude.

The graph in Figure 25.9, and the corresponding picture underneath the graph that shows the diffraction pattern from a laser shining through a single slit, show how much of the wave's energy is concentrated in the forward direction. The secondary peaks have much less intensity than the central maximum. The central maximum is also twice as wide as are the secondary peaks, at least at small angles.

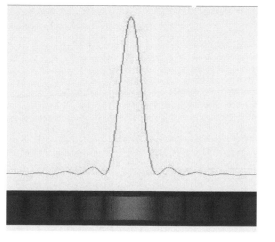

Figure 25.9: A graph of the intensity as a function of angle that corresponds to the diffraction pattern below the graph. The diffraction pattern comes from a laser shining on a single slit.

EXPLORATION 25.3 – An equation for the single slit

Step 1 – *Return to Figure 25.2, which illustrates how waves from two sources can interfere constructively at a particular point if the path length difference is, for instance, one wavelength. Now turn this two-source situation into a single-slit situation by filling in the space between the original two sources with more sources.* Figure 25.10 models a single-slit as being made up of a number of sources of waves laid out across the width of the opening. Note that while we use d to represent the distance between the two sources, we generally use a to represent the width of the single opening.

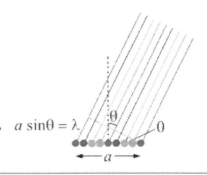

$a \sin\theta = \lambda$

Figure 25.10: A modification of Figure 25.2, turning the double-source situation into a single-slit situation by filling the space between the original sources with additional sources.

Step 2 – *The two sources in red that are at either end of the line of sources constructively interfere, in the situation shown, because their path-length difference is a full wavelength. What is the path-length difference for the source colored red that is in the middle of the set of sources, and the source at the left end of the line? What kind of interference is associated with these two sources?* The path length for the left source is half a wavelength longer than the path length for the middle source, which corresponds to destructive interference.

Step 3 – *What kind of interference results for the two blue sources, or the two orange sources, or the two green sources?* The path-length difference for all these pairs of sources is half a wavelength, corresponding to destructive interference. For the point in question, the waves from the left half of the opening cancel the waves arriving from the right half of the opening.

Step 4 – *If the equation $d \sin\theta = m\lambda$ gives the angles at which constructive interference occurs for two sources, what does the equation $a \sin\theta = m\lambda$ correspond to for the single slit?* The equation $a \sin\theta = m\lambda$ gives the angles at which <u>destructive</u> interference occurs for the single slit.

$$a \sin\theta = m\lambda \,,$$ (Equation 25.5: **destructive interference for a single slit**)

where m is an integer, and a is the width of the slit.

Key idea: The diffraction of waves passing through a single opening can be understood in terms of interference between waves leaving all points on the opening. The narrower the opening, the more the waves spread out. **Related End-of-Chapter Exercises: 19 – 21.**

Diffraction for sound waves

Diffraction takes place for other waves, such as water waves and sound waves, just as it does for light waves, with Equation 25.5 even applying for these other kinds of waves. Horn speakers, such as those shown in Figure 25.11, are often shaped to exploit diffraction, causing the waves to spread out in a particular dimension when they emerge from the end of the speaker.

Essential Question 25.3: As you are walking toward the open door to a room, you can hear the conversation between two people inside, even though you can't see the people. Explain why the sound waves are diffracted by the doorway, while the light is not.

Figure 25.11: These speakers are shaped to take advantage of diffraction for sound waves. The narrower the opening, the wider the diffraction pattern. The speakers are designed to diffract sound waves horizontally, where they can be heard by people on a train platform. Photo credit: A. Duffy.

Answer to Essential Question 25.3: A big difference between the sound waves and the light waves is the wavelength. The sound waves have wavelengths that are on the order of a meter, while the wavelengths of the light waves are about six orders of magnitude smaller. The width of the doorway is comparable to the wavelength of the sound waves, and so the sound waves experience significant diffraction. The doorway is so large compared to the wavelength of light, however, that the light goes in a straight line out the door, with negligible diffraction.

25-4 Diffraction: Double Slits and Circular Openings

The bottom graph in Figure 25.12 shows the relative intensity, as a function of position, of the light striking a screen after passing through a double slit. If each slit acted as a source of light, emitting waves uniformly in all directions, we would expect the peaks on the screen to be equally bright, as shown in the "Double Source" picture. Instead, each opening emits a diffraction pattern, as shown in the "Single Slit" picture. The interference between the two diffraction patterns results in the "Double Slit" pattern at the bottom, with the amplitude of the peaks predicted by the double-source equation being reduced by a factor given by the single-slit equation.

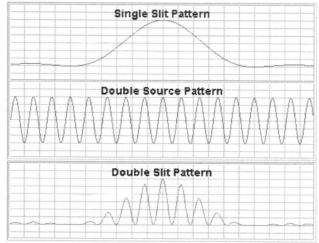

The "Double Slit" pattern exhibits a phenomenon known as **missing orders**. Peaks that are predicted in the pattern by the double-source equation, $d \sin\theta = m_d \lambda$, coincide with zeros from the single-slit equation, $a \sin\theta = m_s \lambda$, and are thus missing from the pattern.

Figure 25.12: When light passes through a double slit, the interference maxima are not equally bright, but drop off quite dramatically in brightness as you move away from the center of the pattern, as shown at the bottom. The double-slit pattern is a combination of the single-slit pattern (at the top) and the double-source pattern (in the middle).

A bit of history

Prior to 1800, there was a big debate in physics about the nature of light. The Dutch scientist Christiaan Huygens (1629 – 1695) came up with a way to explain many optical phenomena (such as refraction) in terms of light acting as a wave. The main proponent of the particle theory, however, was Sir Isaac Newton (1643 – 1727), who called it the corpuscular theory. With the weight of Newton behind it, the particle model of light won out until Thomas Young's double-slit experiment in 1801, followed by the work of the Frenchman, Augustin Fresnel, who studied diffraction in the early 1800's.

In 1818, Siméon Poisson realized that if light acted as a wave, the shadow of a round object should have a bright spot at its center. The light would leave all points on the edge of the object, and constructively interfere to produce a bright spot at the center of the shadow, because that point has a path-length difference of zero. Poisson actually put forward the idea of the bright spot as a way to disprove the wave theory, so he was somewhat taken aback when Dominique Arago did an experiment to show that there really is such a bright spot. These days, it is easy to create the bright spot at the center of a shadow by diverging a laser beam with a lens and then shining the beam onto a smooth metal ball. The shadow produced by such an arrangement is shown in Figure 25.13.

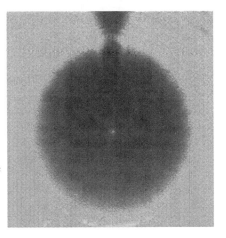

Figure 25.13: The bright spot at the center of the shadow of a ball bearing, demonstrating that light acts as a wave. Photo credit: A. Duffy.

Diffraction by a circular opening

A related and common phenomenon is diffraction by a circular opening (commonly called a circular aperture), such as the one we all look through, the pupil in each of our own eyes. For a circular opening, the angle at which the first zero occurs in a diffraction pattern is given by:

$$\theta_{min} = \frac{1.22\lambda}{D}$$ (Eq. 25.6: **The first zero in a diffraction pattern from a circular aperture**)

where D is the diameter of the opening. Note that the larger the diameter of the opening, the narrower the width of the central peak in the diffraction pattern. This dependence on the diameter of the opening has implications for how close two objects can be before you cannot resolve them. For instance, when you look up at the sky at night, two stars that are very close together may appear to you to be a single star. If you look at the same patch of sky through binoculars, or through a telescope, however, you can easily tell that you're looking at two separate stars. The light enters binoculars or telescopes through an aperture that is much larger than your pupil, and thus experiences much less diffraction.

It turns out that you can just resolve two objects when the first zero in the diffraction pattern associated with the first object coincides with the maximum in the diffraction pattern associated with the second object. Hence, Equation 25.6 gives the minimum angular separation between two objects such that you can just resolve them. Figure 25.14 illustrates the issue, where two objects are too close to be resolved by a human eye in bright sunlight, when the pupil is small, but can be resolved by the same eye when it is dark out, and the pupil has become larger to let in more light.

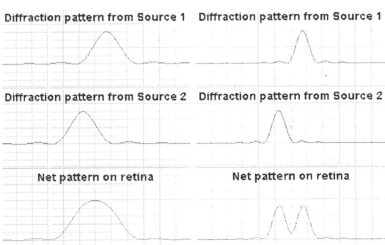

Diffraction pattern from Source 1   Diffraction pattern from Source 1

Diffraction pattern from Source 2   Diffraction pattern from Source 2

Net pattern on retina   Net pattern on retina

Related End-of-Chapter Exercises: 9, 22, 23, and 46.

Figure 25.14: On the left, we see that the angular separation between two objects is too small for them to be resolved, with the patterns overlapping too much on the retina. The images on the left correspond to a human eye in bright sunlight, when the pupil is small. The images on the right correspond to the same situation, but viewed in the dark. In the dark, the pupil expands to let in more light, reducing the spreading associated with diffraction.

Essential Question 25.4: Consider the double-slit pattern in Figure 25.12. Noting the location of the missing orders in the pattern, what is the ratio of d to a for this double slit? That is the ratio of the center-to-center distance between the two openings (d) to the width of each opening (a).

Answer to Essential Question 25.4: One way to answer this question is to set up a ratio of the double-source equation to the single-slit equation:

$$\frac{d\sin\theta}{a\sin\theta} = \frac{m_d\lambda}{m_s\lambda} \quad \Rightarrow \quad \frac{d}{a} = \frac{m_d}{m_s} \text{, for the same angle, } \theta.$$

The position corresponding to $m_s = 1$ is where we find the first zero on one side of the central maximum in the single-slit pattern. Looking at the double-slit pattern in Figure 25.15, and counting the peak in the center of that pattern as $m_d = 0$, we can see that the peak at $m_d = 5$ lines up with the first zero in the single-slit pattern, and is thus a missing order. With $m_d / m_s = 5/1$, we have $d/a = 5$ in this case.

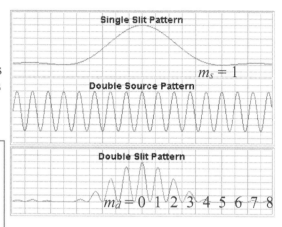

Figure 25.15: In this case, the first zero in the single-slit pattern corresponds to the same position, and therefore the same angle, as the $m_d = 5$ peak in the double-source pattern, leading to a missing order in the double-slit pattern.

25-5 Reflection

As we have discussed in Chapter 21 for waves on a string, when a wave reflects from the fixed end of a string, the reflected wave is inverted. When a wave reflects from the free end of a string, the reflected wave is upright.

What happens when the end of the string is neither perfectly free nor perfectly fixed, such as when a light string is tied to a heavy string? As shown in Figure 25.16 (a) and (b), when a wave is traveling along the light string, the point where the strings meet acts more like a fixed end than a free end. Part of the wave is transmitted onto the heavy string, and the part that reflects back along the light string is inverted. Conversely, as in Figure 25.16 (c) and (d), when a wave is traveling along the heavy string, the point where the strings meet acts more like a free end. Part of the wave is again transmitted into the second medium, while the part that reflects is upright.

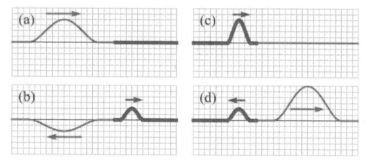

Figure 25.16: (a) When a wave traveling on a light string encounters the boundary between the light string and a heavy string, part of the wave is transmitted onto the heavy string, and part reflects back onto the light string, as in (b). The boundary acts like a fixed end, so the reflected wave is inverted. (c) If the wave is traveling along the heavy string before striking the boundary, the part of the wave that reflects is reflected upright, as in (d). In this situation, the speed of the wave on the light string is three times the speed of the wave on the heavy string.

An analogous process happens for light, or for any other electromagnetic wave. When a light wave traveling in one medium (medium 1) encounters an interface between that medium and a second medium (medium 2) with a different index of refraction, part of the light wave is transmitted into the second medium, and part is reflected back into the first medium. Whether the reflected wave is inverted or not depends on how the indices of refraction compare, as summarized in the box below, and as shown pictorially in Figure 25.17.

A light wave reflecting from a medium with a higher index of refraction than the medium the wave is traveling in ($n_2 > n_1$) is inverted upon reflection. If the second medium has a smaller index of refraction than the first ($n_2 < n_1$), the wave is reflected upright. For a sine wave, inverting the wave has the same effect as shifting the wave by half a wavelength, so we will treat an inversion upon reflection as a half wavelength shift.

Figure 25.17: (a) When light traveling in one medium reflects from a medium with a larger index of refraction, the part of the wave that is reflected is inverted upon reflection. (b) If the second medium has a smaller index of refraction than the first, the reflected part of the wave reflects upright. In both cases, the reflected wave has been shifted to the right to distinguish it from the incident wave.

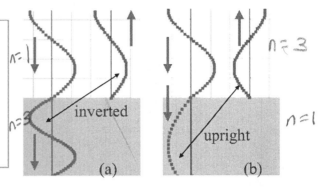

EXPLORATION 25.5 – Double-source interference with a single source

Figure 25.18 shows a situation in which a single source of sound waves is located above the floor. At any point, such as at point A in the figure, waves are received directly from the source, but waves are also received after being reflected from the floor.

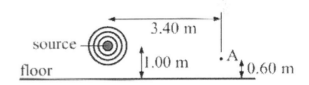

Figure 25.18: A source of sound waves, in air, located above a flat floor.

Step 1 – We can treat this situation as if there are two sources of waves. Where, effectively, is the second source located? The second source is where the image of the first source is located. Treating the floor like a plane mirror, reflecting the first source in the mirror gives the second source at the position shown in Figure 25.19.

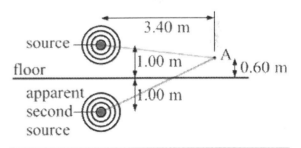

Step 2 – To analyze the interference between the waves from the two sources, we consider the mirror-image source to be 180° out of phase with the first source. Explain why we do this. The 180° phase shift comes from the fact that the wave does not actually originate at the second source. Instead, it originates at the first source, and reflects from the floor, producing an inversion of the wave upon reflection. Inverting the wave is equivalent to shifting the wave half a wavelength, which is equivalent to a 180° phase shift.

Figure 25.19: The floor acts like a plane mirror for sound waves. Effectively, there is a second source of waves where the image of the first source is created by the mirror.

Key idea: Even reflecting sound waves experience an inversion upon reflection.
Related End-of-Chapter Exercises: 8, 50 – 52.

Essential Question 25.5: Return to the situation described in Exploration 25.5, and assume the speed of sound is 340 m/s. Using the geometry of right-angled triangles, we can determine that point A is a distance of $\sqrt{(3.40 \text{ m})^2 + (0.40 \text{ m})^2} = 3.42 \text{ m}$ from the source, and a distance of

$\sqrt{(3.40 \text{ m})^2 + (1.60 \text{ m})^2} = 3.76 \text{ m}$ from the apparent second source. What is the lowest frequency sound wave from the source that will produce completely constructive interference at point A?

Answer to Essential Question 25.5: In this situation, the condition for constructive interference is that the path-length difference is half a wavelength. Using an integer number of wavelengths plus a half-wavelength would also produce constructive interference, but it would also decrease the wavelength. To get the lowest frequency, we need the longest wavelength. The wave that reflects from the floor is inverted upon reflection, and an inversion is equivalent to traveling an additional half wavelength, so the net shift is a full wavelength. This gives a path-length difference of 3.76 m – 3.42 m = 0.34 m. The path-length difference is half a wavelength, so a full wavelength is 0.68 m, corresponding to a frequency of v/λ = 340 m/s / 0.68 m = 500 Hz.

25-6 Thin-Film Interference: The Five-Step Method

The photograph in Figure 25.20 shows some colorful soap bubbles. The beautiful colors of the bubbles are caused by thin-film interference, interference between light reflecting from the outer surface of a soap bubble and light reflecting from the inner surface of the bubble. The colors we see are directly related to the thickness of the bubble wall. The basic process of thin-film interference is illustrated in Figure 25.21.

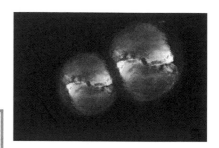

Figure 25.20 The colors in these soap bubbles are produced by thin-film interference - interference between light reflecting from the outer and inner surfaces of a bubble. Photo credit: George Horan, from publicdomainpictures.net

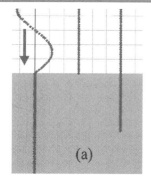 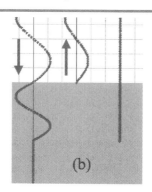 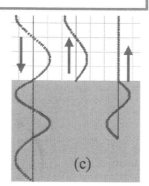 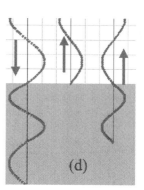

    (a)   (b)   (c)   (d)

Figure 25.21: These successive images are separated in time by one period of the wave. The thin film, shown in pink, is characterized by an index of refraction n_2. (a) A wave traveling in medium 1 is incident on the interface separating media 1 and 2 along the normal to the interface. (b) Part of the wave is transmitted into medium 2, while part reflects back into medium 1. The reflected wave is shifted to the right for clarity. The thin film happens to be exactly one wavelength thick. In general, the wave reflecting from the top surface of the film can be inverted upon reflection, or reflect without being inverted. In this case, the reflected wave is inverted because $n_2 > n_1$. (c) At the interface separating media 2 and 3, the wave is also partly reflected and partly transmitted. This reflected wave is shown on the far right of the diagram. In this case, the reflected wave is not inverted upon reflection because $n_3 < n_2$. (d) The two reflected waves interfere with one another in medium 1. By adjusting the thickness of the thin film, this interference can be completely constructive, completely destructive, or something in between.

    Note that, in Figure 25.21, the wave that reflects off the bottom surface of the film travels a total extra distance of 2 wavelengths, compared to the wave that reflects off the film's top surface. What kind of interference occurs between the two reflected waves? As we can see from Figure 25.21(d), the waves interfere destructively. The extra path length is an integer number of wavelengths, but the inversion upon reflection at the top surface introduces a half wavelength shift that causes peaks in one reflected wave to align with troughs in the other, and vice versa.

Let's take a systematic approach to analyzing a thin-film situation. The basic idea is to determine the effective path-length between the wave reflecting from the top surface of the film and the wave reflecting from the bottom surface. For a film of thickness t, and with a wave incident along the normal, the effective path-length difference accounts for the extra distance of $2t$ traveled by the wave that reflects from the bottom surface of the film, as well as for any inversions that occur when a wave reflects from a higher-n medium.

For a wave that does get inverted by reflecting from a higher-n medium, we will treat the inversion as an extra half-wavelength contribution to the wave's path-length. We do this because for a sine wave, inverting the wave is equivalent to shifting the wave by half a wavelength. However, a general thin-film situation involves three different media, and hence three different wavelengths! Which wavelength is it that matters? The wave that reflects from the bottom surface of the film is the one that travels the extra distance. Because the extra distance traveled is in the thin film, the wavelength that matters is the wavelength in the thin film. It is helpful to remember the relationship between the wavelength in the film and the wavelength in vacuum:

$$\lambda_{\text{film}} = \frac{\lambda_{\text{vacuum}}}{n_{\text{film}}}.$$ (Eq. 25.7: **An expression for the wavelength of light in the thin film**)

The Five-Step Method for analyzing thin films
Our approach will assume that the wave starts in medium 1, and is normally incident on a thin film (medium 2) of thickness t that is on a third medium (medium 3), as shown in Figure 25.21.

Step 1 – Determine Δ_t, the shift for the wave reflecting from the top surface of the film. This contribution to the path length is non-zero only if the wave is inverted upon reflection. If $n_2 > n_1$, $\Delta_t = \lambda_{\text{film}} / 2$. If $n_2 < n_1$, $\Delta_t = 0$.

Step 2 – Determine Δ_b, the shift for the wave reflecting from the bottom surface of the film. This contribution to the path length is at least $2t$, because that wave travels an extra distance t down through the film, and t back up through the film. There is an extra half-wavelength contribution if the wave is inverted upon reflection. If $n_3 > n_2$, $\Delta_b = 2t + \lambda_{\text{film}} / 2$. If $n_3 < n_2$, $\Delta_b = 2t$.

Step 3 – Determine Δ, the effective path-length difference. Simply subtract the results from the previous steps to find the relative shift between the two waves. $\Delta = \Delta_b - \Delta_t$.

Step 4 – Bring in the interference condition appropriate to the situation. If the interference is constructive, such as when we see a particular color reflecting from the thin film, we set the effective path-length difference equal to an integer number of wavelengths ($\Delta = m\lambda_{\text{film}}$, where m is an integer). If the interference is destructive, $\Delta = (m+0.5) \lambda_{\text{film}}$.

Step 5 – Solve the resulting equation. In general, the equation relates the thickness of the thin film to the wavelength of the light.

Related End-of-Chapter Exercises: 24 – 28.

Essential Question 25.6: Fill in Table 25.1, which summarizes the various possibilities for what Δ_t and Δ_b can be.

	$n_2 > n_1$	$n_2 < n_1$
$n_3 > n_2$	$\Delta_t =$ $\Delta_b =$	$\Delta_t =$ $\Delta_b =$
$n_3 < n_2$	$\Delta_t =$ $\Delta_b =$	$\Delta_t =$ $\Delta_b =$

Table 25.1: A table for summarizing the various results for Δ_t and Δ_b.

	$n_2 > n_1$	$n_2 < n_1$
$n_3 > n_2$	$\Delta_t = \lambda_{film}/2$ $\Delta_b = 2t + (\lambda_{film}/2)$	$\Delta_t = 0$ $\Delta_b = 2t + (\lambda_{film}/2)$
$n_3 < n_2$	$\Delta_t = \lambda_{film}/2$ $\Delta_b = 2t$	$\Delta_t = 0$ $\Delta_b = 2t$

Answer to Essential Question 25.6: The shift for the wave reflecting from the top surface, Δ_t, depends on how n_2 compares to n_1. The shift for the wave reflecting from the bottom surface, Δ_b, depends on how n_3 compares to n_2.

Table 25.2: Summarizing the various results for Δ_t and Δ_b.

25-7 Applying the Five-Step Method

EXPLORATION 25.7 – Designing a non-reflecting coating
High-quality lenses, such as those for binoculars or cameras, are often coated with a thin non-reflecting coating to maximize the amount of light getting through the lens. We can apply thin-film ideas to understand how such a lens works. Explaining why such lenses generally look purple will also be part of our analysis. In this example, we will assume light is traveling through air before it encounters the non-reflective coating ($n = 1.32$) that is on top of the glass ($n = 1.52$). Figure 25.22 shows the arrangement. The coating is completely non-reflective for just one wavelength, so we will design it to be non-reflective for light with a wavelength in vacuum of 528 nm, which is close to the middle of the visible spectrum.

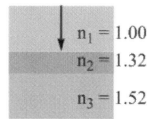

Figure 25.22: The arrangement of air (top), coating (middle), and glass (bottom) for a typical situation of a non-reflective coating on a glass lens.

Step 1 – *Determine Δ_t, the shift for the wave reflecting from the air-coating interface.* Because the coating has a higher index of refraction than the air, this wave is inverted upon reflection, giving $\Delta_t = \lambda_{film}/2$.

Step 2 – *Determine Δ_b, the shift for the wave reflecting from the coating-glass interface.* The glass has a higher index of refraction than the coating, so this wave is also inverted upon reflection. For a coating of thickness t, $\Delta_b = 2t + (\lambda_{film}/2)$.

Step 3 – *Determine Δ, the effective path-length difference.* $\Delta = \Delta_b - \Delta_t = 2t$.

Step 4 – *Bring in the appropriate interference condition.* In this situation, we do not want light to reflect from the coating. We can accomplish this by having the reflected waves interfere destructively. Setting the effective path-length difference equal to $(m + 1/2)$ wavelengths gives:

$$2t = (m + 1/2)\lambda_{film} \, .$$

Step 5 – *Solve for the minimum possible coating thickness.* To solve for the smallest possible coating thickness, we choose the smallest value of m that makes sense, remembering that m is an integer. In this case, $m = 0$ gives the smallest coating thickness.

$$2t_{min} = (0 + 1/2)\lambda_{film} \quad \Rightarrow \quad t_{min} = \frac{\lambda_{film}}{4} = \frac{\lambda_{vacuum}}{4n_{film}} = \frac{528 \text{ nm}}{4 \times 1.32} = 100 \text{ nm} \, .$$

Step 6 – *If a 100-nm-thick film produces completely destructive interference for 528 nm green light, what kind of interference will it produce for the violet end of the spectrum (400 nm) and the red end of the spectrum (700 nm)? Why does this make the lens look purple in reflected light?* In the coating, 400 nm violet light has a wavelength of 400 nm / 1.32 = 303 nm. Thus, an effective path-length difference of $2t$ = 200 nm shifts one reflected violet wave relative to another by 200 nm / 303 nm, a shift of about 2/3 of a wavelength. The interference is partly destructive, so some violet light reflects from the coating. For red light of 700 nm, with a wavelength in the film of 700 nm / 1.32 = 530 nm, the relative shift is 200 nm / 530 nm = 0.38 wavelengths. Again, this produces partly destructive interference, so some red light reflects. When white light shines on the film, therefore, almost no green light is reflected, small amounts of yellow and blue are reflected, a little more orange and indigo are reflected, and even more red and violet are reflected. Thus, the reflected light is dominated by red and violet, which makes the film look purple.

Key ideas: The five-step method can be applied in all thin-film situations, to help us relate the film thickness to the wavelength of light.   **Related End-of-Chapter Exercises: 31, 32, 54.**

EXAMPLE 25.7 – A soap film

A ring is dipped into a soap solution, creating a round soap film. (a) When the ring is held vertically, explain why horizontal bands of color are observed, as seen in Figure 25.23(a). (b) As time goes by, the film gets progressively thinner. Where the film is very thin, no light reflects from the film, so it looks like the film is not there anymore, as in the top right of Figure 25.23(b). Apply the first three steps of the five-step method to explain why, in the limit that the film thickness approaches zero, the two reflected waves interfere destructively.

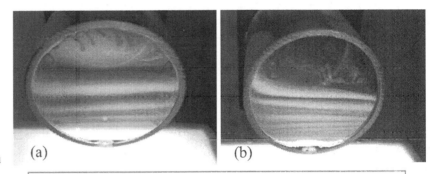

Figure 25.23: (a) A vertical soap film generally has horizontal bands. (b) When the film gets very thin, it does not reflect any light whatsoever, as is happening at the top right of the film in this case. Photo credit: A. Duffy.

SOLUTION

(a) The film thickness is approximately constant at a given height, with that thickness corresponding to constructive interference for a particular wavelength (color). Gravity pulls the fluid down toward the bottom of the film, so the film thickness decreases as the vertical position increases, changing the wavelength (color) associated with a particular height.

(b) The index of refraction of the soap film is essentially that of water (n = 1.33), with the film being surrounded by air (n = 1.00). The wave reflecting from the front surface of the film is in air, reflecting from the higher-n film, so it experiences a half-wavelength shift: $\Delta_t = \lambda_{film}/2$. The wave reflecting from the back surface of the film reflects from a lower-n medium, so the effective path-length is simply $\Delta_b = 2t$, where t is the film thickness. The effective path-length difference is therefore $\Delta = \Delta_b - \Delta_t = 2t - \lambda_{film}/2$. In the limit that the film thickness t approaches zero, the effective path-length difference has a magnitude of half a wavelength. Shifting one wave with respect to the other by half a wavelength produces destructive interference, and the interference is destructive for all wavelengths, so no light is reflected when the film is very thin.

Related End-of-Chapter Exercises: 10, 12.

Essential Question 25.7: For the situation shown in Exploration 25.7, the non-reflective coating on glass, what kind of interference results as the thickness of the coating approaches zero?

Answer to Essential Question 25.7: Returning to the result of Step 3 in Exploration 25.7, the effective path-length difference between the two waves is $\Delta = 2t$. Thus, in the limit that the thickness, t, approaches zero, the effective path-length difference approaches zero and the interference is constructive.

Chapter Summary

Essential Idea: Interference and Diffraction.
In many situations, light acts as a wave. In general, waves diffract through narrow openings, and waves interfere with one another. Examples of this behavior with light occur when a laser beam is incident on one or more narrow openings, when light passes through the pupil of your eye, and when light interacts with thin films such as those in soap bubbles.

Constructive Interference – from Double Slits to Diffraction Gratings
For a wave of wavelength λ that is incident on a number of equally spaced narrow openings, where the number of openings is at least two, the angles at which constructive interference occurs are given by

$d \sin\theta = m\lambda$,   (Equation 25.3: **constructive interference, for $N > 1$ sources**)
where m is an integer, and d is the distance between neighboring openings.

Destructive Interference – Single and Double Slits
For a wave of wavelength λ that is incident on a single slit of width a, the angles at which destructive interference occurs are given by

$a \sin\theta = m\lambda$,   (Equation 25.5: **diffraction minima for a single slit**)
where m is an integer, and a is the width of the slit.

For a double slit, the interference minima occur at angles given by

$d \sin\theta = (m + 1/2)\lambda$,   (Equation 25.4: **destructive interference, for two sources in phase**)
where m is an integer.

Limits imposed by diffraction
For a circular opening, the angle at which the first zero occurs in a diffraction pattern is given by

$\theta_{min} = \dfrac{1.22\lambda}{D}$,   (Eq. 25.6: **The minimum angle between two sources to be resolvable**)

where D is the diameter of the opening. This equation can be applied to our own eyes.

Thin-film interference
The colorful patterns exhibited by thin films, such as soap bubbles, can be understood by following the five-step method outlined in Section 25.6. Such patterns result from the wave reflecting from one surface of the film interfering with the wave reflecting from the other surface of the film. A key part of the analysis is accounting for the fact that when waves in one medium reflect from a second medium that has a lower index of refraction, the reflected wave is upright, while if the second medium has a higher index of refraction, the reflected wave is inverted. This inversion upon reflection is like an extra half-wavelength distance traveled by the wave.

End-of-Chapter Exercises

Exercises 1 – 12 are conceptual questions designed to see whether you understand the main concepts in the chapter.

1. Red laser light shines on a double slit, creating a pattern of bright and dark spots on a screen some distance away. State whether the following changes, carried out separately, would increase, decrease, or produce no change in the distance between the bright spots on the screen, and justify each answer. (a) Replace the red laser with a green laser. (b) Decrease the spacing between the slits. (c) Decrease the distance between the slits and the screen. (d) Immerse the entire apparatus in water.

2. Light of a single wavelength shines onto a double slit. A particular point on the opposite side of the double slit from the light source happens to be 1800 nm farther from one slit than the other. Assume that the point receives some light from each slit, and that the beams arriving at the point from each slit are of equal intensity. For the following wavelengths, determine whether the interference at the point is constructive, destructive, or something in between. (a) 400 nm violet light, (b) 500 nm green light, (c) 600 nm orange light, (d) 700 nm red light. Explain each of your answers.

3. The graph in Figure 25.24 shows $\sin\theta$ as a function of wavelength for different orders of light. The red line corresponds to the first-order ($m = 1$) spectrum. (a) What does the blue line correspond to? (b) Copy the graph and draw the line corresponding to the third-order spectrum. (c) What is the largest wavelength for which there is a third-order spectrum for this grating?

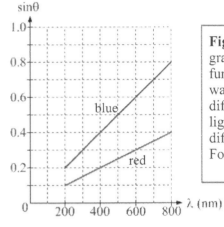

Figure 25.24: The graph shows $\sin\theta$ as a function of wavelength for different orders of light shining on a diffraction grating. For Exercise 3.

4. The pattern in Figure 25.25 represents the interference pattern set up in a room by two speakers (the red circles) broadcasting identical single-frequency sound waves in phase with one another. (a) If you walk slowly along the line shown in yellow, from one end to the other, what will you hear? Explain your answer. (b) If the wavelength of the sound waves is 1.5 m, how far apart are the speakers?

Figure 25.25: An interference pattern set up in a room by two speakers broadcasting identical single-frequency waves. For Exercise 4.

5. A red laser shining on something creates the pattern shown in Figure 25.26. Is the laser shining on a single slit, double slit, or a diffraction grating? Explain your answer.

Figure 25.26: The pattern at the center of a screen, produced by a red laser beam shining on a single slit, a double slit, or a diffraction grating. The distance between neighboring tick marks, shown on the screen below the pattern, is 8.00 mm. For Exercise 5.

6. A beam of white light strikes a glass prism, as shown in Figure 25.27(a). The white light is made up of only three colors. These are, in alphabetical order, green, red, and violet. The graph of the index of refraction vs. wavelength for the glass is shown below to the right of the prism. For each of the three rays labeled (a) – (c) on the diagram, label the ray with its color. Use W for white, G for green, R for red, and V for violet. (b) The prism is now replaced by a diffraction grating with a grating spacing of $d = 1300$ nm. The three colors in the beam of white light have, in order of increasing wavelength, wavelengths of 400 nm, 500 nm, and 700 nm. For the seven rays labeled (d) – (j) in Figure 25.27(b), label the ray with its color. Use W for white, G for green, R for red, and V for violet.

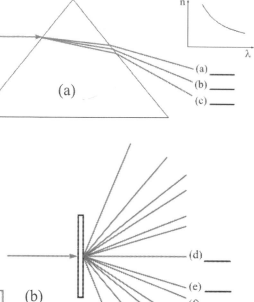

(a)

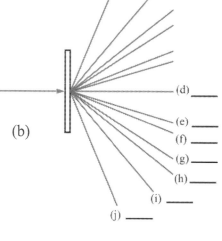

(b)

Figure 25.27: (a) A beam of violet, green, and red light is incident on a prism. (b) The same beam is incident on a diffraction grating. For Exercise 6.

7. Figure 25.28 shows the $m = 0$ through $m = 2$ lines that result when green light is incident on a diffraction grating. The squares on the grid in the figure measure 10 cm × 10 cm. The horizontal blue line at the top of the figure represents a screen, which is 1.0 m long and 90 cm from the grating. Approximately how far from the grating should the screen be located so that the two second-order green lines are just visible at the left and right edges of the screen?

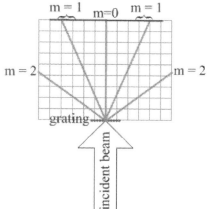

Figure 25.28: The $m = 0$ through $m = 2$ lines that result when green light with a wavelength of 540 nm is incident on a diffraction grating. The horizontal line at the top represents a screen, which is 1.0 m long. The squares on the grid measure 10 cm × 10 cm. For Exercise 7.

8. Two speakers send out identical single-frequency sound waves, in phase, that have a wavelength of 0.80 m. As shown in Figure 25.29, the speakers are separated by 3.6 m. Three lines, labeled A through C, are also shown in the figure. Line A is part of the perpendicular bisector of the line connecting the two sources. If you were to walk along these lines, would you observe completely constructive interference, completely destructive interference, or something else? Answer this question for (a) line A, (b) line B, and (c) line C. Briefly justify each of your answers.

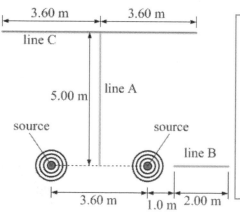

Figure 25.29: Three lines located near two speakers (each labeled "source") that are broadcasting identical single-frequency sounds, for Exercise 8.

9. Figure 25.30 shows, at the top, the pattern resulting from light with a wavelength of 480 nm passing through only one slit of a double slit. In the middle of the figure is the pattern that would result if both slits were illuminated and the slits sent out light uniformly in all directions. At the bottom of the figure is the actual pattern observed when the light illuminates both slits. What is the ratio of the distance between the slits to the width of one of the slits?

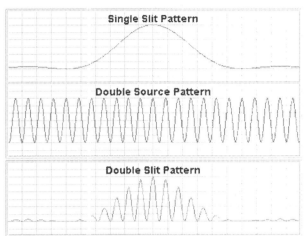

Figure 25.30: The pattern on a screen that results from 480 nm light illuminating a double slit is shown at the bottom. This pattern is a combination of the single slit pattern (top) and double source pattern (middle). For Exercise 9.

10. Figure 25.31 shows four situations in which light is incident perpendicularly on a thin film (the middle layer in each case). The indices of refraction are $n_1 = 1.50$ and $n_2 = 2.00$. In the limit that the thickness of the thin film approaches zero, determine whether the light that reflects from the top and bottom surfaces of the film interferes constructively or destructively in (a) case A, (b) case B, (c) case C, and (d) case D.

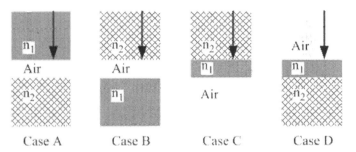

Figure 25.31: Four thin-film situations involving different arrangements of the same three media, for Exercise 10.

11. A soap film, surrounded by air, is held vertically so that, from top to bottom, its thickness varies from a few nm to a few hundred nm. The film is illuminated by white light. Which is closer to the top of the film, the location of the first band of red light, produced by completely constructive thin-film interference, or the first band of blue light? Explain.

12. Figure 25.32 illustrates a phenomenon known as Newton's rings, in which a bull's–eye pattern is created by thin-film interference. The film in this case is a thin film of air that is between a piece of glass with a spherical surface (such as a watch glass) that is placed on top of a flat piece of glass. Two possible patterns, one with a dark center and one with a bright center, are shown in the figure. (a) Which pattern would you see when you look down on the rings from above, and which would you see when you look up at them from below? Explain.

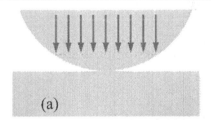

Figure 25.32: The phenomenon of Newton's rings comes from light shining through an object with a spherical surface that rests on a flat surface (a). Interference between light reflecting from the top surface of the film (between the spherical surface and the flat surface) and light reflecting from the bottom surface produces a bull's-eye pattern. Two possible patterns are shown in (b) and (c). For Exercise 12.

Exercises 13 – 15 deal with the interference from two sources.

13. Two speakers broadcasting identical single-frequency sound waves, in phase with one another, are placed 4.8 m apart. The speed of sound is 340 m/s. You are located at a point that is 10.0 m from one speaker, and 8.4 m from the other speaker. What is the lowest frequency for which you observe (a) completely constructive interference? (b) completely destructive interference?

14. Two speakers broadcasting identical single-frequency sound waves, in phase with one another, are placed 6.5 m apart. The wavelength of the sound waves is 3.0 m. You stand directly in front of the speaker on the left (along the dashed line in Figure 25.33), at a distance of 4.5 m from it. Your friend than changes the wavelength of the identical waves being emitted by the speakers. What are the two largest wavelengths that, at your location, result in (a) completely constructive interference, and (b) completely destructive interference?

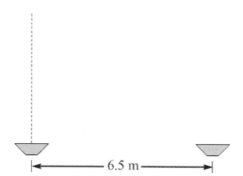

15. Two speakers broadcasting identical single-frequency sound waves, in phase with one another, are placed 6.5 m apart. The wavelength of the sound waves is 3.0 m. You stand directly in front of the speaker on the left (along the dashed line in Figure 25.33), but some distance from it. How far are you from that speaker if the interference at your location is (a) completely constructive? (b) completely destructive? Find all the possible answers in each case.

Figure 25.33: Two speakers broadcasting identical single-frequency waves, for Exercises 14 and 15.

Exercises 16 – 18 involve double slits and diffraction gratings.

16. When the beam of a red laser is incident on a particular diffraction grating, the $m = 1$ bright fringe is observed at an angle of 28.0°. At what angle is the (b) $m = 2$ bright fringe, and (c) the $m = 3$ bright fringe?

17. Light with a wavelength of 540 nm shines on two narrow slits that are 4.40 μm apart. At what angle does the fifth dark spot occur on a screen on the far side of the slits from the light source?

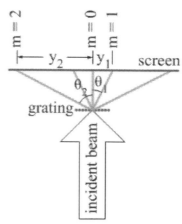

18. Laser light shines onto a diffraction grating, creating the pattern of bright lines shown in Figure 25.34. The lines strike a screen (in blue in the figure) that is a distance L away from the grating, creating some bright spots on the screen. The distance between the central spot and the mth bright spot to either side is denoted y_m. (a) What is the relationship between θ_m (the angle between the mth bright line and the $m = 0$ line) and y_m? (b) Show that, in the limit that θ_m is small, y_m is given by $y_m = \dfrac{m\lambda L}{d}$.

Figure 25.34: Single-frequency light shining on a double slit, for Exercise 18.

Exercises 19 – 23 involve single slits and diffraction by circular openings.

19. Light of a particular wavelength is incident on a single slit that has a width of 2.00 μm. If the second zero in the diffraction pattern occurs at an angle of 30°, what is the wavelength of the light?

20. The label on a green laser pointer states that the wavelength of the laser is 532 nm. You shine the laser into an aquarium filled with water, with an index of refraction of 1.33, and onto a single slit (in the water) that has a width of 1.60 μm. At what angle is the first zero in the diffraction pattern?

21. Light with a wavelength of 600 nm shines onto a single slit, and the diffraction pattern is observed on a screen 2.5 m away from the slit. The distance, on the screen, between the dark spots to either side of the central maximum in the pattern is 25 mm. (a) What is the distance between the same dark spots when the screen is moved so it is only 1.5 m from the slit? (b) What is the width of the slit?

22. On a dark night, you watch a car drive away from you on a long straight road. If the car's red tail lights are LED's emitting a wavelength of 640 nm, the distance between the lights is 1.50 m, and your pupils are 6 mm in diameter, what is the maximum distance the car can get away from you before the two individual lights look like one light to you?

23. A spy satellite takes in light through a circular opening 2.0 m in diameter. (a) If the wavelength of the light is 540 nm, and the satellite is 250 km above the ground, how close together can two small objects be on the ground for the satellite to be able to resolve them? (b) If the pupils in your eyes are 4.0 mm in diameter, how far above the ground would you be to achieve the same resolution as the satellite?

Exercises 24 – 28 are designed to give you practice with applying the five-step method for thin-film interference. For each of these problems, carry out the following steps. (a) Determine Δ_t, the shift for the wave reflecting from the top surface of the film. (b) Determine Δ_b, the shift for the wave reflecting from the bottom surface of the film. (c) Determine Δ, the effective path-length difference. (d) Bring in the interference condition appropriate to the situation. (e) Solve the resulting equation to solve the problem.

24. When you shine red light, with a wavelength of 640 nm, straight down through air onto a thin film of oil that coats a water surface, the film looks dark because of destructive interference. The index of refraction of the oil is 1.60, while that of water is 1.33. The goal of the problem is to determine the smallest non-zero film thickness. Carry out the five-step method as outlined above.

25. A ring is dipped into a soap solution, resulting in a circular soap film in the ring. When the plane of the ring is horizontal, the film looks green to you when you look straight down onto the film from above. The soap film is surrounded on both sides by air, and the index of refraction of the film is that of water, 1.33. If the film thickness is such that it produces completely constructive interference for green light with a wavelength, in vacuum, of 532 nm, what is the minimum non-zero thickness of the film? Carry out the five-step method, as outlined above, to solve the problem.

26. A thin film of glass, with an index of refraction of 1.5, is used to coat diamond, which has an index of refraction of 2.4. The thickness of the thin film is 200 nm. Light, traveling through air, shines down along the normal to the film, as shown in Figure 25.35. If we define the visible spectrum as extending from 400 nm to 700 nm (measured in air), for which wavelength in the visible spectrum (measured in air) does the film produce completely constructive interference? Carry out the five-step method, as outlined above, to solve the problem.

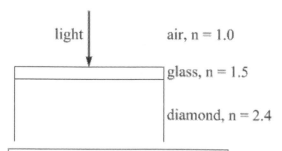

Figure 25.35: A 200-nm-thick film of glass is placed on top of diamond. For Exercises 26 and 27.

27. Return to the situation shown in Figure 25.35 and described in Exercise 26. Now determine for which wavelength in the visible spectrum (measured in air) the film produces completely destructive interference. Carry out the five-step method, as outlined above, to solve the problem.

28. A thin film with an index of refraction of 1.70 is used as a non-reflective coating on a glass lens that has an index of refraction of 1.50. What are the three smallest non-zero thicknesses of the film that will produce completely destructive interference for light that has a wavelength of 510 nm in vacuum? Assume the light is traveling in air before encountering the film, and that it strikes the film at normal incidence. Carry out the five-step method, as outlined above, to solve the problem.

Exercises 29 – 33 involve practical applications of the interference of light.

29. Understanding a particular spectrum is important in many areas of science, including physics and chemistry, where it can be used to identify a gas, for instance. To create a spectrum, light is generally sent through a diffraction grating, splitting the light into the various wavelengths that make it up. (a) The first step in the process is to calibrate the grating, so we know the grating spacing. Sodium has two yellow lines that are very close together in wavelength at 590 nm. When light from a sodium source is passed through a particular diffraction grating, the two yellow lines overlap, looking like one line at an angle of 33.7° in the first-order spectrum. What is the grating spacing? (b) The hydrogen atom is the simplest atom there is, consisting of one electron and one proton, and it has thus been well studied. When hydrogen gas is excited by means of a high voltage, three of the prominent lines in the spectrum are found at wavelengths of 658 nm, 487 nm, and 435 nm. When the light is passed through the diffraction grating we calibrated with sodium, at what angles will these three lines appear in the first-order spectrum? Scientists observing these lines can be confident that the source of the light contains hydrogen.

30. Return to the situation described in Exercise 29. Another application of spectra produced by a diffraction grating is in astrophysics, where the Doppler shift of a particular galaxy can be measured to determine the velocity of the galaxy with respect to us. (a) If the red hydrogen line in the galaxy's spectrum is observed at 690 nm instead of 658 nm, is the galaxy moving toward us or away from us? (b) Recalling that the Doppler equation for electromagnetic waves states that the magnitude of the shift in frequency associated with relative motion between a source and observer is $|f' - f| = vf / c$, determine v, the relative speed of the galaxy with respect to us.

31. Thin coatings are often applied to materials to protect them. In a particular manufacturing process, a company wants to deposit a 200–nm–thick coating onto glass mirrors to protect the mirrors during shipping. The coating material has an index of refraction of 1.30, while that of the glass is 1.53. White light in air is incident on the film along the normal to the surface, and the film looks the color of the wavelength that is experiencing completely constructive interference. If the technician observing the coating as it is being deposited, and gradually increasing in thickness, views the light reflecting from the coating, at which of the following points should the technician stop the deposition process? When the reflected light is violet (400 nm), green (520 nm), orange-red (612 nm), or none of these? Explain.

32. As shown in Figure 25.36(a), two flat pieces of glass are touching at their left edges, and are separated at their right edges by a cylindrical wire. This apparatus can be used to determine the diameter of the wire. When the apparatus is illuminated from above with yellow light with a wavelength of 590 nm, you see the thin-film interference pattern shown in Figure 25.36(b) when you look down on the apparatus from above. Note that the third dark fringe from the left is exactly halfway between the left and right edges of the pieces of glass. At the point where the third dark film from the left appears, (a) how many wavelengths thick is the film, and (b) how thick is the film? (c) How is the diameter of the wire related to the answer to part (b)? (d) What is the diameter of the wire?

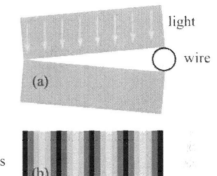

Figure 25.36: (a) A thin film of air is trapped between two flat pieces of glass. The pieces of glass are in contact at their left edges, and are separated at their right edges by a thin wire. (b) The interference pattern you observe when you look down on the film from above, when the film is illuminated from above with 590-nm light, for Exercise 32.

33. In a compact disk (CD) player, to read the information on a CD an infrared laser, with a wavelength of 780 nm in air, reflects from flat-topped bumps and the flat surroundings (known as the land) on the CD. When the laser beam reflects solely from the top of a bump, or solely from the land, a significant signal is reflected back. However, when the beam is moving from a bump to the land, or vice versa, destructive interference between the two parts of the beam, one part which travels a shorter distance than the other, results in a low signal. Thus, music can be encoded as a binary (two-state) signal. What is the height of the bumps on a CD, if the transparent polycarbonate coating on the CD has an index of refraction of 1.55? The bump height is designed to be the smallest needed to produce completely destructive interference between waves reflecting from the bumps and waves reflecting from the land.

General problems and conceptual questions

34. Christiaan Huygens made a number of important contributions to our understanding of the wave nature of light. Do some research about him and his contributions, and write a couple of paragraphs about what you find.

35. The graph in Figure 25.37 shows $\sin\theta$ as a function of wavelength for different orders of light. Let's say that the red line corresponds to the first-order ($m = 1$) spectrum. At what angle is (a) the third-order spot for 500 nm light? (b) the fourth-order spot for 400 nm light?

36. The graph in Figure 25.37 shows $\sin\theta$ as a function of wavelength for different orders of light. What is the grating spacing if the red line corresponds to (a) the first-order ($m = 1$) spectrum? (b) the fifth-order ($m = 5$) spectrum?

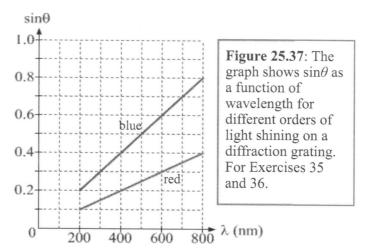

Figure 25.37: The graph shows $\sin\theta$ as a function of wavelength for different orders of light shining on a diffraction grating. For Exercises 35 and 36.

37. A laser with a wavelength of 600 nm is incident on a pair of narrow slits that are separated by a distance of $d = 3.00 \times 10^{-5}$ m. The resulting interference pattern is projected onto a screen 2.00 m from the slits. (a) How far is one of the first-order bright spots from the central bright spot on the screen (measuring from the center of each spot)? Note that for small angles $\sin\theta \approx \tan\theta$. (b) Does the answer change if the entire apparatus is immersed in water, which has an index of refraction of 4/3? If so, how does it change?

38. Figure 25.38 shows the $m = 0$ through $m = 2$ lines that result when green light with a wavelength of 540 nm is incident on a diffraction grating. Also shown, as dashed lines, are the two $m = 1$ lines for a second wavelength. The squares on the grid in the figure measure 10 cm × 10 cm. (a) What is the second wavelength? (b) What is the grating spacing? (c) Will there be $m = 3$ lines for either the green light or the second wavelength in this situation? Explain.

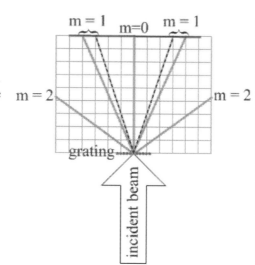

Figure 25.38: The $m = 0$ through $m = 2$ lines that result when green light with a wavelength of 540 nm is incident on a diffraction grating. The two $m = 1$ lines (dashed lines) are for a second wavelength. The horizontal line at the top represents a screen, which is 1.0 m long. The squares on the grid measure 10 cm × 10 cm. For Exercise 38.

39. Light with a wavelength of 400 nm shines onto a double slit. A particular point on the far side of the double slit from the light source happens to be exactly 6 wavelengths farther from one slit than the other. (a) At this particular point, do we expect to see constructive interference or destructive interference? (b) For which wavelengths in the visible spectrum (400 – 700 nm) will the interference be completely constructive at the point? (c) For which wavelengths in the visible spectrum will the interference be completely destructive at the point?

40. Figure 25.39 shows the $m = 0$ and $m = 1$ lines coming from a red laser beam, with a wavelength of 632 nm, that shines on a diffraction grating. The squares in the grid measure 10 cm × 10 cm. Duplicate the figure, and show all the lines resulting from 450 nm blue light shining on the same grating.

41. Return to the situation described in Exercise 40, and shown in Figure 25.39. (a) Determine the grating spacing. (b) For what range of grating spacings would there be three, and only three, orders to either side of the central maximum with 632 nm red light?

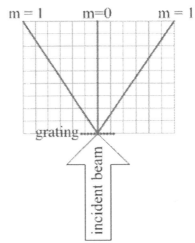

Figure 25.39: The $m = 0$ and $m = 1$ lines that result when red light with a wavelength of 632 nm is incident on a diffraction grating. The squares on the grid measure 10 cm × 10 cm. For Exercises 40 and 41.

42. A red laser, shining on a double slit, creates the pattern shown in Figure 25.40 at the center of a screen placed 2.00 m on the opposite side of the double slit from the laser. If the laser wavelength is 632 nm, what is the distance between the two slits in the double slit?

43. Return to the situation described in Exercise 42, and shown in Figure 25.40. When the red laser is replaced by a second laser, exactly 5 dots are observed within a distance of 24.0 mm at the center of the screen, instead of having exactly 7 dots in that distance, as in Figure 25.33. What is the wavelength of the second laser?

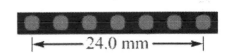

Figure 25.40: The pattern of dots created on a screen a distance 2.00 m from a double slit, when a red laser shines on the slits. For Exercises 42 and 43.

44. Laser light with a wavelength of 632 nm is incident on a pair of identical slits that are 5.60 μm apart. (a) If the slits are very narrow, how many bright fringes would you expect to see on one side of the central maximum? (b) In the pattern on a screen, you notice that, instead of a bright fringe where you expect the fourth bright fringe to be, there is a dark spot. The first three bright fringes are where you expect them to be, however. What is the width of each slit? (c) Are there any other fringes missing, in addition to the fourth one on each side?

45. Repeat Exercise 44, but, this time, use a wavelength of 440 nm.

46. Red light, with a wavelength of 650 nm, is incident on a double slit. The resulting pattern on the screen 1.2 m behind the double slit is shown in Figure 25.41. If the slits are 2.40 μm apart, what is the width of each of the slits?

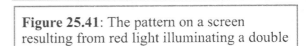

Figure 25.41: The pattern on a screen resulting from red light illuminating a double slit. For Exercise 46.

47. Light with a wavelength of 440 nm illuminates a double slit. When you shine a second beam of light on the double slit, you notice that the 4th-order bright spot for that light occurs at the same angle as the 5th-order bright spot for the 440 nm light. (a) What is the wavelength of the second beam? (b) If the angle of these beams is 40.0°, what is the distance between the two slits in the double slit?

48. When light from a red laser, with a wavelength of 632 nm, is incident on a diffraction grating, the second-order maximum occurs at an angle of 15.4°. (a) What is d, the grating spacing for the diffraction grating? (b) At what angle is the second-order maximum if the red laser is replaced by a green laser with a wavelength of 532 nm?

49. Figure 25.42 shows the pattern at the center of a screen, produced by a red laser beam shining on either a single slit, a double slit, or a diffraction grating. If the laser has a wavelength of 632 nm, and the screen is 1.4 m away, determine the width of the single slit (if the laser shines on a single slit) or the distance between the slits in the double slit (if the laser shines on a double slit) or the grating spacing (if the laser shines on a diffraction grating).

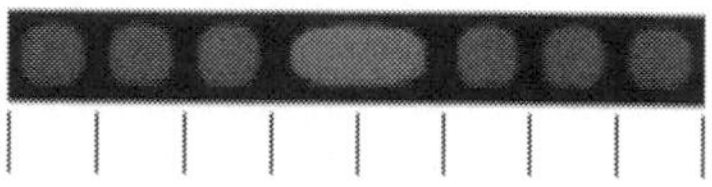

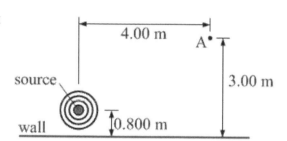

Figure 25.42: The pattern at the center of a screen, produced by a red laser beam shining on a single slit, a double slit, or a diffraction grating. The distance between neighboring tick marks, shown on the screen below the pattern, is 8.00 mm. For Exercise 49.

50. Figure 25.43 shows a source of sound that is located 0.800 m from a wall. The source is emitting waves of a single frequency. Point A is located some distance from the source, as shown. If the speed of sound in air is 340 m/s, find the three lowest frequencies that produce, at A, (a) completely constructive interference, and (b) completely destructive interference.

51. Return to the situation described in Exercise 50, and shown in Figure 25.43. If the source is emitting the lowest frequency sound wave to produce completely destructive interference at point A, what is the minimum distance the source can be moved, directly toward the wall, so the interference becomes completely constructive at A? It is acceptable to answer this by approximating that A is a long way from the source.

Figure 25.43: A source of single-frequency waves is located 0.800 m from a wall, for Exercises 50 and 51.

52. Two speakers send out identical single-frequency sound waves, in phase, that have a wavelength of 0.80 m. As shown in Figure 25.44, the speakers are separated by 3.6 m. Three lines, labeled A through C, are also shown in the figure. Line A is part of the perpendicular bisector of the line connecting the two sources. (a) How many points are there along line C at which the interference between the waves from the two speakers is completely destructive? (b) Relative to the midpoint of line C, approximately where are the points of destructive interference?

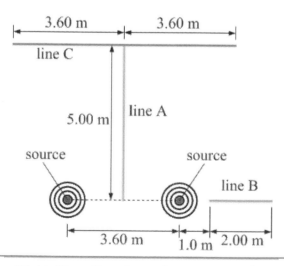

Figure 25.44: Three lines located near two speakers (each labeled "source") that are broadcasting identical single-frequency sounds, for Exercise 52.

53. Figure 25.45 shows four situations in which light is incident perpendicularly on a thin film (the middle layer in each case). The indices of refraction are $n_1 = 1.50$ and $n_2 = 2.00$. (a) Determine the minimum non-zero thickness of the film that results in constructive interference for 450 nm light (measured in vacuum) that reflects from the top and bottom surfaces of the film in case C. (b) Does this film thickness also produce constructive interference for 450 nm light in any of the other cases? Explain why or why not.

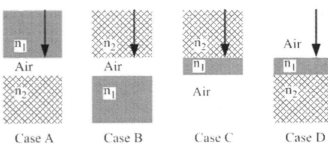

Figure 25.45: Four thin-film situations involving different arrangements of the same three media, for Exercise 53.

54. A thin piece of glass with an index of refraction of $n_2 = 1.50$ is placed on top of a medium that has an index of refraction $n_3 = 2.00$, as shown in Figure 25.40. A beam of light traveling in air ($n_1 = 1.00$) shines perpendicularly down on the glass. The beam contains light of only two colors, blue light with a wavelength in air of 450 nm and orange light with a wavelength in air of 600 nm. What is the minimum non-zero thickness of the glass that gives completely constructive interference for (a) the blue light reflecting from the film? (b) BOTH the blue and orange light simultaneously?

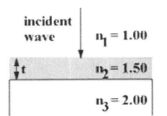

Figure 25.46: A thin film of glass on top of a medium that has an index of refraction of $n_3 = 2.00$. For Exercise 54.

55. Light traveling in air is incident along the normal to a thin film of unknown material that sits on a thick piece of glass ($n = 1.50$). The index of refraction of a typical medium is in the range $1.0 - 2.4$. Confining ourselves to this range, what is the index of refraction of the unknown material if the film is 120 nm thick, and it produces completely destructive interference for light, in air, with a wavelength of 540 nm? Find all the possible answers.

56. Sound waves traveling in air encounter a mesh screen. Some of the waves reflect from the screen, while the rest pass through and reflect from a wall that is 30.0 cm behind the mesh screen. You observe completely destructive interference between the two reflected waves when the frequency of the sound waves is 275 Hz. (a) Do the sound waves experience an inversion when they reflect from the mesh screen and from the wall? Explain. (b) What is the speed of sound in this situation?

57. Return to the situation described in Exercise 56. What are the next two frequencies, above 275 Hz, that will also produce completely destructive interference?

58. It is somewhat ironic that the phenomenon of Newton's rings (see Exercises 12 and 59), which provide evidence for the wave behavior of light, are named after Newton, because Sir Isaac Newton was a firm believer in the particle model of light. Do some research on Newton's contributions to optics, and write a couple of paragraphs about it.

59. Figure 25.47 illustrates a phenomenon known as Newton's rings, in which a bull's-eye pattern is created by thin-film interference. The film in this case is a thin film of air that is between a piece of glass with a spherical surface (such as a watch glass) that is placed on top of a flat piece of glass. The spherical surface of the top piece of glass has a radius of curvature of 500 cm. (a) How many wavelengths of 500 nm light (measured in air) fit in the film of air at a point 1.00 cm from the point where the top piece of glass makes

contact with the bottom piece? (b) Would you expect constructive or destructive interference to occur at this point? (c) Would your answers change if the air was replaced with a fluid with an index of refraction of 1.25? If so, how? Assume that the two pieces of glass have indices of refraction of about 1.5.

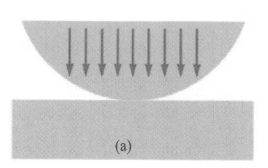  

(a)

> **Figure 25.47**: The phenomenon of Newton's rings comes from light shining through an object with a spherical surface that rests on a flat surface. Interference between light reflecting from the top surface of the film (between the spherical surface and the flat surface) and light reflecting from the bottom surface produces a bull's-eye shaped pattern. For Exercise 59.

60. A particular metal ruler has thin lines on it every half millimeter. As shown in Figure 25.48, laser light is incident on the ruler at an angle of $\alpha = 4.00°$ with respect to the ruler. (a) For the situation shown in the figure, find the relationship between the wavelength of the incident light and the angles (β values) at which constructive interference occurs. Use d to represent the spacing between the lines on the ruler. (b) If the first-order maximum occurs at an angle of $\beta = 4.93°$, what is the wavelength of the laser light?

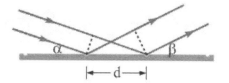

> **Figure 25.48**: Laser light is incident on a metal ruler. The light is incident at an angle of 4.00°, measured from the ruler. After interacting with the ruler, the rays of light leaving the ruler interfere constructively with one another when the rays make an angle β with the ruler surface. The dashed lines on the left and right are perpendicular to the incoming and outgoing rays, respectively. For Exercise 60.

61. Three students are working together on a problem involving thin-film interference. Comment on the part of their conversation that is reported below.

Evan: Do you know the equation for constructive interference in a thin-film situation?

Alison: There isn't one equation that works all the time – it depends on how the different indices of refraction compare.

Christian: Here's one, though, 2t equals m plus a half wavelengths. That's what we worked out in class when we did the soap film.

Evan: Isn't m plus a half for destructive interference?

Christian: Usually, it is, but with thin films you always get one of the waves flipping upside down when it reflects, which is like shifting it half a wavelength.

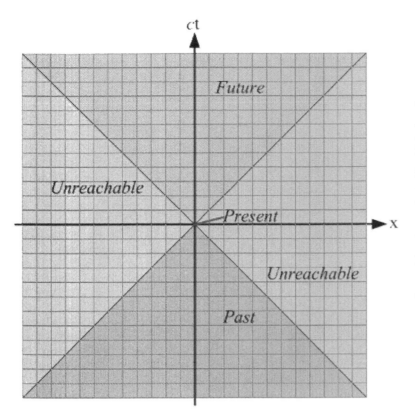

This spacetime diagram shows a light cone, which divides spacetime into past, present (the origin), future, and the points in spacetime that either can not affect the present, or can not be affected by you. The limiting factor is the finite speed of light. If you send a light pulse, traveling in vacuum, from the present in the positive or negative x-direction, it travels along the boundary between the future (the group of points that can be affected by you in the present) and the points in spacetime that are unreachable by you from the present. We will learn about spacetime and spacetime diagrams in this chapter.

Chapter 26 – Special Relativity

CHAPTER CONTENTS

Albert Einstein (1879 – 1955) is almost certainly the most famous physicist of all time. Among Einstein's many contributions to science is the theory of relativity. Einstein and others (notably Lorentz and Minkowski) first came up with the Special Theory of Relativity, which we will discuss in this chapter. Special Relativity deals with the special case of observers in inertial reference frames, which we can take to mean constant-velocity reference frames, with gravitational interactions ignored. Later, Einstein developed the General Theory of Relativity, which includes the effects of gravity.

Special Relativity deals, for the most part, with the physics of the very fast. By very fast, we mean a reasonable fraction of the speed of light. Light travels through vacuum at the speed of 3×10^8 m/s, which is much faster than any of us are likely to travel in our lifetimes. The key point is that we are used to observing objects which travel at modest speeds (compared to the speed of light in vacuum, at least) with respect to us. Special Relativity tells us that space and time can behave quite differently from how we are used to them behaving. However, given that our experience with space and time is limited to low relative speeds, maybe we should not be too surprised that space and time exhibit interesting behaviors at high relative speeds.

On the other hand, the behavior of space and time, as we will learn in this chapter, is so different from the way most people think about them that you may decide not to believe it. If so, you should be aware that there is ample experimental evidence to support the implications of Special Relativity, some of which we will discuss in the chapter.

26-1 Observers

Although Special Relativity is often thought of as applying to fast-moving objects, we can see some effects of Special Relativity at low speeds, too. Let's explore this, and get comfortable with the idea of how the same situation looks to different observers.

EXPLORATION 26.1 – Who needs magnetism?

Consider one observer, Jack, who is watching two pairs of charges, as shown in Figure 26.1. The charges are identical, and we will assume that the charges within each pair repel one another, but that the pairs of charges are far enough apart that one pair of charges does not influence the other pair. One pair of charges is initially at rest, with respect to Jack, while each charge in the other pair of charges has an initial velocity v directed to the right. There is also a second observer, Jill, who has a constant velocity v directed to the right with respect to Jack.

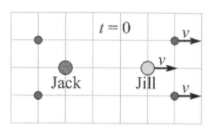

Figure 26.1: The situation at $t = 0$ involving two pair of charges, as seen from Jack's frame of reference. According to Jack, the charges in the left pair are released from rest, while each charge in the right pair have an initial velocity directed right. There is also a second observer, Jill, who is moving right at constant velocity with respect to Jack.

Step 1 – *Jack observes that the charges that were initially at rest move apart more quickly than the charges that were initially moving with respect to him. Figure 26.2 illustrates what Jack sees after a time T has passed. Using principles of electricity and magnetism, explain Jack's observations.*

First, consider the electrostatic interaction between the charges. The charges are identical, so by Coulomb's law, Jack expects the charges that were initially at rest to repel and accelerate away from one another. Jack expects the charges that were initially moving to repel each other, too, but he also expects those charges to interact magnetically, because they are like two parallel currents. When Jack reviews Chapter 19, he recalls that two parallel currents attract one another. When Jack does the calculations, he finds that the repulsive electrostatic force is larger than the attractive magnetic force, so he expects the charges on the right to move apart, but not as fast as the charges on the left. This is exactly what he observes.

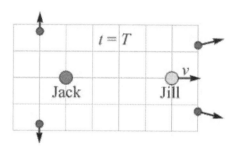

Figure 26.2: The situation at $t = T$, where the charges on the left have separated more than the charges on the right, according to Jack.

Step 2 – *Now draw two more pictures, one showing the initial situation from Jill's frame of reference, and the other showing the situation at a time T after the charges are released, again from Jill's frame of reference. Once again, use principles of physics to explain what Jill observes.*
What Jill observes is a mirror image of what Jack sees. Initially, at $t' = 0$, where t' denotes times measured by Jill, Jill sees the charges on the right at rest and the charges on the left, and Jack, moving to the left with velocity v.

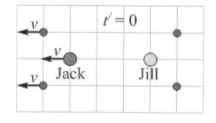

Figure 26.3: The initial situation from Jill's reference frame.

At a time $t' = T$, according to Jill, Jill's view of the situation is again a mirror image of what Jack sees. As shown in Figure 26.4, Jill sees the charges on the right, which were initially at rest with respect to her, move apart more quickly than the charges on the left, because of the attractive magnetic force associated with the charges on the left.

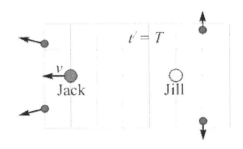

Step 3 – *What, if anything, about the previous steps does not make sense to you?* Possibly, it all makes sense. However, you may be bothered by the fact that the two observers disagree on which charges move apart more quickly. Jack observes that the charges on the left move apart more quickly, while Jill observes that the charges on the right move apart more quickly. For most people, that does not make sense – shouldn't everyone agree on what is happening? This is a key part of relativity – observers in different reference frames observe different things.

Figure 26.4: The situation at $t' = T$, where the charges on the right have separated more than the charges on the left, according to Jill.

Step 4 – *Come up with an alternate explanation for what Jack and Jill observe. Instead of using magnetism to explain why the initially moving charges move apart more slowly, come up with an alternate explanation involving how time works in different reference frames.* This is an unfair question, because we're asking you to be Einstein here. However, the relativistic argument is that there is no need to resort to magnetism. The charges simply repel one another, and move apart. However, observers observe time passing differently (moving more slowly) in reference frames that are moving with respect to that observer.

In this situation, the charges on the left are in Jack's reference frame, while the charges on left are in Jill's reference frame, which is moving at a constant velocity v to the right with respect to Jack's reference frame. Thus, Jack observes Jill's wristwatch running slowly, and everything in her reference frame (such as the charges on the right) to be moving in slow motion. Jill, on the other hand, observes her wristwatch to be running just fine, and the charges on the right to move apart as she expects, but she observes Jack's wristwatch to be running slowly, and the left-hand charges, which are in Jack's reference frame, to be moving in slow motion.

Key ideas: We investigated a variety of ideas in this Exploration. First, magnetism is a relativistic effect, which means that you can change the strength of a magnetic interaction simply by changing the relative velocity between you and a set of interacting charges. Second, observers in different reference frames can disagree on what happens in certain cases. However, any constant-velocity reference frame is just as good as any other reference frame in terms of making observations. In the situation above, for instance, Jill's reference frame is no better or worse than Jack's. Third, time works in a manner that is quite different from what we're used to based on our past experience, in that it runs at different rates in different reference frames (we will explore this in more detail in the sections that follow).   **Related End-of-Chapter Exercises: 1, 2.**

Postulates of special relativity: Relativity is based on two simple ideas.
   1. The speed of light in vacuum is the same for all observers.
   2. There is no preferred reference frame. The laws of physics apply equally in all reference frames.
One implication of these postulates is that nothing can travel faster than the speed of light in vacuum, which is $c = 3.00 \times 10^8$ m/s.

Essential Question 26.1: (a) How fast are you moving right now? (b) What is your speed associated with being on the Earth while the Earth is spinning around its axis? (c) What is your speed associated with being on Earth while the Earth is orbiting the Sun?

Answer to Essential Question 26.1: (a) The obvious answer is that you are at rest. However, the question really only makes sense when we ask what the speed is measured with respect to. Typically, we measure our speed with respect to the Earth's surface. If you answer this question while traveling on a plane, for instance, you might say that your speed is 500 km/h. Even then, however, you would be justified in saying that your speed is zero, because you are probably at rest with respect to the plane. (b) Your speed with respect to a point on the Earth's axis depends on your latitude. At the latitude of New York City (40.8° north), for instance, you travel in a circular path of radius equal to the radius of the Earth (6380 km) multiplied by the cosine of the latitude, which is 4830 km. You travel once around this circle in 24 hours, for a speed of 350 m/s (at a latitude of 40.8° north, at least). (c) The radius of the Earth's orbit is 150 million km. The Earth travels once around this orbit in a year, corresponding to an orbital speed of 3×10^4 m/s. This sounds like a high speed, but it is too small to see an appreciable effect from relativity.

26-2 Spacetime and the Spacetime Interval

We usually think of time and space as being quite different from one another. In relativity, however, we link time and space by giving them the same units, drawing what are called spacetime diagrams, and plotting trajectories of objects through spacetime. A spacetime diagram is essentially a position versus time graph, with the position axes and time axes reversed.

EXPLORATION 26.2 – A spacetime diagram

We can convert time units to distance units by multiplying time by a constant that has units of velocity. The constant we use is c, the speed of light in vacuum.

Step 1 – *Plot a graph with position on the x-axis and time, converted to distance units, on the y-axis. This graph is a spacetime diagram. We are at rest in this coordinate system, which means that the spacetime diagram is for our frame of reference. At t = 0 and x = 0, we send a pulse of light in the positive x-direction. On the graph, show the trajectory of this light pulse, which travels at the speed of light in vacuum. How far does the pulse travel in 5 meters of time?* The spacetime diagram is shown in Figure 26.5. Because the pulse of light travels at the speed of light, the pulse's trajectory is a straight line with a slope of 1. The pulse travels 5 m of distance in 5 m of time, which makes things easy to plot.

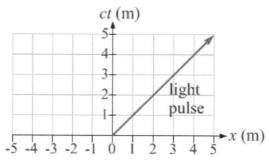

Figure 26.5: A spacetime diagram, showing the trajectory of a light pulse that travels in the +x-direction.

Step 2 – *On the spacetime diagram, plot the trajectory (this is called a worldline) of a superfast mosquito that passes through x = 0 at t = 0. With respect to us, the mosquito moves in the positive x-direction at half the speed of light. What is the connection between the slope of the worldline and the velocity of the mosquito?* This spacetime diagram is shown in Figure 26.6. Because the mosquito's velocity is constant, the slope of its worldline is constant. The mosquito is always half the distance from the origin that the light pulse is (at least according to us!).

The slope of an object's worldline is the inverse of the velocity, if the velocity is expressed as a fraction of c. In this case, the mosquito has a velocity of +0.5, so the slope of its worldline is +2.

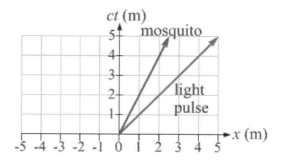

Figure 26.6: The spacetime diagram, showing the worldline of a mosquito traveling in the +x-direction at half the speed of light.

Step 3 – Add two events to the spacetime diagram. *According to us, Event 1 occurs at x = 0 at t = 0, and Event 2 occurs at x = +2 m at t = +4 m of time. According to us, what is the time interval between the two events, and what is their spatial separation?*

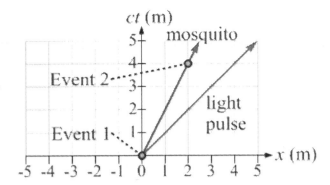

Note that we define an event as something that takes place at a particular point in space and at a particular instant in time. The amended spacetime diagram with the events marked on it is shown in Figure 26.7. The time interval, converted to distance units, between the two events is $c\Delta t = 4$ m. The spatial separation is $\Delta x = 2$ m.

Figure 26.7: The spacetime diagram, now showing the space and time coordinates, as measured from our reference frame, of two events.

The spacetime interval: It turns out that observers in different constant-velocity reference frames always agree on the value of the spacetime interval between two events, as defined by

$$(c\Delta t)^2 - (\Delta x)^2 = (c\Delta t')^2 - (\Delta x')^2 = \text{(spacetime interval)}^2, \qquad \text{(Eq. 26.1: The spacetime interval)}$$

where $c\,\Delta t$ and $c\,\Delta t'$ are the time intervals, converted to distance units, between the two events as measured from two different frames of reference, and Δx and $\Delta x'$ are the spatial separations between the two events, as measured in the same two reference frames. Note that, if the left-hand side of the equation gives a negative number, it is appropriate to reverse the order of the terms.

Step 4 – What is the spatial separation $\Delta x'$ between Event 1 and Event 2 as measured by the mosquito? Knowing this, use Equation 26.1 to find the time interval $c\Delta t'$ between the events, as measured by the mosquito. Both of the events lie on the worldline of the mosquito, so the mosquito thinks that they both take place at $x' = 0$. Thus, the spatial separation between them, as measured by the mosquito, is $\Delta x' = 0 - 0 = 0$. Let's now work out the value of the spacetime interval between these events, as measured by us.

$$\text{(spacetime interval)}^2 = (c\Delta t)^2 - (\Delta x)^2 = (4\text{ m})^2 - (2\text{ m})^2 = 12\text{ m}^2.$$

In the mosquito's reference frame, then, the time difference is given by

$$(c\Delta t')^2 = \text{(spacetime interval)}^2 - (\Delta x')^2 = 12\text{ m}^2 - 0 = 12\text{ m}^2.$$

Thus, we observe that 4 m of time pass between the events, while the mosquito observes only $\sqrt{12}$ m $= 3.5$ m of time passing between them. The main point here is that observers in different reference frames measure different amounts of time passing between the same two events.

Key ideas: In relativity, the emphasis is often on what is different in two different reference frames. However, as we have learned earlier (such as with energy conservation) the parameters that everyone agrees on are generally most important. In relativity, all observers agree on the value of the spacetime interval. In addition, in relativity we treat space and time as different components of a spacetime coordinate system, rather than treating them as completely different things, as we are more used to doing. **Related End-of-Chapter Exercises: 9 – 18.**

Essential Question 26.2: Return to Exploration 26.2. Add your worldline to the spacetime diagram shown in Figure 26.7.

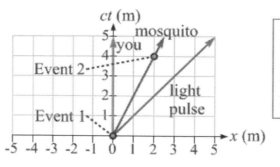

Answer to Essential Question 26.2: The spacetime diagram is drawn for your reference frame, so you remain at *x* = 0 at all times. Thus, your worldline is a straight line up the *ct* axis, as shown in Figure 26.8.

Figure 26.8: The spacetime diagram from Figure 26.7, updated to show your worldline.

26-3 Time Dilation – Moving Clocks Run Slowly

In Sections 26-1 and 26-2, we looked at situations in which time passes at different rates in different reference frames. This may be at odds with your own experience of time. However, we are not used to dealing with speeds that are significant fractions of the speed of light, so perhaps we should not be too surprised that time has such an interesting behavior when we compare reference frames that are moving at high speed with respect to one another.

The time-honored statement summarizing how time behaves is that "moving clocks run slowly." In this section, we will investigate how a clock that uses light pulses behaves when it is viewed from a moving reference frame. We will also discuss some of the experimental evidence supporting this behavior of time.

EXPLORATION 26.3 – A light clock

Figure 26.9 shows a light clock, which has a source (or emitter) of light pulses (at the bottom), a mirror at the top to reflect the light, and a detector at the bottom to record the pulses. The clock runs at a rate equal to the rate at which the pulses are received by the detector.

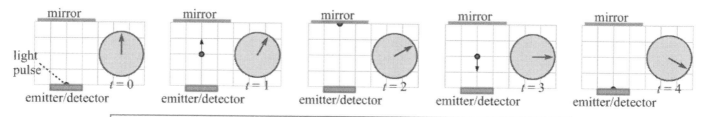

Figure 26.9: A clock that uses light pulses to measure the passage of time.

Step 1 – *Let's say the clock above belongs to Jack. Relative to Jack, Jill moves to the right at a constant speed of 60% of the speed of light in vacuum. Jill's light clock is identical to Jack's. Sketch a diagram, similar to that in Figure 26.9, showing how the light pulses in Jill's clock travel from emitter to mirror to detector, according to Jack.* According to Jack, Jill's clock is moving, so a light pulse in Jill's clock travels farther to reach the mirror after leaving the emitter than does a light pulse in Jack's clock. When Jack's clock reads 1 unit, Jill's clock is reading 0.8 units, according to Jack. This pattern continues, with Jill's clock continuing to show time passing by at 80% of the rate at which Jack's clock measures time passing, according to Jack.

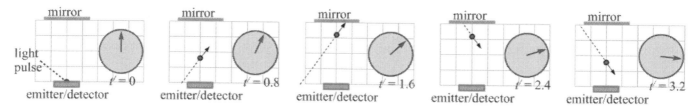

Figure 26.10: What Jill's light clock looks like, according to Jack. Compare this figure to Figure 26.9 to see what Jack sees on his clock and Jill's clock at the same instants.

Step 2 – *Sketch a diagram showing how the light pulses in Jack's clock travel from emitter to mirror to detector, according to Jill.* According to Jill, her clock is working fine (she sees her clock as in Figure 26.9), but Jack's clock is running slowly. All observers see light traveling at the speed of light, so Jill sees Jack's clock running slowly because the light pulses in Jack's clock, shown in Figure 26.11, travel farther to reach the mirror than do the pulses in Jill's clock.

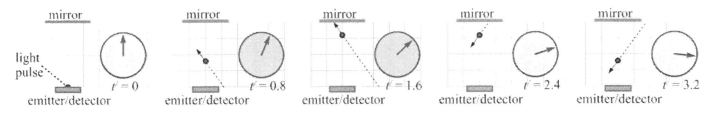

> **Figure 26.11**: What Jack's light clock looks like, according to Jill. Compare this figure to Figure 26.9 to see what Jill sees on her clock and Jack's clock at the same instants.

Proper time and time dilation: An observer measures the proper time interval Δt_{proper} between two events when that observer is present at the location of both events. Observers for whom the events take place in different locations measure a longer time interval Δt between the events.

$$\Delta t = \frac{\Delta t_{proper}}{\sqrt{1 - \dfrac{v^2}{c^2}}} = \gamma\,\Delta t_{proper},\qquad\text{(Equation 26.2: \textbf{Time dilation})}$$

where v is the relative speed between the two reference frames.

Key ideas: The faster a clock moves with respect to an observer, the more slowly it ticks off time, according to that observer. All time-keeping devices, including beating hearts, act this way, not just the light clocks we explored here. **Related End-of-Chapter Exercises: 19 – 22, 34.**

Experimental evidence for time dilation

In 1941, Bruno Rossi and David Hall compared the rate at which muons (essentially heavy electrons) entering the Earth's atmosphere passed through their detectors when they were at the top of Mt. Washington in New Hampshire, at an altitude of 6300 feet. They also measured the rate at which muons passed through the detectors at sea level. For an observer at rest on the Earth, these muons take 6.4 µs to cover 6300 feet, considerably longer than the 2.2 µs average lifetime of muons created in the laboratory. Thus, one would expect almost all the muons to decay before reaching sea level. Rossi and Hall's measurements showed that significantly more muons reached sea level than would be expected. This can be explained by time dilation. According to an Earth-based observer, time passes more slowly for these muons, which travel at over 99% of the speed of light. When 6.4 µs has elapsed for us, we observe clocks in the muons' frame of reference to be running more slowly, explaining why they last longer than we expect.

Another experiment was carried out in 1971 by J. C. Hafele and R. E. Keating. They flew four atomic clocks eastward around the world, and then flew the clocks westward around the world. After each trip, the clocks were compared to an identical clock that remained at the United States Naval Observatory. Both the effects of general and special relativity were important in this experiment, but the observed results (a loss of 59 ± 10 ns for the eastward trip, and a gain of 273 ± 7 ns for the westward trip) were in agreement with the predictions of relativity.

Essential Question 26.3: Return to Exploration 26.3. Who is aging more slowly, Jack or Jill?

Answer to Essential Question 26.3: Who ages more slowly depends on who you ask. According to Jack, time passes more slowly for Jill, who is moving with respect to Jack, and thus Jack says that Jill is aging more slowly. According to Jill, time passes more slowly for Jack, who is moving with respect to Jill, and thus Jill says that Jack is aging more slowly. This brings us to the famous **twin paradox**, in which one twin goes off at high speed to explore a distant galaxy, and returns some years later to find that the twin who remained behind on Earth is considerably older than the traveling twin. The apparent paradox is why there is no symmetry in this situation. The twin who stayed behind should see the traveling twin to be aging more slowly, but the traveling twin should see the twin who stayed behind to be aging more slowly. The resolution of the paradox is that the situation is not symmetric – the traveling twin changed reference frames halfway through, from a frame of reference in which she travels away from the Earth to a new frame of reference in which she travels toward the Earth. In both reference frames, interestingly, she does observe the twin who remained on Earth to be aging more slowly than she is, but she also observes the age of the twin who remained behind to jump when the traveling twin switches reference frames.

26-4 Length Contraction

Time is not the only thing that behaves in an unusual way, space does too. Observers who view objects moving with respect to them, for instance, measure the length of the object to be contracted (shorter) along the direction of motion.

Let's now work through an example that ties together all of the ideas we have discussed thus far in the chapter

---

Length contraction: Assume that two points separated by some distance are in the same reference frame. An observer measures the proper length L_{proper} between these two points when the points are at rest with respect to that observer (that is, the observer is in the same frame of reference as the points). For an observer in a different reference frame, when the displacement between the points has a component parallel to the velocity of the points, the observer measures a contracted length between the points. This length contraction is most pronounced when the displacement from one point to the other is parallel to the velocity, in which case the length L measured by the observer is

$$L = \sqrt{1 - \frac{v^2}{c^2}}\, L_{proper} = \frac{L_{proper}}{\gamma}, \qquad \text{(Equation 26.3: Length contraction)}$$

where v is the relative speed between the two reference frames.

EXAMPLE 26.4 – Isabelle's travels

Let's say that Planet Zorg is at rest with respect to the Earth, and that in the reference frame of the Earth, the distance between Earth and Zorg is 40 light-years. Isabelle is passing by the Earth in her rocket ship traveling toward Zorg at a constant velocity of 0.8 c. At the instant she passes by, you, who are on Earth, send a light pulse toward Zorg.

   (a) According to you, how long does the light pulse take to reach Zorg?
   (b) According to you, how long does Isabelle take to reach Zorg?
   (c) According to Isabelle, what is the spatial distance between the event of Isabelle passing Earth and the event of Isabelle arriving at Zorg?
   (d) Use Equation 26.1 to find the time it takes Isabelle to reach Zorg, according to Isabelle.
   (e) What is the distance between Earth and Zorg, according to Isabelle?
   (f) How long does the light pulse take to reach Zorg, according to Isabelle?

SOLUTION

(a) According to you, the planets are separated by a distance of 40 light-years. Because light covers 1 light-year every year, it takes the light pulse 40 years to reach Zorg, according to you.

(b) According to you, Isabelle travels at 4/5 the speed that light does, so Isabelle should take 5/4 the time that light does.

$$t = \frac{40 \text{ light-years}}{0.8\,c} = 50 \frac{\text{light-years}}{c} = 50 \text{ years} .$$

Note that light-years divided by the speed of light in vacuum, c, gives years. Also, note that we don't have to use a relativistic equation to get the answer. We are, instead, using the basic idea that for motion with constant speed, time is simply the distance divided by the speed.

(c) For Isabelle, these two events happen at the same location, right outside Isabelle's rocket. Thus, their spatial separation, according to Isabelle, is $\Delta x' = 0$.

(d) Equation 26.1 is the equation for the spacetime interval. According to you, the time between the events is 50 years which, when multiplied by c, gives $c\Delta t = 50$ light-years, and the events are separated spatially by $\Delta x = 40$ light-years. The spacetime interval is given by:

$$(\text{spacetime interval})^2 = (c\Delta t)^2 - (\Delta x)^2 = (50 \text{ light-years})^2 - (40 \text{ light-years})^2 = (30 \text{ light-years})^2 .$$

Solving for the time interval from Isabelle's perspective, we get:

$$(c\Delta t')^2 = (\text{spacetime interval})^2 - (\Delta x')^2 = (30 \text{ light-years})^2 - 0 = (30 \text{ light-years})^2 .$$

In Isabelle's reference frame, we have $c\Delta t' = 30$ light-years. Dividing by c gives a time interval of 30 years, so it only takes Isabelle 30 years to go from Earth to Zorg, according to Isabelle.

Note that 30 years is also the proper time between the two events, because Isabelle is present at both events. The proper length between Earth and Zorg, on the other hand, is the 40 light-years measured by you, because the planets are at rest in your reference frame. At this point, you may be concerned that it looks like Isabelle travels faster than light, since she travels to Zorg in 30 years while light travels there in 40 years, but we are about to resolve that.

(e) We could use the length contraction equation to determine the distance between the planets, according to Isabelle. A simpler method is that, according to Isabelle, Zorg travels toward her at a constant velocity of $0.8\,c$, and Zorg passes her 30 years after Earth passes her. Thus, Zorg must have covered a distance of $0.8\,c \times 30$ years, which is 24 light-years, in that time, which represents the distance between the planets, according to Isabelle.

(f) According to Isabelle, the planets are separated by a distance of 24 light-years. Isabelle sees the light pulse traveling away from her at c, and Zorg coming toward her at $0.8c$, for a relative speed between them of $1.8c$, according to Isabelle. Thus, according to Isabelle, the pulse and Zorg would meet at a time of 24 light-years divided by $1.8c$, or 13.3 years. Both you and Isabelle agree that nothing travels faster than light in this situation, by the way.

Related End-of-Chapter Exercises: 6, 23 – 25.

Essential Question 26.4: If your clock and Isabelle's clock both read zero when Isabelle passes Earth, what will the clocks read, according to both you and Isabelle, when Isabelle reaches Zorg?

Answer to Essential Question 26.4: According to you, 50 years elapses on your clock during Isabelle's trip to Zorg. Using the time dilation equation with a relative speed of 0.8*c*, we can show that Isabelle's clock is running at 60% of the rate of yours, according to you. Thus, when 50 years passes by on your clock, only 30 years passes by on Isabelle's clock, according to you. This is consistent with the result of part (d) in Example 26-4.

As we showed in part (d), according to Isabelle, the time interval between Earth passing her and Zorg passing her to is 30 years. However, if you see her clock running at 60% of the rate of yours, Isabelle sees your clock running at 60% of the rate of hers. 60% of 30 years is 18 years, so, according to Isabelle, your clock only reads 18 years when Isabelle and Zorg meet.

26-5 The Breakdown of Simultaneity

Let's explore the behavior of time in more detail, beginning with extending our understanding of spacetime diagrams. From our work earlier in the book, we are used to being able to transform from one coordinate system to another by simply rotating the *x* and *y* axes through the same angle, as demonstrated in Figure 26.12. The two coordinate systems share the same origin, and a point such as *x* = +3 m, *y* = +4 m (a distance of 5 m from the origin) transforms into *x'* = +5 m, *y'* = 0 (not coincidentally also 5 m from the origin). For any point a distance *r* from the origin, we have the equivalent of the spacetime interval equation,

$$x^2 + y^2 = (x')^2 + (y')^2 = r^2.$$

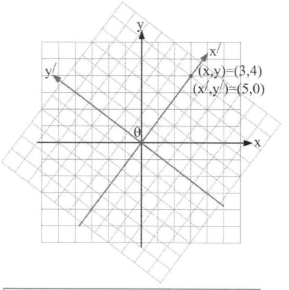

Figure 26.12: Rotating the *x* and *y* axes through the same angle transforms coordinate systems in the *x-y* plane.

We can something similar with the spacetime diagram, except that the space and time axes rotate in opposite directions. The angle of rotation depends on the relative velocity between one reference frame and the other. With the super-fast mosquito we looked at in Exploration 26-2, the mosquito's time axis, as viewed from your frame of reference, was rotated as it was because the mosquito was moving at half the speed of light with respect to you. Similarly, the mosquito's *x*-axis (as seen from your reference frame) is rotated through the same angle, but in the opposite direction, so that the slope of the mosquito's *x*-axis is equal to the mosquito's speed, with respect to you, expressed as a fraction of the speed of light. The mosquito's coordinate system, according to you, is shown in Figure 26.13.

Examining events 1 and 3 in Figure 26.13, we can see that events which are simultaneous in one reference frame may not be simultaneous in another. Events 1 and 3 are on the mosquito's *x*-axis, which means they take place at the same instant (at *t'* = 0, in this case) and are thus simultaneous in the mosquito's frame of reference. In your reference frame, however, Event 1 occurs before Event 3. Simultaneity is relative! In a situation like this, events are simultaneous for observers in different reference frames only when they occur at the same time and place.

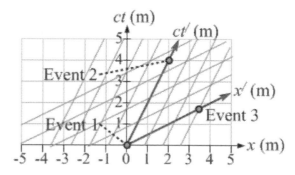

Figure 26.13: Transforming coordinate systems on a spacetime diagram is accomplished by rotating the time and space axes in opposite directions, by an angle determined by the relative velocity between the two coordinate systems.

Let's construct a spacetime diagram, from your frame of reference, for Isabelle's travels, from Example 26.4. Isabelle is traveling at $0.8c$ with respect to you so, in your reference frame, Isabelle's coordinate system is rather distorted. This is shown in Figure 26.14. If you and an observer called Yan, who is in your frame of reference but located on Zorg, synchronize your clocks, according to you and Yan, all four of the following events are simultaneous:

- Isabelle passes Earth
- Your clock reads zero
- Isabelle's clock reads zero
- Yan's clock reads zero

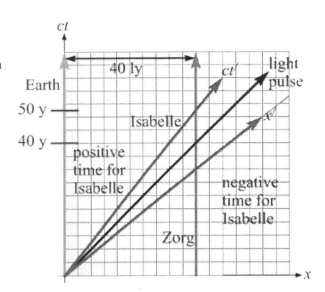

Figure 26.14: A spacetime diagram, from your frame of reference, for the situation of Isabelle's travels between Earth and Zorg. The boxes on the grid measure 4 ly × 4 ly.

Isabelle agrees that the first three events are simultaneous, but she disagrees that Yan's clock reads zero at that instant. From the spacetime diagram, we can divide spacetime into locations at which Isabelle records a positive time on her clock, and locations where Isabelle records a negative time. According to Isabelle, Yan's clock reads zero long before your clock reads zero. In fact, according to Isabelle, Yan's clock is set 32 years ahead of your clock! This helps explain a mystery regarding Essential Question 26.4. You think Isabelle takes 50 years to reach Zorg, and Yan sends a message to you that, when Isabelle passed Zorg, Yan's clock read 50 years and Isabelle's read 30 years. This makes complete sense to you. Isabelle's message to you, however, states that, when Zorg passed her, Yan's clock read 50 years, her clock read 30 years, but that your clock read only 18 years. This is consistent with Isabelle observing both your clock and Yan's clock to be running slow, but with Yan's clock 32 years ahead of your clock. During her trip, according to Isabelle, 18 years ticked by on both your clock and Yan's clock.

Differences in clock readings are NOT associated with the travel time of light
You should be clear that the different times and lengths recorded by observers in different reference frames are not caused by the observers being separated by some distance and light taking some time to travel from one observer to another. When we say things like "When Isabelle reaches Zorg, from her reference frame your clock reads 18 years," that is not the reading that Isabelle would see on your clock at that instant, looking back many light-years through a powerful telescope to Earth. Instead, we imagine multiple observers at different locations in Isabelle's reference frame, their clocks synchronized with Isabelle's. Each of these observers, when they pass Earth, can send a message to Isabelle to tell her what their clock reading was, and what your clock was reading, when they passed Earth. Many years after passing Zorg, Isabelle will finally get the message confirming what your clock was reading when she passed Zorg, according to an observer in her reference frame. Thus, the statement "When Isabelle reaches Zorg, from her reference frame your clock reads 18 years" is interpreted as "When Isabelle reaches Zorg, the observer in Isabelle's frame of reference who is next to your clock records that your clock reads 18 years." The observer is right there, so light travel time is not an issue.

Related End-of-Chapter Exercises: 44 – 46.

Essential Question 26.5: According to Isabelle, how many years after passing Zorg would she get the message stating what your clock read when she passed Earth, from the observer in Isabelle's frame of reference who sees Earth pass by when Isabelle sees Zorg pass by?

Answer to Essential Question 26.5: According to Isabelle, she and the relevant observer in her reference frame are separated by 24 light-years, the distance between Earth and Zorg in her reference frame. The fastest the message could be sent would be at light speed, in which case the message would take 24 years to reach Isabelle.

Chapter Summary

Essential Idea: Special Relativity.
The closer relative speeds get to the speed of light, the more time and space exhibit unusual behaviors. This includes (but is not limited to) clocks in different reference frames running at different rates, lengths measured to be different by different observers, and events that are simultaneous for one observer taking place at different times for other observers.

The Postulates of Special Relativity
Relativity is based on two simple ideas, or postulates. One implication of these postulates is that nothing can travel faster than the speed of light in vacuum, which is $c = 3.00 \times 10^8$ m/s.
1. The speed of light in vacuum is the same for all observers.
2. There is no preferred reference frame. The laws of physics apply equally in all reference frames.

The Spacetime Interval
Observers in different constant-velocity reference frames always agree on the value of the spacetime interval between two events, as defined by

$$(c\Delta t)^2 - (\Delta x)^2 = (c\Delta t')^2 - (\Delta x')^2 = \text{(spacetime interval)}^2, \quad \text{(Eq. 26.1: \textbf{the spacetime interval})}$$

where $c\,\Delta t$ and $c\,\Delta t'$ are the time intervals, converted to distance units, between the two events as measured from two different frames of reference, and Δx and $\Delta x'$ are the spatial separations between the two events in the same two reference frames. If the left-hand side of the equation gives a negative number, it is appropriate to reverse the order of the terms.

Proper time and time dilation
An observer measures the proper time interval Δt_{proper} between two events when that observer is present at the location of both events. Observers for whom the events take place in different locations measure a longer time interval Δt between the events.

$$\Delta t = \frac{\Delta t_{proper}}{\sqrt{1 - \dfrac{v^2}{c^2}}} = \gamma\,\Delta t_{proper}, \qquad \text{(Equation 26.2: \textbf{Time dilation})}$$

where v is the relative speed between the observers.

Length contraction
An observer measures the proper length L_{proper} between two points when the points are at rest with respect to that observer (that is, the observer is in the same frame of reference as the points). For an observer in a different reference frame, the observer measures a contracted length L between the points along the direction of motion. v is the relative speed between the observers.

$$L = \sqrt{1 - \frac{v^2}{c^2}}\, L_{proper} = \frac{L_{proper}}{\gamma}. \qquad \text{(Equation 26.3: \textbf{Length contraction})}$$

End-of-Chapter Exercises

Exercises 1 – 8 are primarily conceptual questions designed to see whether you understand the main concepts of the chapter.

1. Return to the Jack and Jill situation in Exploration 26.1. Let's now add a third observer, Martin, who, according to Jack, is traveling in the same direction as Jill but at a different speed. From Martin's point of view, the charges in Jack's reference frame move apart at the same rate that the charges in Jill's reference frame move apart. Explain qualitatively how this is possible.

2. If magnetism is a relativistic phenomenon, then we should be able to explain effects that we previously attributed to magnetic interactions without resorting to magnetism. An example is the situation shown in Figure 26.15, in which an object with a negative charge has an initial velocity v directed parallel to a long straight wire. The current in the long straight wire is directed to the left. (a) First, use the principles of physics we discussed in Chapter 19 to explain how to determine the direction of the force exerted by the wire on the moving charge, and state the direction of that force. (b) Now, we'll work through the process without magnetism. Imagine that the current in the wire is associated with electrons flowing to the right with the same speed, v, that the charged object has. The wire, when no current flows, has an equal amount of positive and negative charge. We can imagine the positive charges to be at rest in the original frame of reference, shown in Figure 26.15(b). If we look at the situation from the point of view of the charged object, however, which charges in the wire are moving, the positive charges or the negative charges? (c) From the frame of reference of the charged object, will the distance between the electrons in the wire be length-contracted? What about the distance between the positive charges in the wire? (d) Again from the frame of reference of the charged object, will the wire appear to be electrically neutral, or will it have a net positive or net negative charge? (e) Looking at the electric interactions only, will the wire exert a net force on the charged object? If so, in what direction is it, and is this direction consistent with the answer to part (a)?

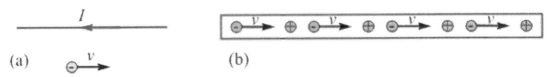

(a)

(b)

Figure 26.15: (a) A negatively charged object near a long straight current-carrying wire, and (b) a close-up of the wire showing the moving electrons and stationary positive charges making up the wire, for Exercise 2.

3. According to Michael, two events take place at the same location, but are separated in time by 5 minutes. Jenna is moving at a reasonable fraction of the speed of light with respect to Michael. According to Jenna, (a) do the events take place at the same location, and (b) is the time interval between them longer than, equal to, or shorter than 5 minutes?

4. When you and your rocket are at rest on the Earth, you measure the length of your rocket to be 60 m. When you and your rocket are traveling at 80% of the speed of light with respect to the Earth, how long do you measure your rocket to be?

5. Consider events A and B on the spacetime diagram shown in Figure 26.16. Do the events occur at the same time, or at the same location, for any of the four observers? Explain.

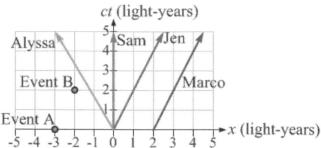

6. You are on a train that is traveling at 70% of the speed of light relative to the ground. Your friend is at rest on the ground, a safe distance from the track, watching the train go past. Which of you measures the proper length for the following? (a) The length of the train. (b) The distance between the two rails the train is riding on. (c) The distance between the two cities the train is traveling between.

Figure 26.16: This spacetime diagram shows worldlines for four observers, whose motion is confined to the x-axis, as well as the spacetime coordinates of two events, in Sam's frame of reference. For Exercise 5.

7. Return to the Earth, Zorg, you and Isabelle situation from Exploration 26-4, and the spacetime diagram for this situation that was drawn for it, from your frame of reference, in Figure 26.14. Now sketch the spacetime diagram for this situation from Isabelle's frame of reference.

8. One way that observers in the same reference frame can synchronize their clocks is to use a light pulse. For instance, you and your friend, who is located some distance from you, might agree to both set your clocks to read zero when you each observe a light pulse that is emitted from a source exactly halfway between you and which spreads out uniformly in all directions. Comment on this possible alternate method of clock synchronization. Your friend can bring his clock to where you are, you can synchronize them, and then your friend can carry his clock back to his original location. What, if anything, is wrong with that method?

Exercises 9 – 13 involve the spacetime interval.

9. According to you, two events take place at the same location, with a time interval of 60 meters of time between them. According to a second observer, the time interval between the events is 75 meters of time. (a) How many nanoseconds does 60 meters of time represent? (b) What is the distance between the locations of the two events, according to the second observer?

10. Four observers watch the same two events. The time interval and spatial separation between the events, according to each of the observers, is shown in Table 26.1. Fill in the blanks in the table.

Table 26.1: The time interval and spatial separation between the same two events, according to four different observers, for Exercise 10.

Observer	Time interval	Spatial separation
Anna		7.0 m
Bob		14.0 m
Caroline	35.7 meters of time	21.0 m
Dewayne	50.0 meters of time	

11. According to you, two events take place simultaneously at locations that are 100 m apart. According to a second observer, the spatial separation between the locations of the events is 150 m. What is the time interval between the events, according to the second observer?

12. According to you, two events take place 100 ns apart with a spatial separation between them of 18 m. (a) What is the spacetime interval for the two events? (b) For a second observer, the two events occur at the same location. What is the time interval between them, in nanoseconds, according to the second observer?

13. The spacetime diagram in Figure 26.17 shows three events. Determine the spacetime interval between (a) Event 1 and Event 2, (b) Event 1 and Event 3, and (c) Event 2 and Event 3.

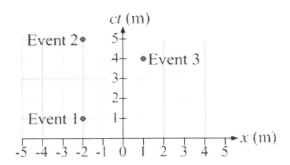

Figure 26.17: A spacetime diagram showing three events, for Exercise 13.

Exercises 14 – 18 involve drawing and interpreting spacetime diagrams.

14. Consider the spacetime diagram shown in Figure 26.18, for four people whose motion is confined to the x-axis. The squares on the grid measure 1 m × 1 m. (a) Which worldline is impossible? Why? (b) What is Keith's velocity with respect to Erica? (c) What is Sai's velocity with respect to Erica?

15. Consider the spacetime diagram shown in Figure 26.18. (a) What are the coordinates of events A and B, according to Erica? (b) What is the spacetime interval for these two events? (c) According to Sai, what is the spatial separation between events A and B? (d) According to Sai, what is the time difference between the events?

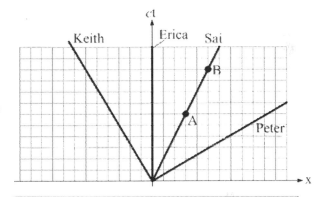

Figure 26.18: A spacetime diagram showing worldlines for four people and two events, A and B. For Exercises 14 and 15.

16. Consider the spacetime diagram in Figure 26.19, which shows worldlines for you and for two other astronauts. All the motion is confined to the x-axis. (a) What is Sasha's velocity with respect to you? (b) What is Michel's velocity with respect to you? (c) What is your velocity with respect to Michel?

17. Consider the spacetime diagram shown in Figure 26.19. (a) Add two events to the spacetime diagram that, according to you, happen at a distance of 5 light-years from you, but which are separated by a time interval of 4 years. (b) Add two events to the spacetime diagram that happen at different times, but at the same location, according to Sasha.

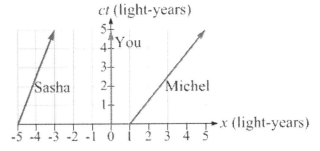

Figure 26.19: A spacetime diagram showing worldlines for three astronauts, for Exercises 16 and 17.

18. Draw a spacetime diagram, with units of meters, showing (a) your worldline, and (b) the worldline of a rocket that, according to you, passes through $x = +100$ m at $t = 0$ and travels in the positive x-direction at 60% of the speed of light. (c) According to you, what is the rocket's location at $t = +600$ meters of time?

Exercises 19 – 22 involve time dilation.

19. According to you, an astronaut's trip from Earth to a distant planet takes 50 years. The distant planet is in the same frame of reference as the Earth, and the astronaut travels at constant velocity between the two planets. (a) How long does the trip take, according to the astronaut, if the astronaut's speed with respect to you is (i) 5% of the speed of light, (ii) 50% of the speed of light, and (iii) 95% of the speed of light? (b) What is the distance between the planets, according to you, for the three different cases in part (a)?

20. Light takes 500 s to travel from the Sun to Earth. In the frame of reference of the light, how much time does the trip take?

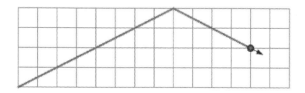

21. Figure 26.20 shows the path followed by a light pulse in Randi's light clock. Randi and her clock are moving at constant velocity to the right with respect to you. What is Randi's speed with respect to you?

Figure 26.20: The path followed by a light pulse in Randi's clock, as observed by you. For Exercise 21. The grid on the diagram is square, according to you.

22. According to you, the distance between the Earth and a distant star is 1000 light-years. You and the star are in the same reference frame as the Earth. (a) According to you, what is the shortest time a message could be sent from the Earth to the star? (b) In the frame of reference of Megan, who is traveling at constant velocity between the Earth and the star, could the trip take only 10 years? If not, explain why not. If so, determine Megan's speed with respect to you. (c) Which observer, you or Megan, measures the proper time between the two events of Megan passing Earth and Megan passing the distant star? Explain.

Exercises 23 – 25 involve length contraction.

23. According to you, the distance between the Earth and a distant star is 1000 light-years. You and the star are in the same reference frame as the Earth. (a) In the frame of reference of Rajon, who is traveling at constant velocity between the Earth and the star, could the Earth and the star be separated by a distance of only 5 light-years? If not, explain why not. If so, determine Rajon's speed with respect to you. (b) Which observer, you or Rajon, measures the proper length between the Earth and the star? Explain.

24. A rocket is passing by the Earth traveling at 80% of the speed of light in a direction parallel to the length of the rocket. As the rocket passes, you measure the rocket's length to be precisely 100 m, using a tape measure in which lines are marked every millimeter. According to an observer moving with the rocket, (a) what is the length of the rocket, and (b) how far apart are the marks on your tape measure?

25. If you examine Figures 26.9 and 26.10 carefully, you will notice that the images of the mirror, and the images of the emitter/detector, are shorter in Figure 26.10 than they are in Figure 26.9. (a) What is the explanation for this? (b) For the particular situation described in Exploration 26.3, how much shorter are the mirrors in Figure 26.10 compared to those in Figure 26.9?

Exercises 26 – 31 involve ideas from General Relativity. In case you are intrigued by Einstein's ideas and you would like to know more, these exercises will give you some starting points for further reading.

26. These days, GPS (Global Positioning System) units can be carried by hikers and sailors, and are built into cars and airplanes, to provide accurate information about someone or something's position on Earth. Do some research regarding how and why the clocks on the GPS satellites are corrected for effects associated with General Relativity, and write a couple of paragraphs about this.

27. In 1960, Robert Pound and Glen Rebka did an interesting experiment at Harvard University to prove that the frequency of light is affected by gravity. Do some reading about their experiment, describing how gravity affects the frequency of light, and how the Pound and Rebka experiment verified the effect.

28. Joseph Taylor and Russell Hulse won the Nobel Prize for Physics in 1993. Do some reading about their research, and write a couple of paragraphs about it. Be sure to mention how their work is connected to General Relativity.

29. LIGO and LISA are acronyms for two large physics experiments that relate to General Relativity. Do some research about them, first to determine what these acronyms stand for, and then so you can write up a couple of paragraphs about how the experiments are designed to work and what they are trying to find.

30. A famous experiment in 1973 to test the predictions of relativity theory involved atomic clocks, which are incredibly accurate time-keepers. One atomic clock was placed on an airplane that circled the Earth going eastward. A second clock was placed on a plane that went westward around the Earth. A third clock was left at the US Naval Observatory. When the clocks were all brought back together, they showed that slightly different time intervals had gone by. Could the principles of physics be used to explain the observations? Was it Special Relativity or General Relativity that was more important in the experiment? Do some reading about the experiment, and write a couple of paragraphs describing the implications of the results.

31. In 1919, Arthur Eddington led an expedition to the island of Príncipe in an effort to test one of Einstein's predictions regarding how light is influenced by gravity. What did the expedition determine, and why did they have to go to Príncipe to do this?

General problems and conceptual questions

32. The spacetime interval between two events is 40 m. The events have a spatial separation of (1) 18 m, according to Gary, (2) 8 m, according to Megan, and (3) 0 m, according to Shawn. What is the time interval, in meters of time, between the events, according to (a) Gary, (b) Megan, and (c) Shawn.

33. Review the discussion of the Rossi and Hall experiment in Section 26.3, and assume that the muons are traveling at 99% of the speed of light. (a) From the point of view of an observer at rest on the Earth, how much time elapses on the clock of an observer who is in the reference frame of the muons? (b) From the point of view of an observer in the reference frame of the muons, what is the vertical distance between the top of Mt. Washington and sea level, and how long does it take a muon to traverse that distance?

34. Let's investigate a light clock that is moving with respect to you, but moving in a direction perpendicular to the mirror instead of parallel to it as in Exploration 26.3. The clock is shown in Figure 26.21. You observe the emitter and the mirror to be separated by 90 cm, and the clock to be moving at 50% of the speed of light. Note that this situation parallels the Brandi and Mia situation from Chapter 4, in which Mia ran on a moving sidewalk. (a) How long does a light pulse take to travel from the emitter to the mirror, according to you? (b) How long does the pulse take to travel back from the mirror to the detector, according to you? (c) What is the total round trip time for the light pulse in the clock, according to you? (d) If you have an identical clock, what is the round trip time for a light pulse in your clock? (e) Is the time dilation equation (Equation 26.2) valid for this situation? Explain why or why not.

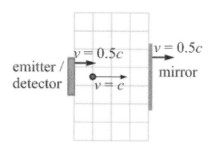

Figure 26.21: A light clock that is moving in a direction perpendicular to the plane of its mirror. For Exercise 34.

35. A spacetime diagram, showing worldlines for four observers whose motion is confined to the x-axis, is shown in Figure 26.22. (a) Which two observers have the same velocity? What is the spatial separation between the two observers with the same velocity, according to (b) Sam, (c) Jen, and (d) Marco?

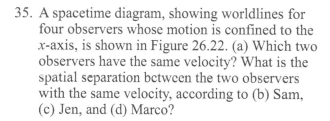

36. (a) What is the spacetime interval between events A and B on the spacetime diagram in Figure 26.22? According to Jen, what is (b) the spatial separation between the two events, and (c) the time interval between the two events? (d) Do any of the other three observers agree with Jen on the answers to (b) and (c)? If so, who agrees with Jen?

Figure 26.22: This spacetime diagram shows worldlines for four observers, whose motion is confined to the x-axis, as well as the spacetime coordinates of two events, in Sam's frame of reference. For Exercises 35 – 36.

37. The spacetime diagram in Figure 26.23 shows the spacetime coordinates of three events, as measured by a particular observer. Determine the spacetime interval between (a) events A and B, (b) events A and C, and (c) events B and C.

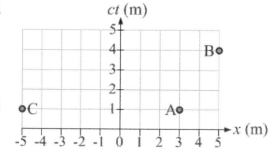

38. The spacetime diagram in Figure 26.23 shows the spacetime coordinates of three events, as measured by a particular observer whose worldline follows the ct axis. (a) According to a second observer, who is traveling at constant velocity along the x-axis with respect to the first observer, events A and B occur at the same location. What is the speed of the second observer with respect to the first observer? (b) According to a third observer, who is traveling at constant velocity along the x-axis with respect to the first observer, events C and B occur simultaneously. What is the speed of the third observer with respect to the first observer?

Figure 26.23: A spacetime diagram showing the spacetime coordinates of three events, as measured by a particular observer. For Exercises 37 and 38.

39. You are standing on a train platform that has a length of 150 m, according to you. According to a passenger on the train, the train has a length of 250 m. As the train passes through the station at very high speed, however, you observe the length of the train to be exactly the same as the length of the platform, 150 m. If we define event A to be the front of the train passing one end of the platform, and event B as the rear of the train passing the other end of the platform, you observe that the two events are simultaneous. (a) What is the speed of the train, according to you? (b) What is the spacetime interval between the events A and B? (c) What is the time interval between the two events, according to the passenger on the train?

40. You are on a train that is traveling at 70% of the speed of light relative to the ground. Your friend is at rest on the ground, a safe distance from the track, watching the train go past. According to you, the train has a length of 300 m, and the distance between the two cities the train is traveling between is 1200 km. What does your friend measure for (a) the length of the train, and (b) the distance between the two cities the train is traveling between.

41. Return to Exercise 40. Assuming the velocity of the train is constant, what is the time taken by the train to travel between the two cities according to (a) you, and (b) your friend?

42. *An introduction to the pole-and-barn paradox.* A particular pole is 10 m long, according to Paul, who is at rest with respect to the pole. According to you, the distance between the front and back doors of a barn is only 5 m. However, Paul runs fast enough, while holding the pole parallel to the ground, that the pole appears to be just under 5 m long, according to you. Thus, with the barn's front door open and back door closed, Paul can run through the front door of the barn, and you can close the front door and your assistant can simultaneously open the back door so that Paul passes through the barn without any trouble. The apparent paradox comes when the situation is looked at from Paul's perspective. According to Paul, (a) what is the length of the pole, and (b) what is the distance between the two barn doors? (c) Is it possible for Paul's pole, according to Paul, to be entirely inside the barn, as you saw it to be? (d) According to you, Paul can pass through the barn without the pole hitting either door. According to Paul, is this possible? Either explain this apparent paradox yourself, or do some background reading about the barn and pole paradox to resolve it.

43. Table 26.2 gives the readings on the clocks of three observers, as recorded by one of the three observers. The relative velocities between the observers remain constant at all times. (a) Complete the table, filling in the missing readings. (b) Which of the three observers is recording the clock readings? Explain. (c) What is Sarah's speed with respect to Josh?

	Josh	Rachel	Sarah
Event A	2 hours	0	2 hours
Event B	5 hours		6 hours
Event C	11 hours	3 hours	

Table 26.2: A table of clock readings for various events, as recorded by one of the three observers.

44. Let's return to the Isabelle, you, and Yan situation that was introduced in Section 26-4, and elaborated on in Section 26-5. You and Yan, who are in the same reference frame but separated by 40 light-years, according to you, agree that you will set your clocks to zero when a light pulse, emitted from the point midway between you and Yan, reaches you. The light travels at the speed of light in vacuum. Isabelle, who is traveling at 80% of the speed of light while traveling from Earth, where you are, to Zorg, where Yan is, happens to pass you at the same time the light pulse reaches you, so she sets her clock to read zero at the same time you do. (a) According to you, how long does it take the light pulse to reach you after it is emitted? (b) According to Isabelle, how far is it from the point where the light pulse was emitted to you? Isabelle agrees that the point where the light pulse is emitted is halfway between you and Yan. (c) According to Isabelle, how long does the light pulse take, after being emitted, to reach Yan? Note that, according to Isabelle, the light pulse and Yan are both moving. (d) According to Isabelle, how long does the light pulse take, after being emitted, to reach you? (e) On Isabelle's clock, how much time elapses between the light pulse reaching Yan and the light pulse reaching you? (f) According to Isabelle, how much time elapses on Yan's clock between the light pulse reaching Yan and the light pulse reaching you? Note that this answer should agree with the information in Section 26-5.

45. Return to Exercise 44. Sketch a spacetime diagram, from your frame of reference, for the situation described in Exercise 44. Draw worldlines for you, Yan, Isabelle, the light pulse that travels from the midpoint between you and Yan to you, and the light pulse that travels from the midpoint between you and Yan to Yan.

46. Repeat Exercise 45, but now sketch the spacetime diagram for Isabelle's frame of reference.

47. You are traveling along the intergalactic freeway in your personal rocket. A large transporter in the next lane, which has the same velocity as you, is exactly 4 times the length of your rocket. An identical transporter in the oncoming lane, however, appears to be the same length as your rocket. Relative to you, at what speed is transporter in the oncoming lane traveling?

48. Two students are considering a particular question. Comment on the part of their conversation that is reported below.

Erin: The question says "As the rocket goes by at 75% of the speed of light, observers in the Earth's reference frame mark the locations of the tip and tail of the rocket at the same time. They then measure the distance between the marks to be 60 meters. What is the length of the rocket?" Uh, 60 meters, right?

Katie: Doesn't the speed have something to do with it? Don't we have to use the length contraction equation to find the length of the rocket in the rocket's frame of reference?

Erin: Well, the question doesn't really say, right? It asks for the length of the rocket, but according to who? Who's the observer?

Katie: What if we did two answers, one 60 meters, and the other the length that's seen by somebody in the rocket?

Erin: OK, I'll go with that. Now, length gets contracted, so the other length should be less than 60 meters, right?

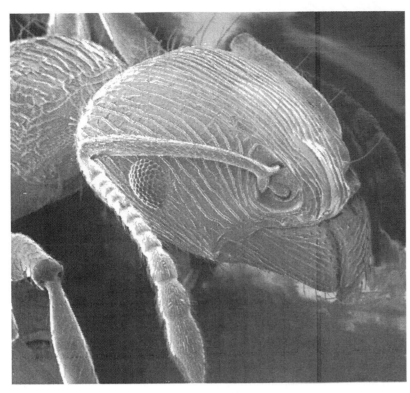

This image of the head of an ant was obtained with an electron microscope. Although we often think of electrons as particles, electron microscopes exploit the wave nature of electrons to obtain incredibly detailed pictures like this. One of the key ideas of this chapter is wave-particle duality – everything exhibits both a wave and a particle nature.

Image credit: Photo courtesy of the U. S. Geological Survey.

Chapter 27 – The Quantum World

CHAPTER CONTENTS

Our modern world is full of applications of quantum physics, including the laser that reads the disk in your DVD player or scans the bar codes on items you buy at the store, the solar cells that power your calculator, and the light-emitting diodes (LEDs) that are being increasingly used in traffic lights, car brake lights, and home lighting. It is amazing how much we depend on the quantum nature of matter when, just 100 years ago, the study of quantum physics was just beginning.

Quantum physics is, essentially, the study of the behavior of tiny things such as electrons, and is interesting because this behavior is so different from the behavior of the everyday objects we are used to dealing with. When we throw a baseball, we can predict very accurately what it will do. If we throw the baseball the same way over and over, the path it follows should be similar every time. The behavior of electrons is not nearly so predictable. Given the same set of initial conditions, an electron may do many different things. We can calculate the various probabilities for the different electron behaviors, but in a particular situation we may not be able to know exactly what the electron will do.

As with Special Relativity, quantum physics will bring with it some surprising results. Not the least of these is that we will need a new model of light. It turns out that light, in certain cases, acts as if it is made up of particles, which is very much at odds with the wave model of light we spent all of Chapter 25 discussing.

27-1 Planck Solves the Ultraviolet Catastrophe

By the end of the 19th century, most physicists were confident that the world was well understood. Aside from a few nagging questions, everything seemed to be explainable in terms of basic physics such as Newton's laws of motion and Maxwell's equations regarding electricity, magnetism, and light. This confidence was soon to be shaken, however.

One of the nagging questions at the time concerned the spectrum of radiation emitted by a so-called **black body**. A perfect black body is an object that absorbs all radiation that is incident on it. Perfect absorbers are also perfect emitters of radiation, in the sense that heating the black body to a particular temperature causes the black body to emit radiation with a spectrum that is characteristic of that temperature. Examples of black bodies include the Sun and other stars, light-bulb filaments, and the element in a toaster. The colors of these objects correspond to the temperature of the object. Examples of the spectra emitted by objects at particular temperatures are shown in Figure 27.1.

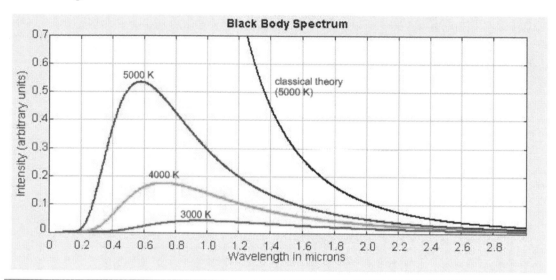

Figure 27.1: The spectra of electromagnetic radiation emitted by hot objects. Each spectrum corresponds to a particular temperature. The black curve represents the predicted spectrum of a 5000 K black body, according to the classical theory of black bodies.

At the end of the 19th century, the puzzle regarding blackbody radiation was that the theory regarding how hot objects radiate energy predicted that an infinite amount of energy is emitted at small wavelengths, which clearly makes no sense from the perspective of energy conservation. Because small wavelengths correspond to the ultraviolet end of the spectrum, this puzzle was known as the **ultraviolet catastrophe**. Figure 27.1 shows the issue, comparing the theoretical predictions to the actual spectrum for an object at a temperature of 5000 K. There is clearly a substantial disagreement between the curves.

The German physicist Max Planck (1858 – 1947) was able to solve the ultraviolet catastrophe through what, at least at first, he saw as a mathematical trick. This trick, which marked the birth of quantum physics, also led to Planck being awarded the Nobel Prize for Physics in 1918. Planck determined that if the vibrating atoms and molecules were not allowed to take on any energy, but instead were confined to a set of equally-spaced energy levels, the predicted spectra matched the experimentally determined spectra extremely well. Planck determined that, for an atom oscillating with a frequency f, the allowed energy levels were integer multiples of the base energy unit hf, where Planck's constant h has the value 6.626×10^{-34} J s.

$E = nhf$,   (Equation 27.1: **Allowed energy levels for an oscillator in a blackbody**)

where n is an integer.

Thus was born the idea of quantization, as applied to energy. If a quantity is quantized, it can take on only certain allowed values. Charge, as we discussed in chapter 18, is an example of something that is quantized, coming in integer multiples of the electronic charge e. Money is an example of an everyday item that is quantized, with quantities of money coming in integer units of a base unit, such as the penny in the United States and Canada.

Let us turn now to a second physical phenomenon that was puzzling scientists at the end of the 19th century. This phenomenon is called the **photoelectric effect**, and it describes the emission of electrons from metal surfaces when light shines on the metal. The photoelectric effect, or similar effects, have a number of practical applications, including the conversion of sunlight into electricity in solar panels, as well as the image-sensing systems in digital cameras.

Let's put the photoelectric effect experiment into context. First, recall that, beginning in 1801 with Thomas Young's double-slit experiment, physicists carried out a whole sequence of experiments that could be explained in terms of light acting as a wave. All these interference and diffraction experiments showed that light was a wave, and this view was supported theoretically by the prediction of the existence of electromagnetic waves, via Maxwell's equations. Then, in 1897, J.J. Thomson demonstrated that electrons exist and are sub-atomic particles. The stage was set for an explanation of the photoelectric effect in terms of light acting as a wave.

Predictions of the wave model of light regarding the photoelectric effect
The explanation for how light, as a wave, might interact with electrons in a metal to knock them out of the metal is fairly straightforward, based on the absorption of energy from the electromagnetic wave by the metal. Note that all metals have what is known as a **work function**, which is the minimum energy required to liberate an electron from the metal. Essentially, the work function represents the binding energy for the most weakly bound electrons in the metal.

Remember that the intensity of an electromagnetic wave is defined as the wave's power per unit area. Predictions based on the wave model of light include:
- Light (that is, electromagnetic waves) of any intensity should cause electrons to be emitted. If the intensity is low, it will just take longer for the metal to absorb enough energy to free an electron.
- The frequency of the electromagnetic waves should not really matter. The key factor governing electron emission should be the intensity of the light.
- Increasing intensity means more energy per unit time is incident on a given area, and thus we might expect both more electrons to be emitted and that the emitted electrons would have more kinetic energy.

Amazingly, despite a century of success in explaining many experiments, the predictions of the wave model of light are completely at odds with experimental observations Again, as we will discover in Section 27-2, it took the intellect of Albert Einstein to explain what was going on.

Related End-of-Chapter Exercises: 1, 2, 36, 37.

Essential Question 27.1: In Figure 27.1, we can see that the intensity of light emitted by an object at 5000 K has a maximum at a wavelength of about 0.6 microns (600 nm). (a) What frequency does this correspond to? (b) What is difference between energy levels at this frequency?

Answer to Essential Question 27.1: (a) Assuming the wavelength is measured in vacuum, we can use the wave equation $f = c / \lambda$ to find that the frequency corresponding to a wavelength of 600 nm is $f = c / \lambda = (3.00 \times 10^8 \text{ m/s}) / (6 \times 10^{-7} \text{ m}) = 5 \times 10^{14} \text{ Hz}$. (b) Applying equation 27.1, with $n = 1$, we find that the difference between energy levels is extremely small, being $E = hf = (6.626 \times 10^{-34} \text{ J s}) \times (5 \times 10^{14} \text{ Hz}) = 3 \times 10^{-19} \text{ J}$.

27-2 Einstein Explains the Photoelectric Effect

On the previous page, we introduced the photoelectric effect (the emission of electrons from a metal caused by light shining on the metal), and discussed how the wave theory of light led to predictions about the experiment that simply did not fit the experimental observations. It was at this point, in 1905, that Albert Einstein stepped in. First, Einstein built on Planck's explanation of the spectrum of a black body. Planck had theorized that oscillators (such as atoms) in a black body could only take on certain energies, with the energy levels separated by an energy hf, where f is the oscillation frequency. Einstein went on to propose that when such an oscillator dropped from one energy level to the next lowest level, losing an energy hf, the missing energy was given off as light, but given off as a packet of energy. Such packets of energy now go by the name *photon*. In some sense, then, a photon is like a particle of light, with an energy given by

$$E = hf \, , \qquad \text{(Equation 27.2: \textbf{Energy of a photon})}$$

where f is the frequency of the electromagnetic wave corresponding to the photon. Note how the wave and particle properties of light are brought together in this equation – the energy of a particle of light depends on the light's frequency.

In applying the photon concept to the photoelectric effect, Einstein modeled the process not as a wave interacting with a metal, but as interactions between single photons and single electrons. If the light incident on the metal has a frequency f, then the beam of light can be thought of as being made up of a stream of photons, each with an energy of hf. If each photon is absorbed by a single electron, giving up its energy to the electron, electrons are emitted from the metal as long as the energy an electron acquires from a photon exceeds the metal's work function, W_0, which represents the minimum binding energy between the electrons and the metal.

Predictions of the particle model of light regarding the photoelectric effect

For the predictions of the wave model, refer to the previous page. The particle model makes quite different predictions for the photoelectric effect than does the wave model, including:
- The frequency corresponding to $hf = W_0$ is a critical frequency (known as the *threshold frequency*) for the experiment. Below this frequency, the energy of the photons is not enough for the electrons to overcome the work function. No electrons are emitted when the frequency of the incident light is less than the threshold frequency. Electrons are emitted only when the frequency of the light exceeds the threshold frequency.
- Treating the process as a single photon – single electron interaction leads to a straightforward equation governing the process that is based on energy conservation. When the frequency of the light exceeds the threshold frequency, part of the photon energy goes into overcoming the binding energy between the electron and the metal, with the energy that remains being carried away by the electron as its kinetic energy. Thus,

$$K_{\max} = hf - W_0 \, . \qquad \text{(Equation 27.3: \textbf{The photoelectric effect})}$$

- Increasing the intensity of the incident light without changing its frequency means that more photons are incident per unit time on a given area. If the light frequency is below the threshold frequency, no electrons are emitted no matter what the intensity is. If the frequency exceeds the threshold frequency, increasing the intensity causes more electrons to be emitted, but the maximum kinetic energy of the emitted electrons does not change.

It took several years for these predictions to be verified; however, by 1915 experiments showed clearly that Einstein was correct, confirming that light has a particle nature. For his explanation of the photoelectric effect, Einstein was awarded the Nobel Prize in Physics in 1921.

EXPLORATION 27.2 – The photoelectric effect experiment
Let's look at one method for carrying out the photoelectric effect experiment. This will involve a review of some of the concepts from Chapter 17, such as electric potential and the workings of a capacitor.

Step 1 – *A diagram of the experimental apparatus is shown in Figure 27.2. Light of a frequency higher than the work function shines on a metal plate (plate 1), causing electrons to be emitted. These electrons are collected by a second plate (plate 2), and the electrons travel back through a wire from plate 2 to plate 1. An ammeter measures the current in the wire, while an adjustable battery, with a voltage set initially to zero, is also part of the circuit. Explain, using conservation of energy and concepts from Chapter 17, how adjusting the battery voltage enables us to measure the maximum kinetic energy of the emitted electrons. Neglect gravity.*

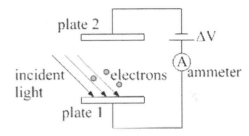

Figure 27.2: A diagram of the experimental apparatus for carrying out the photoelectric effect experiment.

If the battery is connected so plate 2 is negative and plate 1 is positive, electrons emitted from plate 1 do not reach plate 2 unless the kinetic energy they have when they leave plate 1 exceeds the change in electric potential energy, $e\,\Delta V$, associated with electrons crossing the gap from plate 1 to plate 2. The electrons that do not make it across the gap return to plate 1. Thus, as the battery voltage increases from zero, fewer electrons cross the gap and the ammeter reading drops. The smallest battery voltage ΔV_{min} needed to bring the current (and the ammeter reading) to zero is directly related to the maximum kinetic energy of the emitted electrons. By energy conservation, the potential energy of the electrons just below plate 2 is equal to the kinetic energy they have at plate 1 (defining plate 1 as the zero for potential energy). In equation form, we have $K_{max} = e\,\Delta V_{min}$.

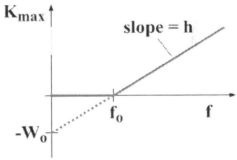

Step 2 – *Sketch a graph showing the maximum kinetic energy of the electrons as a function of the frequency of the incident light, both above and below the threshold frequency f_0. This graph, which matches Equation 27.3, is shown in Figure 27.3.*

Figure 27.3: A graph of the maximum kinetic energy of electrons emitted when light shines on a metal, as a function of the frequency of the light. Below the threshold frequency, no electrons are emitted. Above the threshold frequency, the maximum kinetic energy of the emitted electrons increases linearly with frequency.

Key idea: To explain the photoelectric effect experiment, we treat light as being made up of particles called photons.   **Related End-of-Chapter Exercises: 4, 5, 42, 46.**

Essential Question 27.2: Plot a graph like that in Figure 27.3, but for a different metal that has a larger threshold frequency. Comment on the similarities and differences between the two graphs.

The new graph, corresponding to a larger threshold frequency, is shown as the rightmost graph in Figure 27.4. For frequencies above the threshold frequency, the slopes are identical because the slope of this section of the graph is Planck's constant. Also, the new line has a y-intercept that is more negative, consistent with the larger work function in that case.

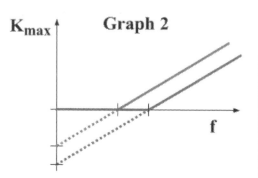

Figure 27.4: Two graphs showing the maximum kinetic energy for electrons emitted by light incident on two different metals. For the rightmost graph, the threshold frequency is larger, as is the work function of the metal, compared to that for the original graph.

27-3 A Photoelectric Effect Example

In this section, we will do an example of a photoelectric effect problem. Let us begin, though, by discussing the electron volt (eV), which is a unit of energy that is often used in settings involving the photoelectric effect.

The electron volt

How much kinetic energy does an electron gain, or lose, when it is accelerated through a potential difference of 1 volt? The electron's change in kinetic energy is equal in magnitude, and opposite in sign, to the change in electric potential energy it experiences. Thus, the electron's change in kinetic energy is $\Delta K = -(-e)\Delta V = e\Delta V$. If we used the electronic charge in coulombs

($e = 1.60 \times 10^{-19}$ C) and kept the potential difference in volts, we would obtain an energy in joules, but it would be a very small number.

As an alternative, we can define a different energy unit, the electron volt, such that an electron accelerated through a potential difference of 1 volt experiences a change in kinetic energy that has a magnitude of 1 electron volt (eV). If the potential difference is 500 volts, the electron's change in kinetic energy has a magnitude of 500 eV, etc. The conversion factor between joules and electron volts is

$1 \text{ eV} = 1.60 \times 10^{-19}$ J .   (Eq. 27.4: **Conversion factor between joules and electron volts**)

We will make use of the electron volt in the following example. It is also helpful to express Planck's constant in eV s, which can be done as follows:

$$h = \frac{6.626 \times 10^{-34} \text{ J s}}{1.602 \times 10^{-19} \text{ J/eV}} = 4.136 \times 10^{-15} \text{ eV s} .$$ (Eq. 27.5: **Planck's constant in eV s**)

EXAMPLE 27.3 – Solving problems involving the photoelectric effect
Using the experimental apparatus shown in Figure 27.5, when ultraviolet light with a wavelength of 240 nm shines on a particular metal plate, electrons are emitted from plate 1, crossing the gap to plate 2 and causing a current to flow through the wire connecting the two plates. The battery voltage is gradually increased until the current in the ammeter drops to zero, at which point the battery voltage is 1.40 V.

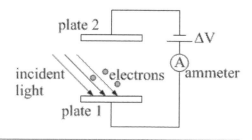

Figure 27.5: A diagram of the experimental apparatus for carrying out the photoelectric effect experiment.

(a) What is the energy of the photons in the beam of light, in eV?
(b) What is the maximum kinetic energy of the emitted electrons, in eV?
(c) What is the work function of the metal, in eV?
(d) What is the longest wavelength that would cause electrons to be emitted, for this particular metal?
(e) Is this wavelength in the visible spectrum? If not, in what part of the spectrum is this light found?

SOLUTION
(a) Assuming that the wavelength corresponds to the wavelength in vacuum, we can first convert the wavelength to the frequency using

$$f = \frac{c}{\lambda} = \frac{3.00 \times 10^8 \text{ m/s}}{2.40 \times 10^{-7} \text{ m}} = 1.25 \times 10^{15} \text{ Hz} .$$

Now, we can use Equation 26.2 to find the photon energy.

$$E = hf = (4.136 \times 10^{-15} \text{ eV s}) \times (1.25 \times 10^{15} \text{ Hz}) = 5.17 \text{ eV} .$$

(b) As we discussed in Exploration 27.2, the maximum kinetic energy of the emitted electrons is related to the minimum voltage across the two plates needed to stop the electrons from reaching the second plate (this is known as the **stopping potential**). In this case, the stopping potential is 1.40 V, so the maximum kinetic energy of the electrons is 1.40 eV.

(c) Now that we know the photon energy and the maximum kinetic energy of the electrons, we can use Equation 27.3 to find the work function of the metal.
$$W_0 = hf - K_{max} = 5.17 \text{ eV} - 1.40 \text{ eV} = 3.77 \text{ eV} .$$

(d) The maximum wavelength that would cause electrons to be emitted corresponds to the threshold frequency for this situation. Let's first determine the threshold frequency, f_0.
$$W_0 = hf_0, \quad \text{so} \quad f_0 = \frac{W_0}{h} = \frac{3.77 \text{ eV}}{4.136 \times 10^{-15} \text{ eV s}} = 9.12 \times 10^{14} \text{ Hz} .$$

Converting the threshold frequency to wavelength, assuming the light is traveling in vacuum, gives
$$\lambda_{max} = \frac{c}{f_0} = \frac{3.00 \times 10^8 \text{ m/s}}{9.12 \times 10^{14} \text{ Hz}} = 3.29 \times 10^{-7} \text{ m} .$$

(e) This wavelength is 329 nm, less than the 400 nm (violet) wavelength that marks the lower bound of the visible spectrum. This light is beyond violet, in the ultraviolet.

Related End-of-Chapter Exercises: 13 – 16, 43, 44.

Essential Question 27.3: With a particular metal plate, shining a beam of red light on the metal causes electrons to be emitted. (a) If we replace the red light by blue light, do we know that electrons will be emitted? (b) If the two beams have the same intensity and are incident on equal areas of the plate, do we get the same number of electrons emitted per second in the two cases? Assume that in both cases the probability that a photon will cause an electron to be emitted is the same in both cases (e.g., for every two photons incident on the plate, one electron is emitted).

Answer to Essential Question 27.3: (a) Blue light has a higher frequency than red light, and thus the photons in the blue light have a higher energy than the photons in the red light. We know that the photons in the red light have an energy larger than the metal's work function, because electrons are emitted, so the photons in the blue light have more than enough energy to cause electrons to be emitted, too. (b) We actually get fewer electrons emitted with the blue light. The energy per photon is larger for the blue light, so to achieve the same intensity there are fewer photons per second per unit area incident on the plate with the blue light. Fewer photons produce fewer electrons, although the electrons are emitted with, on average, more kinetic energy.

27-4 Photons Carry Momentum

In Chapter 6, we defined an object's momentum as the product of the object's mass and its velocity. Photons have no mass, so we might expect them to also have no momentum. However, we have already discussed the fact that light does carry momentum – recall the discussion of radiation pressure and solar sailboats in Chapter 22. Thus, photons do have momentum, but we need an equation that gives momentum for massless particles, keeping in mind that the units must be the familiar units of momentum that we have worked with throughout the book.

$$p = \frac{E}{c} = \frac{hf}{c} = \frac{h}{\lambda}.$$ (Equation 27.6: **Momentum of a photon**)

One of the key pieces of evidence supporting the photon model of light is an experiment involving light interacting with matter. When light of a particular frequency is incident on matter, the light can change both direction and frequency. The shift in frequency cannot be explained in terms of the wave model of light, but the particle (photon) model provides quite a straightforward explanation. The phenomenon is known as the Compton effect after its discoverer, the American physicist A. H. Compton (1892 – 1962), for which he won the 1927 Nobel Prize in Physics.

The explanation of the Compton effect is very similar to that of the photoelectric effect. In both cases, a single photon interacts with a single electron. With the Compton effect, the analysis is similar to our analysis of a two-dimensional collision in Section 7.6, except that, in this situation, one of the objects (the electron) is initially at rest. As always in a collision situation, momentum is conserved. In a collision like this, involving a photon and an electron, there is nothing to transfer energy out of the system, so energy is also conserved. A diagram of the collision is shown in Figure 27.6.

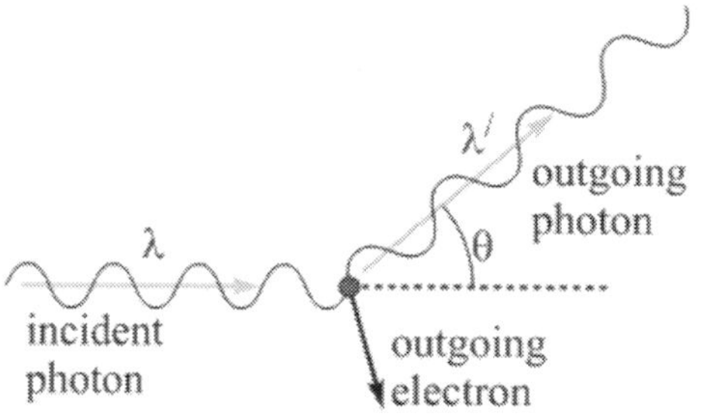

Figure 27.6: In the Compton effect, an incident photon of wavelength λ collides with an electron (the gray sphere). After the collision, the wavelength of the photon is λ′.

The photon is incident with a wavelength λ, and the photon transfers some of its energy to the electron. Thus, after the collision, the photon has lower energy, which means its frequency is lower while its wavelength λ′ is higher. The photon's direction changes by an angle θ. Applying energy conservation to this situation gives one equation, while applying momentum conservation in two dimensions gives us two more equations. Combining these equations is somewhat involved, but it leads to a simple relationship relating the change in wavelength experienced by the photon to the angle through which the photon has been scattered.

$$\Delta\lambda = \lambda' - \lambda = \frac{h}{m_e c}(1 - \cos\theta),$$ (Equation 27.7: **The Compton Effect**)

where m_e is the mass of the electron, and the quantity $h/(m_e c) = 2.43 \times 10^{-12}$ m is known as the Compton wavelength. Because $\cos\theta$ varies from +1 to −1, the quantity $(1- \cos\theta)$ varies from 0 to 2. Thus, the shift in the photon's wavelength from the Compton effect varies from 0 (for a scattering angle of 0, which is essentially no collision) to two Compton wavelengths, for a scattering angle of 180°.

EXAMPLE 27.4 – Working with the Compton effect
A photon with a wavelength of 4.80×10^{-11} m collides with an electron that is initially stationary. As shown in Figure 27.7, the photon emerges traveling in a direction that is at 120° to the direction of the incident photon.

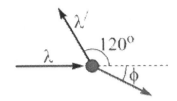

(a) What is the wavelength of the photon after the collision with the electron?
(b) What is the magnitude of the momentum of the incident photon? What is the magnitude of the momentum of the outgoing photon?
(c) What is the magnitude of the electron's momentum after the collision?

Figure 27.7: The geometry of the Compton effect situation described in Example 27.4.

SOLUTION
(a) To find the wavelength of the outgoing photon, we apply Equation 27.7. This gives:

$$\lambda' = \lambda + \frac{h}{m_e c}(1-\cos\theta) = (4.80\times10^{-11} \text{ m}) + (2.43\times10^{-12} \text{ m})\times(1+0.5) = 5.16\times10^{-11} \text{ m}.$$

(b) To find a photon's momentum, we can apply Equation 27.6, $p = h/\lambda$.

For the incident photon, $p = \dfrac{h}{\lambda} = \dfrac{6.626\times10^{-34} \text{ J s}}{4.80\times10^{-11} \text{ m}} = 1.380\times10^{-23}$ kg m/s .

For the outgoing photon, $p' = \dfrac{h}{\lambda'} = \dfrac{6.626\times10^{-34} \text{ J s}}{5.16\times10^{-11} \text{ m}} = 1.284\times10^{-23}$ kg m/s

(c) Momentum is conserved in the collision, so we can apply momentum conservation to find the x and y components of the electron's momentum.

x-direction : $p = p'_x + p_{e,x}$ so

$p_{e,x} = p - p'_x = (1.380\times10^{-23} \text{ kg m/s}) - (1.284\times10^{-23} \text{ kg m/s})\cos(120°) = 2.022\times10^{-23}$ kg m/s.

y-direction : $0 = p'_y - p_{e,y}$ so

$p_{e,y} = p'_y = (1.284\times10^{-23} \text{ kg m/s})\sin(120°) = 1.112\times10^{-23}$ kg m/s.

Applying the Pythagorean theorem, we can find the magnitude of the electron's momentum.

$p_e = \sqrt{(p_{e,x})^2 + (p_{e,y})^2} = \sqrt{(2.022\times10^{-23} \text{ kg m/s})^2 + (1.112\times10^{-23} \text{ kg m/s})^2} = 2.308\times10^{-23}$ kg m/s

Related End-of-Chapter Exercises: 6, 7, 17 – 20, 49 – 53.

Essential Question 27.4: Return to Example 27.4. (a) What is the electron's speed after the collision? (b) If the direction of the electron's velocity after the collision makes an angle φ with the direction of the incident photon, what is φ?

Answer to Essential Question 27.4: (a) To find the electron's speed, we can divide the magnitude of the electron's momentum by the electron's mass.

$$v_e = \frac{p_e}{m} = \frac{2.308 \times 10^{-23} \text{ kg m/s}}{9.11 \times 10^{-31} \text{ kg}} = 2.53 \times 10^7 \text{ m/s}.$$

This speed is less than 10% of the speed of light in vacuum, so we are safe applying the non-relativistic equations that we used in Example 27.4.

(b) We can find the direction of the electron's momentum from the components of the momentum.

$$\tan\phi = \frac{p_{e,y}}{p_{e,x}} = \frac{1.112 \times 10^{-23} \text{ kg m/s}}{2.022 \times 10^{-23} \text{ kg m/s}} \qquad \text{which gives } \phi = 28.8°.$$

27-5 Particles Act Like Waves

As we have learned in the previous two sections, light, which in many instances acts like a wave, also exhibits a particle nature. This was a rather surprising result, but the surprises kept coming. Recall that equation 27.6 relates the momentum of a photon to the wavelength of the light, $p = h/\lambda$. In 1923, Louis de Broglie (1892 – 1987) proposed turning the equation around and applying it to objects that we normally think of as particles.

The de Broglie wavelength: de Broglie proposed that Equation 27.6 could be applied to objects we think of as particles, such as electrons and neutrons, in the form:

$$\lambda = \frac{h}{p} = \frac{h}{mv}.$$ (Equation 27.8: **the de Broglie wavelength**)

In other words, de Broglie proposed that everything that moves has an associated wavelength. When de Broglie's idea was verified, de Broglie was awarded the Nobel Prize in Physics in 1929 for the idea.

For objects that we are used to dealing with in our daily lives, such as balls and cars and people, their de Broglie wavelength is so small that there is, effectively, no wave behavior. For instance, when you pass through a door you simply go through as a particle, rather than diffracting into the next room. For a person traveling at about 1 m/s, for instance, the de Broglie wavelength is about 1×10^{-35} m. This wavelength is so many orders of magnitude smaller than the objects and openings that we encounter everyday that our particle nature dominates.

In contrast, for tiny objects that we generally think of as particles, such as electrons and protons, their tiny mass produces a much larger wavelength. An electron with a kinetic energy of 10 eV, for instance, has a de Broglie wavelength of 3.9×10^{-10} m. That sounds small, but that wavelength is comparable in size to objects that an electron encounters. For instance, the spacing between atoms in a solid object is similar to the de Broglie wavelength of a 10 eV electron, and thus a crystal, with its regular array of atoms, can act as a diffraction grating for electrons. A sample diffraction pattern obtained by diffracting electrons from a crystal is shown in Figure 27.8. Similar patterns are obtained from diffraction with light.

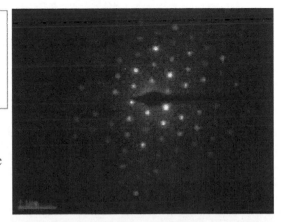

Figure 27.8: An electron diffraction image obtained from electrons incident on a sample of anthophyllite asbestos. From this pattern, the crystal structure, including the spacing between the atoms, can be deduced. Image credit: California Department of Public Health.

Experimental evidence for the de Broglie wavelength

In addition to the electron diffraction images that we just discussed, another persuasive piece of evidence supporting the idea of the de Broglie wavelength is the interference pattern obtained when electrons are incident on a double slit. Light of a particular wavelength produces a particular interference pattern when incident on a double slit. Replacing the light with electrons, which have a de Broglie wavelength equal to that of the wavelength of the light, results in the same interference pattern.

One application of the wave nature of electrons is in microscopy. One factor that limits the resolution of a light microscope is the wavelength of the light used. The same rules apply to an electron microscope, but the de Broglie wavelength of the electrons in an electron microscope can be 1000 times less than that of visible light. Such an electron microscope can resolve features 1000 times smaller than those resolved in a light microscope. An example of what an electron microscope can do is shown in Figure 27.9.

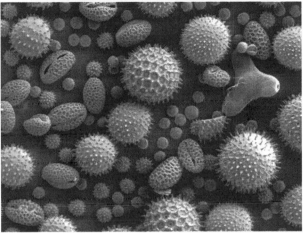

Figure 27.9: An electron microscope image showing various types of pollen. Image credit: Dartmouth Electron Microscope Facility.

Experiments with other particles, including protons, neutrons, and hydrogen and helium atoms, have also been carried out, all of which have verified that such objects exhibit both a particle nature and a wave nature, with a wavelength given by the de Broglie equation.

Wave-particle duality

The fact that everything, including ourselves, exhibits both a particle nature and a wave nature is known as wave-particle duality. Typically, to explain the result of a particular experiment, we use either the wave model or the particle model. However, recent experiments have shown interesting mixes of both. When electrons are incident on a double slit, for example, an interference pattern is produced on a screen beyond the slits – this shows the wave nature of electrons. If you add a detector that tells you which slit the electron passed through in every case, however, the pattern changes to that expected for particles. If your detector is faulty, however, and only tells you 30% of the time which slit an electron passed through, the resulting pattern is a mix of that expected from the wave nature (contributing 70% to the pattern, in this case) and that expected from the particle nature (the other 30%). This experiment also demonstrates how the act of making an observation can change the results of an experiment.

Related End-of-Chapter Exercises: 10, 21 – 25, 57.

Essential Question 27.5: Protons have a mass approximately 1800 times larger than the mass of the electron. If electrons traveling at a speed v produce a particular interference pattern when they encounter a particular double slit, with what speed would the protons have to travel to produce exactly the same interference pattern using the same double slit?

Answer to Essential Question 27.5: The interference pattern produced depends on the de Broglie wavelength of the particles incident. To produce the same pattern as the electrons, the protons must have the same wavelength. Examining Equation 27.8, we see that wavelength is determined by momentum, so the protons need to have the same momentum as the electrons. To have the same momentum with a factor of 1800 in the mass, the protons must have a speed of $v/1800$.

27-6 Heisenberg's Uncertainty Principle

The interplay between the uncertainty in an object's position, Δx, and the uncertainty in its momentum, Δp, was first quantified by the German physicist Werner Heisenberg (1901 – 1976). Heisenberg won the Nobel Prize in Physics in 1932 for his contributions to quantum theory. Heisenberg's uncertainty principle states that you cannot know something to infinite accuracy. More specifically, the uncertainty principle states that for two linked quantities, such as the position and momentum of an object, the more accurately you know one of those quantities, the less accurately you can know the other.

$$\Delta x\, \Delta p \geq \frac{h}{4\pi} = 5.273 \times 10^{-35}\ \text{J s}\,. \qquad \text{(Equation 27.9: \textbf{Heisenberg's uncertainty principle})}$$

The uncertainty principle also applies to other linked quantities, such as energy and time, or to two components of angular momentum.

For any object that we are used to dealing with on a daily basis, Heisenberg's uncertainty principle is essentially irrelevant. For instance, if you could measure the position of a 1 kg water bottle, which you see as being at rest on a table, with an uncertainty of 1 nm, Heisenberg's uncertainty principle tells us that the uncertainty in the water bottle's momentum must be at least 5.273×10^{-26} kg m/s. With its mass being 1 kg, the water bottle's velocity must therefore be uncertain by at least 5.273×10^{-26} m/s, a number so small that it is essentially meaningless.

For tiny objects like electrons, however, the limitations associated with Heisenberg's uncertainty principle are quite important. For the electron bound to the nucleus of a hydrogen atom, for instance, making a reasonable assumption about the uncertainty of the electron's momentum leads to an uncertainty in position that is similar in size to the atom itself. Thus, for the hydrogen atom, we can say that the electron is in the atom, but we cannot say exactly where in the atom it is at any point in time. We will investigate this idea further in Chapter 28.

Applying the Uncertainty Principle to the single-slit experiment
As we discussed earlier in this chapter, electrons interact with single and double slits in much the same way that light does. In both cases, the resulting diffraction pattern or interference pattern can be understood in terms of the wave nature of the electrons or of the light. However, applying the uncertainty principle can also give us some insight into the experiment.

Imagine a beam of electrons traveling toward a wide slit. We have a good idea of the speed and direction of the electrons in the beam – in other words, the momentum of the electrons is well known. The slit is so wide that, while we know that the electrons pass through the slit, we do not have much information about their position. With a large uncertainty in position, according to the Uncertainty Principle, the uncertainty in momentum can be small. Thus, the momentum of each electron is essentially unchanged from what it was before encountering the slit, and the electrons pass through the slit in a straight line.

The narrower we make the slit, however, the more knowledge we have about the position of an electron when it passes through the slit. By the Uncertainty Principle, the more accurately we know an object's position, the less accurately we can know its momentum. In the experiment, this uncertainty in position manifests itself as a larger spread in the beam after the beam passes through the slit. As we discussed with light in Chapter 25, the smaller the width of the slit, the more spread there is in the beam. This concept is shown in Figure 27.10.

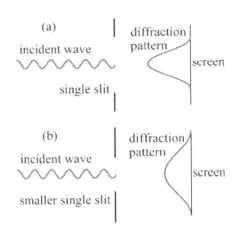

Figure 27.10: The single slit that an electron beam passes through in (a) is larger than that in (b), but the spread in the beam is smaller in (a). This is a consequence of the uncertainty principle.

EXAMPLE 27.6 – Investigating the Uncertainty Principle
Table 27.1 shows the mass, momentum, uncertainty in position, and uncertainty in momentum (assuming we are minimizing this uncertainty, according to the Uncertainty Principle) for three objects, a baseball, a virus, and an electron. In each case, the object's velocity is 10 m/s, and the uncertainty in position is 1 angstrom (Å). 1 Å = 1 × 10⁻¹⁰ m. (a) Some of the values in Table 27.1 are missing. Complete the table to fill in the missing data. (b) Comment on the size of the uncertainty in momentum in the three cases.

Object	Mass (kg)	Momentum (kg m/s)	Δx (m)	Δp (kg m/s)
Baseball	0.15	1.5	1×10^{-10}	5.27×10^{-25}
Virus	2.0×10^{-17}		1×10^{-10}	
Electron	9.11×10^{-31}		1×10^{-10}	

Table 27.1: The mass and the uncertainty in position for a particular situation involving three objects of very different mass. In each case, the velocity is 10 m/s. The momentum and momentum uncertainty are not shown for the virus and the electron.

SOLUTION
(a) The missing data are shown in Table 27.2. Note that, with the uncertainty in position being the same in each case, the uncertainty in momentum is also the same in each case.

Object	Mass (kg)	Momentum (kg m/s)	Δx (m)	Δp (kg m/s)
Baseball	0.15	1.5	1×10^{-10}	5.27×10^{-25}
Virus	2.0×10^{-17}	2.0×10^{-16}	1×10^{-10}	5.27×10^{-25}
Electron	9.11×10^{-31}	9.11×10^{-30}	1×10^{-10}	5.27×10^{-25}

Table 27.2: The mass, momentum, position uncertainty, and momentum uncertainty in a particular situation involving three objects of very different mass.

(b) For a baseball, the limitations imposed by the uncertainty principle are so small as to be meaningless. Even for a tiny object like a virus, the momentum uncertainty is much smaller than the momentum. However, for the electron, the momentum uncertainty is orders of magnitude larger than its momentum, giving us little confidence in the stated value of the electron's velocity.

Related End-of-Chapter Exercises: 26 – 30.

Essential Question 27.6: What if Planck's constant, *h*, had a value around 1 J s instead of its actual value of 6.6 x 10⁻³⁴ J s? Would this change how we interact with the world?

Answer to Essential Question 27.6: The fact that Planck's constant is tiny means that the uncertainty principle is important for objects around the size of an atom or less. If Planck's constant was a great deal larger, such as 1 J s, the uncertainty principle (and the de Broglie wavelength) would be relevant for all objects we deal with on an everyday basis. The world would be a much stranger place than it already is. Cars would diffract from tunnels, batters in baseball would have to deal with the wave nature of the baseball, etc.

Chapter Summary

Essential Idea: The Quantum World.
    The quantum world deals with interactions between very small particles like electrons, protons, and atoms. Because of the tiny value of Planck's constant ($h = 6.626 \times 10^{-34}$ J s), the behavior of such particles is very different from the behavior of everyday objects. Among other things, these tiny objects exhibit both a wave nature and a particle nature.

Black body radiation
Black body radiation is the radiation, in the form of electromagnetic waves, which emanates from a warm object. An example is the red-orange glow given off by a toaster element. Black body radiation is historically significant, because Max Planck explained it in terms of quantized energy levels, the first time quantized energy levels had been used.

The Photoelectric effect
When light is incident on a metal, electrons can be emitted from the metal, in a process known as the photoelectric effect. Electrons are emitted only if the frequency of the light is larger than a particular threshold frequency that depends on the metal. Increasing the intensity of the incident light does not change the maximum kinetic energy of the emitted electrons. The photoelectric effect cannot be explained in terms of light acting as a wave. Instead, it is explained in terms of light being made up of particles, which we call photons, with the electrons being emitted because of single photon – single electron interactions.

$$E = hf ,$$ (Equation 27.2: **Energy of a photon**)
where f is the frequency of the electromagnetic wave corresponding to the photon.

$$K_{max} = hf - W_0 ,$$ (Equation 27.3: **The photoelectric effect**)
where K_{max} is the maximum kinetic energy of the electrons and W_0 is the work function.

The Compton effect
    Despite having no mass, photons carry momentum, given by

$$p = \frac{E}{c} = \frac{hf}{c} = \frac{h}{\lambda} .$$ (Equation 27.6: **Momentum of a photon**)

    The fact that photons have momentum is demonstrated by the Compton effect, a single photon – single electron collision in which both momentum and energy are conserved.

$$\Delta\lambda = \lambda' - \lambda = \frac{h}{m_e c}(1 - \cos\theta) ,$$ (Equation 27.7: **The Compton Effect**)

where m_e is the mass of the electron, and the quantity $h/(m_e c) = 2.43 \times 10^{-12}$ m is the Compton wavelength. θ is the angle between the directions of the incident and outgoing photons, while λ' is the wavelength of the outgoing photon and λ is the wavelength of the incident photon.

Wave-particle duality

Light is not the only thing that exhibits both a wave nature and a particle nature – everything exhibits such wave-particle duality. The wavelength of an object is inversely proportional to its momentum.

$$\lambda = \frac{h}{p} = \frac{h}{mv}.$$ (Equation 27.8: **the de Broglie wavelength**)

To explain the results of a particular experiment, usually either the wave nature or the particle nature is used.

Heisenberg's uncertainty principle

Quantum physics actually puts a limit on how accurately we can know something. More specifically, the uncertainties in two related quantities, such as the position and momentum of an object, are related in such a way that the smaller the uncertainty in one of the quantities, the larger the uncertainty has to be in the other quantity.

$$\Delta x \, \Delta p \geq \frac{h}{4\pi} = 5.273 \times 10^{-35} \text{ J s}.$$ (Equation 27.9: **Heisenberg's uncertainty principle**)

End-of-Chapter Exercises

Exercises 1 – 12 are conceptual questions that are designed to see if you have understood the main concepts of the chapter.

1. Astronomers can determine the temperature at the surface of a star by looking at the star's color. Explain how the color of a star corresponds to its temperature, and comment on whether a blue star has a higher surface temperature than does a red star, or vice versa.

2. An incandescent light bulb gives off a bright yellow-white glow when it is connected to a wall socket. If the potential difference across the bulb is reduced, however, not only does the bulb get dimmer, the emitted light takes on a distinct orange hue. Explain this.

3. With a particular metal plate, shining a beam of blue light on the metal causes electrons to be emitted via the photoelectric effect. If we reduce the intensity of the light shining on the metal, without changing its wavelength, what happens? Explain your answer.

4. The work functions of gold, aluminum, and cesium are 5.1 eV, 4.1 eV, and 2.1 eV, respectively. If light of a particular frequency causes photoelectrons to be emitted when the light is incident on an aluminum surface, explain if we know whether this means that photoelectrons are emitted from a gold surface or a cesium surface when the light is incident on those surfaces.

5. With a particular metal plate, shining a beam of green light on the metal causes electrons to be emitted. (a) If we replace the green light by blue light, do we know that electrons will be emitted? (b) If the two beams have the same intensity and are incident on equal areas of the plate, do we get the same number of electrons emitted per second in the two cases? Assume that the probability that a photon will cause an electron to be emitted is the same in both cases (e.g., for every two photons incident on the plate, one electron is emitted).

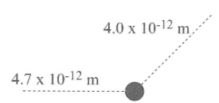

4.0 x 10⁻¹² m

4.7 x 10⁻¹² m

6. The diagram in Figure 27.11 represents the Compton effect collision between a photon and an electron. The diagram shows the paths followed by the incident and outgoing photons, and the wavelengths of these photons. Which is the incident photon and which is the outgoing photon?

Figure 27.11: The diagram shows a representation of a particular Compton effect collision. The blue circle represents the initial position of the electron, which is at rest before the collision. The dashed lines represent the direction of the incident and outgoing photons – the lines are labeled with the photon wavelengths. For Exercise 6.

7. A photon is incident on an electron that is initially at rest. The photon experiences a Compton effect collision with the electron such that the photon, after the collision, is traveling in a direction exactly opposite to that of the incident photon. In what direction is the electron's velocity after the collision? Explain your answer.

8. One problem with solar sailboats is that the force they experience, because of photons from the Sun reflecting from the sails, cannot be directed toward the Sun. In an effort to overcome this problem, an inventor proposes a solar sailboat with a built-in light source. Instead of using sunlight to drive the sailboat, the inventor proposes attaching a high-power laser to the sailboat, and shining the light from the laser onto the sails. According to the inventor, by adjusting the angle of the sails, the sailboat can be made to turn in any direction, and, once pointed in the correct direction, the light from the laser can then propel the sailboat in the desired direction. What, if anything, is wrong with this idea?

9. Electrons are accelerated from rest and are then incident on a double slit. The pattern the electrons make on the screen is shown in Figure 27.12. If the potential difference through which the electrons are accelerated is reduced and a new pattern is observed on the screen, describe how the new pattern differs from that shown in Figure 27.12.

Figure 27.12: The pattern of dots created by electrons striking a screen a distance 2.40 m from a double slit, when an electron beam is incident on the slits. For Exercise 9.

10. The interference pattern shown in Figure 27.12 for electrons that pass through a double slit is something of an idealization. If we take a close up view of the pattern on the screen, we see that it is really more like the pattern shown in Figure 27.13. How does this figure

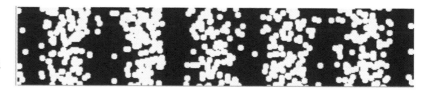

Figure 27.13: A close up view of dots created by electrons striking a screen after passing through a double slit, for Exercise 10.

support the idea of wave-particle duality? In particular, comment on whether any part of the pattern is associated with the wave nature of electrons, and whether any part is associated with the particle nature of electrons.

11. Two experiments are carried out, using the same double slit. In the first experiment, a beam of electrons is incident on the double slit. In the second experiment, the electron beam is replaced by a beam of protons. In which experiment are the peaks in the interference patterns farther apart if the electrons and protons have equal (a) de Broglie wavelengths? (b) momenta? (c) speed? (d) kinetic energy?

12. To create a beam of fast-moving electrons, you accelerate electrons from rest by placing them between two metal plates that have a large potential difference across them. The electrons emerge from this system by means of a tiny hole you have drilled in the plate the electrons accelerate toward. You find, however, that instead of a narrow, well-defined beam, that the electrons are spread over a range of angles. (a) Come up with an explanation, based on the principles of physics covered in this chapter, to explain your observations. (b) Will the problem (the spread in the beam) get better or worse when you make the hole smaller? Explain. (c) Will the problem get better or worse when you increase the potential difference across the metal plates? Explain.

Exercises 13 – 16 involve the photoelectric effect.

13. Gold has a work function of 5.1 eV. (a) What is the threshold (minimum) frequency of light needed to cause electrons to be emitted from a gold plate via the photoelectric effect? (b) If the frequency of the light incident on the gold plate is twice the threshold frequency, what is the maximum kinetic energy of the emitted electrons?

14. When ultraviolet light with a wavelength of 290 nm is incident on a particular metal surface, electrons are emitted via the photoelectric effect. The maximum kinetic energy of these electrons is 1.23 eV. (a) What is the work function of the metal? (b) What is the threshold frequency for this particular metal?

15. Iron has a work function of 4.5 eV. Plot a graph of the maximum kinetic energy (in eV) of photoelectrons emitted when light is incident on an iron plate as a function of the frequency of the light, which can vary from 0 to 2×10^{15} Hz.

16. An incomplete graph of the maximum kinetic energy (in eV) of photoelectrons emitted when incident light is incident on a particular metal is shown in Figure 27.14. Assume the two points shown are accurate. (a) What is the work function of the metal? (b) What is the threshold frequency in this case? (c) Complete the graph, to show the maximum kinetic energy at all frequencies up to 2×10^{15} Hz.

Figure 27.14: An incomplete graph of maximum kinetic energy of photoelectrons emitted from a particular metal plate, as a function of the frequency of the incident light. For Exercise 16.

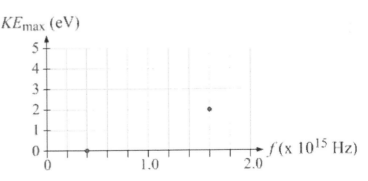

Exercises 17 – 20 involve the Compton effect.

17. A photon with a wavelength of 5.0×10^{-12} m is incident on an electron that is initially at rest. If the photon that travels away from this collision is traveling in a direction that is at 120° to that of the incident photon, what is its wavelength?

18. A photon with a wavelength of 6.14×10^{-12} m is incident on an electron that is initially at rest. If the photon experiences a Compton effect collision with the electron, what is (a) the minimum possible, and (b) the maximum possible, wavelength of the photon after the collision? (c) In which direction does the electron travel after the collision in the situation described in (a) and (b)?

19. A photon is incident on an electron that is initially at rest. The photon experiences a Compton effect collision with the electron such that the photon, after the collision, is traveling in a direction perpendicular to that of the incident photon. (a) How does the wavelength of the photon after the collision compare to that of the photon before the collision? (b) Can the electron after the collision be traveling in a direction exactly opposite to that of the photon after the collision? Explain why or why not.

20. A photon collides with an electron that is initially at rest. After the collision, the photon has a wavelength of 4.0×10^{-12} m, and it is traveling in a direction that is at 45° to that of the incident photon. What is the wavelength of the incident photon?

Exercises 21 – 25 involve the de Broglie wavelength and wave-particle duality. Review Chapter 25 for some relevant concepts and equations.

21. When photons of a certain wavelength are incident on a particular double slit, the angle between the central maximum and one of the first-order maxima in the interference pattern is 10°. If the photons are replaced by electrons, and the electrons have a de Broglie wavelength less than that of the photon wavelength, what (if anything) will happen to the angle between the central maximum and one of the first-order maxima?

22. When light from a red laser, with a wavelength of 632 nm, is incident on a certain double slit, a particular interference pattern is observed. The laser light is replaced by a beam of electrons, all with the same energy, and exactly the same interference pattern is observed. What is (a) the wavelength, (b) frequency, and (c) energy of the electrons?

23. Electrons of a particular energy are incident on a double slit, producing an interference pattern with interference maxima and minima. When the electron energy is reduced, do the interference maxima get closer together, farther apart, or remain the same? Briefly justify your answer.

24. Figure 27.15 shows the $m = 0$ and $m = 1$ lines coming from an electron beam, in which the electrons have a de Broglie wavelength of 54 nm, that is incident on a double slit. The squares in the grid measure 20 cm × 20 cm. Determine the distance between the two slits in the double slit.

Figure 27.15: The $m = 0$ and $m = 1$ lines that result when electrons with a wavelength of 54 nm are incident on a double slit. The squares on the grid measure 20 cm × 20 cm. For Exercise 24.

25. Electrons of a particular energy are incident on a double slit. When the electrons are detected by a detector 3.0 m beyond the double slit, the distance between the central maximum and one of the first-order maxima in the interference pattern is found to be 8.0 mm. The distance between the slits in the double slit is 120 nm. What is the (a) wavelength, and (b) magnitude of the momentum of the electrons?

Exercises 26 – 30 involve the Heisenberg uncertainty principle. In all cases below, assume that the motion is one-dimensional.

26. What is the minimum uncertainty in an object's position if the object has a speed of 20 m/s and a mass of (a) 100 g? (b) 1×10^{-10} kg? (c) 1×10^{-25} kg?

27. According to the Heisenberg uncertainty principle, what is the minimum uncertainty in a proton's speed if the proton has an uncertainty in position of (a) 50 mm? (b) 500 nm? (c) 5.0×10^{-10} m?

28. Repeat Exercise 27, with an electron instead of a proton.

29. Imagine that you live in a strange world where Planck's constant h is 66.3 J s. You park your bike, which has a mass of 20 kg, in a location such that the uncertainty in the bike's location is 10 cm. What is the minimum uncertainty in the bike's speed?

30. You place an electron on a wire that has a length of 6.0 nm, so that you know the electron is on the wire but you don't know exactly where it is. What is the minimum uncertainty in the electron's momentum?

Exercises 31 – 34 involve applications of quantum physics.

31. In 2007, a team led by researchers at the University of Delaware announced a new world record for efficiency by a solar cell, at 42.8%. This means that the solar cell transformed 42.8% of the incident light energy into electrical energy. Estimate what area of these world-record solar cells would be required to supply the energy needs of a household of four people in the United States, based on an estimate of 2×10^{12} J of energy used annually by such a household. You can assume that the intensity of sunlight falling on the cells is 1000 W/m², and that, on average, the sun is shining on the solar cells for 6 hours per day.

32. Many luxury automobiles are now using xenon high-intensity discharge headlights. Compared to incandescent light bulbs, in which the filament is around 2800 K, the light given off by the high-intensity headlights has a spectrum similar to that of a black body at 4200 K. Does this help explain why the high-intensity discharge headlights have a blue tinge in comparison to the light from a standard incandescent bulb? Explain based on the principles of physics covered in this chapter.

33. One of the hot topics in research these days is the development of light-emitting diodes (LEDs) as a potential replacement for the incandescent bulbs that are commonly used for household lighting. The advantage of LEDs over incandescent bulbs is that LEDs are very efficient at transforming electrical energy into visible light, while it is commonly said that "incandescent bulbs give off 90% of their energy as heat." (a) Explain, using principles of physics, how an incandescent light bulb works. (b) Typically, the filament in an incandescent bulb is made from tungsten. Why is tungsten used instead of a cheaper, more readily available metal such as aluminum? (c) What is meant by the phrase "incandescent bulbs give off 90% of their energy as heat?"

34. Ernst Ruska was awarded a 50% share of the Nobel Prize in Physics in 1986. Write a couple of paragraphs regarding the work he did, mentioning in particular how the work is connected to the principles of physics discussed in this chapter.

35. An infrared thermometer is a thermometer that can determine an object's temperature without the thermometer needing to make contact with the object. For instance, you could aim the thermometer at the hot water flowing out of the faucet in your kitchen, and the thermometer would read the temperature of the hot water. (a) Briefly explain, using the principles of physics covered in this chapter, how such a thermometer works. (b) Explain why the thermometer is known as an infrared thermometer.

General problems and conceptual questions

36. One morning, as you wait for the toaster to finish toasting a couple of slices of bread, you measure the spectrum of light being emitted by the glowing toaster elements. The peak wavelength of this light is 600 nm. What is the temperature of the toaster elements?

37. You have probably heard the phrase "white hot" before. Approximately what temperature is a black body that is hot enough to look white?

38. A typical helium-neon laser pointer, emitting light with a wavelength of 632 nm, has a beam with an intensity of 800 W/m² and a diameter of 3.00 mm. How many photons are emitted by the laser pointer every second?

39. The work functions of gold, aluminum, and cesium are 5.1 eV, 4.1 eV, and 2.1 eV, respectively. When light of a particular frequency shines on a cesium surface, the maximum kinetic energy of the emitted photoelectrons is 2.5 eV. What is the maximum kinetic energy of the photoelectrons emitted when the same light shines on (a) an aluminum surface? (b) a gold surface?

40. We consider the visible spectrum to run from 400 nm to 700 nm. (a) What is the equivalent energy range, in eV, of photons for light in the visible spectrum? If white light, composed of wavelengths covering the full range of the visible spectrum but no more, is incident on a surface, will photoelectrons be emitted if the surface is (b) carbon (work function = 4.8 eV)? (c) potassium (work function = 2.3 eV)? Explain your answers to parts (b) and (c).

41. Zinc has a work function of 4.3 eV. A standard lecture demonstration involves attaching a zinc plate to an electroscope, and then charging the plate by rubbing it with a negatively charged rod. At that point, the electroscope indicates that it is charged, and we know that the charge is negative. (a) If light in the visible spectrum, from a bright incandescent light bulb, shines on the plate, do we expect photoelectrons to be emitted from the plate, causing the electroscope to discharge? Why or why not? (b) What is the maximum wavelength of light that would cause the electroscope to discharge? What part of the spectrum is this wavelength in?

42. Table 27.3 gives a set of data for a photo-electric effect experiment, showing the maximum kinetic energy of the emitted electrons at a number of different frequencies of the light incident on a particular metal plate. Plot a graph of the maximum kinetic energy as a function of frequency and determine, from your graph, (a) the threshold frequency, (b) the work function of the metal, and (c) the slope of that part of the graph above the threshold frequency.

Frequency ($\times 10^{15}$ Hz)	K_{max} (eV)
0.5	0
1.0	0
1.5	1.9
2.2	4.8
2.9	7.7
3.7	11.0

Table 27.3: A set of data from a photo-electric effect experiment, showing the maximum kinetic energy of the emitted electrons at a number of different photon frequencies, for Exercise 35.

43. In a particular photoelectric effect experiment, photons with an energy of 5.0 eV are incident on a surface, producing photoelectrons with a maximum kinetic energy of 2.2 eV. If the energy of the photons doubles, does the maximum kinetic energy of the photoelectrons also double? Explain your answer.

44. In a particular photoelectric effect experiment, photons with an energy of 5.6 eV are incident on a surface, producing photoelectrons with a maximum kinetic energy of 3.3 eV. (a) What is the work function of the metal? (b) What is the minimum photon energy necessary to produce a photoelectron in this situation? (c) What is the corresponding threshold frequency?

45. With a particular metal plate, shining a beam of green light on the metal causes electrons to be emitted. (a) If we replace the green light by red light, do we know that electrons will be emitted? (b) If the two beams have the same intensity and are incident on equal areas of the plate, do we get the same number of electrons emitted per second in the two cases, assuming that photons are emitted in both cases? Assume that the probability that a photon will cause an electron to be emitted is the same in both cases (e.g., for every two photons incident on the plate, one electron is emitted).

46. Figure 27.16 shows a graph of the maximum kinetic energy of emitted photoelectrons as a function of the energy of the photons that are incident on a particular surface. From this graph, determine (a) the work function of the surface, and (b) the threshold frequency.

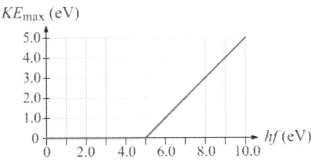

Figure 27.16: This graph shows the maximum kinetic energy of emitted electrons as a function of the energy of the incident photons, for Exercise 46.

47. A green helium-neon laser pointer emits light with a wavelength of 532 nm. The beam has an intensity of 900 W/m^2 and a diameter of 4.00 mm. (a) How many photons are emitted by the laser pointer every second? (b) What is the magnitude of the momentum of each of these photons? How much momentum is imparted to an object, in a 1.00-second interval, that (c) completely absorbs all these photons? (d) reflects all these photons straight back toward the laser pointer?

48. A laser beam that is completely absorbed by a black surface exerts a force of 2.5×10^{-8} N on the surface. (a) What is the net momentum transferred to the surface by the beam every second? (b) If the wavelength of light emitted by the laser is 632 nm, what is the magnitude of the momentum of each photon in the beam? (c) How many photons strike the surface every second?

49. A photon with a wavelength of 6.00×10^{-12} m is incident on an electron that is initially at rest. If the photon that travels away from this collision is traveling in a direction that is at 45° to that of the incident photon, what is its wavelength?

50. Return to the situation described in Exercise 49. Relative to the direction of the incident photon, in what direction does the electron travel after the collision?

51. Photons with a particular wavelength experience Compton-effect collisions with electrons that are at rest. Figure 27.17 shows a graph of the wavelength of the photons that travel away from the various collisions as a function of the cosine of the scattering angle (this is the angle between the direction of the incident photon and the direction of the photon after the collision). For the incident photons, determine the (a) wavelength, (b) frequency, and (c) energy, in eV.

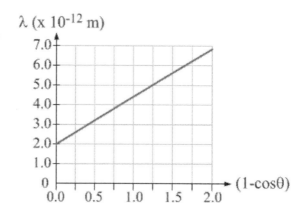

52. Photons with a particular wavelength experience Compton-effect collisions with electrons that are at rest. Figure 27.17 shows a graph of the wavelength of the photons that travel away from the collision as a function of 1 minus the cosine of the scattering angle. We repeat the experiment with incident photons that have wavelengths twice as large as that of the original incident photons. If we plot a new graph, like that in Figure 27.17, how will the graphs compare in terms of their (a) slopes, and (b) y-intercepts?

Figure 27.17: A graph of the wavelength of the photon after colliding with a stationary electron in a Compton effect collision, as a function of 1 minus the cosine of the scattering angle. The scattering angle is the angle between the direction of the incident photon and the direction of the photon after the collision. For Exercises 51 and 52.

53. A photon is incident on an electron that is initially at rest. The photon and electron experience a Compton-effect collision such that the electron, after the collision, is traveling in the same direction that the photon was before the collision. The electron's momentum after the collision is 5.0×10^{-22} kg m/s. (a) Write down an expression to show how the wavelength of the photon after the collision is related to that of the photon after the collision. (b) Write down an expression to show how the momentum of the photon before the collision is related to the momentum of the photon after the collision and the electron's momentum after the collision. Now solve your two equations to find the wavelength of the photon (c) before the collision, and (d) after the collision.

54. Electrons with an energy of 8.00 eV are incident on a double slit in which the two slits are separated by 400 nm. What is the angle between the two second-order maxima in the resulting interference pattern?

55. A beam of electrons, shining on a double slit, creates the pattern shown in Figure 27.18 at the center of a screen placed 2.40 m on the opposite side of the double slit from the electron source. If the de Broglie wavelength of the electrons is 135 nm, what is the distance between the two slits in the double slit?

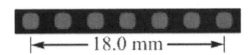

Figure 27.18: The pattern of dots created by electrons striking a screen a distance 2.40 m from a double slit, when an electron beam is incident on the slits. For Exercises 55 – 56.

56. A beam of electrons, shining on a double slit, creates the pattern shown in Figure 27.18 at the center of a screen placed 1.50 m on the opposite side of the double slit from the electron source. (a) If the distance between the two slits is 500 nm, what is the de Broglie wavelength of the electrons? (b) If the electrons achieved this wavelength by being accelerated from rest through a potential difference, what is the magnitude of the potential difference?

57. A baseball has a mass of 150 g. With what speed would a baseball have to be moving so that its de Broglie wavelength is 632 nm, the same as the wavelength of light from a red helium-neon laser?

58. Consider the opening image of this chapter, showing the head of an ant as imaged by an electron microscope. To be able to resolve tiny details, with a size much less than the wavelength of visible light, the de Broglie wavelength of the electrons also needs to be tiny. (a) What is the speed of an electron that has a de Broglie wavelength of 10 nm? (b) If the electron achieved this speed by being accelerated from rest through a potential difference, what was the magnitude of the potential difference?

59. To create a beam of fast-moving electrons, you accelerate electrons from rest by placing them between two metal plates that have a large potential difference across them. The electrons emerge from this system by means of a tiny hole you have drilled in the plate the electrons accelerate toward. You find, however, that instead of a narrow, well-defined beam, that the electrons are spread over a range of angles. To help solve this problem, make use of Equation 25.6, which states that the angle of the first minimum in the diffraction pattern from a circular opening of diameter D is $\theta_{min} = 1.22\lambda / D$. (a) If the electrons are accelerated from rest through a potential difference of 500 V, what is their de Broglie wavelength? (b) What diameter hole is required for the beam to diffract through the hole so that $\theta_{min} = 10°$? (c) If the hole diameter is reduced by a factor of 2, what is θ_{min}? (d) If, instead, the potential difference is doubled, what is θ_{min}?

60. Two students are having a conversation. Comment on the part of their conversation that is reported below. This one may require a little background reading.

Maggie: I really don't buy this wave-particle duality stuff. I mean, let's say we do this. We send an electron beam through a single slit. When the slit is wide, the electrons go straight through. The narrower we make the slit, the greater the probability that the electrons hit one of the edges of the slit, changing the direction. The narrower the slit, the greater the spread in the beam after passing through the slit – who needs waves, it's all particles.

Dipesh: That sounds sensible. Except, when you look at the pattern carefully, there are some angles where you don't get any electrons going – can you explain that just with particles?

Maggie: Oh, good point. OK, let's think about the double slit a little. What if, instead of firing a whole beam of electrons at the double slit, you just did one electron at a time. Wouldn't the electron have to choose one slit or the other? Plus, we shouldn't get interference any more - one electron wouldn't be able to interfere with itself, right?

Dipesh: What a cool idea. Let's google it and see if anyone has done an experiment like that. I'll type in "single electron interference double slit" and see what comes up.

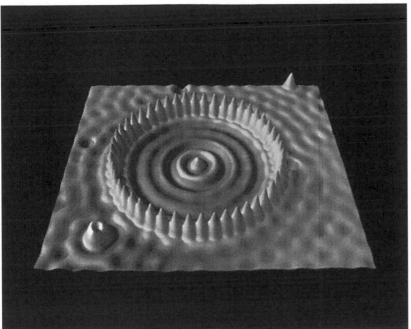

An image from a scanning tunneling microscope (STM), showing electrons trapped inside a corral of iron atoms. We now have the technology to manipulate individual atoms, moving them around one-by-one to create structures such as the one shown here. This also gives us insight into the behavior of electrons.

Image credit: Courtesy of Mike Crommie (UC – Berkeley), for research funded by IBM Research Division.

Chapter 28 – The Atom

CHAPTER CONTENTS

Our understanding of the atom has grown tremendously over the last 100 years or so. From having only a fuzzy picture of what exactly an atom is, and how atoms of one element differ from atoms of another element, we now have a clear understanding of what an atom is and what it is made of.

In contrast to the situation of 100 years ago, we now know that:
- Atoms are made up of a nucleus, generally consisting of neutrons (which have no net charge) and protons (each proton has a charge of $+e$), and electrons (each with a charge of $-e$).
- In a neutral atom, both the number of protons and the number of electrons is equal to Z, the atom's atomic number.
- The chemical properties of an atom are determined primarily by the electrons.
- Electrons are not little balls that orbit the nucleus in well-defined orbits, like planets orbiting the Sun. Instead, electrons are generally thought of as corresponding to charge clouds around the nucleus, with the density of the charge cloud at a particular location corresponding to the probability of finding the electron at that location.

Carrying on from Chapter 27, these insights into the atom have come from applying the theory of quantum physics to the atom. In this Chapter, we will focus mainly on the electrons in the atom. In Chapter 29, we will delve into the nucleus.

28-1 Line Spectra and the Hydrogen Atom

Figure 28.1 gives some examples of the line spectra emitted by atoms of gas. The atoms are typically excited by applying a high voltage across a glass tube that contains a particular gas. By observing the light through a diffraction grating, the light is separated into a set of wavelengths that characterizes the element.

Figure 28.1: Line spectra from hydrogen (top) and helium (bottom). A line spectrum is like the fingerprint of an element. Astronomers, for instance, can determine what a star is made of by carefully examining the spectrum of light emitted by the star.

The spectrum of light emitted by excited hydrogen atoms is shown in Figure 28.1(a). The decoding of the hydrogen spectrum represents one of the great scientific mystery stories. First on the scene was the Swiss mathematician and schoolteacher, Johann Jakob Balmer (1825 – 1898), who published an equation in 1885 giving the wavelengths in the visible spectrum emitted by hydrogen. The Swedish physicist Johannes Rydberg (1854 – 1919) followed up on Balmer's work in 1888 with a more general equation that predicted all the wavelengths of light emitted by hydrogen:

$$\frac{1}{\lambda} = R\left(\frac{1}{n_1^2} - \frac{1}{n_2^2}\right),$$ (Eq. 28.1: **The Rydberg equation for the hydrogen spectrum**)

where $R = 1.097 \times 10^7 \text{ m}^{-1}$ is the Rydberg constant, and the two n's are integers, with n_2 greater than n_1.

Neither Balmer nor Rydberg had a physical explanation to justify their equations, however, so the search was on for such a physical explanation. The breakthrough was made by the Danish physicist Niels Bohr (1885 – 1962), who showed that if the angular momentum of the electron in a hydrogen atom was quantized in a particular way (related to Planck's constant, in fact), that the energy levels for an electron within the hydrogen atom were also quantized, with the energies of the electrons being given by:

$$E_n = \frac{-13.6 \text{ eV}}{n^2},$$ (Eq. 28.2: **Energies of the electron levels in the hydrogen atom**)

where n is any positive integer.

When the electron in a hydrogen atom drops down from a higher energy state to a lower energy state, a photon is given off that has an energy equal to the difference between the electron energy levels – thus, energy is conserved. Because the differences between the electron energy levels are limited, the photons that are emitted by excited hydrogen atoms are emitted at specific wavelengths, giving the few bright lines shown in Figure 28.1(a).

According to Bohr's model of the atom, the lines in Figure 28.1(a) correspond to photons emitted when electrons drop down to the second-lowest energy level in hydrogen from the levels with $n = 3, 4, 5, 6$ or 7. Bohr predicted, however, that photons should be observed at wavelengths corresponding to electrons dropping down from an excited state to the ground ($n = 1$) state. We don't see these wavelengths with our eyes because they are in the ultraviolet region of the spectrum. When scientists using detectors sensitive in the ultraviolet region found light emitted by hydrogen at the exact wavelengths predicted by Bohr, it was a tremendous validation of the Bohr model of the hydrogen atom. Bohr received the Nobel Prize in Physics in 1922 for his work.

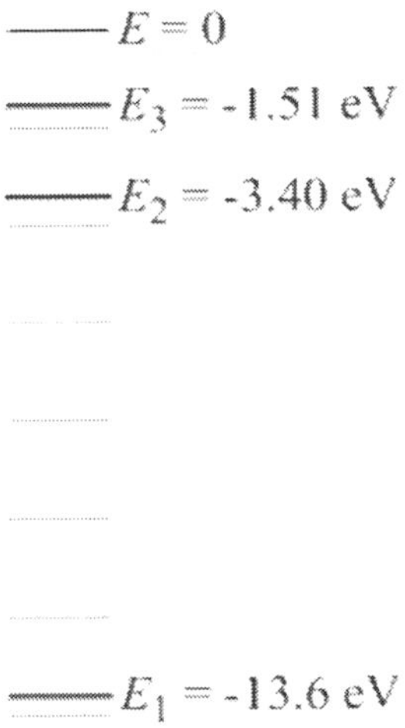

———— $E = 0$

———— $E_3 = -1.51$ eV

———— $E_2 = -3.40$ eV

EXPLORATION 28.1 – Building the energy-level diagram for hydrogen
Creating a diagram of the energy levels for hydrogen can help explain how the photon energies arise. We will start with the three lowest levels.

Step 1 – Using Equation 28.2, determine the energies of the three lowest energy levels for hydrogen. Then create an energy-level diagram, which looks like a ladder with rungs that are unequally spaced. The three lowest energy levels correspond to $n = 1$, 2, and 3. Substituting these values of n into Equation 28.2 gives $E_1 = -13.6$ eV, $E_2 = -3.40$ eV, and $E_3 = -1.51$ eV. The corresponding energy-level diagram is shown in Figure 28.2. Note that all the energy levels for $n > 3$ fall between $E = 0$ and the -1.51 eV of the $n = 3$ level.

———— $E_1 = -13.6$ eV

Figure 28.2: An energy-level diagram, showing the three lowest energy levels for hydrogen.

Step 2 – Let's confine ourselves to electrons that make transitions between only the three energy levels shown in Figure 28.2. Mark these transitions on the energy-level diagram with downward-pointing arrows from one level to a lower level. How many different photon energies are associated with these transitions? Determine the energies of these photons. With three energy levels, we can get three different electron transitions, and thus three different photon energies. The transitions are shown on the energy-level diagram in Figure 28.3. In decreasing order, by photon energy, the photon energies are:

———— $E = 0$

$E_3 = -1.51$ eV

$E_2 = -3.40$ eV

$$E_{3\to1} = E_3 - E_1 = -1.51 \text{ eV} - (-13.6 \text{ eV}) = 12.1 \text{ eV} ;$$

$$E_{2\to1} = E_2 - E_1 = -3.40 \text{ eV} - (-13.6 \text{ eV}) = 10.2 \text{ eV} ;$$

$$E_{3\to2} = E_3 - E_2 = -1.51 \text{ eV} - (-3.40 \text{ eV}) = 1.89 \text{ eV} .$$

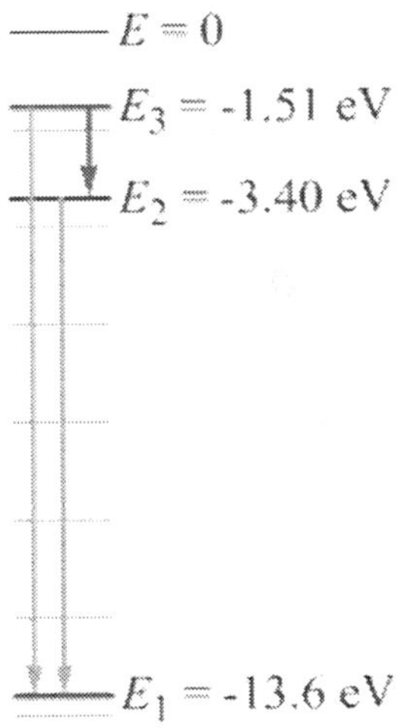

Figure 28.3: The energy-level diagram is modified to show the electron transitions that are possible between the lowest three energy levels in hydrogen. The photons emitted in the two transitions that end at the $n = 1$ level are in the ultraviolet region, while the photon associated with the $n = 3$ to $n = 2$ transition is red.

———— $E_1 = -13.6$ eV

Key idea: The energy of a photon emitted by an electron that drops down from one energy level to a lower energy level is equal to the difference in energy between those two energy levels.
Related End-of-Chapter Exercises: 1, 2, 4, 13 – 18, 33 – 38.

Note that atoms can also absorb energy, in the form of photons, but they only absorb photons with an energy equal to the difference in energy between two of the atom's electron energy levels. In this case, the electrons make a transition from a lower energy level to a higher level.

Essential Question 28.1: Imagine that there is an atom with electron energy levels at the following energies: –31 eV, –21 eV, –15 eV, and –12 eV. Assuming that electron transitions occur between these levels only, (a) how many different photon energies are possible? (b) what is the (i) minimum and (ii) maximum photon energy?

Answer to Essential Question 28.1: (a) In general, four energy levels give six photon energies. One way to count these is to start with the highest level, –12 eV. An electron starting in the –12 eV level can drop to any of the other three levels, giving three different photon energies. An electron starting at the –15 eV level can drop to either of the two lower levels, giving two more photon energies. Finally, an electron can drop from the –21 eV level to the –31 eV level, giving one more photon energy (for a total of six). (b) The minimum photon energy corresponds to the 3 eV difference between the –12 eV level and the –15 eV level. The maximum photon energy corresponds to the 19 eV difference between the –12 eV level and the –31 eV level.

28-2 Models of the Atom

It is amazing to think about how far we have come, in terms of our understanding of the physical world, in the last century or so. A good example of our progress is how much our model of the atom has evolved. Let's spend some time discussing the evolution of atomic models.

Ernest Rutherford probes the plum-pudding model

J. J. Thomson, who discovered the electron in 1897, proposed a plum-pudding model of the atom. In this model, electrons were thought to be embedded in a ball of positive charge, like raisins are embedded in a plum pudding. Ernest Rutherford (1871 - 1937) put this model to the test by designing an experiment that involved firing alpha particles (helium nuclei) at a very thin film of gold. The experiment was carried out in Rutherford's lab by Hans Geiger and Ernest Marsden. If the plum-pudding model was correct, the expectation was that the alpha particles should make it through the ball of spread-out positive charge with very little deflection. For the most part, this was the case; however, a small fraction of the alpha particles were deflected through large angles, with some even being deflected through 180°. Rutherford made a famous statement about this, which was "It was almost as incredible as if you fired a 15-inch shell at a piece of tissue paper and it came back and hit you." Through a careful analysis of the results, Rutherford determined that the positive charge of the atom was not spread out throughout the volume occupied by the atom, but was instead concentrated in a tiny volume, orders of magnitude smaller than that of the atom, which we now call the **nucleus**.

Niels Bohr provides a theoretical framework for Rydberg's equation

The next major advance in our understanding of the atom came from the Danish physicist, Niels Bohr, who incorporating ideas from Rutherford and quantum ideas. In Bohr's model, electrons traveled in circular orbits around a central nucleus, similar to the way planets travel around the Sun. It is important to understand that the Bohr model does not reflect reality, but it provides a basis for our understanding of the atom. With an analysis based on principles of physics we have discussed earlier in this book, such as the attractive force between charged particles, Bohr was able to show that the quantized energy levels in hydrogen were completely consistent with Rydberg's equation (see Equation 28.1 in section 28-1) for the wavelengths of light emitted by hydrogen. In Bohr's model, the angular momenta of the electrons are also quantized, a result we also accept today. Where the Bohr model breaks down is in the electron orbits. In the Bohr model, the electrons are confined to planar orbits of very particular radii. This is not at all the modern view of the atom, which we understand using quantum mechanics.

The modern view of the atom

Over the course of the 20th century, many people, Bohr included, contributed to furthering our understanding of the atom. We will spend some time in Chapter 29 exploring the nucleus, so for the moment let us focus our attention on the electrons in the atom. As far as the electrons are concerned, the nucleus can be thought of as a tiny ball of positive charge.

In the Bohr model of the atom, the electrons are found only in certain orbits, with the radii of the orbits being quantized, so an electron will never be found at other distances from the nucleus. Our modern understanding is rather different. Now, we talk about the probability of finding the electron at a particular distance from the nucleus. For an electron in the **ground state** (the lowest-energy state) of the hydrogen atom, for instance, the Bohr model states that the electron is a distance of 5.29×10^{-11} m from the nucleus (this is known as the **Bohr radius**). The modern view of where the electron in hydrogen's ground state is located is illustrated by Figure 28.4. Even in the modern view, the most likely place to find this ground-state electron is at a distance of one Bohr radius from the nucleus. However, as the graph shows, the electron can be found at any distance from the nucleus, aside from right at the nucleus or infinitely far away.

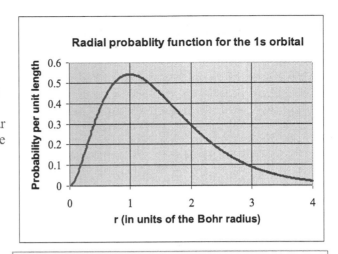

Figure 28.4: A graph of the probability, per unit length, of finding the electron in the hydrogen ground state at various distances from the nucleus. The total area under the curve, when the curve is extended to infinity, is 1 – the electron is 100% likely to be found between $r = 0$ and r = infinity.

Graphs like that in Figure 28.4 come from solving the Schrödinger equation, which is essentially conservation of energy applied to the atom. Solutions to the Schrödinger equation are called wave functions. The square of a wave function gives the probability of finding a particle in a particular location. The Schrödinger equation is named for the Austrian physicist Erwin Schrödinger (1887 – 1961), who shared the 1933 Nobel Prize in Physics for his contributions to quantum mechanics.

Quantum tunneling

If you throw a tennis ball against a solid wall, the ball will never make it to the far side of the wall unless you give it enough kinetic energy to pass over the top of the wall. Such rules do not apply to quantum particles. If the wave function of a quantum particle extends through a barrier to the far side, then there is some probability of finding the particle on the far side of the barrier. Even if the particle's energy is insufficient to carry it over the barrier, the particle will eventually be found on the far side of the barrier. This process, of passing through a barrier, is known as **quantum tunneling**.

Quantum tunneling is exploited in scanning tunneling microscopes (STM's), in which a very sharp tip is scanned over the surface. By measuring the rate at which electrons tunnel across the gap between this tip and the surface, a two-dimensional picture of the surface can be created. The chapter-opening picture shows an image of a quantum corral. First, a corral of iron atoms was created on a surface by pulling the atoms into place with an STM. The STM was then used to scan the surface, to visualize electrons trapped inside the corral. Note how the electrons are wave-like in this situation, and nor particle-like.

Related End-of-Chapter Exercises: 5 – 7.

Essential Question 28.2: Return to the graph in Figure 28.4. Is the ground-state electron in hydrogen more likely to be found at a position closer than 1 Bohr radius from the nucleus, or farther than 1 Bohr radius from the nucleus? Using the graph, estimate the relative probability of finding the electron in these two ranges.

Answer to Essential Question 28.2: The area under the curve for the region beyond 1 Bohr radius from the nucleus is clearly larger than the area for the region less than 1 Bohr radius away from the nucleus – thus, the electron is more likely to be found farther than 1 Bohr radius away. A reasonable estimate for the ratio of the areas is 2:1. In other words, the electron, when in the ground state of hydrogen, is approximately twice as likely to be found at a radius farther than 1 Bohr radius from the nucleus than it is to be found at a distance less than one Bohr radius.

28-3 The Quantum Mechanical View of the Atom

In the Bohr model of the atom, only one quantum number, n, was used. With that single quantum number, Bohr came up with expressions for the energies of quantized energy levels, quantized angular momenta of the electrons, and quantized radii of the electron orbits. Our modern view of the atom, applying the equations of quantum mechanics, is a little more complicated, requiring four quantum numbers to completely specify the various electron energy states. These quantum numbers are:

1. n, the principal quantum number. To a first approximation, the energies of the quantized energy levels in an atom with an atomic number Z and just one electron are given by:

$$E_n = \frac{(-13.6 \text{ eV}) Z^2}{n^2},$$ (Eq. 28.3: **Energies of the electron levels**)

where $n = 1, 2, 3, \ldots$

2. ℓ, the orbital quantum number. This quantum number quantizes the magnitude of the electron's orbital angular momentum (this is somewhat analogous to the angular momentum associated with a planet's orbit around the Sun). The magnitude of the orbital angular momentum is given by:

$$L = \sqrt{\ell(\ell+1)} \, \frac{h}{2\pi},$$ (Eq. 28.4: **Orbital angular momentum**)

where $\ell = 0, 1, 2, \ldots n-1$.

3. m_l, the magnetic quantum number. This quantum number quantizes the direction of the electron's angular momentum (this is known as **space quantization**, and is illustrated in Figure 28.5). Conventionally, we say that this quantum number defines the z-component of the angular momentum. For a given value of the orbital quantum number, the different values of m_l give electron states of the same energy unless a magnetic field is present, in which case the states have different energies (this is called the **Zeeman effect**).

$$L_z = \frac{m_\ell h}{2\pi}, \qquad \text{where } m_l = -\ell, -\ell+1, \ldots \ell-1, \ell.$$

(Eq. 28.5: **z-component of the orbital angular momentum**)

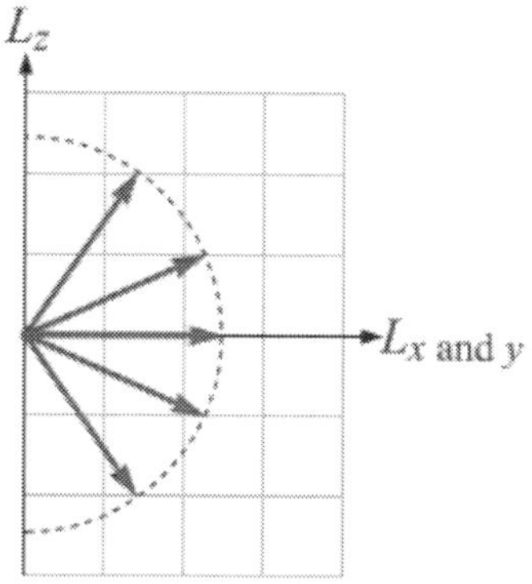

Figure 28.5: The allowed directions of the orbital angular momentum, when $\ell = 2$. The half-circle has a radius of $\sqrt{6}$ units. The units on both axes are angular momentum units of $h/(2\pi)$.

4. m_s, the spin quantum number. The concept of electron spin is somewhat analogous to the angular momentum associated with a planet's rotation about its own axis. The electron has two possible spin states, which we refer to as spin up and spin down. The two possible z-components of the spin angular momentum, S_z, are given by:

$$S_z = m_s \frac{h}{2\pi}, \qquad \text{(Eq. 28.6: Spin angular momentum)}$$

where $m_s = \pm \frac{1}{2}$.

Visualizing the wave functions

Figure 28.6 shows a common way to visualize the various electron wave functions for hydrogen. Remember that there is only one electron in the hydrogen atom, but the electron has an infinite number of states that it can choose between – some of the lower-energy states are shown below. When we look at pictures like this, we again interpret them in terms of probability. The brighter the picture is at a particular point, the more likely it is to find the electron at that point. Although the pictures are two-dimensional, the wave functions are three-dimensional.

The image in the top left of Figure 28.6 corresponds to the graph in Figure 28.4. At first glance, they appear to contradict one another. For instance, the graph in Figure 28.4 shows that the probability of finding the electron at $r = 0$ is zero, while the top left image in Figure 28.6 has the brightest (highest probability) spot right in the center, where $r = 0$. The probability of finding the electron at any particular point, not just the center point, is zero, because a point has no volume. It makes more sense to discuss the probability of finding the electron within a particular volume. Both the graph in Figure 28.4 and all six diagrams below are best interpreted that way. From that perspective, the diagram at the upper left and the graph in Figure 28.4 are consistent.

Figure 28.6: A visualization of the wave functions (or, atomic orbitals) for the $n = 1$, 2, and 3 levels for hydrogen. The brighter it is in a particular region, the more likely it is to find the electron in that region. Conversely, there is no chance of finding the electron in a region that is black. The numbers on the right side of the diagram show values of n, the principal quantum number, for each row. The letters across the top represent the values of the orbital quantum number, ℓ, for a particular column. Only one wave function is shown in the first row because, when $n = 1$, the only possible value of the orbital quantum number is $\ell = 0$. Similarly, there are only two images in the second row because there are two allowed ℓ values when $n = 2$. The wave functions are positive in the red areas and negative in the blue areas. In all cases shown, $m_\ell = 0$.

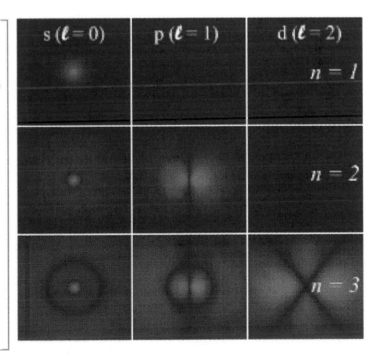

Related End-of-Chapter Exercises: 8, 9, 19 – 23.

Essential Question 28.3: How many different electron states are there that have $n = 3$?

Answer to Essential Question 28.3: If $n = 3$, then the orbital quantum number ℓ can take on 3 possible values, 0, 1, and 2. If $\ell = 0$, $m_\ell = 0$, and there are two possible spin states – thus, there are 2 states with $\ell = 0$. If $\ell = 1$, $m_\ell = -1$, 0, or +1, each with two possible spin states – thus, there are 6 states with $\ell = 1$. If $\ell = 2$, $m_\ell = -2, -1, 0, +1,$ or $+2$, each with two possible spin states – thus, there are 10 states with $\ell = 2$. That is a total of 18 states with $n = 3$.

28-4 The Pauli Exclusion Principle

In Section 28-3, we discussed the fact that it takes four quantum numbers to completely specify the state of an electron. In this section and in Section 28-5, we examine how these four quantum numbers determine the structure of the periodic table. To understand how these four numbers can determine a structure as complex as the periodic table, we begin with the Pauli exclusion principle, which is a simple statement with far-reaching consequences.

The Pauli exclusion principle: No two electrons can simultaneously occupy the same electron state in an atom. In other words, no two electrons in an atom can simultaneously have the same set of four quantum numbers.

The exclusion principle is named for the Austrian physicist Wolfgang Pauli (1900–1958), shown in the picture in Figure 28.7. Pauli was awarded the Nobel Prize in Physics in 1945 for the exclusion principle.

With the exclusion principle in mind, let's examine the ground-state (lowest energy) configurations of various atoms. In general, equation 28.3, in which the energy of an electron state is determined solely by the principal quantum number, is inadequate. The energy is also determined by the electron's orbital angular momentum. In general, for a given value of the principal quantum number, n, the higher the value of the orbital quantum number, ℓ, the higher the energy of that state. Let's begin by looking at the ground state configurations of a few elements in the periodic table. These are shown in Table 28.1. An explanation of what the number-letter-number notation means is given in Figure 28.8.

Figure 28.7: A photograph of Wolfgang Pauli, a colorful character who came up with the exclusion principle. Photo credit: Fermilab employees, via Wikimedia Commons (a public-domain image).

Atomic number	Name (symbol)	Ground-state configuration
2	Helium (He)	$1s^2$
6	Carbon (C)	$1s^2\,2s^2\,2p^2$
10	Neon (Ne)	$1s^2\,2s^2\,2p^6$
15	Phosphorus (P)	$1s^2\,2s^2\,2p^6\,3s^2\,3p^3$
20	Calcium (Ca)	$1s^2\,2s^2\,2p^6\,3s^2\,3p^6\,4s^2$
26	Iron (Fe)	$1s^2\,2s^2\,2p^6\,3s^2\,3p^6\,4s^2\,3d^4$

Table 28.1: Ground-state configurations for selected atoms.

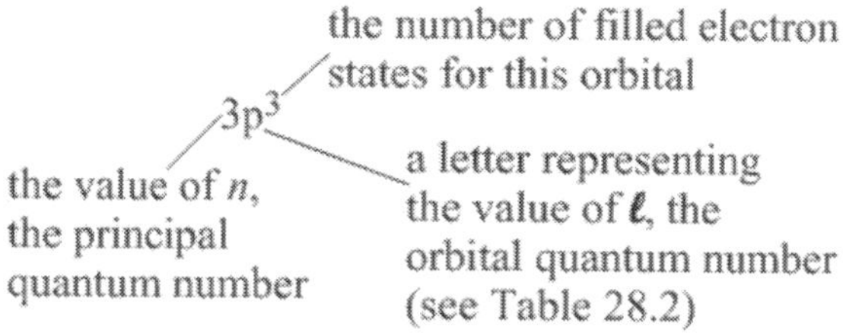

$3p^3$ —
the number of filled electron states for this orbital

the value of n, the principal quantum number

a letter representing the value of ℓ, the orbital quantum number (see Table 28.2)

Figure 28.8: Explaining the number-letter-number notation.

Referring to Table 28.1, we see that helium, with only two electrons, can have both electrons in the $n = 1$ shell, which can only fit two electrons. For atoms that have more than 2 electrons, two electrons are in the $n = 1$ level, and then the others are in states that have larger n values. The $n = 2$ level can fit eight electrons (two with $\ell = 0$ and six with $\ell = 1$). Thus, for elements up to and including neon, which has 10 electrons in its ground-state configuration, the electrons are in the $n = 1$ and $n = 2$ levels. Beyond this, electron state with higher n values come into play.

The layout of the ground-state configurations in Table 28.1 needs further explanation. This is provided by Table 28.2, which shows what value of ℓ the various letters correspond to, and by Figure 28.9, which shows how the various orbitals compare to one another, in terms of their energy level. As you wind your way through Figure 28.9 from top to bottom, the energy of a particular orbital increases. This figure shows the order in which electrons fill states in the ground-state configuration, because the ground-state configuration minimizes the total energy of the electrons.

Value of ℓ	Letter (stands for)
0	s (sharp)
1	p (principal)
2	d (diffuse)
3	f (fundamental)
4	g (letter after f)
5	h (letter after g)
6	i (letter after h)

Table 28.2: Letters for various ℓ values.

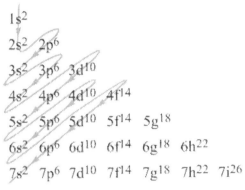

Figure 28.9: The order in which electrons fill the different subshells.

Terminology

An **atomic orbital** is both the mathematical function that describes where an electron can be, and the corresponding region in space. It takes three quantum numbers (n, ℓ, and m_ℓ) to define an orbital. Because there are two possible values of the spin quantum number, each atomic orbital can contain two electrons.

A **subshell** is the set of orbitals with the same values of n and ℓ.

A **shell** is the set of orbitals with the same value of n.

Figure 28.9 is consistent with what is known as the **aufbau principle**, derived from the German word *aufbauprinzip*, which means "building-up principle." The basic rule is that subshells are filled in order of lowest $n + \ell$ values. In cases of equal $n + \ell$ values, the subshells with the lower n value are filled first. The diagonal lines in Figure 28.9 represent subshells that have equal $n + \ell$ values, with those with the lower n values coming first.

Related End-of-Chapter Exercises: 10, 11, 57.

Essential Question 28.4: What is the minimum atomic number an atom should have before $4d^x$, where x represents a positive integer, appears in the atom's ground-state configuration? What is the range of values x can take on?

Answer to Essential Question 28.4: The atom should have an atomic number of at least 39. Following the twists and turns of Figure 28.9, we can count that there are 38 electron states up to and including the 5s subshell. Thus, the 39th electron should go into the 4d subshell. Because the 4d subshell can contain as many as 10 electrons, the x in $4d^x$ is an integer between 1 and 10.

28-5 Understanding the Periodic Table

In the previous two sections, we have laid the groundwork for understanding the periodic table. An idealized periodic table is laid out in Figure 28.10. Table 28.2 and Figure 28.9 are repeated here from Section 28-4 so you can more easily see the connection between the diagram that goes with the aufbau principle (Figure 28.9) and the periodic table.

Value of ℓ	Letter (stands for)
0	s (sharp)
1	p (principal)
2	d (diffuse)
3	f (fundamental)
4	g (letter after f)
5	h (letter after g)
6	i (letter after h)

Table 28.2: Letters for various ℓ values.

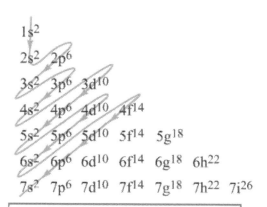

Figure 28.9: The order in which electrons fill the different subshells.

Figure 28.10 shows an idealized periodic table, showing the state of the last few electrons in each element's ground-state configuration. Atoms with completely filled subshells, or half-filled subshells, are particularly stable, and atoms with similar last-electron states generally have similar chemical properties. Note that the true ground-state configurations of elements with atomic numbers 105 and above are not yet known.

In reality, there are some deviations from the idealized behavior shown in Figure 28.10. These deviations can be quite instructive. Some of those deviations include:

- In column 11, the last two terms in the ground-state configurations for copper (Cu), silver (Ag), and gold (Au) are actually $4s^1\,3d^{10}$, $5s^1\,4d^{10}$, and $6s^1\,5d^{10}$, respectively. Each of these configurations ends with a half-full s subshell and a full d subshell, rather than a full s subshell and an almost-full d subshell, demonstrating that full and half-full subshells are particularly stable (lower energy) states.
- In column 6, the last two terms in the ground-state configurations for chromium (Cr) and molybdenum (Mo) are actually $4s^1\,3d^5$ and $5s^1\,4d^5$, respectively. Each of these configurations ends with a half-full subshell and a half-full d subshell, rather than a full s subshell and partly-filled d subshell, demonstrating that half-full subshells are particularly stable (lower energy) states.
- The last two terms in the ground-state configurations for gadolinium (Gd, element 64) and curium (Cm, element 96) are actually $4f^7\,5d^1$ and $5f^7\,6d^1$, respectively. Each of these configurations has a half-full f subshell and one electron in a d subshell, rather than partly-filled f subshell, again demonstrating that half-full subshells are particularly stable (lower energy) states.

Periodic table (Figure 28.10), with columns labeled 1, 2, then transition-metal groups 3–12, and main-group columns 13–18.

1	2	3	4	5	6	7	8	9	10	11	12	13	14	15	16	17	18
1 H $1s^1$																	2 He $1s^2$
3 Li $2s^1$	4 Be $2s^2$											5 B $2p^1$	6 C $2p^2$	7 N $2p^3$	8 O $2p^4$	9 F $2p^5$	10 Ne $2p^6$
11 Na $3s^1$	12 Mg $3s^2$											13 Al $3p^1$	14 Si $3p^2$	15 P $3p^3$	16 S $3p^4$	17 Cl $3p^5$	18 Ar $3p^6$
19 K $4s^1$	20 Ca $4s^2$	21 Sc $3d^1$	22 Ti $3d^2$	23 V $3d^3$	24 Cr $3d^4$	25 Mn $3d^5$	26 Fe $3d^6$	27 Co $3d^7$	28 Ni $3d^8$	29 Cu $3d^9$	30 Zn $3d^{10}$	31 Ga $4p^1$	32 Ge $4p^2$	33 As $4p^3$	34 Se $4p^4$	35 Br $4p^5$	36 Kr $4p^6$
37 Rb $5s^1$	38 Sr $5s^2$	39 Y $4d^1$	40 Zr $4d^2$	41 Nb $4d^3$	42 Mb $4d^4$	43 Tc $4d^5$	44 Ru $4d^6$	45 Rh $4d^7$	46 Pd $4d^8$	47 Ag $4d^9$	48 Cd $4d^{10}$	49 In $5p^1$	50 Sn $5p^2$	51 Sb $5p^3$	52 Te $5p^4$	53 I $5p^5$	54 Xe $5p^6$
55 Cs $6s^1$	56 Ba $6s^2$	*	72 Hf $5d^2$	73 Ta $5d^3$	74 W $5d^4$	75 Re $5d^5$	76 Os $5d^6$	77 Ir $5d^7$	78 Pt $5d^8$	79 Au $5d^9$	80 Hg $5d^{10}$	81 Tl $6p^1$	82 Pb $6p^2$	83 Bi $6p^3$	84 Po $6p^4$	85 At $6p^5$	86 Rn $6p^6$
87 Fr $7s^1$	88 Ra $7s^2$	**	104 Rf $6d^2$	105 Db $6d^3$	106 Sg $6d^4$	107 Bh $6d^5$	108 Hs $6d^6$	109 Mt $6d^7$	110 Ds $6d^8$	111 Rg $6d^9$	112 Uub $6d^{10}$	113 Uut $7p^1$	114 Uuq $7p^2$	115 Uup $7p^3$	116 Uuh $7p^4$	117 Uus $7p^5$	118 Uuo $7p^6$

* Lanthanides

57 La $4f^1$	58 Ce $4f^2$	59 Pr $4f^3$	60 Nd $4f^4$	61 Pm $4f^5$	62 Sm $4f^6$	63 Eu $4f^7$	64 Gd $4f^8$	65 Tb $4f^9$	66 Dy $4f^{10}$	67 Ho $4f^{11}$	68 Er $4f^{12}$	69 Tm $4f^{13}$	70 Yb $4f^{14}$	71 Lu $5d^1$

** Actinides

89 Ac $5f^1$	90 Th $5f^2$	91 Pa $5f^3$	92 U $5f^4$	93 Np $5f^5$	94 Pu $5f^6$	95 Am $5f^7$	96 Cm $5f^8$	97 Bk $5f^9$	98 Cf $5f^{10}$	99 Es $5f^{11}$	100 Fm $5f^{12}$	101 Md $5f^{13}$	102 No $5f^{14}$	103 Lr $6d^1$

Legend:

☐ Alkali metals   ☐ Transition metals   ☐ Non-metals   ☐ Halogens
☐ Alkaline earth metals   ☐ Poor metals   ■ Metalloids   ■ Noble gases

Figure 28.10: This figure lays out an idealized version of the periodic table, to show the predicted term in the ground-state configuration for each element. When viewed this way, one gets a better understanding of why the table is laid out as it is. In a given column, for instance, the elements have similar chemical properties because they have similar electron configurations. In the 17th column, for instance, these elements (known as halogens) have similar properties because in their ground-state configurations, each of the elements has 5 electrons in a p orbital, that p orbital being the highest-energy orbital that contains electrons for those elements. For most of the elements, the last term in their ground-state configurations match what is shown here, but some elements differ from the idealized version shown here, as detailed on the previous page. Also, note that helium is generally shown in the 18th column, because it is just as unreactive as the other noble gases, but you could make a strong argument for helium belonging at the top of column 2, based on its ground-state configuration.

Related End-of-Chapter Exercises: 55, 56, 59.

Essential Question 28.5: In Figure 28.10, the final term in the ground-state configuration of the element with the chemical symbol Po (element 84) is shown to be $6p^4$. What is the common name for this element? What is the complete ground-state configuration for this element, showing all the terms?

Answer to Essential Question 28.5: The common name for the element with the chemical symbol Po is polonium (not potassium, which has a chemical symbol of K). To write out the complete ground-state configuration of polonium, we can wind our way through Figure 28.9, starting from the top: $1s^2 2s^2 2p^6 3s^2 3p^6 4s^2 3d^{10} 4p^6 5s^2 4d^{10} 5p^6 6s^2 4f^{14} 5d^{10} 6p^4$.

28-6 Some Applications of Quantum Mechanics

Lasers

The word **laser** is an acronym, coming from the phrase light amplification by the stimulated emission of radiation. In our modern world, there are many applications of lasers. Such applications include bar-code readers in stores; surgery, particularly in eye surgery (including LASIK, Laser-Assisted in Situ Keratomileusis), where it is absolutely critical to do precision cutting; laser pointers, laser printers, and CD and DVD players; and fiber-optic communications.

The vast majority of lasers manufactured worldwide are diode lasers, in which light is produced from carefully constructed layers of semiconductors. In this section, we will focus on a different kind of laser, the helium-neon (or HeNe) laser that you may have seen in class or even used in a physics lab experiment. It is quite common to do demonstrations or experiments showing diffraction or interference of light using HeNe lasers.

As the name suggests, a helium-neon laser contains a mixture of helium and neon gas, at a relatively low pressure, with many more helium atoms than neon atoms. Coincidentally, the difference in energy between two of the electron energy levels in helium is almost the same as the difference in energy between two of the electron energy levels in neon – this is why these two elements are used. A high-voltage electrical discharge through the gas will excite helium atoms from their ground state to one of the $n = 2$ states, requiring an energy difference of 20.61 eV. These $n = 2$ states are **metastable**, which means that the electron will not immediately drop down to the ground state – it will remain in the excited state for a while.

With an extra 0.05 eV worth of kinetic energy, the excited helium atoms, when they collide with neon atoms that are in the ground state, can transfer 20.66 eV of energy to the neon atoms, just what is required to boost an electron in neon from the ground state to one of the $n = 3$ states (specifically, the 3s state). This level is also metastable, but electrons in some of these excited states will spontaneously drop down to the 2p state, emitting a photon of 632.8 nm (in air), corresponding to the wavelength of light emitted by a typical red HeNe laser. These photons interact with the excited neon atoms, which encourages them also to make the 3s to 2p transition. This part of the process is the *stimulated emission* that is part of what laser stands for. An energy-level diagram for the HeNe laser is shown in Figure 28.11.

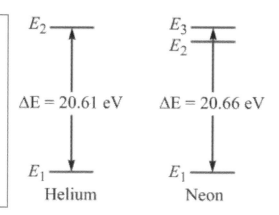

Figure 28.11: An energy-level diagram for the HeNe laser. Because the energy difference between the two lowest energy levels in helium almost exactly equals the energy difference between the $n = 1$ and $n = 3$ levels in neon, excited helium atoms can transfer energy to ground-state neon atoms via collisions. Note that, despite what it looks like in the diagram, the E_1 levels in helium and neon are at completely different energies. What is critical, however, is that the difference between two of the helium levels almost exactly matches the difference between two of the neon levels.

Once the light has been created, by the process of stimulated emission inside the laser, a laser beam must then be created. The photons and the low-pressure mixture of helium and neon are contained in a tube that is typically 15 – 50 cm in length. At one end of the tube is a highly reflective mirror, while at the other end of the tube (the end that the beam emerges from) is a mirror that reflects most of the light, but which allows a little light (about 1%) to pass through. Generally, the photons emerge after bouncing back and forth many times between the mirrors, resulting in a beam of light that has very little spread.

The design of the mirrors inside the laser is also interesting, because the mirrors exploit thin-film interference for a wavelength corresponding to the wavelength of light emitted by the laser. There are several different electron transitions associated with helium and neon, so a HeNe laser can actually emit several different wavelengths (a different wavelength for each transition between electron energy levels). By adjusting the thin films on the mirrors inside the laser, the laser can be optimized for the emission of the standard 632.8 nm red light, or a different wavelength. The first HeNe laser ever made, in the 1960's, for instance, emitted ultraviolet light with a wavelength of 1150 nm.

Fluorescence and phosphorescence

Two more applications of quantum mechanics, again associated with electron energy levels and the photons that are emitted when electrons make a transition from a higher-energy state to a lower-energy state, are fluorescence and phosphorescence. These two phenomena are similar, in that exposing a fluorescent or phosphorescent material to (usually) ultraviolet light will excite electrons from lower-energy states to higher-energy states. When the electrons drop back toward the lower-energy state, however, they do so by dropping down a smaller step in energy, to an intermediate level. If this smaller energy causes photons to be emitted in the visible spectrum, we can then see them. In fluorescent materials, the phenomenon is present only when the light source that excites the upward transitions is present. In phosphorescent materials, it takes a long time, on average, before the excited electrons make a transition to a lower level. Because the electron transitions occur over a long time period, visible light continues to be emitted long after the light source that excites the upward transitions is removed. A photograph of a variety of fluorescent minerals is shown in Figure 28.12.

Figure 28.12: A variety of fluorescent minerals, photographed while they are being exposed to ultraviolet light. In general, these minerals are much less colorful when they are not fluorescing. Image credit: Hannes Grobe, via Wikimedia Commons.

Related End-of-Chapter Exercises: 12, 27.

Essential Question 28.6: Consider the information given in Figure 28.11. (a) If the $n = 3$ to $n = 2$ transition in neon produces a photon with a wavelength of 632.9 nm (in vacuum), what is the difference in energy between these two levels? (b) What is the difference in energy between the $n = 2$ and $n = 1$ levels in neon?

Answer to Essential Question 28.6: (a) The photon energy is the difference in energy. Converting the wavelength to energy, we get:

$$E = hc/\lambda = (6.626\times10^{-34} \text{ J s})(3.00\times10^8 \text{ m/s})/(632.9\times10^{-9} \text{ nm}) = 3.141\times10^{-19} \text{ J}.$$

Converting this energy to electron volts gives 3.141×10^{-19} J$\times\dfrac{1 \text{ eV}}{1.602\times10^{-19} \text{ J}} = 1.96$ eV .

(b) If the $n = 3$ and $n = 1$ levels are separated by 20.66 eV of energy, and the energy difference between the $n = 3$ and $n = 2$ levels is 1.96 eV, the difference between the $n = 2$ and $n = 1$ levels must be 20.66 eV – 1.96 eV = 18.70 eV.

Chapter Summary

Essential Idea: The Atom.
   In our modern view of the atom, the nucleus is modeled as a point particle with a positive charge. The chemical properties of the atom are dominated by the behavior of the atom's electrons. In contrast to the popular representation of an atom being made up of a positive nucleus orbited by electrons in well-defined circular or elliptical orbits, our modern view is that an electron behaves more like a cloud of negative charge, with the density of the charge cloud at a point corresponding to the probability of finding the electron at that point.

Atomic spectra
Exciting a gas causes light to be emitted. The resulting emission spectrum is the fingerprint of the gas – lines are seen at specific wavelengths. A photon produced by an atom comes from an electron that makes a transition from a higher energy level to a lower energy level within the atom. The energy of the photon corresponds to the difference in energy between these electron energy levels.

The four quantum numbers
   One of the early models of the atom, the Bohr model, used a single quantum number to define the quantization of the energy levels, allowed radii, and angular momenta of the electron orbits. Our modern view, the quantum mechanical view of the atom, four quantum numbers are used to completely describe the state of a particular electron. Each quantum number is associated with a physical property of the electron.

   The energy of the electron is quantized by the principal quantum number, n. Allowed values of n are $n = 1, 2, 3, \ldots$ For the hydrogen atom, the energy levels have energies given by:

$$E_n = \frac{-13.6 \text{ eV}}{n^2}, \qquad \text{(Eq. 28.2: \textbf{Energies of the electron levels in the hydrogen atom})}$$

   The magnitude of the orbital angular momentum of the electron is quantized by the orbital quantum number, ℓ, which can take on values $\ell = 0, 1, \ldots, n-1$. The magnitude of the orbital angular momentum is given by:

$$L = \sqrt{\ell(\ell+1)}\,\frac{h}{2\pi}. \qquad \text{(Eq. 28.4: \textbf{Orbital angular momentum})}$$

   The direction of the orbital angular momentum is also quantized. In general, we say that the z-component can take on a limited number of directions, determined by the magnetic quantum number, m_ℓ, which can take on the values $m_\ell = -\ell, -\ell +1,....., \ell -1, \ell$. The z-component of the orbital angular momentum is given by:

$$L_z = \frac{m_\ell h}{2\pi},$$ (Eq. 28.5: **z-component of the orbital angular momentum**)

Finally, the spin angular momentum can take on one of only two values, conventionally referred to as "spin up" and "spin down." The spin angular momentum is characterized by the spin quantum number, which can take on values of $+1/2$ or $-1/2$.

Understanding the periodic table of elements

One key to understanding the periodic table is the Pauli exclusion principle – no two electrons in an atom can have the same set of four quantum numbers.

The layout of the periodic table of the elements has to do with the highest-energy filled or partly filled subshell in the ground-state electron configuration of an atom. Elements that have similar ground-state configurations (such as four electrons in a p subshell) are grouped in a column in the periodic table, and generally have similar chemical properties.

Complete ground-state configurations are written with triplets of numbers and letters, in the form $3p^4$. The first number (3, in this case) represents the value of n, the principal quantum number, for the orbital. The letter (p, in this case) stands for the value of the orbital quantum number, ℓ. The second number (4, in this case) is the number of electrons in the subshell in this configuration.

Applications of atomic physics

Two applications of atomic physics include some lasers (such as the helium-neon laser) and the properties of fluorescence and phosphorescence. These applications involve the emission of photons from electrons in atoms dropping down from one energy level to another.

End-of-Chapter Exercises

Exercises 1 – 12 are conceptual questions that are designed to see if you have understood the main concepts of the chapter.

1. In a particular line spectrum, photons are observed that have an energy of 2.5 eV. Choose the phrase below that best completes the following sentence. The element producing the light definitely has …
    (i)… an electron energy level at an energy of 2.5 eV.
    (ii)… an electron energy level at an energy of –2.5 eV.
    (iii)… two electron energy levels that are 2.5 eV apart in energy.

2. Imagine that there is an atom with electron energy levels at the following energies: –88 eV, –78 eV, and –60 eV. Confining ourselves to electron transitions between these levels only, would we expect to see photons emitted from this atom with the following energies? Explain why or why not. (a) 60 eV, (b) –60 eV, (c) 18 eV, (d) 138 eV.

3. To be visible to the human eye, photons must have an energy between about 1.8 eV and 3.1 eV. In a particular atom, one of the electron energy levels is at an energy of –60.0 eV. Can electron transitions associated with this energy level produce photons that are visible to the human eye? Explain why or why not.

4. In Exploration 28.1, we determined that the energies of the three lowest energy levels in hydrogen have energies of $E_1 = -13.6$ eV, $E_2 = -3.40$ eV, and $E_3 = -1.51$ eV. Photons are produced by electrons that transition from the $n = 3$ level to the $n = 1$ level, as well as from the $n = 2$ level to the $n = 1$ level. Which of these photons have the larger (a) energy? (b) frequency? (c) wavelength?

5. Figure 28.13 shows a graph of the probability, per unit distance, of finding the electron in the *3s* orbital of hydrogen, at various distances from the nucleus. (a) At approximately what distance from the nucleus is the electron most likely to be found? Compare this to the value of 9 Bohr radii that is predicted by the Bohr model for the $n = 3$ state. (b) If we determined the area under the curve for this graph, for the region from $r = 0$ to $r = $ infinity, what value would that area turn out to be?

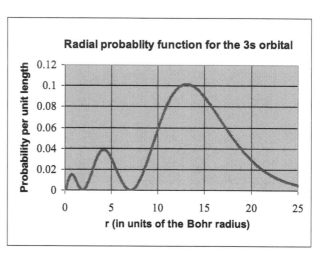

Figure 28.13: A graph of the probability, per unit length, of finding the electron in the *3s* orbital of hydrogen at various distances from the nucleus. For Exercise 5.

6. Figure 28.14 shows the relative probability per unit length as a function of position for a particle confined to a line. The particle is definitely located somewhere between $x = 0$ and $x = d$. (a) At what location(s) between $x = 0$ and $x = d$ is the particle most likely to be found? (b) At what location(s) between $x = 0$ and $x = d$ is the particle least likely to be found?

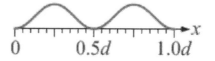

Figure 28.14: A graph of the relative probability per unit length of finding a particle confined to line at a particular position, as a function of position. For Exercise 6.

7. The Bohr model of the atom was an important stage along the path to our understanding of the atom, but it is important to recognize that the Bohr model is quite different from our modern view of the atom. Describe at least two ways in which the Bohr model differs from the modern view.

8. Each of the four quantum numbers is associated with a physical property of the electron. What physical property is associated with (a) the principal quantum number, n? (b) the orbital quantum number, ℓ? (c) the magnetic quantum number, m_ℓ? (d) the spin quantum number, m_s?

9. If the value of the principal quantum number is $n = 2$, what are the allowed values of (a) ℓ, the orbital quantum number, (b) m_ℓ, the magnetic quantum number, and (c) m_s, the spin quantum number? (d) How many different states are there that have $n = 2$?

10. Explain why electrons tend to fill the $4s^2$ orbital before the $3d^{10}$ orbital.

11. Are the following electron configurations valid or invalid? Note that the configurations do not have to be ground-state configurations. If a configuration is valid, state which element it represents. If a configuration is invalid, explain why. (a) $1s^2 2s^3 2p^6$. (b) $1s^2 2s^1 2p^6 3s^2 3p^3$. (c) $1s^2 2s^2 2p^4 3s^2 3p^3$. (d) $1s^2 2s^2 2p^8 3s^1$.

12. Choose the phrase below that best completes the following sentence. A mixture of helium and neon was chosen to create a laser because …

   (i) … helium and neon are from the same column in the periodic table.
   (ii) … there is an energy level in helium that has almost the same energy as an energy level in neon.
   (iii) … the difference between two energy levels in helium is almost the same as the difference between two energy levels in neon.

Exercises 13 – 18 involve line spectra.

13. Two energy levels in a particular atom are at energies of −23.4 eV and −25.6 eV. When an electron makes a transition from one of these levels to the other, a photon is emitted. For the photon, find the (a) energy, (b) frequency, and (c) wavelength. (d) For the transition described here, which level does the electron transition to?

14. In Exploration 28.1, we calculated the $n = 2$ energy level for hydrogen to have an energy of −3.40 eV. (a) What are the energies of the $n = 4$ and $n = 5$ levels for hydrogen? (b) If the electron in the hydrogen atom makes a transition from the $n = 4$ level to the $n = 2$ level, what is the energy of the emitted photon? (c) Repeat part (b), with the electron transitioning from $n = 5$ to $n = 2$, instead.

15. In Exploration 28.1, we sketched an energy-level diagram with the three lowest energy levels of hydrogen. The lines in the visible region of the spectrum that are emitted by hydrogen correspond to electron transitions that come from a higher level to the $n = 2$ level. Because of this, let's focus on the part of the energy-level diagram between an energy of 0 and the $n = 2$ level. (a) Determine the energies of the $n = 4$, $n = 5$, $n = 6$, and $n = 7$ levels for hydrogen. (b) Sketch an energy-level diagram to show the energy levels for $n = 2$ through $n = 7$ for hydrogen. (c) Determine the energies of the photons emitted when electrons transition from the $n = 3$ through $n = 7$ levels down to the $n = 2$ level. (d) Calculate the wavelengths for these photons.

16. A hypothetical atom has electron energy levels at the following energies: −14 eV, −30 eV, −52 eV, and −80 eV. Assume that electron transitions occur between these levels only. (a) How many different photon energies are possible? (b) List the photon energies, from smallest to largest.

17. Figure 28.15 shows an energy-level diagram for a hypothetical atom. Sketch the corresponding emission spectrum for this atom, showing the energies of the photons associated with electron transitions between the four levels shown in the figure.

   $E = 0$
   $E_4 = -1.00$ eV
   $E_3 = -2.50$ eV
   $E_2 = -4.50$ eV

18. Figure 28.15 shows an energy-level diagram for a hypothetical atom. Sketch the corresponding emission spectrum for this atom, showing the wavelengths of the photons associated with electron transitions between the four levels shown in the figure.

   $E_1 = -9.00$ eV

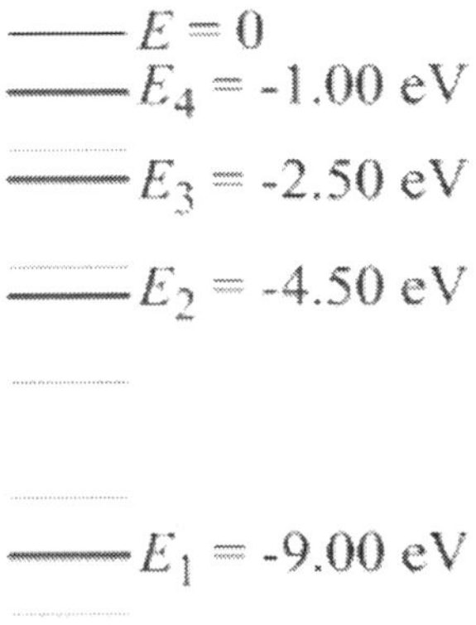

Figure 28.15: An energy-level diagram for a hypothetical atom. For Exercises 17 and 18.

Exercises 19 – 24 involve the four quantum numbers that are associated with the quantum-mechanical view of the atom.

19. If the value of the orbital quantum number is $\ell = 3$, what are the allowed values of (a) m_ℓ, the magnetic quantum number, and (b) m_s, the spin quantum number? (d) How many different states are there that have $\ell = 3$?

20. If the value of the principal quantum number is $n = 4$, what are the allowed values of (a) ℓ, the orbital quantum number, (b) m_ℓ, the magnetic quantum number, and (c) m_s, the spin quantum number? (d) How many different states are there that have $n = 4$?

21. If the value of the orbital quantum number is $\ell = 4$, what are the allowed values of the principal quantum number, n?

22. Figure 28.16 shows the allowed directions of the orbital angular momentum when the orbital quantum number, ℓ, has a particular value. (a) What is the value of ℓ in this case? (b) Re-draw the picture, and label each of the vectors with its corresponding m_ℓ value. (c) In units of $h/(2\pi)$, what is the length of each of the vectors?

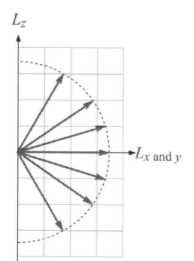

Figure 28.16: The allowed directions of the orbital angular momentum, when ℓ has a particular value. The units on both axes are angular momentum units of $h/(2\pi)$.

For Exercise 22.

23. The value of the magnetic quantum number for a particular electron in the 3d orbital happens to be $m_\ell = +1$. What, if anything, can we say about the values of the other three quantum numbers for this electron?

Exercises 24 – 28 involve applications of atomic physics.

24. Astronomers, particularly astronomers who use radio telescopes, have been able to examine a number of interesting features of the universe through measurements related to what is known as the "21-centimeter line" of hydrogen. The 21-centimeter line refers to photons that have a wavelength of about 21 cm, associated with a transition between two different states of hydrogen. (a) How does the energy of a photon with a wavelength of 21 cm compare to that of a photon in the visible spectrum? (b) Do some research regarding the transition in hydrogen that produces these 21-cm photons, and write a couple of paragraphs explaining the transition, and how the 21-cm line is used in astronomy.

25. One way to view emission spectra is to excite a tube of gas with high voltage, and view the emitted light through a diffraction grating that is held so that the plane of the grating is perpendicular to the direction in which the light travels from the tube to the grating. Recall that we discussed diffraction gratings in Chapter 25. The spectrum of helium gas includes lines at the following wavelengths: 447.1 nm, 501.6 nm, 587.6 nm, and 667.8 nm. When viewed through a diffraction grating in which the openings in the grating are separated by 1500 nm, at what angle (relative to the light coming from the tube) is the first-order line for the (a) 447.1 nm wavelength? (b) 667.8 nm wavelength? (c) How many complete spectra (confining ourselves to the four lines listed here) can be seen when viewing the tube through the diffraction grating?

26. One half of the 1986 Nobel Prize in Physics was awarded to Binning and Rohrer. Do some research about these two scientists, and write a couple of paragraphs regarding what they won the Nobel Prize for, and how it relates to the material covered in this chapter.

27. A neon sign, which is a glass tube filed with excited neon gas, gives off a red-orange glow, as illustrated by the word "Coffee" in the photograph in Figure 28.17. If the color is not red-orange, the tube is generally not filled with neon. Explain qualitatively what the observation about the color of the neon sign tells us about the energy levels and electron transitions for neon atoms.

28. The second (the unit of time) is defined as the duration of 9 192 631 770 cycles of the radiation emitted by an electron making a transition between two ground-state energy levels of the cesium-133 atom. What is the difference between these two energy levels, in electron volts?

Figure 28.17: A neon sign, for Exercise 27. Photo credit: Tom Genovese, from publicdoamainpictures.net

General problems and conceptual questions

29. Niels Bohr, for whom the Bohr model is named, had an interesting life. Do some research about him, and write a couple of paragraphs about him. Include some descriptions about Bohr as a person as well as about his contributions to science.

30. The spectrum emitted by excited sodium gas includes two very closely spaced lines in the yellow part of the visual spectrum, at wavelengths of 588.995 nm and 589.592 nm. The photons associated with this light come from electron transitions that start from one of two closely-spaced 3p levels, and which end at the same 3s level. What is the energy difference of these two 3p levels?

31. When an electron in a particular atom transitions from one energy level to another, the atom emits a photon that has a wavelength of 623 nm. One of the energy levels associated with this transition has an energy of –30.28 eV. What is the energy of the other level?

32. Start with the electron in the ground state in the hydrogen atom, and use Equation 28.2 to calculate energy. How much energy is required to (a) ionize the atom (removing the electron completely)? (b) excite the electron from the ground state to the $n = 9$ level?

33. Which requires more energy, to excite the electron in hydrogen from the ground state to the $n = 2$ level, or to excite the electron from the $n = 2$ level to the $n = 100$ level? Explain.

34. A particular atom emits photons with wavelengths of 485 nm and 607 nm. Which wavelength is associated with the larger difference in energy between two energy levels in that atom? Briefly justify your answer.

35. The spectrum of helium gas includes lines at the following wavelengths: 447.1 nm, 501.6 nm, 587.6 nm, and 667.8 nm. Find the corresponding energy for these photons.

36. If you have the energy-level diagram for a particular atom, you can predict the emission spectrum for that atom (as we did in Exercises 17 and 18). If you have the emission spectrum for an atom, can you determine the atom's energy-level diagram? Explain.

37. When a spectral tube containing gas from a particular element is excited by applying high voltage, the spectrum shown in Figure 28.18 is observed. (a) What is the minimum number of energy levels that can be used to obtain this three-line spectrum? (b) Using the minimum number of levels, sketch an energy-level diagram that is consistent with Figure 28.18.

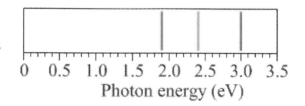

Figure 28.18: The lines in the visible spectrum that are seen when a particular gas is excited with high voltage, for Exercise 37.

38. When a spectral tube containing gas from a particular element is excited by applying high voltage, photons with the following energies are emitted: 3 eV, 13 eV, 16 eV, 25 eV, 38 eV, and 41 eV. No other photons in this range are observed. One of the electron energy levels in the corresponding energy-level diagram is at an energy of –20 eV. (a) What is the minimum number of energy levels that can be used to obtain the given six-line spectrum? (b) Using the minimum number of levels, sketch an energy-level diagram that is consistent with the information given here. (c) Is there only one possible answer to (b)? Explain.

39. Equation 28.3 can be used to find the energies of the energy levels for atoms that have only one electron. Let's use Equation 28.3 to examine the emission spectrum of doubly-ionized lithium, Li^{2+} (that is, lithium atoms with just one electron). (a) The energies of the electron energy levels in hydrogen are equal to the energies of some of the electron energy levels of Li^{2+}. Which n values, for Li^{2+}, give the same energies as those of the $n = 1$, $n = 2$, and $n = 3$ levels in hydrogen? (b) With matching energy levels, the energies of the photons emitted by excited hydrogen could also be produced by excited Li^{2+} atoms. Is it possible to distinguish between spectra from hydrogen and Li^{2+}? If so, how?

40. In a hypothetical atom, the energies of the electron energy levels are given by $E_n = -(144 \text{ eV})/n^2$, where n is a positive integer. (a) What are the energies of the four lowest energy levels in this atom? (b) What are the energies of the photons emitted by electrons that transition between any two of these four levels?

41. In a hypothetical atom, the energies of the electron energy levels are given by $E_n = -(60 \text{ eV})/n$, where n is a positive integer. Consider the lowest six energy levels.

For transitions between the lowest six energy levels only, determine how many different ways there are to produce photons with an energy of (a) 10 eV, and (b) 5 eV. Specify the initial and final values of n for each of these transitions.

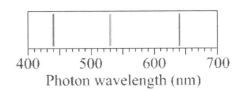

Photon wavelength (nm)

42. The wavelengths of visible photons that are emitted by a tube of excited gas are shown in Figure 28.19. (a) Calculate the energies of these photons. (b) If these photons are emitted in electron transitions that end at an energy level of -16.5 eV, what are the energies of the energy levels that the electrons start from to produce these photons?

Figure 28.19: The wavelengths of photons in the visible spectrum that are seen when a particular gas is excited with high voltage, for Exercise 42.

43. The "Balmer series" is the name given to the set of photon energies (or, equivalently, wavelengths) that correspond to electrons making the transition from higher-n levels to the $n = 2$ level in hydrogen. For the Balmer series, what is the (a) smallest photon energy? (b) largest photon energy?

44. The "Lyman series" is the name given to the set of photon energies (or, equivalently, wavelengths) that correspond to electrons making the transition from higher-n levels to a specific lower level in hydrogen. (a) What is the value of n that characterizes the lower level for the Lyman series? (b) Who was Lyman? Do some research about Lyman and write a paragraph describing this scientist.

45. Return to the situation described in Exercise 44. For the Lyman series, what is the (a) largest photon wavelength? (b) smallest photon wavelength?

46. One of the wavelengths emitted by excited hydrogen gas in a tube that is at rest on the Earth is found to be 656.28 nm. When you carefully measure the corresponding wavelength emitted from the hydrogen in a distant star, however, you find the wavelength to be 658.73 nm. Having learned about the Doppler effect for electromagnetic waves, in Chapter 22, you realize that, with this data, you can determine the velocity of the distant star with respect to the Earth (at least, you can find the component of the velocity that is along the line joining the Earth and the star). Find the value of this velocity.

47. The Hubble Space Telescope is named after the American astronomer Edwin Hubble. Do some research about Edwin Hubble and write a couple of paragraphs about what he is most famous for, and how his work relates to the principles of physics discussed in this chapter. Note that an image of part of the Eagle Nebula, as seen by the Hubble Space Telescope, is shown in Figure 28.20.

Figure 28.20: An image of a vast cloud of gas and dust that is part of the Eagle Nebula, as imaged by the Hubble Space Telescope, for Exercise 47. Image credit: Jeff Hester and Paul Scowen (Arizona State University), and NASA/ESA.

48. We discussed the photoelectric effect experiment in Chapter 27. One way the photoelectric effect experiment is often carried out is to use a mercury light source to illuminate a particular surface. The emission spectrum of mercury has strong lines at wavelengths of 365.5 nm (ultraviolet), 404.7 nm (violet), 435.8 nm (blue), and 546.074 nm (green). The light can be passed through a diffraction grating to separate the light into these different wavelengths. Assume that photons of all these wavelengths have enough energy to overcome the work function of the surface. (a) Which of these wavelengths will produce electrons with the largest kinetic energy? (b) If the maximum kinetic energy of the emitted electrons is 1.27 eV when the green light illuminates the surface, what will the maximum kinetic energy of the emitted electrons be when the violet light illuminates the surface? (c) What is the work function of this surface?

49. Figure 28.21 shows a graph of the probability, per unit distance, of finding the electron in the *2s* orbital of hydrogen, at various distances from the nucleus. (a) At approximately what distance from the nucleus is the electron most likely to be found? Compare this to the value of 4 Bohr radii that is predicted by the Bohr model for the $n = 2$ state. (b) Use the graph to estimate the probability of finding the electron at a distance of less than 2 Bohr radii from the nucleus.

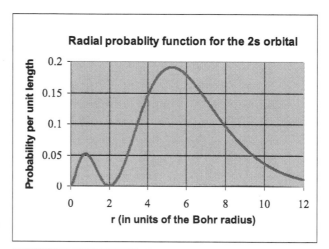

50. For the graph in Figure 28.21, the area under the curve for the region from $r = 0$ to $r = 5.8$ Bohr radii is equal to 0.5. (a) What is the area under the curve from the region from $r = 5.8$ Bohr radii to $r = $ infinity? (b) What does this mean, in terms of the probability of finding the electron at particular distances from the nucleus?

Figure 28.21: A graph of the probability, per unit length, of finding the electron in the *2s* orbital of hydrogen at various distances from the nucleus. For Exercises 49 and 50.

51. A particular electron in an atom is in a state such that its orbital quantum number is $\ell = 3$. Express your answers in terms of h (Planck's constant) and π. (a) What is the magnitude of the electron's orbital angular momentum? (b) What are the possible values of the z-component of the electron's orbital angular momentum?

52. A particular electron in an atom is in a state such that the z-component of its orbital angular momentum is $+h/2\pi$. What, if anything, does this tell us about the value of (a) the electron's orbital quantum number, ℓ? (b) the electron's principal quantum number, n?

53. In Figures 28.5 and 28.16, we sketched vector diagrams to show the allowed directions of the orbital angular momentum vector for two specific values of ℓ. (a) Draw a similar diagram that applies when $\ell = 1$, labeling each vector with the corresponding value of m_ℓ. (b) What is the magnitude of each of the vectors on the diagram?

54. In the absence of an external magnetic field, states with the same values of n and ℓ have the same energy. For instance, with no external magnetic field, electrons in the 3p orbital in a particular atom all have the same energy. If such electrons drop from the 3p level to the 2s level, we would see a single line in the emission spectrum, corresponding to photons with an energy equal to the energy difference between the 3p and 2s levels. When an external magnetic field is applied, however, electrons within an orbital will have slightly different energies, experiencing a shift in energy from the zero-field value that is proportional to the strength of the magnetic field multiplied by the electron's quantum number. This splitting of the energy levels is known as the **Zeeman effect**. (a) For the example mentioned here, electrons transitioning from the 3p to 2s levels, explain why we only have to worry about the splitting of energy levels for the 3p states, not for the 2s states. (b) Sketch, qualitatively, an energy-level diagram showing the 3p and 2s energy levels in both the absence and presence of an external magnetic field. (c) What affect would applying an external magnetic field have on the line in the emission spectrum associated with the 3p to 2s transition?

55. Use the chart in Figure 28.9 to help you answer this question. Write out the complete ground-state configuration for the element with an atomic number of (a) 16. (b) 34. (c) Briefly explain why these elements are found in the same column in the periodic table.

56. A short-hand method for writing out the ground-state configuration of an element is to start from the configuration of a column-18 element, and then add whatever is missing. For instance, the ground-state configuration of zirconium (Zr) can be written as [Kr] $5s^2 4d^2$, which means that the initial terms match those of krypton (Kr), and the additional terms are $5s^2 4d^2$. (a) Write out the complete ground-state configuration for zirconium. (b) Why might a column-18 element be used as the starting point for the short-hand method, as opposed to, say, a column-16 element? Using this short-hand method, write out the ground-state configuration of (c) Nickel (Ni), and (d) Lead (Pb).

57. The ground-state configuration of a particular element is $1s^2 2s^2 2p^6 3s^2 3p^6 4s^2 3d^{10} 4p^6 5s^2 4d^{10} 5p^5$. (a) What is the atomic number of this element? (b) What is the name of this element?

58. Reading up on the properties and applications of different elements can be quite fascinating. Do some research about the element antimony, and write up a couple of paragraphs to tell the story of antimony.

59. Elements in column 1 of the periodic table generally make strong bonds with elements of column 17. An example of this is NaCl, table salt. To form the bond, the column-1 element essentially gives up one of its electrons to the column-17 element. Explain why this generally leads to a stable structure.

60. Two students are having a conversation. Comment on each of the statements that they make below.

Julia: I'm trying to figure out how to draw the energy-level diagram from this spectrum we were given. I converted the six wavelengths to electron volts – is the energy-level diagram just that set of six energies?

Kristin: No – those are all the photon energies. You have to think about where the photons come from – how do the photons relate to the energy levels?

Julia: OK, I remember – a photon gets released when an electron changes levels, and the photon energy equals the difference between the electron levels. So, like, this 2.5 eV photon could come from an electron dropping from a 12.5 eV level to a 10 eV level.

Kristin: That sounds right. Except, how do you know those are the right energies? You could also get 2.5 eV by dropping from 10 eV to 7.5 eV.

Julia: You're right – we could get 2.5 eV an infinite number of ways. Maybe we need to look at the other photons to narrow down the possibilities. Look, I see a pattern here for these three photons, we have 2.5 eV and 4.1 eV, and they add up to 6.6 eV, and we also have a 6.6 eV photon. So, we could have energy levels at 16.6, 12.5, and 10, and that gives us all those photons.

Kristin: Don't we need 14.1 eV, too, to get the 4.1 eV photon?

Julia: No, we've got that already, because the electron can go 16.6 to 12.5 – that's 4.1. OK, so, for those three photons we had three energy levels – if we can do three more energy levels for the other photons we'll be all set. Those other photon energies are 13.4, 17.5, and 20.0. Huh, if you add 13.4 and 17.5 you don't get 20 – how do you do those ones?

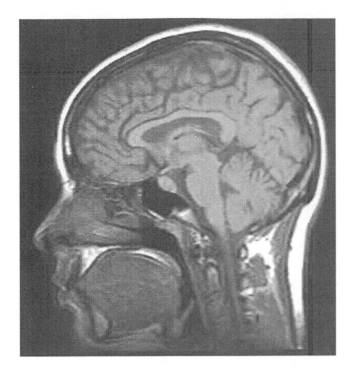

This image shows shows structures in a particular plane inside a person's head, obtained with a magnetic resonance imaging system. Such images are commonly used in hospitals for diagnostic purposes.

Image credit: public-domain image.

Chapter 29 – The Nucleus

CHAPTER CONTENTS

To understand chemical reactions, we can generally neglect the nucleus and focus on the electrons in the atom. To understand things like where the Sun's energy comes from, the production of nuclear power, and what radioactivity is all about, we need to go farther down into the very heart of the atom, into the nucleus.

In this final chapter of the book, we will investigate puzzles such as what holds a nucleus together; what does the most famous equation in physics, $E = mc^2$, have to do with the nucleus; and, how can we use signals from nuclei in our bodies to obtain an MRI, as in the picture above.

Finally, note that, although this chapter marks the end of the book, there is a vast universe out there, and we have only just scratched the surface of all there is to know about how the universe works. This is an exciting time to be alive, because there are plenty of interesting physics questions still to be solved. What is dark energy? Is string theory a good model of how the universe is constructed? Will anyone invent a material that is superconducting at room temperature? What did the early universe look like? How can physics be applied most effectively to treat cancer? It is also an interesting time to be alive because there are a number of important experiments coming on-line that will allow us to probe many of these questions more deeply than ever. With the base of physics knowledge and understanding you have built up now, you should be well-prepared to keep reading about new developments in physics as they come about during your lifetime.

29-1 What Holds the Nucleus Together?

Before answering the question of what keeps the nucleus together, let's go over the basics of what a nucleus is. A nucleus is at the heart of the atom. It is positively charged, because it contains the atom's protons. It also contains the neutrons, which have no net charge. We actually have a word, **nucleons**, for the particles in a nucleus. The atomic mass number, A, for a nucleus is the total number of nucleons (protons and neutrons) in a nucleus. The atomic number, Z, for a nucleus is the number of protons it contains.

The protons and neutrons can be modeled as tiny balls that are packed together into a spherical shape. The radius of this sphere, which represents the nucleus, is approximately:

$$r \approx (1.2 \times 10^{-15} \text{ m}) A^{1/3}. \quad \text{(Eq. 29.1: \textbf{The radius of a nucleus})}$$

This is almost unbelievably small, orders of magnitude smaller than the radius of the atom.

An atom of a particular element contains a certain number of protons. For instance, in the nucleus of a carbon atom there are always 6 protons. If the number of protons changes, we get a new element. There can be different numbers of neutrons in the nucleus, however. Carbon atoms, for instance, are stable if they have six neutrons (by far the most abundant form of natural carbon), seven neutrons, or eight neutrons. These are known as **isotopes** of carbon – they all have six protons, but a different number of neutrons.

The general notation for specifying a particular isotope of an element is shown in Figure 29.1. The notation is somewhat redundant – by definition, aluminum is the 13th element in the periodic table, and thus has an atomic number of $Z = 13$. A little redundancy is not a bad idea, however – it will help us keep things straight later when we write equations for radioactive decay processes and for nuclear reactions.

Figure 29.1: The general notation for specifying a particular isotope is shown at left, where X represents the chemical symbol for the element, A is the atomic mass number (the number of protons + neutrons) and Z is the atomic number (the number of protons). A specific case, the isotope aluminum-13. is shown at right.

$$^A_Z X \qquad\qquad ^{27}_{13}Al$$

The masses and charges of the three basic building blocks of atoms are shown below in Table 29.1. In addition to specifying the mass in kilograms, the mass is also shown in atomic mass units (u). By definition, 1 atomic mass unit is 1/12th the mass of a neutral carbon-12 atom.

$$1\,\text{u} = 1.660540 \times 10^{-27}\ \text{kg}.$$

Particle	Charge	Mass (kg)	Mass (u)
neutron	0	1.674929×10^{-27} kg	1.008664
proton	$+e$	1.672623×10^{-27} kg	1.007276
electron	$-e$	9.109390×10^{-31} kg	0.00054858

Table 29.1: The charge and mass of the neutron, proton, and electron. Recall that $e = 1.602 \times 10^{-19}$ C.

Related End-of-Chapter Exercises: 1, 41, 42.

EXPLORATION 29.1 – Holding the nucleus together
    Let's explore what it is that holds a nucleus together, starting by considering whether what holds the nucleus together is a force we already know about.

Step 1 – *First, apply Coulomb's Law to calculate the electrostatic force between two protons that are separated by 1 × 10⁻¹⁵ m. Can this force hold the nucleus together?* Substituting the appropriate values into Coulomb's Law gives:

$$F = \frac{k(+e)(+e)}{r^2} = \frac{(9.0 \times 10^9)(1.602 \times 10^{-19} \text{ C})^2}{(1.0 \times 10^{-15} \text{ m})^2} = 231 \text{ N}.$$

This is a repulsive force, so it certainly cannot be the force holding the nucleus together – the electrostatic repulsion between the protons is trying to spread the protons apart.

Step 2 – *Let's use Newton's Law of Universal Gravitation to find the gravitational force between the two protons. Is the gravitational force sufficient to keep the nucleus together?* Substituting the appropriate values into the law of gravitation gives:

$$F = \frac{Gmm}{r^2} = \frac{(6.67 \times 10^{-11} \text{ N m}^2/\text{kg}^2)(1.6726 \times 10^{-27} \text{ kg})^2}{(1.0 \times 10^{-15} \text{ m})^2} = 1.87 \times 10^{-34} \text{ N}.$$

This gravitational force is an attractive force, but at about 36 orders of magnitude less than the electrostatic force, it is negligible compared to the electrostatic force.

Step 3 – *Are any of the other forces we have encountered already in this book responsible for holding the nucleus together?* Other forces we have dealt with include forces of tension and friction, and normal forces. These forces are macroscopic manifestations of forces between charges, however, and they do not apply at the microscopic level of the nucleus. Another force we looked at is the magnetic force applied on charged objects but, as we discussed in Chapter 26, magnetism can actually be interpreted as another manifestation of the electrostatic force.

Step 4 – *What do you conclude about the force responsible for holding the nucleus together?* Our conclusion must be that there is a force we have not yet discussed that is responsible for holding the nucleus together. The basics of this force are given in the box below.

The force that holds a nucleus together is the **nuclear force**, a short-range force between nucleons. At very small separations, the nuclear force is repulsive, keeping the protons and neutrons from getting too close to one another. In general, however, protons and neutrons exert attractive forces on other protons and neutrons via the nuclear force, up to a separation distance of about 1.3 × 10⁻¹⁵ m (at larger separations the nuclear force is negligible). To keep a nucleus together, the net attraction from all the nuclear forces between nucleons must balance the net repulsion from all the protons mutually repelling one another.

Key idea: Nuclei are held together, against the electrostatic repulsion trying to tear them apart, by the nuclear force, a (generally) attractive, but short-range, force between nucleons (neutrons and protons).   **Related End-of-Chapter Exercise: 16.**

Essential Question 29.1: Your friend says that two isotopes of carbon are specified by $^{12}_{6}\text{C}$ and $^{14}_{8}\text{C}$. Do you agree with your friend? Explain why or why not.

Answer to Essential Question 29.1: The isotope $^{12}_{6}\text{C}$ is certainly correct – it is the isotope of carbon that makes up 98.9% of naturally occurring carbon on Earth. On the other hand, $^{14}_{8}\text{C}$ can not be correct. The lower number represents the number of protons in the nucleus, and carbon atoms all have 6 protons. The correct notation for carbon-14 is $^{14}_{6}\text{C}$.

29-2 $E = mc^2$

As we discussed in Section 29-1, the atomic mass unit is defined as 1/12th the mass of a neutral carbon-12 atom. Table 29.1, copied from Section 29-1, shows the mass of a neutron, proton, and electron in both kilograms and atomic mass units. The data seems to be contradictory. We will explore, and resolve, that apparent contradiction in Exploration 29.2.

Table 29.1: The charge and mass of the neutron, proton, and electron.

Particle	Charge	Mass (kg)	Mass (u)
neutron	0	1.674929×10^{-27} kg	1.008664
proton	$+e$	1.672623×10^{-27} kg	1.007276
electron	$-e$	9.109390×10^{-31} kg	0.00054858

EXPLORATION 29.2 – Nuclear binding energy

Step 1 – *How many neutrons, protons, and electrons are in a neutral carbon-12 atom?* A neutral carbon-12 atom has six neutrons, six protons, and six electrons.

Step 2 – *Find the total mass of the individual constituents of a neutral carbon-12 atom. Compare this total mass to the mass of a neutral carbon-12 atom.*
The total mass of six neutrons, six protons, and six electrons is:

$$6 \times 1.008664 \text{ u}$$
$$+ 6 \times 1.007276 \text{ u}$$
$$+ 6 \times 0.00054858 \text{ u}$$
$$= 12.098931 \text{ u}$$

The total mass of a carbon-12 atom is, by definition, exactly 12 u. This seems to be a contradiction – how can the masses of the individual constituents of the atom add up to more than the mass of the atom itself?

To resolve the contradiction, we apply the most famous equation in physics, $E = mc^2$. The idea is that the missing mass, amounting to 0.098931 u, is converted to energy, in the form of the binding energy that holds the atom together.

The difference in mass between the individual constituents of an atom and the atom itself is known as the **mass defect**. The mass that is "missing" from the atom is the atom's binding energy (almost of this binding energy is associated with the nucleus, rather than the electrons). The conversion from mass to energy is done using Einstein's famous equation:

$$E = mc^2,$$ (Equation 29.2: **The equivalence of mass and energy**)

where c = 2.998×10^8 m/s is the speed of light in vacuum.

Step 3 – *Use Equation 29.2 to determine the energy, in MeV (mega electron volts) associated with a mass of 1 u.* Let's first convert 1 u, expressed in units of kilograms, to joules.

$$E = mc^2 = (1.660539 \times 10^{-27} \text{ kg})(2.998 \times 10^8 \text{ m/s})^2 = 1.492493 \times 10^{-10} \text{ J}.$$

We can now convert to MeV using the conversion factor 1 MeV = 1.602176×10^{-13} J.

$$1.492493 \times 10^{-10} \text{ J} \times \frac{1 \text{ MeV}}{1.602176 \times 10^{-13} \text{ J}} = 931.5 \text{ MeV}.$$

Step 4 – *Use the conversion factor derived in step 3 to find the binding energy, in MeV, of a carbon-12 atom, as well as the binding energy per nucleon.* The conversion factor we just derived is that 1 u is equivalent to 931.5 MeV of energy. Using this conversion factor to convert our value for the mass defect of carbon-12 from part (b), 0.098931 u, we get

$$0.098931 \text{ u} \times \frac{931.5 \text{ MeV}}{1 \text{ u}} = 92.15 \text{ MeV}.$$

To more easily compare different atoms, the binding energy for an atom is often expressed in terms of the binding energy per nucleon. Carbon-12 has 12 nucleons, so dividing the 92.15 MeV binding energy by 12 nucleons gives 7.68 MeV per nucleon.

Key idea: The binding energy for an atom is its mass defect (the difference between the total mass of the atom's neutrons, protons, and electrons and the mass of the atom), converted to energy. Electron binding energies are measured in electron volts, whereas atomic binding energies are measured in millions of electron volts – almost all the binding energy for an atom is associated with the energy holding the nucleus together.
Related End-of-Chapter Exercises: 2, 3, 13 – 15, 17, 18, 43.

Comparing chemical energy to nuclear energy

Much of the energy we use on a daily basis comes from chemical energy, such as through the burning of fossil fuels. Burning a gallon of gasoline in a car engine, for instance, typically produces about 1.3×10^8 J of energy. However, the energy obtained comes from changes in bonds between atoms, and this is associated with the electron energy levels. Breaking and forming bonds (generally associated with carbon atoms, when we're talking about fossil fuels) typically frees up a few electron volts at a time.

The relatively modest energy output, per unit mass, available from chemical energy contrasts with the much larger energy output, per unit mass, available from nuclear energy sources. Converting mass directly to energy gives enormous amounts of energy, but it is not feasible to convert all mass to energy. It is possible to convert a fraction of the mass associated with a nucleus into energy - as we will discuss in more detail later in this chapter, this can be done through the processes of nuclear fusion or nuclear fission (nuclear power plants make use of nuclear fission). Changes in the nucleus through these processes generally produce millions of electron volts per nucleus, several orders of magnitude more than is obtained from sources of chemical energy. Tapping into the vast sources of energy associated with the nucleus is why nuclear power is so attractive.

Essential Question 29.2: The power output of the Sun is about 4×10^{23} J/s. This energy comes from converting mass into energy. How much mass is the Sun losing every second?

Answer to Essential Question 29.2: Solving for the mass converted to energy within the Sun every second gives $m = E/c^2 = (4 \times 10^{23} \text{ J})/(9 \times 10^{16} \text{ m}^2/\text{s}^2) \approx 4 \times 10^6 \text{ kg}$. This is a huge mass, but it represents a tiny fraction of the Sun's mass of 2.0×10^{30} kg.

29-3 Radioactive Decay Processes

In general, there are three types of radioactive decay processes, named after the first three letters of the Greek alphabet, alpha, beta, and gamma. In the alpha and beta decay processes, a nucleus emits a particle, or a collection of particles, turning into a nucleus of a different element. The gamma decay process is more analogous to what happens in an atom when an electron drops from a higher energy level to a lower energy level, emitting a photon. Gamma decay occurs when a nucleus makes a transition from a higher energy level to a lower energy level, emitting a photon in the process. Because nuclear energy levels are generally orders of magnitude farther apart than are electron energy levels, however, the photon released in a gamma decay process is very high energy, and falls in the gamma ray region of the electromagnetic spectrum.

Radioactive decays can happen spontaneously when the products resulting from the decay process are more stable than the original atom or nucleus. In any kind of radioactive decay process, a number of conservation laws are satisfied, as explained in the box below.

All nuclear reactions and decays satisfy a few different conservation laws. First of all, the process can generally be viewed as a super-elastic collision, and thus linear momentum is conserved. Kinetic energy is generally not conserved, but any excess or missing kinetic energy can be explained in terms of a conversion of mass into kinetic energy. Charge must also be conserved in a reaction or a decay. In addition to the preceding guidelines, the number of nucleons (the number of protons plus neutrons) must also be conserved, a law known as conservation of nucleon number.

Alpha decay
An alpha particle is a helium nucleus, two protons and two neutrons, which is particularly stable. Heavy nuclei can often become more stable by emitting an alpha particle – this process is known as **alpha decay**. Equation 29.3 describes alpha decay, in which a nucleus with a generic chemical symbol X_1, with atomic mass number A and atomic number Z, transforms into a second nucleus, X_2, with an atomic number of A–4 and atomic number Z–2. The number of neutrons, protons, and electrons (assuming all three atoms are neutral) is the same on both sides of the equation.

$$^A_Z X_1 \Rightarrow \, ^{A-4}_{Z-2} X_2 + \, ^4_2\text{He}. \qquad \text{(Equation 29.3: \textbf{General equation for alpha decay})}$$

A particular example of alpha decay is the transformation of uranium-238 into thorium-234.

$$^{238}_{92}\text{U} \Rightarrow \, ^{234}_{90}\text{Th} + \, ^4_2\text{He}. \qquad \text{(Equation 29.4: \textbf{An alpha decay example})}$$

Table 29.3 in Section 29-8 gives the masses of a number of isotopes. The atomic masses of uranium-238, thorium-234, and helium-4 are 238.050786 u, 234.043596 u, and 4.002603 u, respectively. The total mass on the right side of Equation 29.4 is 238.046199 u, which is lower in mass, by 0.004587 u, than the mass of the uranium-238. How do we explain this mass difference?

The missing mass is converted to kinetic energy, which is shared by the two atoms after the decay. Using our conversion factor of 931 MeV/u, 0.004587 u corresponds to 4.273 MeV of kinetic energy, most of which is carried away by the helium nucleus after the decay.

Beta-minus decay

There are two kinds of beta decay, beta-plus and beta-minus. A beta-minus particle is familiar to us – it is an electron – so let's examine beta-minus decay first. The general equation for beta-minus decay, which takes the nucleus one step up the periodic table, is

$$_Z^A X_1 \Rightarrow {}_{Z+1}^{A} X_2^+ + {}_{-1}^{0}e^- + \overline{\nu}_e. \qquad \text{(Eq. 29.5: General equation for beta-minus decay)}$$

The beta-minus decay process can be viewed as one of the neutrons in the nucleus decaying into a proton and an electron (the electron is symbolized by $_{-1}^{0}e^-$). The last term on the right-hand side of Equation 29.5 represents an electron anti-neutrino. In the early 20th-century, analysis of beta-minus decay processes seemed to indicate a violation of energy conservation and of momentum conservation. In 1930, Wolfgang Pauli proposed that the missing energy and momentum was being carried away by a particle that was very hard to detect, which Enrico Fermi called the **neutrino** (little neutral one). Pauli was proven correct, and we now know that the Sun emits plenty of neutrinos, which interact so rarely that the majority of neutrinos incident on the Earth pass right through without interacting at all! Note that, when comparing the masses on the two sides of the decay, you can neglect the mass of the anti-neutrino, and looking up the mass of the neutral version of the nucleus on the right accounts for the electron, because the atom on the right is positively charged.

An example of beta-minus decay is the decay of thorium-234 into protactinium-234.

$$_{90}^{234}\text{Th} \Rightarrow {}_{91}^{234}\text{Pa}^+ + {}_{-1}^{0}e^- + \overline{\nu}_e. \qquad \text{(Eq. 29.6: A specific example of beta-minus decay)}$$

Beta-plus decay

A beta-plus particle is a positron, which is the antimatter version of the electron. It has the same mass and the same magnitude charge as the electron, but the sign of its charge is positive. A beta-plus decay takes the nucleus one step down the periodic table.

$$_Z^A X_1 \Rightarrow {}_{Z-1}^{A} X_2^- + {}_{+1}^{0}e^+ + \nu_e. \qquad \text{(Eq. 29.7: General equation for beta-plus decay)}$$

The neutrino in this case is an electron neutrino (there are two other kinds of neutrino, each with an antimatter version). In this case, when comparing the masses on the two sides of the decay, you can neglect the mass of the neutrino. In addition to the mass of the neutral version of the nucleus on the right, you need to add two electron masses, one for the extra electron (the atom is negatively charged) and one for the positron, which has the same mass as the electron.

An example of beta-plus decay is the decay of astatine-210 into polonium-210.

$$_{85}^{210}\text{As} \Rightarrow {}_{84}^{210}\text{Po}^- + {}_{+1}^{0}e^+ + \nu_e. \qquad \text{(Eq. 29.8: A specific example of beta-plus decay)}$$

Gamma decay

In gamma decay, the atom does not turn into anything different, as the nucleus simply decays from a higher-energy state to a lower-energy state. Using an asterisk to denote the higher state, the general equation for a gamma decay is

$$_Z^A X_1^* \Rightarrow {}_Z^A X_1 + \gamma. \qquad \text{(Equation 29.9: General equation for gamma decay)}$$

Related End-of-Chapter Exercises: 4 – 6, 23, 45.

Essential Question 29.3: If a carbon-13 atom ($_6^{13}\text{C}$) experienced alpha decay, what would it decay into? Use the atomic mass data in Table 29.3 in Section 29-8 to help you explain why carbon-13 will not spontaneously undergo alpha decay.

Answer to Essential Question 29.3: If carbon-13 experienced alpha decay, the process would be written $^{13}_{6}C \Rightarrow {}^{9}_{4}Be + {}^{4}_{2}He$. There are 13 nucleons and 6 protons on both sides of the reaction.

Looking up the relevant masses in Table 29.3, we find that the total mass after the reaction is larger than the mass of the carbon-13 atom. Spontaneous reactions occur when the total mass of the decay products is less than the mass of the initial atom – in that case, the missing mass shows up as the kinetic energy of the decay products. Thus, carbon-13 will not spontaneously exhibit alpha decay, because it does not make sense from the perspective of energy conservation.

29-4 The Chart of the Nuclides

A nuclide is an atom that is characterized by what is in its nucleus. In other words, it is characterized by the number of protons it has, and by the number of neutrons it has. Figure 29.2 shows the chart of the nuclides, which plots, for stable and radioactive nuclides, the value of Z (the atomic number), on the vertical axis, and the value of N (the number of neutrons) on the horizontal axis.

If you look at a horizontal line through the chart, you will find nuclides that all have the same number of protons, but a different number of neutrons. These are all nuclides of the same element, and are known as isotopes (equal proton number, different neutron number).

On a vertical line through the chart, the nuclides all have the same number of neutrons, but a different number of protons. These nuclides are known as isotones (equal neutron number, different proton number).

It is interesting to think about how the various decay processes play a role in the chart of the nuclides. Only the nuclides shown in black are stable. The stable nuclides give the chart a line of stability that goes from the bottom left toward the upper right, curving down and away from the Z = N line as you move toward higher N values. There are many more nuclides that are shown on the chart but which are not stable - these nuclides decay by means of a radioactive decay process.

In general, the dominant decay process for a particular nuclide is the process that produces a nucleus closer to the line of stability than where it started. In an alpha decay process, the resulting nuclide (usually referred to as the daughter nuclide), has a Z value that is 2 less than the parent nuclide, and the N value is also 2 less than that of the parent nuclide. This is because the alpha particle takes away two protons and two neutrons. On the chart of the nuclides, therefore, an alpha decay process moves the nucleus two spaces down and two spaces left. Thus, nuclides that decay via alpha decay are found mainly at the upper right of the chart. Because the line of stability curves down as the number of nucleons increases, as well as because, above a certain number of nucleons, there are no stable nuclides, a decay process that moves a nucleus down and to the left on the chart tends to bring that nucleus closer to the line of stability.

Beta-plus decay, on the other hand, increases the neutron number by one and decreases the proton number by one. In other words, a beta-plus decay moves a nucleus down and to the right on the chart. Nuclides that decay via beta-plus decay, therefore, are generally found above and to the left of the line of stability.

The beta-minus decay is, in many ways, the opposite of beta-plus. Beta-minus decay decreases the neutron number by one and increases the proton number by one, moving a nucleus up and to the left on the chart. Nuclides that decay via beta-plus decay, therefore, are generally found below and to the right of the line of stability.

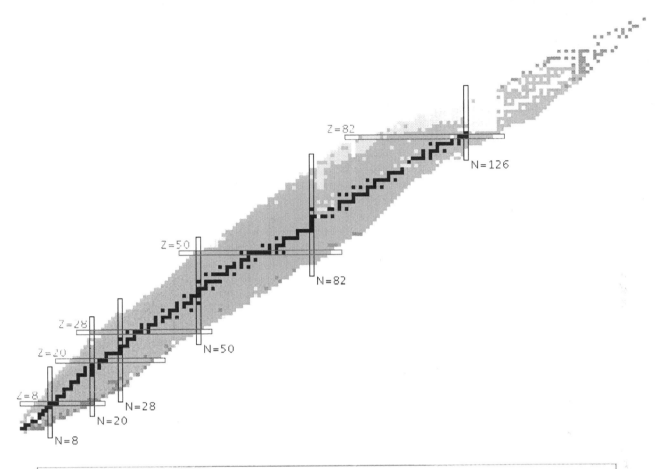

Figure 29.2: The chart of the nuclides, plotted with Z (the atomic number) on the vertical axis, and N (the number of neutrons) on the horizontal axis. Nuclides in black are stable - all other nuclides exhibit radioactive decay. The stable nuclides start out following the line Z = N (at bottom left), but then, as the number of nucleons increases, stable nuclides gradually increase the number of neutrons they have relative to their number of protons. Viewed in color, the chart is color-coded according to the nuclides primary decay mode. Above the line of stability, the chart is mainly light blue - those nuclides decay via beta-plus decay. Below the line of stability, the chart is mainly purple - those nuclides decay via beta-minus decay. At the top right of the chart, many of the nuclides are shown in yellow - those nuclides decay via the alpha decay process. In general, a radioactive decay will produce another nuclide that is closer to the line of stability than the original nuclide was. Also in general, the closer the nuclide is to the line of stability, the more stable it is, and the longer its half-life is. See Section 29-5 for an explanation of half-life. Chart of Nuclides from National Nuclear Data Center, using information extracted from the Chart of Nuclides database, http://www.nndc.bnl.gov/chart/

Related End-of-Chapter Exercises: 60 – 62.

Essential Question 29.4: Sodium-23, which has 11 protons and 12 neutrons, is stable. Sodium-22, however, is unstable. What type of radioactive decay process would you expect sodium-22 to undergo? What is the daughter nuclide that is produced in this decay?

Answer to Essential Question 29.4: Sodium-22 is to the left of sodium-23 on the chart of the nuclides, so we expect it to decay via beta-plus decay, taking the nucleus down and to the right on the chart. In fact, sodium-22 is a well-known positron emitter; it does decay via the beta-plus process. Through beta-plus decay, sodium-22, with 11 protons and 11 neutrons, produces the nuclide with 10 protons and 12 neutrons. That is neon-22, which happens to be stable.

29-5 Radioactivity

A **radioactive nucleus** is a nucleus that will spontaneously undergo radioactive decay. For an individual radioactive nucleus, it is not possible to predict precisely when the nucleus will decay. However, the statistics of radioactive decay are well understood, so for a sample of radioactive material containing a large number of nuclei, we can accurately predict the fraction of radioactive nuclei that will decay in a particular time interval.

For a given isotope, we can generally look up the **half-life**. The half-life is the time it takes for half of a large number of nuclei of this particular isotope to decay. Half-lives can vary widely from isotope to isotope. For instance, the half-life of uranium-238 is 4.5 billion years; the half-life of carbon-14 is 5730 years; and the half-life of oxygen-15 is about 2 minutes.

For a particular sample of material, containing one isotope of radioactive nuclei, the number of radioactive decays that occur in a particular time interval is related to the half-life of that isotope, but it is also proportional to the number of radioactive nuclei in the sample. The time rate of decay is given by

$$\frac{\Delta N}{\Delta t} = -\lambda N \,,$$ (Equation 29.10: **Decay rate for radioactive nuclei**)

where N is the current number of radioactive nuclei, and the **decay constant** λ is related to the half-life, $T_{1/2}$, by the equation

$$\lambda = \frac{\ln(2)}{T_{1/2}} = \frac{0.693}{T_{1/2}} \,.$$ (Eq. 29.11: **The connection between decay constant and half-life**)

A process in which the rate that a quantity decreases is proportional to that quantity is characterized by exponential decay. The exponential equation that describes the number of a particular radioactive isotope that remain after a time interval t is

$$N = N_i e^{-\lambda t} \,,$$ (Equation 29.12: **The exponential decay of radioactive nuclei**)

where N_i is a measure of the initial number of radioactive nuclei (the number at $t = 0$).

EXAMPLE 29.5 – Calculations for exponential decay

A particular isotope has a half-life of 10 minutes. At $t = 0$, a sample of material contains a large number of nuclei of this isotope. How much time passes until (a) 80% of the original nuclei remain? (b) 1/8th of the original nuclei remain? See if you can answer part (b) in your head, without using Equation 29.12.

SOLUTION

(a) Let's first determine approximately what the answer is. We know that in 10 minutes (one half-life), 50% of the nuclei would decay. To get 20% of the nuclei to decay must take less than half of a half-life, so the answer should be something like 3 or 4 minutes. In Equation 29.12, we can use a number of different measures for N and N_i, as long as we are consistent. For instance, we can measure them both in terms of mass (e.g., grams), or use the actual number of nuclei, or express them as a percentage or fraction. It makes sense to use fractions or percentage here, because the question was posed in terms of percentage. Thus, we're looking for the time when $N = 80\%$ of N_i, or when $N = 0.8\,N_i$. Note that the units on λ and t must match one another, but they can be completely different from the units on N and N_i. Applying Equation 29.12, we get:

$$0.8N_i = N_i\,e^{-\lambda t}, \quad \text{which becomes} \quad 0.8 = e^{-\lambda t}.$$

The inverse function of the exponential function is the natural log, so to bring down the t term we can take the natural log of both sides. Thus, $\ln(0.8) = -\lambda t$.

Bringing in the half-life via Equation 29.11, we get:

$$t = \frac{-\ln(0.8)}{\lambda} = \frac{-\ln(0.8)}{\ln(2)}\,T_{1/2} = \frac{-\ln(0.8)}{\ln(2)}\,(10\ \text{min}) = 3.2\ \text{min}.$$

(b) If the fraction of nuclei remaining can be expressed as 1 divided by a power of 2, the time taken is an integer multiple of the half life (the integer being equal to the power to which 2 is raised). Table 29.2 illustrates the idea.

Table 29.2: The table shows the relationship between the number of half-lives that have passed and the fraction of radioactive nuclei remaining. In general, if the fraction remaining is expressed as $1/(2^x)$ (where x does not have to represent an integer), the amount of time that has passed is x multiplied by the half-life.

Number of half-lives	Fraction of radioactive nuclei remaining
0	$1/(2^0) = 1$
1	$1/(2^1) = 1/2$
2	$1/(2^2) = 1/4$
3	$1/(2^3) = 1/8$
4	$1/(2^4) = 1/16$
5	$1/(2^5) = 1/32$

In our example, if $1/8^{\text{th}}$ of the nuclei remain, three half-lives must have passed. Three half-lives is 30 minutes, in this case.

A graph of Equation 29.11 is shown in Figure 29.3. This figure, as it must be, is also consistent with the data in Table 29.2.

Related End-of-Chapter Exercises: 7–9, 48–54.

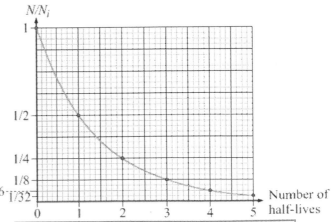

Figure 29.3: A graph of the exponential decay of radioactive nuclei, showing the fraction of nuclei remaining as a function of the number of half-lives that have elapsed.

Essential Question 29.5: Equation 29.10 can be re-arranged to $\Delta N = -\lambda N(\Delta t)$. This form of the equation gives a reasonable approximation of the number of decayed nuclei as long as Δt is much smaller than what?

Answer to Essential Question 29.5: It is best to use Equation 29.12 to find the number of nuclei remaining after a particular time interval. Equation 29.10 is designed to give the decay rate at a particular instant in time; however, it will give a good approximation of the number of decays that have taken place in a time interval if the time interval is much smaller than the half life.

29-6 Nuclear Fusion and Nuclear Fission

In Section 29-2, we calculated the mass defect for carbon-12, and used that information to determine the binding energy per nucleon for carbon-12. The basic process is:
- Calculate the total mass of the individual neutrons, protons, and electrons in an atom.
- Calculate the mass defect for the atom by subtracting the atomic mass from the total mass of the individual constituents.
- Convert the mass defect from mass to energy, using the conversion factor 931.5 MeV / u. This represents the atom's binding energy, which is almost all in the nucleus.
- Divide the binding energy by the number of nucleons (neutrons plus protons) to find the average binding energy per nucleon.

Following the procedure above, we obtain the graph in Figure 29.4, showing the average binding energy per nucleon for a variety of common isotopes. The most stable isotope, having the largest binding energy per nucleon, is nickel-62, followed closely by iron-58 and iron-56.

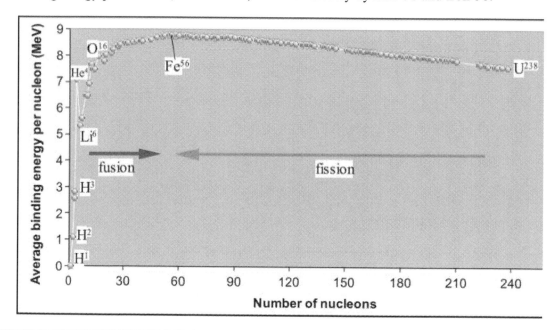

Figure 29.4: A graph of the average binding energy per nucleon for most common isotopes.

Nuclear fusion

The larger the average binding energy per nucleon, the more stable a particular isotope is. Thus, light elements (less than 50 nucleons, say) can, in general, become more stable by joining together to form a nucleus that has more binding energy per nucleon. This process is known as **nuclear fusion**, and it releases a significant amount of energy. Nuclear fusion is the process by which the Sun generates its energy, for instance. Currently, the Sun is made up mostly of hydrogen, which is gradually fusing together to become helium. When the hydrogen is used up, the helium atoms will fuse together with one another, or with other light atoms, to form heavier nuclei. All of these fusion reactions, resulting in atoms with more binding energy per nucleon, produce energy. The process continues for billions of years until the Sun's atoms fuse together to become nickel and iron, which are the most stable elements of all. At this point, the fusion reactions will cease, because the peak of stability will have been reached, and the Sun will, essentially, die of old age.

An example fusion reaction is the fusion of deuterium and tritium, which are both isotopes of hydrogen, into helium. This reaction may well form the basis, in future, of controlled fusion reactions that generate energy in a fusion reactor.

$$\,^2_1H + \,^3_1H \Rightarrow \,^5_2He \Rightarrow \,^4_2He + \,^1_0n.$$

The products of this reaction, the helium-4 atom and the neutron, carry away 17.6 MeV between them, 3.5 MeV for the helium atom and 14.1 MeV for the neutron. This is an enormous amount of energy compared to the several eV of energy produced in a typical chemical reaction.

Nuclear fission
At the heavy end of the scale in Figure 29.4, those nuclei with a large number of nucleons can reach a more stable state by splitting apart into smaller pieces that have a higher average energy per nucleon. This process is known as **nuclear fission**, and it is used in nuclear reactors, in well-controlled reactions, to generate energy. In uncontrolled reactions, a chain of fission reactions can occur so quickly that nuclear meltdown occurs, or a nuclear bomb explodes.

An example of a fission reaction that may occur in a nuclear reactor is

$$\,^1_0n + \,^{235}_{92}U \Rightarrow \,^{236}_{92}U \Rightarrow \,^{141}_{56}Ba + \,^{92}_{36}Kr + 3\left(\,^1_0n\right).$$

Note that the reaction is triggered by bombarding the uranium-235 atom with a neutron, temporarily creating a uranium-236 atom that quickly splits into krypton, barium, and three more neutrons. These neutrons can go on to cause more uranium-235 atoms to split apart. In a controlled reaction, the rate at which reactions occur should be constant, so if it takes one neutron to start a reaction, only one of the product neutrons (on average) are allowed to go on to produce further reactions. The other neutrons are absorbed by a moderator inside a reactor, which could be heavy water or a boron control rod embedded in the reactor core.

A typical fission reaction, like the one shown, above releases on the order of 200 MeV. As with typical fusion reactions, this is millions of times larger than the energy released by burning oil or gas in a chemical reaction, explaining why nuclear power is so appealing in comparison to the burning of fossil fuels. That huge advantage has to be weighed against the negative aspects of nuclear energy, including the fact that nuclear reactors produce radioactive waste products that must be handled carefully and stored securely for rather long times.

Related End-of-Chapter Exercises: 10, 11, 29 – 33.

Essential Question 29.6: The fission reaction shown above is just one possible way that a uranium-235 atom, bombarded with a neutron, can split up. What is the missing piece in another of the many possible reactions, which is shown here?

$$\,^1_0n + \,^{235}_{92}U \Rightarrow \,^{236}_{92}U \Rightarrow ? + \,^{94}_{38}Sr + 2\left(\,^1_0n\right).$$

Answer to Essential Question 29.6: The reaction must satisfy conservation of nucleon number (there are 236 nucleons) and conservation of charge (in this case, there are 92 protons). For both sides of the reaction to have 236 nucleons and 92 protons, the missing piece must have 140 nucleons and 54 protons. The element with 54 protons is xenon so the missing piece is $^{140}_{54}Xe$.

29-7 Applications of Nuclear Physics

Radiocarbon dating

One well-known application of nuclear physics is the use of carbon-14, which is radioactive, to determine the age of artifacts made from materials that used to be alive, such as bowls made of wood. The idea is that when the tree from which the wood was taken was alive, it was exchanging carbon with the atmosphere, and the ratio of carbon-14 to non-radioactive carbon-12 in the wood matched the corresponding ratio in the atmosphere. This ratio is maintained at an approximately constant value of 1 carbon-14 atom to every 10^{12} carbon-12 atoms by cosmic rays that turn nitrogen-14 in the atmosphere into carbon-14. After a tree dies or is cut down, the carbon-14 in the wood decays, decreasing the carbon-14 to carbon-12 ratio. The more time passes, the smaller the carbon-14 to carbon-12 ratio, so the carbon-14 to carbon-12 ratio can be used to estimate the time that has passed since the tree was cut down. This process is called **radiocarbon dating**.

A famous artifact dated using radiocarbon dating is the Shroud of Turin, which was long believed to be the burial cloth of Jesus Christ. Tests on small samples of the fabric indicate that the fabric dates not from 2000 years ago, however, but from around 700 years ago instead.

Carbon-14 decays via the beta-minus process, with a half-life of 5730 years, so radiocarbon dating works well for artifacts with an age ranging from several hundred years old to about 60000 years old, an age range corresponding to a reasonable fraction of the carbon-14 half-life to several times the carbon-14 half-life. To date much older artifacts, such as dinosaur bones, which are about 100 million years old, carbon-14 would not be appropriate, because so many carbon-14 half-lives would have passed in 100 million years that the amount of carbon-14 in the sample would be negligibly small. A similar process to radiocarbon dating can be carried out, however, using an isotope with a much longer half-life than carbon-14.

EXAMPLE 29.7 – Dating a bowl

While on an archeological dig, you uncover an old wooden bowl. With your radiation detector, you measure 560 counts per minute coming from the bowl, each of these counts corresponding to the beta-minus decay of a carbon-14 atom into a nitrogen-14 atom (your detector picks up the fast-moving electron that is emitted from each of these decays). Using the same kind of wood, but from a tree that was recently cut down, you make a similar bowl of the same shape and mass as the one you unearthed at the excavation site. This bowl registers 800 counts per minute in your detector. Estimate the age of the old bowl you unearthed.

SOLUTION

Our working assumption is that the old bowl would also have emitted 760 counts per minute originally, when it was first made from wood from a tree that had been recently cut down. This assumes the ratio of carbon-14 to carbon-12 in our atmosphere has remained constant over time, which is approximately true. Accurate dating involves correcting for effects such as the fluctuation of carbon-14 to carbon-12 in the atmosphere in the past, but we can get a good estimate of the age of the bowl without these corrections. Applying Equation 29.12 gives:

$$560 = 800e^{-\lambda t}, \quad \text{which becomes} \quad \frac{560}{800} = 0.7 = e^{-\lambda t}.$$

Taking the natural log of both sides gives $\ln(0.7) = -\lambda t$. Bringing in the half-life via Equation 29.11, we get:

$$t = \frac{-\ln(0.7)}{\lambda} = \frac{-\ln(0.7)}{\ln(2)} T_{1/2} = \frac{-\ln(0.7)}{\ln(2)}(5730 \text{ y}) = 2900 \text{ y} \; .$$

Thus, the site you are exploring contains artifacts from approximately 3000 years ago.

Radiation therapy

A common method of treating cancer is with radiation therapy, which involves treatment with ionizing radiation associated with nuclear decays. For instance, x-rays can be directed at a tumor within the body, damaging the DNA of the cancer cells with the energy deposited by the photons. It is hard to avoid damaging healthy cells with this process, but if the tumor is targeted from various directions the energy deposited in the tumor can be maximized while the damage to surrounding healthy cells is minimized.

Another example of radiation therapy is in the treatment of prostate cancer, in which small radioactive rods, called seeds, are embedded in the prostate, damaging the DNA of the cancer cells with the energy that comes from the decay. The seeds may contain, for example, iodine-125 or palladium-103, which are emitters of gamma rays or x-rays.

Medical imaging – PET and MRI

Two more applications of nuclear physics are **positron emission tomography** (PET) and **magnetic resonance imaging** (MRI). If you get a PET scan (like that in Figure 29.5), you must first take in an isotope that is a positron emitter. A common example is to use fluorodeoxyglucose (FDG), in which the fluorine atom is fluorine-18, which decays via the beta-plus (positron) process with a half-life of about two hours – this short half-life minimizes your exposure to radiation. FDG is taken up by cells that use glucose, so FDG is useful for studying cells with significant glucose uptake, such as those in the brain or in a cancer tumor.

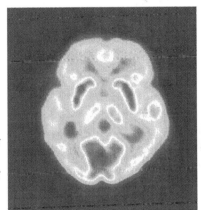

When a fluorine-18 atom decays, it emits a positron. The positron and a nearby electron annihilate one another, turning into two high-energy photons (gamma rays), which exit the body in almost exactly opposite directions. If the photons are detected by detectors surrounding the body, the path of the photons can be determined. After many such photon pairs have been detected, areas of high glucose uptake in the body can be reconstructed.

Figure 29.5: An image of a human brain obtained with positron emission tomography. The red and blue areas correspond to high and low positron activity, respectively. Image credit: Jens Langner, via Wikimedia Commons.

A different process is at work in MRI, in which hydrogen nuclei in our bodies (in water and lipid molecules, in particular) are excited by strong magnetic fields. Signals from these nuclei are then detected, and the signals can be used to create an image of what is going in inside a body being scanned. Such scans are often used to diagnose soft-tissue injuries. The image that opens this chapter shows such an MRI scan.

Related End-of-Chapter Exercises: 12, 26, 27, 35, 58, 59.

Essential Question 29.7: The ratio of carbon-14 to carbon-12 in a wooden implement, unearthed at an archeological dig, is only ¼ as large as it is in a growing tree today. Estimate its age.

Answer to Essential Question 29.7: To reduce the level to ¼ = ½ × ½ of its original value, two half-lives must have passed. Two half-lives for carbon-14 represents 11000 – 12000 years.

29-8 A Table of Isotopes

In doing calculations of mass defect, or nuclear decays and reactions, we need to know the atomic mass(es) of the atom or atoms involved. This information is shown in Table 29.3.

Z, Element	Isotope : Atomic mass	Z, Element	Isotope : Atomic mass
1, Hydrogen	$^{1}_{1}H$: 1.00782503 u	15, Phosphorus	$^{31}_{15}P$: 30.97376163 u
	$^{2}_{1}H$: 2.01410178 u	16, Sulfur	$^{32}_{16}S$: 31.97207100 u
	$^{3}_{1}H$: 3.01604928 u, (β-)	17, Chlorine	$^{35}_{17}Cl$: 34.96885268 u
2, Helium	$^{4}_{2}He$: 4.00260325 u		$^{37}_{17}Cl$: 36.96590259 u
3, Lithium	$^{6}_{3}Li$: 6.01512279 u	18, Argon	$^{40}_{18}Ar$: 39.96238312 u
	$^{7}_{3}Li$: 7.01600455 u	19, Potassium	$^{39}_{19}K$: 38.96370668 u
4, Beryllium	$^{9}_{4}Be$: 9.0121822 u		$^{41}_{19}K$: 40.96182576 u
5, Boron	$^{10}_{5}B$: 10.0129370 u	20, Calcium	$^{40}_{20}Ca$: 39.96259098 u
	$^{11}_{5}B$: 11.0093054 u	21, Scandium	$^{45}_{21}Sc$: 44.9559119 u
6, Carbon	$^{12}_{6}C$: 12.00000000 u		$^{46}_{21}Sc$: 45.9551719 u, (β-)
	$^{13}_{6}C$: 13.00335484 u	22, Titanium	$^{46}_{22}Ti$: 45.9526316 u
	$^{14}_{6}C$: 14.00324199 u, (β-)		$^{48}_{22}Ti$: 47.9479463 u
7, Nitrogen	$^{14}_{7}N$: 14.00307400 u	23, Vanadium	$^{51}_{23}V$: 50.9439595 u
8, Oxygen	$^{15}_{8}O$: 15.0030656 u, (β+)	24, Chromium	$^{52}_{24}Cr$: 51.9405075 u
	$^{16}_{8}O$: 15.99491462 u		$^{53}_{24}Cr$: 52.9406494 u
	$^{18}_{8}O$: 17.9991610 u	25, Manganese	$^{55}_{25}Mn$: 54.9380451 u
9, Fluorine	$^{18}_{9}F$: 18.0009380 u, (β+)	26, Iron	$^{56}_{26}Fe$: 55.9349375 u
	$^{19}_{9}F$: 18.99840322 u		$^{58}_{26}Fe$: 57.9332756 u
10, Neon	$^{20}_{10}Ne$: 19.99244018 u	27, Cobalt	$^{59}_{27}Co$: 58.9331950 u
11, Sodium	$^{22}_{11}Na$: 21.9944364 u, (β+)	28, Nickel	$^{58}_{28}Ni$: 57.9353429 u
	$^{23}_{11}Na$: 22.98976928 u		$^{62}_{28}Ni$: 61.9283451 u
12, Magnesium	$^{24}_{12}Mg$: 23.98504170 u	29, Copper	$^{63}_{29}Cu$: 62.9295975 u
13, Aluminum	$^{27}_{13}Al$: 26.98153863 u		$^{65}_{29}Cu$: 64.9277895 u
14, Silicon	$^{28}_{14}Si$: 27. 97692653 u	30, Zinc	$^{64}_{30}Zn$: 63.9291422 u

Table 29.3: A table of selected isotopes and their atomic masses, taken from data made available by the Lawrence Berkeley Laboratory. Note that the mass, in atomic mass units, of an electron or a positron is 0.00054858 u. The neutron mass is 1.008664 u, and the proton mass is 1.007276 u.

Z, Element	Isotope : Atomic mass	Z, Element	Isotope : Atomic mass
31, Gallium	$^{69}_{31}$Ga : 68.9255736 u	50, Tin	$^{120}_{50}$Sn : 119.902195 u
	$^{71}_{31}$Ga : 70.9247013 u	51, Antimony	$^{121}_{51}$Sb : 120.9038157 u
32, Germanium	$^{74}_{32}$Ge : 73.9211778 u	52, Tellurium	$^{130}_{52}$Te * : 129.9062244 u, (β⁻)
	$^{76}_{32}$Ge : 75.9214026 u, (β⁻)	53, Iodine	$^{127}_{53}$I : 126.904473 u
33, Arsenic	$^{75}_{33}$As : 74.9215965 u		$^{131}_{53}$I : 130.9061246 u, (β⁻)
34, Selenium	$^{80}_{34}$Se : 79.9165213 u	54, Xenon	$^{132}_{54}$Xe : 131.9041535 u
35, Bromine	$^{79}_{35}$Br : 78.9183371 u	55, Cesium	$^{133}_{55}$Cs : 132.90545193 u
	$^{81}_{35}$Br : 80.9162906 u	56, Barium	$^{138}_{56}$Ba : 137.9052472 u
36, Krypton	$^{84}_{36}$Kr : 83.911507 u		$^{141}_{56}$Ba : 140.914411 u, (β⁻)
	$^{85}_{36}$Kr : 84.9125273 u, (β⁻)	82, Lead	$^{208}_{82}$Pb : 207.9766521 u
37, Rubidium	$^{85}_{37}$Rb : 84.91178974 u	83, Bismuth	$^{209}_{83}$Bi : 208.9803987 u, (α)
38, Strontium	$^{88}_{38}$Sr : 87.9056121 u	84, Polonium	$^{210}_{84}$Po : 209.9828737 u, (α)
39, Yttrium	$^{89}_{39}$Y : 88.905848 u	85, Astatine	$^{210}_{85}$At : 209.987148 u, (β⁺)
40, Zirconium	$^{90}_{40}$Zr : 89.9047044 u	86, Radon	$^{222}_{86}$Rn : 222.0175777 u, (α)
	$^{92}_{40}$Zr : 91.9050408 u	87, Francium	$^{223}_{87}$Fr : 223.019736 u, (β⁻)
41, Niobium	$^{93}_{41}$Nb : 92.906378 u	88, Radium	$^{226}_{88}$Ra : 226.0254098 u, (α)
42, Molybdenum	$^{98}_{42}$Mo : 97.9054082 u	89, Actinium	$^{227}_{89}$Ac : 227.027752 u, (β⁻)
43, Technetium	$^{99}_{43}$Tc : 98.9062547 u, (β⁻)	90, Thorium	$^{232}_{90}$Th : 232.0380553 u, (α)
44, Ruthenium	$^{102}_{44}$Ru : 101.9043493 u	91, Protactinium	$^{231}_{91}$Pa : 231.0358840 u, (α)
45, Rhodium	$^{103}_{45}$Rh : 102.905504 u	92, Uranium	$^{235}_{92}$U : 235.0439299 u, (α)
46, Palladium	$^{106}_{46}$Pd : 105.903486 u		$^{238}_{92}$U : 238.0507882 u, (α)
47, Silver	$^{107}_{47}$Ag : 106.905097 u	93, Neptunium	$^{237}_{93}$Np : 237.0481734 u, (α)
	$^{109}_{47}$Ag : 108.904752 u	94, Plutonium	$^{238}_{94}$Pu : 238.0495599 u, (α)
48, Cadmium	$^{112}_{48}$Cd : 111.902758 u		$^{239}_{94}$Pu : 239.0521634 u, (α)
49, Indium	$^{115}_{49}$In : 114.903878 u, (β⁻)	95, Americium	$^{241}_{95}$Am : 241.0568291 u, (α)

Table 29.3, continued: Elements 57 – 81 are emitted to make room for the high-Z elements, which are rather radioactive. Radioactive isotopes are indicated with the decay process in brackets after the mass. *Tellurium-130 actually undergoes double-beta decay, in which two neutrons become protons by emitting electrons.

Related End-of-Chapter Exercises: 19 – 22, 56, 57.

Essential Question 29.8: Table 29.3 shows that krypton-85 experiences beta-minus decay. What does krypton-85 decay into?

Answer to Essential Question 29.8: Beta-minus decay increases the number of protons by 1, without changing the number of nucleons. Thus, beta-minus decay always takes the element one step up the periodic table. In addition to an electron and an anti-neutrino, krypton-85 decays into rubidium-85.

Chapter Summary

Essential Idea: The Nucleus.
Atomic nuclei are almost unimaginably tiny, yet the energy associated with nuclei is orders of magnitude larger than that associated with the electrons in an atom. Nuclear reactions, tapping into the energy of the nucleus, power the Sun (and other stars) as well as nuclear reactors (as well as nuclear bombs).

Holding the nucleus together
To hold a nucleus together, the mutual repulsion between protons in the nucleus is balanced by the nuclear force, which is associated with the interaction between nucleons (neutrons and protons). The nuclear force is a short-range force that is generally attractive.

To find the binding energy for an atom, we first find the mass defect, which is the difference between the total mass of the individual constituents of the atom and the mass of the atom itself. The mass defect is converted to energy via Einstein's famous equation, $E = mc^2$. This results in the mass to energy conversion factor:

$$1 \text{ u is equivalent to } 931.5 \text{ MeV.}$$

Radioactive decay processes
All radioactive decay processes conserve nucleon number and charge. The general equations describing specific processes are:

$$_Z^A X_1 \Rightarrow {}_{Z-2}^{A-4} X_2 + {}_2^4 \text{He.}$$ (Equation 29.3: **General equation for alpha decay**)

$$_Z^A X_1 \Rightarrow {}_{Z+1}^{A} X_2 + {}_{-1}^{0} e^- + \overline{v}_e.$$ (Eq. 29.5: **General equation for beta-minus decay**)

$$_Z^A X_1 \Rightarrow {}_{Z-1}^{A} X_2 + {}_{+1}^{0} e^+ + v_e.$$ (Eq. 29.7: **General equation for beta-plus decay**)

$$_Z^A X_1^* \Rightarrow {}_Z^A X_1 + \gamma.$$ (Equation 29.9: **General equation for gamma decay**)

In the alpha and beta decay processes, the nucleus becomes a nucleus of a different element. In gamma decay, the nucleus simply drops from a higher-energy state to a lower-energy state, in a manner analogous to that of an electron making a transition from one electron energy level to a lower level. An alpha particle is a helium atom; a beta-minus particle is an electron; and a beta-plus particle is a positron.

Radioactivity
The rate at which N radioactive nuclei decay is:

$$\frac{\Delta N}{\Delta t} = -\lambda N.$$ (Equation 29.10: **Decay rate for radioactive nuclei**)

The equation that describes the exponential decay in the number of nuclei of a particular radioactive isotope as a function of time t is

$$N = N_i e^{-\lambda t},$$ (Equation 29.12: **The exponential decay of radioactive nuclei**)

where N_i is a measure of the initial number of radioactive nuclei (the number at $t = 0$).

The decay constant, λ, is related to the half-life, $T_{1/2}$, (the time for half of the nuclei to decay) by

$$\lambda = \frac{\ln(2)}{T_{1/2}} = \frac{0.693}{T_{1/2}}.$$ (Eq. 29.11: **The connection between decay constant and half-life**)

Nuclear fusion and nuclear fission

The most stable nuclei (those with the highest average binding energy per nuclei) are nickel-62, iron-58 and iron-56. Nuclei that are lighter than these most stable nuclei can generally become more stable (increasing the average binding energy per nuclei) by joining together with other light nuclei – this process is known as nuclear fusion.

Very heavy nuclei, in contrast, can generally become more stable by splitting apart, usually into two medium-sized nuclei and a few neutrons. This process is known as nuclear fission. The fission of uranium-235, driven by the bombardment of the uranium-235 atoms with neutrons, is exploited in a nuclear reactor to produce nuclear energy, while the fission of plutonium-239 is what drives the explosion of a nuclear bomb.

End-of-Chapter Exercises

Exercises 1 – 12 are mainly conceptual questions that are designed to see if you have understood the main concepts of the chapter.

1. The symbol for the isotope iron-56 is $^{56}_{26}\text{Fe}$. A neutral iron-56 atom has how many (a) protons? (b) neutrons? (c) nucleons?

2. Which of these numbers is larger, the mass of an iron-56 atom, or the total mass of the individual constituents (neutrons, protons, and electrons) of an iron-56 atom? Briefly explain your answer.

3. If you could convert 1 kg of matter entirely to energy, how much energy would you get?

4. (a) Fill in the blank to complete this decay process: $^{226}_{88}\text{Ra} \Rightarrow$ ____ $+ \, ^{4}_{2}\text{He}$. (b) What kind of decay is this? (c) Based on the fact that this decay process happens spontaneously, which side of the equation do you expect to have more mass? Why?

5. (a) What kind of radioactive decay process gives rise to a positron? (b) What is the electric charge of a positron? (c) Complete this sentence: The mass of a positron is the same as the mass of _____ .

6. Fill in the blanks to complete the following decay processes:
    (a) ____ $\Rightarrow \, ^{15}_{7}\text{N} + \, ^{0}_{+1}e^{+} + v_e$. (b) $^{46}_{21}\text{Sc} \Rightarrow$ ____ $+ \, ^{0}_{-1}e^{-} + \bar{v}_e$. (c) $^{60}_{28}\text{Ni}^{*} \Rightarrow$ ____ $+ \gamma$.

7. A particular sample contains a large number of atoms of a certain radioactive isotope, which has a half life of 1 hour. After 5 hours, approximately what percentage of the original radioactive nuclei remain?

8. You have two samples of radioactive material. At $t = 0$, sample A contains a large number of nuclei that have a half-life of 10 minutes, while sample B contains exactly the same number of a different kind of nuclei, which have a half-life of 40 minutes. (a) In which sample is the rate at which the nuclei decay larger, at $t = 0$? Briefly justify your answer. Find the ratio of the number of nuclei that have decayed in sample A to the number that have decayed in sample B after (b) $t = 40$ minutes, (c) $t = 1$ year (feel free to make a reasonable approximation in part (c)).

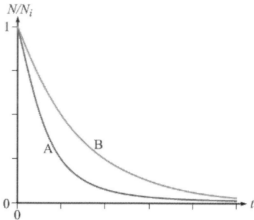

9. Figure 29.6 shows a graph of the ratio of the number of undecayed nuclei to the initial number of nuclei for samples of two different radioactive isotopes, labeled A and B. Which isotope has the larger (a) decay constant? (b) half-life?

Figure 29.6: A graph of the ratio of the number of undecayed nuclei to the initial number of nuclei for samples of two different radioactive isotopes, labeled A and B, for Exercise 9.

10. In stars that are more massive than the Sun, one of the primary methods of fusing hydrogen into helium is the CNO (carbon-nitrogen-oxygen) cycle, which is a sequence of fusion reactions. One of the reactions in the CNO cycle, which liberates about 5 MeV of energy, is shown below. Identify the isotope represented by the question mark.

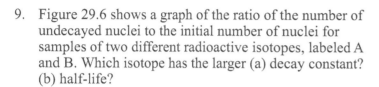

$$_0^1\text{n} + {}_7^{15}\text{N} + {}_1^1\text{H} \Rightarrow ? + {}_2^4\text{He}.$$

11. How many neutrons are produced in the following fission reaction for a uranium-235 atom that combines with a neutron?

$$_0^1\text{n} + {}_{92}^{235}\text{U} \Rightarrow {}_{92}^{236}\text{U} \Rightarrow {}_{52}^{139}\text{Te} + {}_{40}^{94}\text{Zr} + ?\left({}_0^1\text{n}\right).$$

12. Modern human activities have altered the ratio of carbon-14 to carbon-12 in the atmosphere. A good example of this is the burning of fossil fuels, which contain carbon that is millions of years old. The half-life of carbon-14 is about 5700 years, while carbon-12 is stable. Based on this information, has all of our burning of fossil fuels increased or decreased the carbon-14 to carbon-12 ratio in the atmosphere? Explain your answer.

Exercises 13 – 18 involve applications of $E = mc^2$. For some of these exercises you will probably want to make use of the data in Table 29-3 in Section 29-8.

13. For an oxygen-16 atom, calculate (a) the mass defect, in atomic mass units, (b) the total binding energy, in MeV, and (c) the average binding energy per nucleon.

14. (a) What is the difference, in terms of constituents, between a carbon-12 atom and a carbon-13 atom? (b) Compare the mass of a carbon-13 atom to the mass of a carbon-12 atom plus the mass of the extra particle(s) that makes up a carbon-13 atom compared to a carbon-12 atom. Is there a difference? If so, why?

15. In almost all nuclei, there are more than 2 nucleons, and thus there is more than one pair of nucleons interacting via the nuclear force. The situation is much simpler in a deuterium ($_1^2\text{H}$) atom, in which the binding energy is associated with the attractive nuclear force between the single neutron and the single proton. For the deuterium atom, calculate (a) the mass defect, in atomic mass units, (b) the binding energy, in MeV, which is almost entirely associated with the nuclear force between the proton and neutron.

16. In general, for a nucleus to be stable, the attractive forces between nuclei must balance the repulsive forces associated with the charged protons. A helium-4 atom is particularly stable. To help understand why this is, consider how many pairs of (a) interacting protons, and (b) interacting nucleons there are in a helium-4 atom. (c) Briefly explain why your answers to (a) and (b) support the idea that the helium-4 atom is particularly stable.

17. Lead-208 is stable, which is relatively rare for a high-mass nucleus. For $_{82}^{208}\text{Pb}$, calculate (a) the mass defect, in atomic mass units, (b) the total binding energy, in MeV, and (c) the average binding energy per nucleon.

18. Nickel-62 has the largest average binding energy per nucleon of any isotope. Calculate the average binding energy per nucleon for nickel-62.

Exercises 19 – 24 involve radioactive decay processes. Make use of the data in Table 29-3 in Section 29-8.

19. Plutonium-239 ($_{94}^{239}\text{Pu}$) decays via alpha decay. (a) Write out the decay equation for plutonium-239. (b) Calculate the energy released in the decay of one Pu-239 atom.

20. Scandium-46 ($_{21}^{46}\text{Sc}$) decays via the beta-minus process. (a) Write out the complete decay equation for scandium-46. (b) Calculate the energy released in the decay of one scandium-46 atom.

21. A certain isotope decays via the beta-plus process to neon-22 ($_{10}^{22}\text{Ne}$). (a) Write out the complete decay equation for this situation. (b) The mass of a neon-22 atom is 21.99138511 u. Calculate the energy released in this beta-plus decay process.

22. (a) According to Table 29.3, what does krypton-85 decay into? (b) Write out the decay equation for krypton-85. (c) Calculate the energy released in this decay process.

23. Silver-111, $_{47}^{111}\text{Ag}$, which has an atomic mass of 110.905291 u, is radioactive, spontaneously decaying via either alpha decay, beta-plus decay, or beta-minus decay. The masses of the possible decay products are 110.904178 u for $_{48}^{111}\text{Cd}$, 110.907671 u for $_{46}^{111}\text{Pd}$, and 106.906748 u for $_{45}^{107}\text{Rh}$. (a) Explain how you can use these numbers to determine the spontaneous decay process for silver-111. (b) Write out the complete decay reaction for the spontaneous decay of silver-111.

Exercises 24 – 28 involve radioactivity.

24. Oxygen-15 has a half-life of 2 minutes. Nuclear activity is often measured in units of becquerels (Bq), which are the number of nuclear decays per second. If the activity of a sample of oxygen containing some oxygen-15 is 64×10^6 Bq at $t = 0$, what is the activity level of the sample at (a) $t = 4$ minutes, (b) $t = 16$ minutes, and (c) $t = 20$ minutes?

25. Fluorine-20 decays to neon-20 with a half-life of 11 seconds. At $t = 0$, a sample contains 120 grams of fluorine-20 atoms. How many grams of fluorine-20 atoms remain in the sample at (a) $t = 5.0$ seconds, (b) $t = 30$ seconds, and (c) $t = 1.0$ minutes?

26. A supply of fluorodeoxyglucose (FDG) arrives at a positron emission tomography clinic at 8 am. At 5 pm, at the end of the working day, the activity level of the FDG has dropped significantly because of the 110 minute half-life of the radioactive fluorine. Considering equal masses of FDG at 8 am and 5 pm, by what factor has the activity level been reduced by 5 pm?

27. Four hours after being injected with a radioactive isotope for a positron emission tomography scan, the activity level of the material injected into the patient has been reduced by a factor of 8 compared to its initial value (at the time of the injection). After another four hours elapses, will the activity level be reduced by another factor of 8? Briefly justify your answer.

28. After 3 hours, the activity level of a sample of a particular radioactive isotope, which decays into a stable isotope, has fallen to 20% of its initial value. Calculate the half-life of this isotope.

Exercises 29 – 33 involve nuclear fusion and nuclear fission. Make use of the data in Table 29-3 in Section 29-8.

29. (a) What is the second object produced in the following fusion reaction? $^2_1\text{H} + ^6_3\text{Li} \Rightarrow ^7_3\text{Li} + \underline{\quad}$.
(b) How much energy is produced in this reaction?

30. In Exercise 10, we examined one of the fusion reactions in the CNO (carbon-nitrogen-oxygen) cycle, which is a primary method of fusing hydrogen into helium in massive stars. The primary reactions in the CNO cycle are shown in Figure 29.7. Two more of the fusion reactions in the CNO cycle are $^1_1\text{H} + ^{13}_6\text{C} \Rightarrow ^{14}_7\text{N}$ and $^1_1\text{H} + ^{14}_7\text{N} \Rightarrow ^{15}_8\text{O}$. Calculate the energy released in each of these fusion reactions.

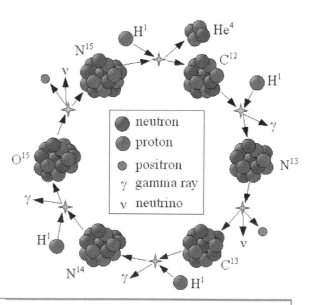

Figure 29.7: The CNO cycle of fusion reactions that produces much of the energy in massive stars. The CNO cycle has the net effect of fusing hydrogen into helium.

31. In relatively low-mass stars, such as our Sun, much of the energy generated by the star comes from the proton-proton chain, which has the net effect of fusing hydrogen into helium. The step that produces the helium-4 atom is: $^{3}_{2}He + ^{3}_{2}He \Rightarrow ^{4}_{2}He + ^{1}_{1}H + ^{1}_{1}H$. The helium-3 atoms are produced from the fusion of hydrogen isotopes in earlier steps in the process. How much energy is released in the step that produced helium-4? Note that the mass of $^{3}_{2}He$ is 3.01602932 u.

32. Complete the following fission reaction: $^{1}_{0}n + ^{235}_{92}U \Rightarrow ^{236}_{92}U \Rightarrow ^{143}_{56}Ba + \underline{\quad} + 3\left(^{1}_{0}n\right)$.

33. How much energy is released in the following fission reaction?
$$^{1}_{0}n + ^{235}_{92}U \Rightarrow ^{236}_{92}U \Rightarrow ^{141}_{56}Ba + ^{92}_{36}Kr + 3\left(^{1}_{0}n\right).$$

Exercises 34 – 38 involve applications of nuclear physics.

34. A common application of nuclear radiation is **food irradiation**, which is the exposure of food products (such as meat, poultry, and fruit) to ionizing radiation to kill bacteria, prolong shelf life, and delay the ripening of fruit. One example of this treatment is the exposure of a meat product to a dose of 3 kilograys of gamma radiation from a cobalt-60 source. One gray (Gy) represents 1 joule of energy exposure per kilogram of food. Cobalt-60 decays by the beta-minus process to an excited form of nickel-60, which then decays to its ground state through the emission of two gamma rays, with energies of 1.17 MeV and 1.33 MeV. (a) How many nuclei are there in 1 gram of cobalt-60, which has an atomic mass of 59.9338171 u? (b) Cobalt-60 has a half-life of 5.27 years. How many nuclei, in a 1-gram sample of cobalt-60, would decay in 1 second? (c) If the two gamma rays associated with each decaying cobalt-60 nuclei deposit all their energy in a 1 kilogram package of meat, how long would the meat have to be exposed to receive a dose of 3 kGy?

35. When radiocarbon dating was carried out on the Shroud of Turin in 1988, the three labs doing the measurements agreed that the shroud dated to approximately the year 1300 AD, providing evidence against the idea that the shroud (a picture of which is shown in Figure 29.8) was the burial cloth of Jesus Christ. If the ratio of carbon-14 to carbon-12 in the atmosphere has remained constant over time at 1.2×10^{-12}, (a) approximately what ratio did the researchers find for the sample of shroud they measured?, and (b) what ratio would they have found if the cloth was 2000 years old?

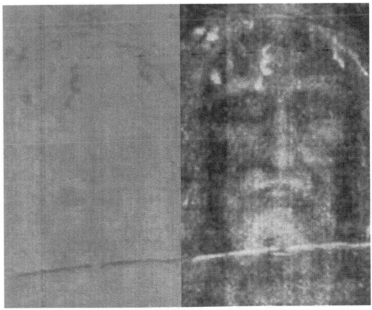

Figure 29.8: A photograph of part of the Shroud of Turin, on the left, and the corresponding negative image, on the right, for Exercise 35. Image credit: Wikimedia Commons.

36. Some smoke detectors have an alpha emitter, americium-241, in them. The alpha particles ionize air molecules in the space between two oppositely charged plates (a parallel-plate capacitor), and the current associated with these ions in the parallel-plate capacitor is measured by the smoke detector. When this current drops, because the ions are blocked by smoke particles, the detector's alarm sounds. (a) When an americium-241 atom decays via alpha decay, what does it decay into? (b) Using the data in Table 29.3 to help you, how much energy (in MeV) is released in the decay process of a single americium-241 atom? For reference, typical ionization energies are tens of electron volts.

37. Single photon emission computed tomography (SPECT) is a medical imaging procedure that makes use of technetium-99m, an excited form of technetium-99 that undergoes gamma decay with a half-life of 6 hours. The gamma rays are relatively low energy, around 140 keV, comparable to standard x-rays. SPECT is thus much like a CT scan, but with photons coming from inside the body rather than passing through the body. Technetium-99m is used in well over half of all medical imaging procedures that use radioactive nuclei, and is especially useful for bone scans and brain scans. By what factor has the activity level of the technetium-99m fallen to 48 hours after it was first administered to the patient?

38. Burning 20 tons of coal provides about 5×10^{11} J, which is approximately the annual energy requirement of an average person in the United States. If this amount of energy was provided by the electricity generated by the fission of uranium-235 in a nuclear reactor, instead, what mass of natural uranium would be required? About 0.7% of natural uranium is uranium-235 (almost all the rest is uranium-238, which does not fission like U-235 does). Assume that the fission of each U-235 atom provides 200 MeV of energy, and that the nuclear power plant has an efficiency of 33% in transforming the energy from the fission reactions into electricity.

General problems and conceptual questions

39. Marie Curie was a pioneer in the field of radioactivity. Do some research about Marie Curie and write a couple of paragraphs about her life and her contributions to nuclear physics.

40. Enrico Fermi made a number of important contributions to our understanding of nuclear physics. Do some research about Fermi and write a couple of paragraphs about his contributions to nuclear physics, describing how his work relates to the principles of physics discussed in this chapter.

41. Equation 29.1 gives the approximate radius of a nucleus, assuming it to be spherical. (a) What is the atomic mass number of a zinc-64 atom? (b) What is the cube root of zinc-64's atomic mass number? (c) Approximately what is the radius of a zinc-64 atom?

42. A particular nucleus has a radius of about 6×10^{-15} m. (a) Approximately what is the atomic mass number of the nucleus? (b) Approximately where would you find such a nucleus on the periodic table, assuming the nucleus is stable?

43. (a) If you convert the mass of an electron entirely to energy, how much energy, in keV, do you get? (b) If you convert the mass of a positron entirely to energy, how much energy, in keV, do you get? When an electron and positron encounter one another, they annihilate one another, and the particles are transformed entirely into two photons. Assume that the electron and positron are initially at rest before the annihilation process. After the annihilation process, what is the (c) energy, and (d) wavelength of each of the two photons?

44. The daily food intake of an average person consists of approximately 10 million joules of food energy. If we could generate this energy internally through the fusion of helium-3 and lithium-6 into two helium-4 atoms plus a proton, which releases about 17 MeV of energy, approximately what mass of helium-3 and lithium-6 would we need to take in every day?

45. Polonium-210 decays via alpha decay into lead-206 and an alpha particle. For the purposes of this exercise, use the approximation that the lead-206 atom has 50 times the mass of the alpha particle. We will also assume that the polonium-210 is at rest before the decay process. (a) Immediately after the decay, how does the momentum of the lead-206 atom compare to the momentum of the alpha particle? (b) Immediately after the decay, what is the ratio of the alpha particle's kinetic energy to the lead-206 atom's kinetic energy?

46. Try this at home. M&M's (the candy) are not radioactive, but a package of M&M's (or something equivalent, like coins) can be used as a model of a system with a half-life. Obtain a package of M&M's and, after first counting how many M&M's there are, do the following. Place all the M&M's in cup (starting with a number of M&M's that is a power of 2, like 64 M&M's, makes this exercise easier), and then shake them out onto a clean surface, like a plate. Remove all the M&M's that have their "m" down – those are the ones that have decayed. Count all the "m" up M&M's, representing undecayed nuclei, and place them back in the cup. Repeat the process until all the M&M's have decayed. (a) Make a table of your results, recording the number of "undecayed nuclei" as a function of the number of throws. (b) Use Equation 29.12 to predict the number of "undecayed nuclei" remaining – add this theoretical data to your table. (c) Plot a graph of your results, and draw the curve representing the theoretical results. (d) Account for any differences between your values and the theoretical values. (e) If we started with a very large number of radioactive nuclei, would we expect to see similar percentage deviations from the theoretical values? Explain. (f) If you repeated the experiment with the M&M's one thousand times, averaged your results, and plotted your averaged results against the theoretical values, would you expect more, less, or the same deviation from the theoretical values compared to the deviation obtained from a single trial? Explain. (g) Eat the M&M's.

47. Much like the M&M's in Exercise 46, a large number of six-sided dice can be used as a model of a system with a half-life. If you shake all the dice, how many throws represent one half-life if you define the "decayed" dice as (a) all those showing an even number, or (b) all those showing a 1?

48. Oxygen-15 has a half-life of 2 minutes. Fluorine-18 has a half-life of 110 minutes. For a sample of oxygen-15 to have the same decay rate as a sample of fluorine-18, how should (a) the number of radioactive nuclei in the two samples compare? (b) the masses of the two samples compare?

49. At $t = 0$, you have two samples of radioactive nuclei. Sample A contains N nuclei of a radioactive isotope that has a half-life of 3 hours. Sample B contains a to-be-determined number of nuclei of a different radioactive isotope, with a half-life of 4 hours. How many nuclei of the second radioactive isotope should sample B contain (at $t = 0$) if, at $t = 12$ hours, (a) the number of undecayed nuclei in the two samples is equal, or (b) the decay rate, measured in nuclei per second, of the two samples is equal.

50. You have two samples of radioactive nuclei. At $t = 0$, sample A contains N nuclei of a particular radioactive isotope, while sample B contains twice as many nuclei of a different radioactive isotope. Also at $t = 0$, the decay rate, measured in nuclei per second, of sample A is three times larger than the decay rate in sample B. How do the half-lives of the two different types of nuclei compare?

51. The graph in Figure 29.9 shows the ratio of the number of undecayed nuclei to the initial number of undecayed nuclei, as a function of time. Use the data in the graph to estimate the half-life of the nuclei.

52. Instead of defining the half-life for a particular isotope, we could define the "one-tenth-life" (or the life for any other fraction less than 1). After a time of a single one-tenth-life has passed, only $1/10^{th}$ of the initial number of radioactive nuclei remain. (a) Determine the ratio of a one-tenth-life to a half-life. (b) The half-life of cesium-137 is 30 years. What is the one-tenth-life for cesium-137? (c) After how many one-tenth-lives does at activity level of a sample of radioactive material drop to 1% of its initial value? (d) How much time elapses until the activity level of a sample of cesium-137 drops to 1% of its initial value?

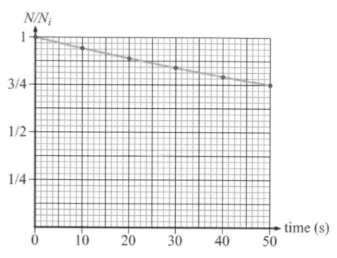

Figure 29.9: A graph of the number of undecayed nuclei to the initial number of undecayed nuclei, as a function of time, for Exercise 51.

53. After 6 days, the activity level of a sample of radioactive material has fallen to 40% of its initial value. After an additional 6 days elapses, what is the activity level of the sample, compared to its initial value? (Hint: there is a very easy way to do this calculation – it is possible to work out the answer without a calculator.)

54. Right now, a sample contains 1.0×10^{20} undecayed nuclei, and a much larger number of decayed nuclei. If the half-life of the undecayed nuclei is 5 months, how many undecayed nuclei did the sample contain a year ago?

55. At $t = 0$ s, a sample of radioactive nuclei contains 1.0×10^{20} undecayed nuclei, which have a half-life of 10 seconds. First, use equation 29.10, with $N = 1.0 \times 10^{20}$ nuclei, to estimate the number of undecayed nuclei remaining at (a) $t = 1$ s, (b) $t = 10$ s, and (c) $t = 20$ s. Second, use equation 29.12 to estimate the number of undecayed nuclei remaining at (d) $t = 1$ s, (e) $t = 10$ s, and (f) $t = 20$ s. (g) Explain which method is the correct method to use to accurately determine the number of undecayed nuclei remaining, and why the other method gives inaccurate results.

56. When a heavy radioactive nucleus decays, it often decays into a nucleus that is also radioactive. The result can be a sequence of decay reactions that continues until a stable nucleus is reached. The sequence of reactions is known as a **decay chain**. An example of a decay chain is shown in Figure 29.10, displaying the sequence of isotopes that make up the chain that starts with the radioactive isotope thorium-232 and ends with the stable isotope lead-208. Note that bismuth-212 has two possible decay modes, so the chain branches from there, but both branches end up at lead-208. (a) Identify the sequence of decays in this particular chain, including the two branches. (b) By comparing the mass of thorium-232 to the mass of lead-208 plus the mass of all the alpha particles (the beta-minus particles are already included) produced in the decay chain, you can find the total energy released by the chain. What is this total energy? Does it matter which branch the chain takes after reaching bismuth-212?

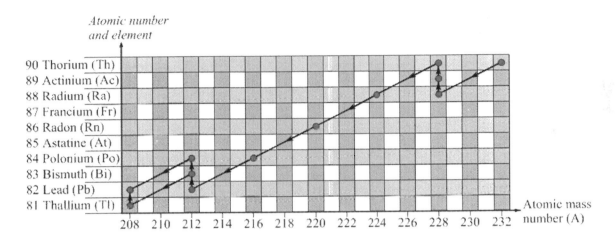

Figure 29.10: The decay chain that starts with the radioactive isotope thorium-232, and ends with the stable isotope lead-208. The radioactive isotopes that the nucleus passes through as it decays are shown with the red circles. Each step in the chain is either an alpha decay, which reduces the atomic number by 2 and the mass number by 4, or a beta-minus decay, which increases the atomic number by 1 without changing the mass number. For Exercise 56.

57. In Exercise 56, we defined the concept of a decay chain. Draw a decay chain diagram, like the one in Figure 29.10, representing the decay chain that starts with radium-226 and ends with lead-206. That sequence proceeds as follows: three alpha decays, followed by two beta-minus decays, followed by one alpha decay, two beta-minus decays, and a final alpha decay. On your decay chain diagram, clearly identify the various isotopes that the chain passes through on the way from the radioactive isotope radium-226 to the stable isotope lead-206.

58. The ratio of carbon-14 to carbon-12 in the shaft of a wooden arrow, unearthed when a foundation was being dug for a new house, is 70% of the same ratio in a growing tree today. Assuming the ratio of carbon-14 to carbon-12 in the atmosphere has been constant, estimate the age of the arrow.

59. In 1991, a body was found by hikers in the Alps, just inside Italy along the border between Italy and Austria. After being removed from the ice that had preserved the body, the body was carefully studied. This individual is now known as Ötzi the Iceman, and one of the methods used to study Ötzi was radiocarbon dating. The radiocarbon dating process carried out at the University of Vienna resulted in a "radiocarbon age" of 4550 years BP. BP stands for "Before Present," where present is defined to be the year 1950 AD, the year when the radiocarbon dating process was first done. In determining a radiocarbon age, the half-life of carbon-14 is taken to be 5568 years. Assuming the ratio of carbon-14 to carbon-12 has held constant in the atmosphere at a value of 1.2×10^{-12}, what was the ratio in Ötzi's body in 1950? Note that correcting Ötzi's age for known historical fluctuations in the carbon-14 to carbon-12 ratio, as well as the correct half-life of carbon-14, led to the conclusion that there was a more than 60% probability that Ötzi died between 3230 and 3100 BC.

60. Nickel has five different stable nuclides, but nickel-66 is not one of them. All of the stable nuclides of nickel have fewer neutrons than nickel-66 has, in fact. (a) What do you think the dominant radioactive decay process is for nickel-66? (b) What is the daughter nuclide produced by this decay process?

61. One of the largest stable nuclides is a nuclide of lead, lead-208, with 82 protons and 126 neutrons. This nuclide can be formed by various radioactive decay processes. (a) If lead-208 is produced by a single beta-plus decay, what was the original nuclide? (b) If lead-208 is produced by a single beta-minus decay, what was the original nuclide? (c) If lead-208 is produced by a single alpha decay, what was the original nuclide? (d) Calculate the amount of energy released in the alpha decay process.

62. Near the low atomic number end of the chart of the nuclides, nuclides with equal numbers of protons and neutrons tend to be stable. As the atomic number increases, however, nuclides need more neutrons than protons to be stable. Consider the nuclide with 50 protons and 50 neutrons. (a) What element is this nuclide? (b) What do you expect the dominant radioactive decay mode to be for this nuclide? Explain your answer.

63. Three students are having a conversation. Comment on how the answers obtained by each student compare to the correct answer to the question they are trying to solve. Explain what, if anything, is wrong with each of their methods.

Mike: OK, the question gives the half-life as 2 years, and it asks for the percentage remaining after just 1 year. Isn't that easy? If half of it decays in 2 years, then 25% decays in 1 year, so the fraction remaining after 1 year is 75%, right?

Jessica: I think you have to use one of the equations. I used Equation 29.12 ($N = N_i e^{-\lambda t}$), and got an answer around 70%.

Debbie: I think Jessica's right, that you need to use an equation. I used Equation 29.10, though, after I re-arranged it to $\Delta N = -\lambda N(\Delta t)$. Then I subtracted what was lost from 100% to get the answer. I got more like 65% left, though.

Answers to selected problems from Essential Physics, Chapter 16

1. This answer is not likely to be correct. Charge is quantized in units of e, which has a value of 1.60×10^{-19} C. Thus, we expect values of charge to be greater than or equal to e, not a small fraction of e, which is what we have in this case.

3. (a) Yes. The simplest way is to have them all start with equal charges. Another one of the many ways is to have A and B start with a total charge of $+10Q$, (such as $+2Q$ on A and $+8Q$ on B), and have C have a charge of $+5Q$, half of the total charge on A and B together. They would all end up with a charge of $+5Q$. (b) No, you can't have all three with different charges. B and C, which touch last, would end up with the same charge as one another. (c) Based on (b), if one sphere has a different sign it must be A. It can be done. One of the many ways is to start with $-8Q$ on A, no charge on B, and $+8Q$ on C. At the end, A has $+2Q$ and B and C each have $-Q$.

5. (a) Yes, the force on B is doubled. (b) This also doubles the force. (c) This increases the force by a factor of 4. (d) This also increases the force by a factor of 4.

9. (a) to the right (b) It could be negative – if so, we have no information about the magnitude of the charge. It could also be positive, but if it is its magnitude must be less than Q.

11. (a) Ball 3 has a positive charge, with a magnitude equal to four times that of the magnitude of the charge on ball 2. Ball 1 has a non-zero charge, so it can only experience no net force when the force on it from ball 2 is exactly balanced by that on ball 3. To have these two forces point in opposite directions requires that the charge on ball 3 be opposite in sign to that on ball 2. To have the magnitudes of the two forces acting on ball 1 be equal, ball 3 must have a larger magnitude charge than ball 2, because ball 3 is farther away. We know that ball 3 is twice as far from ball 1 as ball 2 is, and, in Coulomb's law, that factor of 2 in the distance gets squared to a factor of 4 in the denominator. To make up for this factor of 4 in the denominator, there must be a factor of 4 in the numerator – this is what we get if ball 3 has a charge that has a magnitude four times that of ball 2. (b) Ball 1 has a positive charge, and it must be larger in magnitude than that of ball 3. Ball 2 is attracted to ball 3, because these two charges have opposite signs, so to have a net force on ball 2 that is directed to the left, ball 2 must be attracted to ball 1 (therefore, ball 1 is positive) with a larger force than that applied by ball 3 (thus, the charge on ball 1 is larger than that on ball 3). (c) $1 > 3 > 2$.

13. $3kq^2 / r^2$, in the negative y-direction.

15. There are two solutions. The second charge is located either at $x = +2a$ or at $x = +10a$.

17. (a) You can't set up a situation in which all three balls experience no net force. (b) You can have one of them feel no net force, however. For instance, to have the $+q$ charge feel no net force, place the $+2q$ charge between the $+q$ and $-4q$ charges (to the left of the

−4q charge), in just the right spot that the force it applies to the +q charge is balanced by that from the −4q charge. On the other hand, for the −4q charge to experience no net force, the +2q charge has to be placed at just the right distance to the right of the −4q charge.

21. (a) The acceleration because of the electric field is found from $ma = qE$, so $a = qE/m$, which has a value of about 3.5×10^{13} m/s, directed straight down. This is many orders of magnitude larger than the acceleration due to gravity, so it is quite reasonable to neglect the influence of gravity. (b) 4.3×10^{-11} s (c) 3.2×10^{-8} m

23. There is only one possible solution. There is a charge of $−6.0 \times 10^{-7}$ C at $x = +3$ m.

25. Two possible solutions. The unknown charge is either $+0.5q$ or $+2.5q$.

27. Two possible solutions. The unknown location is either at $x = +a/2$ or at $x = -\dfrac{a}{\sqrt{8}}$.

31. (a) The three net forces add up to zero. This is because we are really adding up three pairs of equal-and-opposite forces, which cancel. (b) Yes, we would always get this result, because we would always be adding up pairs of equal-and-opposite forces.

33. (a) negative (b) more than (c) $−2Q$.

35. This situation is not possible. The unknown charge, if it was positive, would exert a force to the left on the positive test charge. The original charge would exert a downward force on the test charge, however, and the unknown charge could not exert a force on the test charge so that the net force on the test charge had an upward component.

41. 2.2×10^6 m/s

43. 4.08 kg

45. (a) The +q charge at the origin experiences a larger net force than the +q charge on the positive y-axis. In both cases, there is significant cancellation between the individual forces, but the +q charge at the origin has the advantage of being closer, on average, to the charges exerting forces on it, and the two charges on the y-axis exert forces that are in the same direction. (b) The −2q charge on the −y axis experiences a net force of $(0.328)\dfrac{kq^2}{r^2}$ straight down. The −2q charge on the −x axis experiences a net force of $(4.26)\dfrac{kq^2}{r^2}$ at an angle of 84.4 degrees above the negative x-direction. The −2q charge on the +x axis experiences a net force of $(4.26)\dfrac{kq^2}{r^2}$ at an angle of 84.4° above the positive x-direction.

47. (a) $1 > 2 > 3$ (b) $1 > 3 > 2$

49. (a) $\sqrt{17}\,\dfrac{kq}{r^2}$ at an angle of 76° below the negative x-axis   (b) $\dfrac{kq}{r^2}$ to the left

   (c) $\sqrt{13}\,\dfrac{kq}{r^2}$ at an angle of 33.7° below the negative x-axis

51. (a) $+4\sqrt{2}\,q$ (b) $+8\sqrt{2}\,q$ (c) $\dfrac{6kq}{L^2}$ in both cases

53. (a) $\dfrac{4kq}{L^2}$ toward the lower left corner of the square (b) Either make the charge at the bottom right $+5q$, or make the charge at the upper left $+q$.

55. –0.010 C

57. (a) $\dfrac{4\sqrt{2}\,kq}{L^2}$ straight up   (b) $(2.31)\dfrac{kq}{L^2}$ at an angle of 7.3° below the negative x-axis

59. There are two possible solutions. One solution is $+6 \times 10^{-6}$ C, located at $x = -1.0$ m, or the charge can be -6.67×10^{-7} C, located at $x = +3.0$ m.

61. (a) Charge 1 is positive and charge 2 is negative (b) Approximately 22 µC

63. (a) There is one location where the net field is zero, in between the two charges, and closer to the smaller charge. (b) The location is at $x = 0.69$ m.

67. Picture (a)

Answers to selected problems from Essential Physics, Chapter 17

1. (a) No, just because the electric field is zero at a particular point, it does not necessarily mean that the electric potential is zero at that point. A good example is the case of two identical charges, separated by some distance. At the midpoint between the charges, the electric field due to the charges is zero, but the electric potential due to the charges at that same point is non-zero. The potential either has two positive contributions, if the charges are positive, or two negative contributions, if the charges are negative. (b) No, just because the electric potential is zero at a particular point, it does not necessarily mean that the electric field is zero at that point. A good example is the case of a dipole, which is two charges of the same magnitude, but opposite sign, separated by some distance. At the midpoint between the charges, the electric potential due to the charges is zero, but the electric field due to the charges at that same point is non-zero. Both the electric field vectors will point in the direction of the negative charge.

3. (a) Zero. The potential at infinity is zero, and the potential at the midpoint of the dipole, due to the charges on the dipole, is also zero. The potential difference is zero, so no net work is done. (b) Still zero. The path followed does not matter because the electric force is conservative – all that matters is the potential difference between the initial point and the final point, which is zero.

5. 1 is not possible – field lines and equipotentials are perpendicular to one another where they cross. 2 is not possible – for one thing, equipotentials can not cross one another. 3 looks fine – it looks pretty close to a dipole situation.

7. (a) $+4q$ (b) $-6q$ (c) $+4q$

9.

	Potential difference	Capacitance	Charge	Field	Energy
Initially	V_0	C_0	Q_0	E_0	U_0
Dielectric removed	V_0	$C_0/3$	$Q_0/3$	E_0	$U_0/3$
Distance halved	V_0	$2C_0/3$	$2Q_0/3$	$2E_0$	$2U_0/3$

11. (a) The capacitor does work on the dielectric, attracting it inside the capacitor.
    (b) You do work on the dielectric to bring it back out of the capacitor.

13. 3.1×10^6 m/s

15. 7.5 m/s

17. (a) $-\dfrac{6kq^2}{r}$ (b) $-\dfrac{12kq^2}{r}$ (c) $-\dfrac{2kq^2}{r}$

19. 7.3×10^7 m/s

21. There are two possible solutions. The charge is -1.33×10^{-7} C, and located at $x = +8.0$ m, or the charge is -4.44×10^{-8} C, and located at $x = +4.0$ m.

23. (a) and (b) Yes, there can be a place in both of these regions. At some point closer to the $+q$ charge, the smaller distance to the $+q$ charge can make up for the fact that the other charge is larger in magnitude. (c) To the right of the ball that has a charge of $-3q$, the potential will always be negative – being closer to the larger-magnitude charge, that charge will always dominate. (d) The two locations are $x = -2a$ and $x = +a$.

25. +80 volts

27. (a) The forces are equal in magnitude. We can justify this using Newton's third law, or from Coulomb's Law. (b) $\dfrac{3kq^2}{16a^2}$, to the left (c) $\dfrac{kq}{2a^2}$, to the left

The other (c) $+\dfrac{3kq^2}{4a}$ (d) $+\dfrac{2kq}{a}$

31. (a) The ball with charge $+4q$. (b) $+\left(2 + 22\sqrt{2}\right)\dfrac{kq^2}{d} \approx +33\dfrac{kq^2}{d}$

33.

	Potential difference	Capacitance	Charge	Field	Energy
Initially	V_0	C_0	Q_0	E_0	U_0
Distance doubled	V_0	$C_0/2$	$Q_0/2$	$E_0/2$	$U_0/2$
Wires removed	V_0	$C_0/2$	$Q_0/2$	$E_0/2$	$U_0/2$
Dielectric inserted	$V_0/4$	$2C_0$	$Q_0/2$	$E_0/8$	$U_0/8$
Battery re-connected	V_0	$2C_0$	$2Q_0$	$E_0/2$	$2U_0$

35. We can find the capacitance, but to find the energy we need another piece of information, either the potential difference or the charge.

37. (a) The balls repel one another, and because of the symmetry they all accelerate away from the center of the triangle, along lines that go from the center of the triangle through the vertices of the triangle. (b) 550 m/s (c) 780 m/s.

39. (a) This is possible, but only if the charge is negative. The farther away you get from a negative charge, the less negative the potential due to that charge is, so increasing the distance can increase the potential. (b) -8.3×10^{-10} C.

41. (a) 3>1>2 (b) 2>1>3 (c) 2>3>1 (d) 3>1=2

43. (a) Negative. (b) 0.5 s (c) 10 cm

45. (a) 1=2=3 (b) 3>2>1

47. (a), (b), and (c) $+3kq/r$

49. $+(18-\dfrac{9}{\sqrt{2}})\dfrac{kq^2}{L} \approx +11.6\dfrac{kq^2}{L}$

51. (a) Yes, charge 1 must be negative. (b) The magnitude of charge 1 is $3Q$. (c) No, we cannot say what the sign of charge 3 is. (d) We can't say anything about charge 3's magnitude, either.

53. (a) There are two locations a finite distance from the charge where the net potential is zero. One of these places is to the left of the positive charge, on the negative x-axis. The other spot is between the charges, closer to the charge that has a smaller magnitude (closer to the positive charge). (b) At $x = -2.5$ m and at $x = + (5/7)$ m.

55. There are lots of locations off the line at which the net potential is zero – all points that are a distance d from the positive charge and a distance of $(9/5)d$ from the negative charge.

57. (a) at x = -a and at x = +3a (b) also -2q (c) Yes, at x = +a, where the slope of the graph is zero.

59. (a) 2=3>1 (b) 1=2=3 (c) 2=3>1 (d) 1>2=3

Answers to selected problems from Essential Physics, Chapter 18

1. In a circuit, removing a parallel resistor increases the equivalent resistance of the circuit, and reduces the total current in the circuit. By analogy, closing one trail on the ski hill reduces the rate at which skiers make it down the hill. The speed of the lift should be reduced, so the rate at which skiers are carried up the hill is equal to the rate at which they go down the hill.

3. (a) Graph (a) corresponds to the resistors being in series. Graphs (b) and (c) are for the individual resistors. Graph (d) corresponds to the resistors being in parallel. (b) The resistors have resistances of 2 Ω and 3 Ω.

5. (a) A gets dimmer   (b) B gets dimmer (it turns off) (c) C gets brighter
   (d) D gets brighter.

7. (a) D>A=C>B (b) D>A>B=C   (c) C=D>A=B (d) B=C=D>A

9. (a) The current decreases through A. (b) The current increases through B.
   (c) The current decreases through C.   (d) The current decreases through the battery.
   (e) $R_C = 6\ \Omega$

11. (a) 3.0 V   (b) 1.0 A   (c) 0.50 A   (d) The capacitor, which was being charged with a current of 1.0 A at the instant the switch is closed, is being charged with a current of only 0.50 A immediately after the switch is closed. The rest of the current is diverted to resistor B. The capacitor is still charging, it is simply charging more slowly.
    (e) After a long time, the capacitor has a potential difference of 4.0 V across it, and the resistors each have a current of (2/3) of an amp passing through them.

15. (a) Current multiplied by time is charge, so the milliamp-hour is a unit of charge.
    (b) 4⅔ hours

17. (a) The resistance has increased, because of the increased length and the decreased cross-sectional area.   (b) $16R$

19. This requires about 1980 m of copper wire, and just 1.52 m of carbon wire.

21. (a) 288 Ω (b) 144 Ω (c) 96 Ω (d) 150 W

25. (a) $N = 4$ (b) $R = 6\ \Omega$

29. (a) $4R$ (b) 1/3 (c) $I_{2R} > I_{3R} > I_R = I_{5R}$ (d) $\Delta V_{2R} = \Delta V_{3R} > \Delta V_{5R} > \Delta V_R$

31. (a) 96 Ω (b) If we let R represent the resistance of one bulb, the equivalent resistance of the circuit is (6/7)R, and by the power equation P = $(V)^2$/R we find that the power is (7/6) times that of what it is from a single bulb getting the full potential difference. Thus, the total power output is larger than that from a single bulb that has 120 V across it.
(c) Bulbs A and B each have a power of (50/3) W; bulb C has a power of (200/3) W; bulbs D and E each have a power of 37.5 W. The total power is 175 W.

33. (a) 5R   (b) $I_A = 6I_D$ $I_B = 3I_D$ $I_C = 2I_D$ $I_E = I_D$
    (c) $\Delta V_A > \Delta V_B = \Delta V_C > \Delta V_E > \Delta V_D$

37. 28.3 V

39. (a) 1 A down through the battery. (b) 14 V   (c) 8 Ω (d) –3 V

41. $I_1 = +\dfrac{5}{13}$ A $I_2 = +\dfrac{14}{13}$ A $I_3 = -\dfrac{9}{13}$ A

43. (a) $I_1 + I_2 + I_3 = 0$ (b) This junction equation makes it look like all three currents are directed into a junction, and none are directed away (or vice versa, depending on which junction you look at). In reality, there must be at least one current directed into the junction and at least one directed away, which tells us that either one or two of the currents have the wrong sign, and therefore are shown going the wrong way on the diagram.

45. (a) 6.8 V. (b) A is 9.2 V higher than B

47. (a) 0.556 A   (b) 5.44 V   (c) 0.510 A   (d) 4.51 V

49. (a) 100 kW h   (b) 100 kW h
(c) Total energy used = 500 kW h; energy cost = $100; total cost = $102.50
(d) Total energy used = 100 kW h; energy cost = $20; total cost = $24

51. 19.6 V

53. (a) R_1 has the largest current through it. Because R_2 is at least 4 Ω, the total resistance of the branch of the circuit with resistors 2 and 3 is at least 10 Ω, greater than the resistance of the branch containing R_1. Thus, more current will go through R_1.
    (b) The current through R_1 is 1 A no matter what R_2 is. In this circuit, the potential difference across R_1 is always 9 V, so R_1 always has the same current.

55. There are two cases to consider. The 5 Ω resistor can be placed in parallel with the 4 Ω resistor, or it can be placed in series with the 6 Ω resistor. After analyzing both cases, it can be shown that to maximize the current through 3 Ω resistor, the 5 Ω resistor should be placed in parallel with the 4 Ω resistor. This increases the current through the 3 Ω resistor by 0.84 A, from 2.00 A to 2.84 A.

57. $I_A = 0.5$ A, $I_B = 0.22$ A, $I_C = I_D = 0.11$ A, $I_E = 0.33$ A.

59. (a) The 5 Ω resistor has the largest current. All the current that passes through the battery also passes through the 5 Ω resistor, while all the other resistors only get a fraction of the total current. (b) These two resistors have equal currents – both parallel branches of the circuit have 10 Ω of resistance, so the current divides equally between them. (c) The 10 Ω resistor has a larger potential difference than the 8 Ω resistor. Both parallel branches have the same potential difference across them. The 10 Ω resistor has that potential difference across it, while the 8 Ω resistor has to share that potential difference with the pair of 4 Ω resistors. Another way to answer this part is simply to apply Ohm's law to the two resistors – they have the same current, so the one with the largest resistance also has the largest potential difference across it.

61. (a) $+16 \text{ V} - I_2(56 \text{ }\Omega) - I_1(68 \text{ }\Omega) - I_1(47 \text{ }\Omega) = 0$
    (b) $-I_1(47 \text{ }\Omega) + 16 \text{ V} - I_2(56 \text{ }\Omega) - I_1(68 \text{ }\Omega) = 0$. The terms are in a different order, but the equation is equivalent.
    (c) $+I_1(47 \text{ }\Omega) + I_1(68 \text{ }\Omega) + I_2(56 \text{ }\Omega) - 16 \text{ V} = 0$. The terms are in a different order, and every sign is flipped, but the equation is still equivalent to the two above.

63. (a) 3.7 V   (b) $t = 11.4$ s   (c) $t = \infty$

Answers to selected problems from Essential Physics, Chapter 19

1. (a) The magnetic field is perpendicular to the plane of the page.
   (b) The magnetic field is directed into the page.

3. (a) up (b) The field definitely has an upward component, and it may or may not have a component that is directed east or west. It definitely does not have a component directed north or south.

5. (a) Yes – the force is directed out of the page. (b) No, because the magnetic force that acts on a moving charge is always perpendicular to the velocity of the charge.

7. (a) positive (b) negative (c) into the page

9. (a) Yes. Note that there are more field lines entering the region than are leaving it, an indication that there is a net negative charge in the region behind the screen. (b) No. Because magnetic field lines are continuous loops, there are always an equal number of field lines entering a region as there are emerging from a region. That is not the case in this picture.

11. (a) out of the page (b) into the page (c) wire 2

13. (a) 1.75×10^{-4} N, directed in the plane of the page, at an angle of 60° above the positive x-direction.

15. (a) 4.0 N (b) 0 (c) 14.5°

17. (a) $m_2 > m_1 > m_3$ (b) $F_1 = F_2 = F_3$

19. Note that this question was incomplete – you can't answer it unless you know the strength of the magnetic field. Let's assume the magnetic field is $B = 2.0$ T. In that case, (a) 3.5×10^{-3} s (b) 56 m (c) Passing through the negative y-axis at a distance of 56 m from the origin.

21. (a) 1.7×10^{-5} T (b) Yes, at the point 24 cm to the right of wire 2.

23. 6.3×10^{-6} N/m at an angle of 18° below the positive x-axis, assuming the positive x-axis is directed right.

25. (a) 5.0×10^{-6} T out of the page (a) 1.0×10^{-6} T out of the page
    (c) 5.0×10^{-7} T out of the page (a) 3.5×10^{-6} T into the page

27. 4:1

29. (a) a > c > b. (b) 1.0×10^{-3} T

31. 5.3×10^7 m/s

33. (a) The magnets are repelling one another – the field lines at the facing poles curve away from one another, which tells us that the facing poles are like poles. (b) We know like poles are facing one another, but we cannot tell the south poles from the north poles. The iron filings in the picture line up with the field lines, but we can't tell, for instance, whether the right end of the magnet on the right is a north pole, with field lines directed away from it, or a south pole, with field lines directed toward it.

37. (a) The density of field lines passing through the Earth's surface is much higher in the polar regions than it is in other regions of the Earth. Thus, when particles spiral around the field lines, they tend to get carried toward one pole or the other. (b) toward the east.

39. (a) 0 (b) 800 m/s (c) 800 m/s.

41. (a) 3.3×10^4 m. The axis of the spiral is parallel to the x-axis. (b) 5.2 s
(c) 1.6×10^5 m

43. (a) into the page (b) The kinetic energy stays the same. The magnetic force is always perpendicular to the velocity. Any force perpendicular to the velocity can only change the direction of the velocity, not its magnitude. Because the speed is constant, the kinetic energy is also constant. (c) positive (d) $v/2$ (e) see the diagram (f) see the diagram (g) particle 3.

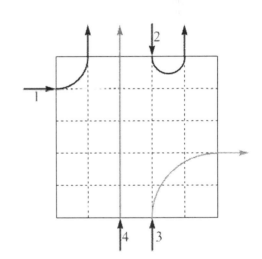

45. (a) into the page. (b) 2.0×10^{-2} T. (c) Faster electrons experience a net force down, so they get deflected down, out of the beam.

47. (a) 4.0×10^4 m/s (b) 1.7 T.

49. (a) $2 > 3 > 1$ (b) wire 1: 4.0×10^{-6} N/m, to the right;
        wire 2: 1.6×10^{-5} N/m, to the left;   wire 3: 1.2×10^{-5} N/m, to the right

51. (a) Zero force (b) Zero net force

53. (a) The loop is experiencing a non-uniform field   (b)
 from the long straight wire. The magnitude of the
field decreases with increasing distance from the wire.

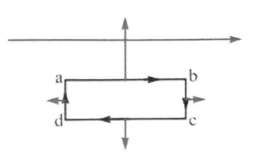

(c) To find the magnetic force on either the left side or the
right side, we would have to account for the fact that
different points on these sides are different distances from

the long straight wire, and so are experiencing different magnetic fields. To calculate these forces would involve doing calculus. Fortunately, we do not have to do this, because the forces on the left and right sides of the loop cancel one another out. (d) 3.6×10^{-6} N, directed up. (e) 1.8×10^{-6} N, directed down. (f) 1.8×10^{-6} N, directed up.

55. There are two possible solutions. Either $I_2 = 2I_1$, directed out of the page, or $I_2 = 6I_1$, directed into the page.

57. (a) $2 > 3 > 1$ (b) to the right.

59. (a) 4.0×10^{-5} N m (b) 4.0×10^{-5} N m (c) zero

61. The magnitude of the angular acceleration is 107 rad/s^2.

1. (a) $3 > 4 > 1 = 2$ (b) $3 > 1 = 2 > 4$

3. (a) Yes, if there continues to be no magnetic field passing through the loop.
   (b) Yes, if the magnetic field through the loop is changing with time, and just happens to be passing through zero at this one instant.

5. There is no induced current.

7. To the right.

9. $2 > 1 > 3$

11. "A non-zero potential difference when the switch closes that quickly drops to zero." In this situation, closing the switch results in the magnetic flux through the transformer coil changing from zero to some non-zero, but constant, value. Thus, there is only an induced emf momentarily, while the flux is changing.

13. (a) 9.0 mV   (b) 0   (c) 7.8 mV

15. $t_C > t_B > t_A = t_D$

17. (a) No change. In this case, the flux depends only on the field and on the area of the loop.   (b) The maximum emf increases by a factor of 3, because moving the loop across the field boundaries 3 times faster increases the rate of change of the magnetic flux through the loop by a factor of 3.   (c) The maximum induced current also increases by a factor of 3. It is proportional to the induced emf, by Ohm's Law.

19. (a) This is possible if the current is increasing in magnitude. To oppose the extra field lines that pass through the loop out of the page, the induced current generates a magnetic field that is into the page – this requires a clockwise current.

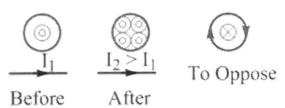

(b) This is possible if the current is decreasing in magnitude. To oppose the loss of field lines that pass through the loop into of the page, the induced current generates a magnetic field that is into the page – this requires a clockwise current.

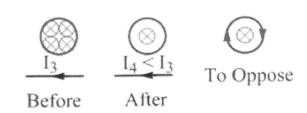

23. (a) The induced emf is the same at all three times. The induced emf depends on the time rate of change of the magnetic flux, which is the slope of the graph. Because the flux vs. time graph has a constant slope, the induced emf has a constant value.
    (b) 5.0×10^{-4} V

25. (a)

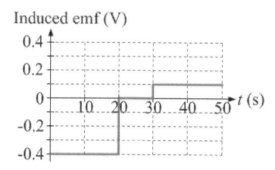

    (b) clockwise

    (c) clockwise

    (d) counterclockwise

29. The rod will be moving faster in the first experiment, when the switch is open. When the switch is closed, in experiment 2, an induced current flows through the rod. The magnetic field acts on this current to exert a force to the left on the moving rod, reducing the rod's acceleration to the right.

31. 90 V

33. 4.8×10^5 rad/s

35. (a) We expect the current associated with the back emf to subtract from the original current. This is consistent with Lenz's law, in which loops tend to oppose whatever is causing the flux to change through them, which in this case is the spin because of the original current.
    (b) The net current drops as the motor speed increases. Thus, the lights in your house dim briefly initially, when there is no induced current associated with the back emf, and the net current through the device is large. When the net current in the device drops after a few seconds, the lights return to normal brightness.

37. (a) 12 A (b) 80 W (c) The input power is 240 W. If the computer used 80 W, the remaining 160 W must be dissipated in the transformer itself, raising the temperature of the transformer.

39. (a) The standard European outlet supplies 240 V, while the standard American outlet supplies 120 V. Devices are generally designed to operate with either one or the other of these voltages, but not both. (b) 2:1

41. (a) 3.0×10^{-4} T m^2 (b) 70.5°

43. (a) 0.80 A counterclockwise (b) 0.80 A clockwise (c) 0.80 A clockwise
    (d) 1.6 A counterclockwise

45. The direction of the induced current is clockwise.

Before   **After**   **To Oppose**

47. (a)

(b) Multiple graphs are possible. What matters is the slope of the graph in the different regions. You can slide the entire graph up or down, to give a much different flux graph, while still producing the same graph of induced emf.

51. (a) Counterclockwise (b) The net force is down, away from the long straight wire (c) The net force causes the loop to move away from the wire, toward a region of reduced flux. This is completely consistent with Lenz's law. The increased current in the long straight wire causes the magnetic flux through the loop to increase. By Lenz's law, the induced current produces a force that attempts to bring the flux through the loop back toward its original value.

53. (a) $I = \dfrac{\varepsilon}{R}$, clockwise (b) $F_{net} = \dfrac{\varepsilon LB}{R}$, to the right (c) $I = 0$ (d) $F_{net} = 0$

   (e) $v = \dfrac{\varepsilon}{BL}$, to the right

55. (a) left-to-right

(c) $v = \dfrac{mgR}{B^2 L^2}$

(b)

$F = ILB$

mg mg

Initially   Terminal velocity

(d) The brightness of the bulb increases as the rod accelerates, and then the bulb maintains a steady glow after the rod has reached its terminal velocity.

57. The train would still slow down, which makes sense from the perspective of energy conservation. Reversing the field reverses the direction of the swirling eddy currents, but reversing both the current and the field keeps the forces in the same direction. Thus, the forces still act to slow the rotation rate of the wheels, which slows the train.

59. In a solid core, the changing magnetic flux in the core tends to produce eddy currents that circulate in planes that are perpendicular to the direction of the magnetic field passing through the core. These currents give rise to a magnetic field that tends to reduce the time rate of change of the magnetic field in the core, reducing the induced emf in the secondary coil. With laminated sheets, the eddy currents are confined to tiny swirls within each sheet, minimizing their negative impact.

Answers to selected problems from Essential Physics, Chapter 21

1.

3.

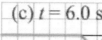

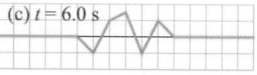

5. (a) The midpoint between the speakers always experiences constructive interference, because the path length difference for that point is zero. The midpoint is the same distance from both speakers, so the waves always interfere constructively.

7.

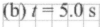

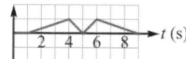

9. There are six different beat frequencies, at 3 Hz, 4, Hz, 5 Hz, 7 Hz, 8 Hz, and 12 Hz.

11. (a) 3.0 cm

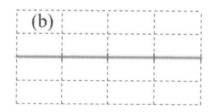

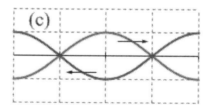

13. (a) 7.0 mm   (b) 4π rad/s   (c) 2.0 Hz   (d) 2.0 m   (e) 4.0 m/s

15. (a) 104 rad/s   (b) 2.4 mm.

17. (a) 25 Hz   (b) 23 m/s   (c) 0.92 m   (d) y = (5.0 mm) cos[(50 π rad/s)t + (6.80 m⁻¹)x]

19. 3.2×10^{-5} W/m²

21. (a) 96 dB   (b) 84 dB

23. (a) 265 Hz   (b) 247 Hz

25. (a) 16.8 m/s   (b) 301 Hz

27. (a) 71.1 kHz (b) 74.3 kHz

29. (a) 3.8 cm (b) 1.9 cm

31. 379 Hz

33. 26 ms

37. (a) C:E:G = 6:5:4 (b) The G pipe is 0.43 m long

39. (a) 24 ms (b) 0.24 m

41. (a) positive x-direction (b) $y = -(4.0 \text{ cm})\sin[(100\pi \text{ rad/s})t - (5\pi \text{ m}^{-1})x]$
    (c) Yes, both the negative signs in the equation switch to positive signs.

43. There are four possibilities, all of which require the second train to also be traveling east. If the second train is behind the first train (both moving east), the second train could either be traveling at 0.9 m/s or at 9.0 m/s. If the second train is ahead of the first train (both moving east), the second train could either be traveling at 0.9 m/s or 9.2 m/s.

45. (a) 3.00 MHz (b) 2.97 MHz (c) 3.04 MHz.

47. (a) 300 kHz (b) 200 kHz

49. (a) to the left (b) The source speed is twice the wave speed. (c) 160 Hz

51. (a) 200 Hz, 400 Hz, and 600 Hz (b) 100 Hz, 300 Hz, and 500 Hz

53.

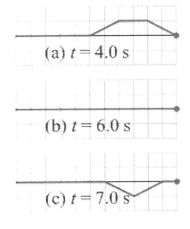

(a) $t = 4.0$ s

(b) $t = 6.0$ s

(c) $t = 7.0$ s

55. displacement

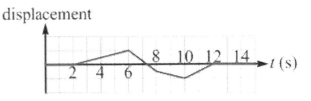

57. displacement

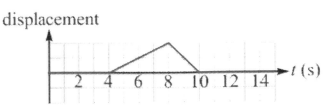

59. 40.9 N

61. (a) 18.3 m/s (b) 0.21 m (c) 62 Hz

Answers to selected problems from Essential Physics, Chapter 22

1. A common way to implement a collision avoidance/mitigation system is to use the Doppler effect for EM waves. If a system mounted on the front of the car emitted pulses consisting of waves of a particular frequency, the time taken to receive the reflected waves back tells the system the distance away that an object ahead of the car is (this requires knowing the speed of light, which we know very accurately). In addition, the frequency of the reflected waves, by the Doppler effect, tells the system the velocity of an object ahead relative to the car. A significant velocity of that object toward the car would indicate an impending collision.

3. There will be no frequency difference.

5. The sky will look darker in case 1. Let's say you are initially looking north, and the Sun is setting in the south, behind you. Light coming from the sky directly above you is polarized along an east-west line. This is exactly what your sunglasses would block if you then, when first facing north, look straight up. In case 2, let's say you were initially facing east, with the Sun setting in the South. Looking up, your sunglasses would block anything polarized along a north-south line. The light coming from above you, however, is polarized along an east-west line, so it gets through your sunglasses in case 2.

7. Graph 2. No matter what the direction of the polarizer's transmission axis is, when unpolarized light passes through the polarizer 50% of the intensity is blocked.

9. It can't be done with only one of these polarizers. With two polarizers, it can be done by passing the light through A and then C, or through C and then A. With three polarizers, polarizer B needs to be either first or last. This gives four possible arrangements: B➔ A➔ C, B ➔ C ➔ A, A➔ C➔ B, or C➔ A➔ B.

11. (a) The sequence could be A➔ B➔ C or C➔B➔A. (b) 67.5 W/m^2

13. It's hard to tell on a logarithmic scale, but the FM band is substantially wider. The AM band is a little wider than 1 MHz. The FM band, in contrast, covers about 20 MHz.

15. (a) VHF = very high frequency (b) 30 MHz – 300 MHz (c) FM radio broadcasts are part of the VHF band, as are some television broadcasts. Communications with airplanes and ships falls under the VHF band, also. Wireless microphones also operate in the VHF band.

17. (a) 4.75×10^{14} Hz (b) 1.58×10^6

19. 15000 V/m. This is a rather large electric field, much larger than you would find in a typical electromagnetic wave.

21. (a) 45° to the incident light   (b)

23. (a) 5.7×10^7 m/s

25. (a) Moving away (b) 1.4×10^7 m/s

27. (a) The man's velocity with respect to the ground was 110 km/h south. He still deserves a speeding ticket, but the verdict should be based on being 10 km/h over the speed limit, as opposed to 60 km/h.

29. (a) 30° (b) 450 W/m^2 (c) at 40° to the vertical (d) 30° (e) 340 W/m^2

31. (a) 360 W/m^2 (b) at 90° to the vertical (c) 60° (d) 90 W/m^2
     (e) at 30° to the vertical (f) 30° (g) 67.5 W/m^2

33. 150 km

35. (a) 2.2×10^7 times larger (b) 1×10^8 W/m^2 (c) 1.9×10^5 V/m

39. 4×10^{26} W

41. (a) About 3 km (b) Every 3 s corresponds to 1 km (c) Yes

43. (a) 8700 W/m^2 (b) 2500 W/m^2 (c) 560 W/m^2 (d) 48 W/m^2

45. The intensity is about 570 W/m^2. This is less than the intensity of sunlight, but within a factor of 2, and hence still very bright. In addition, while the energy in a beam of sunlight is spread out over all wavelengths of the visible spectrum, the energy in the laser beam is all at one wavelength, which increases the danger. Thus, it is not a good idea to look into the laser beam, and you should ask your professor to be more careful.

47. (a) 3.00 MHz (b) 2.97 MHz   (c) 3.04 MHz.

49. 4 cm away from the filament

51. (a) 1170 Hz (b) 1949 Hz   (c) 2924 Hz

53. (a) Toward the officer. (b) 2.268×10^{10} Hz

55. (a) 90 W/m^2 (b) 45°

57. We can't tell. With the correct angle between the transmission axes of the two polarizers, we can get 35% of the intensity if we start with unpolarized light. With polarized light, there are many combinations of angles that will produce a final intensity that is 35% of the initial intensity. So, we can't say whether the incident light is polarized or unpolarized.

Answers to selected problems from Essential Physics, Chapter 23

1. Yes, the image is in the same location for both of you. A plane mirror acts somewhat like a window. Two observers, standing in different locations, looking through a window at the same object agree on the object's location. The same thing happens for a plane mirror. Another way to see this is that the observers see the image because of light rays reflecting from the mirror towards each observer. The image is located where all the reflected rays appear to diverge from.

3. The resulting pattern is a bright vertical line, twice as tall as the slit.

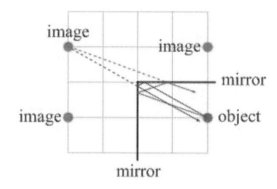

5. (a) You see three images. (b) + (c)

7. Because light rays are reversible, and the image is real, we can't tell which is the object and which is the image.

9. (a) The red ray is drawn incorrectly. The red ray is a parallel ray (parallel to the principal axis), so when it reflects from the mirror it should go directly away from the focal point, F, rather then the center of curvature.

(b)

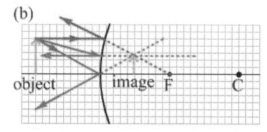

11. (a) Any mirror can produce a virtual image, so this information does not help much. (b) This, also, would happen with any of the three types of mirror. (c) It cannot be a plane mirror, because the height of the image created by a plane mirror is always the same as that of the object. It cannot be a convex mirror, either – moving the object closer to a convex mirror results in an increase in image size. It must be a concave mirror – moving the object closer to a concave mirror when the mirror is creating a virtual image will result in a decrease in the image size.

13. (a) 20° (b) 20°

15. (a) 1/(+12 cm) (b) +12 cm (c) concave mirror

17. (a) 1/(+30 cm) (b) +30 cm (c) concave mirror

19. (a) zero (b) infinity

21. (a) 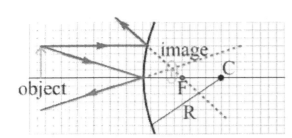   (b) –7.1 cm (c) –0.29

25. (a) –0.56 m (b) 0.070 (c) The result for the magnification is consistent with the warning. The image of the truck is a lot smaller than the real truck, making it appear that the truck is farther away than it really is.

27. (a) 2.5 m (b) inverted

29. (a) 0.49 m (b) Yes, if you can see the clerk, the clerk can, in general, see you. For you to be able to see the clerk in the mirror, light travels in a straight line from the clerk to the mirror, reflects off the mirror (obeying the law of reflection), and travels in a straight line to your eye. Light would travel from you to the clerk along the reverse path so, in general, the clerk could also see you.

31. In conditions where there is lots of light passing through the window from outside, as there is on a sunny day, the light coming through the window into your eye overwhelms the light reflecting from the inside of the window into your eye. When there is very little light passing through the window from outside, on the other hand, as when the train is passing through a dark tunnel, the reflected light dominates.

33. The image maintains its position. Thus, a mirror is much like a window. When you look at a stationary object through a window, the object maintains its position when you move. Similarly, a mirror creates an image of a stationary object that is at a fixed location in space, so looking into the mirror at that image while you are moving is a lot like looking through a window at an object while you are moving.

35. (a) 3.0 m (b) 5.0 m, 8.0 m, and 8.0 m (there are two images 8.0 m from you).

37. (a) 180° (b) 180°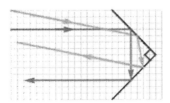

41. (a) –48 cm (b) +16 cm

45. The mirror is in between the object and image, 1.0 m to the right of the object. The mirror is convex, with a focal length of –1.5 m.

47. The LED is at the focal point, so the ray diagram is the opposite of the traditional diagram to show the location of the focal point.

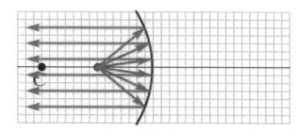

49. The image is located 3.5 cm above the principal axis, and 5.25 cm horizontally to the right of the center of the mirror.

51. (a) 53 cm from the mirror, on the same side as the object (b) –8.3 cm
    (c) real (d) inverted.

53. (a) 12.3 cm from the mirror, on the opposite side as the object (b) 1.9 cm
    (c) virtual (d) upright.

55. The image shifts a little farther from the mirror, and gets a little larger.

57. (a)

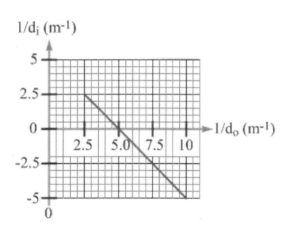

(b) The focal length can be found from the intercept of the graph (either intercept). For instance, the x-intercept is 5.0 m^{-1}. The focal length is the inverse of this, or 1/5 m (= 20 cm).

59. A convex mirror with a focal length of –20 cm.

Answers to selected problems from Essential Physics, Chapter 24

1. This is not possible. If the incident beam comes in along the normal, the part of the beam that is transmitted into the second medium will also follow the normal.

3. 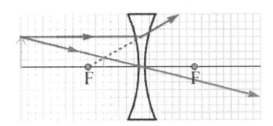 5.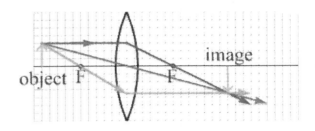

7. The critical angle is larger for red light, because the index of refraction for the glass is smaller for red light than it is for violet light. The smaller the index of refraction of the higher-n medium (the glass), the larger the sine of the critical angle, and the larger the critical angle is.

9. No. The lenses are designed to refract light properly coming from air to glass and back to air. When the light goes from water to glass to water, it will not refract as much, and thus will not compensate correctly for your vision issue.

11. (a) It could be either kind of lens – both can produce virtual images. (b) It must be a diverging lens. Moving an object closer to a diverging lens causes the image to increase in size (this is what is observed), while moving an object closer to a converging lens, when it is giving a virtual image, will cause the image to decrease in size. (c) There's nothing we can say about the focal length, given the information here, aside from the fact that the focal length is negative.

13. 62.8°

15. (a) The refracted beam refracts away from the normal, which happens when light passes from a medium with a higher index of refraction to a medium with a lower index of refraction. Thus, medium 1 has a higher index of refraction than medium 2. (b) Medium 2 has an index of refraction of 1.10, which means that medium 1 has an index of refraction of 2.15. If medium 1 had an index of 1.10, medium 2 would have to have an index of refraction of 0.56, which is not possible – values of the index of refraction are greater than or equal to 1, in general.

17. The index of refraction of medium 1 is less than or equal to $\sqrt{2.0}$.

19. 34.2°, although the light must come from the other side of the interface to experience total internal reflection.

21. (a) $\dfrac{1}{f} = \dfrac{7}{+120 \text{ cm}}$ (b) $f = \dfrac{+120 \text{ cm}}{7} = +17.1 \text{ cm}$ (c) converging lens

23. (a) $\dfrac{1}{f} = +\dfrac{1}{30 \text{ cm}}$ (b) $f = +30$ cm   (c) converging lens

25. (a)

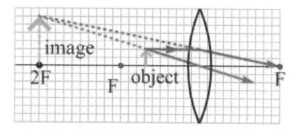

(b) +15 cm   (c) –0.5

27. (a) –30 cm (b) +15 cm   (c)

29. (a) 8.6 cm (b) –0.71   (c) 5.1 cm (d) –0.017

31. What happens with diamond is that the light generally experiences quite a bit of total internal reflection. Total internal reflection occurs when light is traveling in a high-n medium, and it encounters an interface with a low-n medium. If the angle of incidence exceeds the critical angle, the light will be totally internally reflected. Because diamond has such a high index of refraction, the critical angle for a diamond-air interface is particularly small (about 24.6°). Thus, it is very likely that light will be incident on the interface at an angle exceeding the critical angle, and the light will travel a relatively long distance inside the diamond, reflecting internally a number of times, before it emerges from the diamond back to air. It is important that diamond exhibits dispersion, too, which means that the index of refraction of diamond varies with wavelength. The larger the path length traveled, the larger the difference in angle there will be between light of different wavelengths emerging from the diamond. Thus, when white light enters the diamond, it is immediately split into different colors, which bounce around a few times inside the diamond before emerging, at different places on the diamond and at different angles.

33. (a) The wood should be at the focal point of the lens. Thus, it should be a distance equal to the focal length away from the lens. (b) This method only works with a converging lens, which is the type of lens used to correct far-sightedness.

35. Path C is the best path to take. Just like the path that a light beam takes when it travels from a point in a lower-n (higher speed of light) medium to a point in a higher-n (lower speed of light) medium, the lifeguard should not take the straight-line path, but should travel a longer distance (longer than that for the straight-line path) on the beach (the fast medium) and a shorter distance in the water. Path D is not the minimum-time path, however – path C is the one that balances the distances appropriately to minimize the total travel time.

37. 3 > 2 > 1

39.

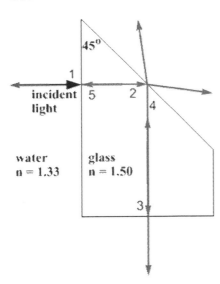

At 1, some light reflects straight back, and some light is transmitted into the glass without changing direction, because the light is incident along the normal.

At 2, some light reflects straight down, and some light refracts back into the air, making an angle of 53° with respect to the normal when it emerges into the air.

At 3, some light reflects straight back, and some light is transmitted into the air without changing direction.

At 4, some light reflects horizontally to the left, and some light refracts back into the air, making an angle of 53° with respect to the normal when it emerges into the air.

At 5, some light reflects straight back, and some light is transmitted into the air without changing direction.

41. (a) The device can only be a converging lens. It is the only kind of lens or mirror that produces a larger image on the same side of the lens or mirror as the object. (b) +24 cm

43. (a) –80 cm (b) +26.7 cm

45. (a) One solution is that the lens is a converging lens, with an object distance of 60 cm and an image distance of +30 cm. The image is real and inverted. (b) The second solution is that the lens is a diverging lens, with an object distance of 20 cm and an image distance of –10 cm. The image is virtual and upright.

(c)

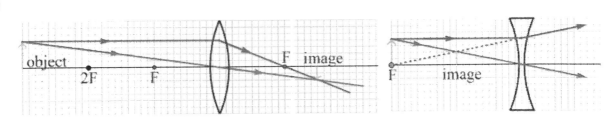

47. The lens is a diverging lens located 150 cm to the right of the object and 90 cm to the right of the image. The lens has a focal length of –225 cm.

49.

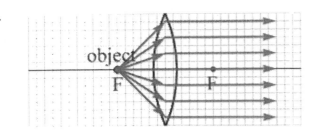

51. (a) The image is 4.8 cm to the left of the lens, and 2.4 cm above the principal axis.

53. (a) $d_i = +45$ cm (b) $h_i = -10$ cm (c) real (d) inverted

55. (a) $d_i = -12.9$ cm (b) $h_i = +2.9$ cm (c) virtual (d) upright

57. The image moves a little closer to the lens, and gets a little smaller.

61. (a)

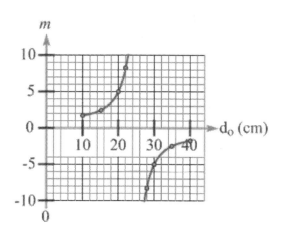

(b) The focal point corresponds to the value of d that gives a magnification of infinity. From the graph, we can see that the magnification approaches infinity as the object distance approaches 25 cm, so $f = +25$ cm.

65. (a) The first image is 10 cm to the right of the lens. It is inverted and the same size as the original object. (b) The mirror creates a second image that is 8 cm to the right of the lens. It is also inverted (in comparison to the original object) and the same size as the original object. (c) The lens creates a third image that is 13.3 cm to the left of the lens. It is upright (in comparison to the original object) and has a height of 6.7 cm. (d) The first image is virtual, because the mirror prevents the light from reaching the image position. The second and third images are real because the light actually passes through the image position.

Answers to selected problems from Essential Physics, Chapter 25

1. (a) decrease (b) increase (c) decrease (d) decrease

3. (a) The second-order ($m = 2$) spectrum. (b)

   (c) 667 nm

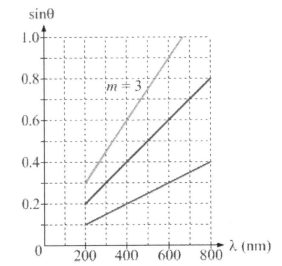

5. The laser is shining on a single slit. We observe that the central maximum is significantly brighter than, and twice as wide as, the other maxima. This happens only with the single slit.

7. About 35 cm.

9. (a) 7:1 (b) 16.2 μm

11. The blue light. The film thickness for completely constructive interference is proportional to the wavelength. The blue light has a smaller wavelength than the red light, so the region of the film that first produces constructive interference for blue light is thinner, and therefore closer to the top, than the region of film that gives constructive interference for red light.

13. (a) 212 Hz (b) 106 Hz

15. (a) 0.52 m or 5.5 m   (b) 2.4 m or 13.3 m

17. 33.5°

19. 500 nm

21. (a) 15 mm (b) 120 μm

23. (a) 8.2 cm   (b) 500 m

25. (a) $\Delta_t = \dfrac{\lambda_{film}}{2}$ (b) $\Delta_b = 2t$ (c) $\Delta = 2t - \dfrac{\lambda_{film}}{2}$ (d) $2t - \dfrac{\lambda_{film}}{2} = m\lambda_{film}$

   (e) $t_{min} = \dfrac{\lambda_{film}}{4} = 100$ nm

27. (a) $\Delta_t = \dfrac{\lambda_{film}}{2}$ (b) $\Delta_b = 2t + \dfrac{\lambda_{film}}{2}$ (c) $\Delta = 2t$ (d) $2t = \left(m + \dfrac{1}{2}\right)\lambda_{film}$

   (e) 400 nm

29. (a) 1060 nm (b) 38.2°, 27.3°, and 24.1°, respectively

31. The technician should stop when the film looks green (520 nm wavelength). Following the five-step method gives $2t = m\lambda_{film}$, which can be solved to find a wavelength in vacuum (or air) of 520 nm.

33. 126 nm

35. (a) 48.6° (b) 53.1°

37. (a) 4.0 cm   (b) Yes, the wavelength decreases to 75% of the original wavelength, so the distance between the spots also decreases to 75% of the original distance. The distance is 3.0 cm.

39. (a) constructive (b) 480 nm and 600 nm (c) 436 nm, 533 nm, and 686 nm

41. (a) 1140 nm (b) $1896 \text{ nm} \leq d < 2528 \text{ nm}$

43. 948 nm

45. (a) 12 (b) 1.4 µm (c) Yes, the 8th and 12th fringes are also missing.

47. (a) 550 nm (b) 2740 nm

49. The width of the single slit is 111 µm.

51. 0.4 m

53. (a) 150 nm. (b) Yes, case D. Cases C and D give the same equation, and involve the same medium as the thin film, resulting in the same thin-film thickness for constructive interference. Cases A and B have a different equation for constructive interference, and the thin film is a different medium, resulting in a different film thickness for constructive interference.

55. One possibility is $n = 1.125$. Another possibility is $n = 2.25$.

57. 825 Hz and 1375 Hz

59. (a) 20 wavelengths   (b) destructive (c) This time there would be 25 wavelengths, but the result would still be destructive interference.

Answers to selected problems from Essential Physics, Chapter 26

3. (a) No. (b) The time interval as measured by Jenna is longer.

9. (a) The time is the distance over the speed of light – the time works out to 200 ns.
   (b) Using the spacetime interval, we can find that the spatial separation is 45 m.

11. Using the spacetime interval, we can find that the time interval is 112 m of time, or, equivalently, 373 ns.

19. (a) (i) 49.94 years. (ii) 43.3 years (iii) 15.6 years
    (b) (i) 2.5 lightyears (ii) 25 lightyears (iii) 47.5 lightyears

21. 0.894 c

23. (a) Yes, the faster Rajon travels, the shorter the distance gets, by length contraction. To contract the distance by a factor of 200, however, Rajon must be traveling at 99.99875% of the speed of light, with respect to you. (b) You measure the proper length – you are at rest with respect to the Earth and the star.

25. (a) The mirrors are length contracted. Remember that contraction only occurs for lengths that are parallel to the velocity. (b) The moving mirrors measure 20% shorter than the stationary mirrors.

33. (a) 0.90 microseconds (b) 889 feet, and 0.90 microseconds

39. (a) 0.8 c (b) 150 m of time, which is equivalent to 0.5 microseconds (c) 200 m of time, which is equivalent to 0.67 microseconds

41. (a) 5.7 milliseconds (b) 8.0 milliseconds

47. 0.968 c

3.   Decreasing the intensity of the light without changing the wavelength means that there are fewer photons per second incident on the metal plate, but the energy of the photons is unchanged. Thus, the emitted electrons have the same maximum kinetic energy that they did before, but the number of electrons emitted per second is reduced.

5.   (a) Yes. Each blue photon has more energy than each green photon, so if the green photons have sufficient energy to cause electrons to be emitted, the blue photons will have more than enough energy to cause electrons to be emitted. (b) There are more electrons emitted with the green light. For the beams to have the same intensity, with the blue light having more energy per photon, there must be fewer blue photons per unit time incident on the plate. Reducing the number of photons produces a corresponding reduction in the number of electrons emitted.

7. To conserve momentum, the electron's velocity must be in the direction of the momentum of the incident photon.

9. The dots in the pattern will be farther apart.

11. (a) same for both (b) same for both (c) the electrons (d) the electrons

13. (a) 1.23×10^{15} Hz (b) 5.1 eV

15.

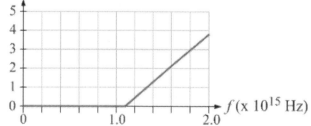

17. 8.6×10^{-12} m

19. (a) After the collision, the photon's wavelength is 2.43×10^{-12} m larger than the wavelength of the incident wavelength. (b) In this case, the electron cannot travel at $180°$ to the outgoing photon. The electron needs to have a component of its velocity in that direction, at $180°$ to the outgoing photon, but it must also have a component of its velocity in the direction of the momentum of the incident photon. This is the only way momentum can be conserved in this collision.

21. With a smaller wavelength, the angle will be smaller.

23. The maxima get farther apart. Reducing the kinetic energy reduces the magnitude of the momentum, which increased the wavelength. By the equation $d\sin\theta = m\lambda$, increasing the wavelength increases the angles between the interference maxima.

27. (a) 6.3×10^{-7} m/s (b) 6.3×10^{-2} m/s (c) 63 m/s

29. 2.6 m/s

31. 600 m^2

33. (a) To turn it on, a light bulb is generally connected to a wall socket that supplies alternating current. As electrons move back and forth through the filament, a substantial amount of energy is transferred from the electrons to the filament. Equilibrium is reached in this process when the filament has become so hot that it glows because of blackbody radiation. At this point, the power transferred to the filament by the electrons is equal to the power radiated by the filament via blackbody radiation. (b) An aluminum filament would melt at the temperatures of a typical light bulb filament. Tungsten, in contrast, has a very high melting point, so it can withstand the high temperatures required to give off light. (c) The temperature of the filament is such that it emits a yellow-white light. At such a temperature, a blackbody emits a relatively small fraction of its energy in the visible spectrum, and a large fraction of its energy in the infrared part of the electromagnetic spectrum. It is the infrared component of the energy that is often referred to as heat.

35. (a) An infrared thermometer exploits the fact that a hot object emits radiation, just like a blackbody, and that the peak wavelength of the emitted radiation is related to the temperature of the object. Thus, by determining the peak wavelength, the thermometer can determine the corresponding temperature of the object. (b) Objects or materials that are hot to the touch but which are not hot enough to glow emit electromagnetic radiation in the infrared part of the spectrum. This kind of thermometer is designed to measure the temperature of such objects, hence the name "infrared thermometer."

37. about 6500 K

39. (a) 0.5 eV (b) The photons do not have enough energy to produce photoelectrons from gold.

41. (a) No, because photons in the visible spectrum have energies of 3.1 eV or less, which, being less than zinc's work function, is insufficient to remove electrons from zinc. (b) 290 nm, which is in the ultraviolet part of the electromagnetic spectrum.

43. No, the maximum kinetic energy does not double, mainly because the work function is constant. If the photon energy increases by 5.0 eV, the maximum kinetic energy of the electrons also increases by 5.0 eV. Thus, doubling the energy of the photons more than triples the maximum kinetic energy of the electrons (7.2 eV, compared to 2.2 eV).

45. (a) No, we do not know that electrons will be emitted for red light. Photons of red light have lower energy than photons of green light, so if the work function of the metal exceeds the energy of the photons of red light, no electrons will be emitted. The work function could be less than that of the energy of the photons of red light, however, in which case electrons would be emitted. (b) No. In this case, there would be more photons emitted per second with the red light than with the green light. With a red photon having less energy than a green photon, there would be more photons incident per second in the beam of red light than in the beam of green light when the beam had the same intensity.

47. (a) 3.0×10^{16} photons/s (b) 1.2×10^{-27} kg m/s (c) 3.8×10^{-11} kg m/s (d) 7.5×10^{-11} kg m/s

49. 6.71×10^{-12} m

51. (a) 2.0×10^{-12} m (b) 1.5×10^{20} Hz (c) 6.2×10^{5} eV

53. (a) $\lambda' = \lambda + 4.86 \times 10^{-12}$ m (b) $\dfrac{h}{\lambda} = (5.0 \times 10^{-22}$ kg m/s$) - \dfrac{h}{\lambda'}$
(c) 1.66×10^{-12} m (d) 6.52×10^{-12} m

55. 108 μm

57. 7.0×10^{-27} m/s

59. (a) 5.5×10^{-11} m (b) 3.8×10^{-10} m (c) 20° (d) 7°

Answers to selected problems from Essential Physics, Chapter 28

1. ... two electron energy levels that are 2.5 eV apart in energy.

3. It is possible for electron transitions involving this level to produce visible photons, if there is another level within 1.8 to 3.1 eV of the –60 eV level. For instance, if there was a level at –58 eV, an electron dropping from the –58 eV level to the –60 eV level would produce a photon of 2 eV, which is in the visible spectrum. Similarly, if there was a level lower than –60.0 eV, say at –62.5 eV, an electron dropping from the –60.0 eV level to the –62.5 eV level would produce a photon of 2.5 eV, which is also in the visible spectrum.

5. (a) The most probable distance is at about 13 Bohr radii from the nucleus, a value that is significantly larger than the value of 9 Bohr radii predicted by the Bohr model.
(b) The total area under the curve = 1, which corresponds to the fact that is 100% certain that the electron will be found at some distance between 0 and infinity from the nucleus.

7. (1) In the Bohr model, only one quantum number is needed to quantize energy, radii, and angular momentum. In the modern quantum-mechanical picture, four quantum numbers are used. (2) In the Bohr model, the electrons travel in circular orbits of well-defined radii. In the modern view, the electrons can be at virtually any distance from the nucleus. (3) In the Bohr model, the electron is treated like a particle. In the modern view, our model of the electron is much more wave-like, with the density of the electron cloud at a particular point corresponding to the probability of finding the electron at that point. There is no equivalent probability interpretation in the Bohr model.

9. (a) 0 or 1   (b) 0, if $\ell = 0$, and –1, 0, or +1, if $\ell = 1$ (c) ±½   (d) 8

11. (a) invalid – there can not be more than 2 electrons in an s-subshell (b) valid – this could be an excited state of silicon (c) valid – this could be an excited state of aluminum (d) valid – there can not be more than 6 electrons in a p-subshell

13. (a) 2.2 eV (b) 5.3×10^{14} Hz   (c) 560 nm (d) the lower level, –25.6 eV

15. (a) –0.850 eV, –0.544 eV, –0.378 eV, –0.278 eV   (b)

(c) 1.89 eV, 2.55 eV, 2.86 eV, 3.02 eV, 3.12 eV

(d) 657 nm, 487 nm, 434 nm, 411 nm, 397 nm

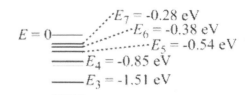

17.

Photon energy (eV)

19. (a) –3, –2, –1, 0, +1, +2, +3 (b) ±½ (c) 14

21. 5, 6, 7, ...

23. n = 3, ℓ = 2, m_ℓ = +1, m_s = ±½

25. (a) 17.34° (b) 26.44° (c) 4 (two on each side)

27. Neon has a number of energy levels that differ in energy from other energy levels by about 2 – 2.3 eV, so that a significant fraction of the visible light emitted by neon consists of photons with energies of 2 – 2.3 eV. These energies correspond to colors in the red and orange part of the visible spectrum.

31. There are two possibilities. The other level could have an energy of –32.27 eV, or it could have an energy of –28.29 eV.

33. It takes more energy to excite the atom from the n = 1 state to the n = 2 state than it does to excite it from the n = 2 state to the n = 100 state. The difference in energy between the n = 2 and n = 1 states is about three times larger than the difference in energy between the n = 100 and n = 2 states.

35. 2.773 eV, 2.472 eV, 2.110 eV, and 1.857 eV, respectively.

37. (a) four (b) There are an infinite number of energy-level diagrams that correspond to this emission spectrum. One of these is shown here.

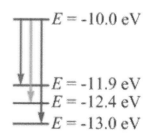

E = -10.0 eV

E = -11.9 eV
E = -12.4 eV
E = -13.0 eV

39. (a) n = 3, n = 6, and n = 9 (b) Yes, it is possible to tell the two spectra apart. In addition to the emission lines it has in common with hydrogen, there will be additional emission lines in the lithium spectrum because of the additional electron levels lithium has, in between the levels it has in common with hydrogen.

41. (a) The six energies are E_1 = –60 eV, E_2 = –30 eV, E_3 = –20 eV, E_4 = –15 eV, E_5 = –12 eV, and E_6 = –10 eV. Thus, there are two transitions that produce a 10 eV photon, $E_6 \rightarrow E_3$ and $E_3 \rightarrow E_2$. (b) There are also two transitions that produce a 5 eV photon, $E_6 \rightarrow E_4$ and $E_4 \rightarrow E_3$.

43. (a) 1.89 eV (b) 3.40 eV.

45. (a) 122 nm. (b) 91 nm.

49. (a) The most likely radius is about 5 Bohr radii, which is somewhat larger than the 4 Bohr radii predicted by the Bohr model. (b) There is about a 5% chance of finding the electron at a radius of less than 2 Bohr radii from the nucleus.

51. (a) $L = \sqrt{12}\,\dfrac{h}{2\pi}$ (b) the possible values are $L_z = -\dfrac{3h}{2\pi}, -\dfrac{2h}{2\pi}, -\dfrac{h}{2\pi}, 0, +\dfrac{h}{2\pi}, +\dfrac{2h}{2\pi}, +\dfrac{3h}{2\pi}$

53. (a)   (b) $L = \sqrt{2}\,\dfrac{h}{2\pi}$

The units on the graph are $h/(2\pi)$

55. (a) The ground state of S is $1s^2\, 2s^2\, 2p^6\, 3s^2\, 3p^4$
(b) The ground state of Se is $1s^2\, 2s^2\, 2p^6\, 3s^2\, 3p^6\, 4s^2\, 3d^{10}\, 4p^4$
(c) These elements are found in the same column in the periodic table because the outermost electrons in both elements have a similar configuration. In both cases, there are 4 electrons in the last sub-shell, which is a p sub-shell. This means that both sulfur (S) and selenium (Se) have similar chemical properties   chemical bonds generally depend on the outermost electrons.

57. (a) 53 (b) Iodine

59. In general, atoms and molecules are most stable when electrons completely fill the sub-shells. The ground-state configuration of a column-1 element is such that it has completely filled sub-shells plus one extra electron. Any column-17 element, on the other hand, is just one electron short of having completely-filled subshells. Thus, when a column-1 element donates its extra electron to a column-17 element within a molecule, both elements essentially have completely filled subshells, producing a stable arrangement.

Answers to selected problems from Essential Physics, Chapter 29

1. (a) 26 (b) 30 (c) 56

3. 9×10^{16} J

5. (a) a beta-plus decay   (b) $+e = +1.6 \times 10^{-19}$ C   (c) an electron

7. About 3%.

9. (a) A   (b) B

11. 3

13. (a) 0.13699 u   (b) 127.61 MeV   (c) 7.9756 MeV/nucleon

15. (a) 0.0023868 u   (b) 2.2233 MeV

17. (a) 1.7566 u   (b) 1636.3 MeV   (c) 7.8668 MeV/nucleon

19. (a) $^{239}_{94}\text{Pu} \rightarrow \, ^{235}_{92}\text{U} + \, ^4_2\text{He}$ (b) 5.245 MeV

21. (a) $^{22}_{11}\text{Na} \rightarrow \, ^{22}_{10}\text{Ne} + \, ^0_{+1}\text{e}^+ + \nu_e$ (b) 1.820 MeV

23. (a) For a spontaneous decay to occur, the total mass of the products must be less than the mass of the original atom, with the missing mass being converted to energy – this energy is the kinetic energy of the products after the decay process. One of the candidate atoms given in the problem has a larger mass than the original silver atom, so the silver atom will not spontaneously decay into that larger mass atoms. The rhodium is quite a bit lower in mass, so let's investigate that possibility – that would be an alpha decay. To verify whether this decay will occur spontaneously, we need to add the mass of the rhodium and the alpha to see whether the total mass after the alpha decay is less than the mass of the original silver atom. The total mass afterwards in that case would be 106.906748 u + 4.00260325 u = 110.909351 u, which is larger than the 110.905291 u of the silver atom – that reaction will not occur spontaneously either. The only possibility left is the cadmium, which would be a beta-minus decay. In a beta-minus decay, the mass of the electron emitted is already included in the mass of the product atom, so all we have to do is to see that the mass of the cadmium atom is less than the mass of the silver atom to know that this is the reaction that the silver atom will undergo.
   (b) $^{111}_{47}\text{Ag} \rightarrow \, ^{111}_{48}\text{Cd} + \, ^0_{-1}\text{e}^- + \overline{\nu}_e$

25. (a) 88 g   (b) 18 g   (c) 2.7 g

27. Yes. That's how radioactive decay works – for equal time intervals, the number of atoms is reduced by the same factor.

29. (a) ^1_1H (b) 5.025 MeV

31. 12.860 MeV

35. (a) 1.1×10^{-12} (b) 0.94×10^{-12}

37. $1/256^{\text{th}}$ of the original activity

41. (a) 64 (b) 4 (c) 4.8×10^{-15} m

43. (a) 511 keV (b) 511 keV (c) 511 keV (d) 2.4×10^{-12} m

45. (a) The momenta are equal-and-opposite. (b) 50 : 1

47. (a) one throw (b) 3.8 throws

49. (a) 0.5 N (b) 2N/3.

51. 120 s

53. 16%

55. (a) 9.3×10^{19} (b) 3.1×10^{19} (c) -3.9×10^{19}
   (d) 9.3×10^{19} (e) 5.0×10^{19} (f) 2.5×10^{19}
   (g) The second method, using Equation 29.12, is correct. The problem with the first method is that it assumes the rate of loss of nuclei is constant, but the rate of loss of nuclei decreases as time goes by, because the number of radioactive nuclei decreases as time goes by.

57.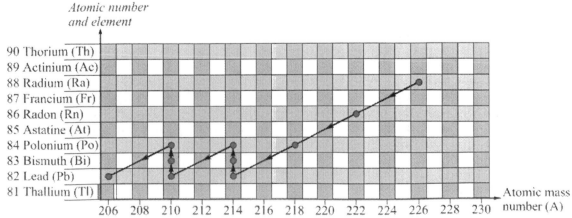

59. 6.8×10^{-13}

Made in the USA
Charleston, SC
10 December 2012